S0-BYH-479

HANDBOOK OF INCINERATION SYSTEMS

Other McGraw-Hill Books of Interest

Baumeister and Marks • MARKS' STANDARD HANDBOOK FOR MECHANICAL ENGINEERS
Brady and Clauser • MATERIALS HANDBOOK
Bralla • HANDBOOK OF PRODUCT DESIGN FOR MANUFACTURING
Corbitt • STANDARD HANDBOOK OF ENVIRONMENTAL ENGINEERING
Elliot • STANDARD HANDBOOK OF POWERPLANT ENGINEERING
Freeman • STANDARD HANDBOOK OF HAZARDOUS WASTE TREATMENT AND DISPOSAL
Ganic and Hicks • THE MCGRAW-HILL HANDBOOK OF ESSENTIAL ENGINEERING INFORMATION AND DATA
Gieck • ENGINEERING FORMULAS
Grimm and Rosaler • HANDBOOK OF HVAC DESIGN
Harris • HANDBOOK OF NOISE CONTROL
Harris and Crede • SHOCK AND VIBRATION HANDBOOK
Hicks • STANDARD HANDBOOK OF ENGINEERING CALCULATIONS
Juran and Gryna • JURAN'S QUALITY CONTROL HANDBOOK
Karassik et al. • PUMP HANDBOOK
Maynard • INDUSTRIAL ENGINEERING HANDBOOK
Parmley • STANDARD HANDBOOK OF FASTENING AND JOINING
Rohsenow, Hartnett, and Ganic • HANDBOOK OF HEAT TRANSFER APPLICATIONS
Rohsenow, Hartnett, and Ganic • HANDBOOK OF HEAT TRANSFER FUNDAMENTALS
Rosaler and Rice • STANDARD HANDBOOK OF PLANT ENGINEERING
Rothbart • MECHANICAL DESIGN AND SYSTEMS HANDBOOK
Shigley and Mischke • STANDARD HANDBOOK OF MACHINE DESIGN
Tuma • ENGINEERING MATHEMATICS HANDBOOK
Tuma • HANDBOOK OF NUMERICAL CALCULATIONS IN ENGINEERING
Wadsworth • HANDBOOK OF STATISTICAL METHODS FOR ENGINEERS AND SCIENTISTS
Young • ROARK'S FORMULAS FOR STRESS AND STRAIN

HANDBOOK OF INCINERATION SYSTEMS

Calvin R. Brunner, P.E., D.E.E.

Incinerator Consultants Incorporated

Reston, Virginia

McGraw-Hill, Inc.

New York St. Louis San Francisco Auckland Bogotá
Caracas Hamburg Lisbon London Madrid
Mexico Milan Montreal New Delhi Paris
San Juan São Paulo Singapore
Sydney Tokyo Toronto

Library of Congress Cataloging-in-Publication Data

Brunner, Calvin R.
Handbook of incineration systems / Calvin R. Brunner.
p. cm.
Includes bibliographical references and index.
ISBN 0-07-008589-7
1. Incinerators. 2. Incineration. I. Title.
TD796.B777 1991
628.4'457—dc20 90-24765
CIP

1 2 3 4 5 6 7 8 9 0 DOC/DOC 9 7 6 5 4 3 2 1

ISBN 0-07-008589-7

The sponsoring editor for this book was Robert Hauserman, the editing supervisor was Joseph Bertuna, and the production supervisor was Suzanne W. Babeuf. It was set in Times Roman by McGraw-Hill's Professional Publishing composition unit.

Printed and bound by R. R. Donnelley & Sons Company.

Portions of this book were first published in 1984 by Van Nostrand Reinhold Co., Inc. They were reprinted in 1988 by Incinerator Consultants Incorporated.

For Claire Helen Brunner, whose support made this book possible.

CONTENTS

PREFACE

Incineration is increasingly looked to as a favorable means of waste treatment and disposal, especially when compared to alternative methods. As presented in this book, the field of incineration encompasses the destruction or processing of solids, sludge and liquid, gaseous, and radioactive wastes.

This handbook has been written to accommodate technical and nontechnical persons alike. It is meant to provide both a broad view of the subject as well as detailed system design techniques of primary interest to the specialist. References and bibliographies appear periodically to direct the reader to relevant publications and other sources of technical information.

The emphasis throughout is on systems design. Before design can begin, however, the applicable statutory requirements must be understood. Two chapters are therefore allotted to the regulations that govern the incineration of hazardous and nonhazardous wastes. Six chapters address analytical methods for system design, from waste characterization to the prediction of air emissions. The various types of incineration systems currently available are discussed in the following eleven chapters, which include design calculations, dimensional data, and other incinerator parameters.

Other chapters include one on energy recovery which presents a method for determining the heat-recovery potential of an incinerator system, along with relevant design examples, and two on air emissions control equipment which contain descriptions of the large variety of control devices available, their capacities, dimensions, design parameters, and systems used for acid gas control. Two chapters present a comprehensive design example, one developed in U.S. units and the other in metric units.

The reader has sufficient information in this single text to determine equipment selection, sizing, and parameters of operation for incineration equipment that burns the vast variety of wastes generated by municipal, commercial, industrial, and institutional sources.

Calvin R. Brunner

ACRONYMS AND ABBREVIATIONS

AQCR	Air quality control region
BAAQMD	Bay Area air quality management district
BACT	Best available control technology
BOD	Biochemical oxygen demand
Btu	British thermal unit
C/A	Cooling air
CAA	Clean Air Act
CAAA	Clean Air Act Amendments
CE	Combustion efficiency
CEQ	Council on Environmental Quality
COD	Chemical oxygen demand
CPI	Chemical process industries
CTG	Control technology guideline
DEC	Department of Environmental Conservation
DEP	Department of Environmental Protection
DG	Dry gas
DOE	United States Department of Energy
DRE	Destruction and removal efficiency
EA	Excess air
EAU	Excess air unit
EIS	Environmental Impact Statement
EOP	Emissions offset policy
EPA	United States Environmental Protection Agency
ESP	Electrostatic precipitator
FBF	Fluid bed furnace
FGD	Flue gas desulfurization
FO	Fuel oil
GEP	Good engineering policy
GLC	Ground-level concentration
HEPA	High-efficiency particulate air
LAER	Lowest achievable emission rate
LEL	Lower explosive limit
LLW	Low-level (radioactive) waste
M/H	Multiple-hearth incinerator

MSW	Municipal solid waste
NA	Nonattainment
NAAQS	National Ambient Air Quality Standards
NEP	National Environmental Policy Act
NESHAPs	National Emission Standard for Hazardous Air Pollutants
NIOSH	National Institute for Occupational Safety and Health
NPDES	National pollutant discharge elimination system
NRC	Nuclear Regulatory Commission
NSPS	New-source performance standards
NSR	New-source review
NTIS	National Technical Information Service
ORD	Office of Research and Development (EPA)
PIC	Product of incomplete combustion
POHC	Principal organic hazardous constituent
POTW	Publicly owned treatment works
PSD	Prevention of significant deterioration
RACT	Reasonable available control technology
RCRA	Resource Conservation and Recovery Act
RDF	Refuse-derived fuel
SAC	Starved air combustion
SAU	Starved air unit
SIP	State implementation plan
SWA	Solid waste administration
TRU	Transuranic material
TSCA	Toxic Substances Control Act
UEL	Upper explosive limit
VOC	Volatile organic compound
WC	Water column

HANDBOOK OF INCINERATION SYSTEMS

CHAPTER 1
INTRODUCTION

The term *waste* is a generalization for myriads of types, sizes, and configurations of materials which have no apparent utility. Household refuse, spent broth from the manufacture of penicillin, concrete chunks from a demolished structure, laboratory animal remains, rejects from a chemical product stream—these are all termed wastes. At one time off-gas generated in the drilling of oil wells in the United States was considered waste and was disposed of by burning on-site (flaring). The price and availability of energy have made the collection and distribution of natural gas economical, and today, instead of classification as a waste, it is considered a fuel source itself. In time, as natural resources are increasingly depleted, more and more means of reclamation and recycling will be developed, removing materials from the category "waste." Industrial reclamation of combustible materials from waste streams; use of refuse and sewage sludge to generate gaseous fuel, hot water, steam, and/or electric power; and recycling of waste paper, used cans, and glass are all steps in the direction of reducing the damaging effect of an industrial society on the earth, the environment, and the people the society serves.

One of the most effective means of dealing with many wastes, to reduce their harmful potential and often to convert them to an energy form, is incineration. In comparing incineration (the destruction of a waste material by the application of heat) to other disposal options such as land burial or disposal at sea or in lagoons, the advantages of incineration are obvious:

- The volume and weight of the waste are reduced to a fraction of its original size.
- Waste reduction is immediate; it does not require long-term residence in a landfill or holding pond.
- Waste can be incinerated on-site, without having to be carted to a distant area.
- Air discharges can be effectively controlled for minimal impact on the atmospheric environment.
- The ash residue is usually nonputrescible, or sterile.
- Technology exists to completely destroy even the most hazardous of materials in a complete and effective manner.
- Incineration requires a relatively small disposal area, not the acres and acres required for lagoons or land burial.
- By using heat-recovery techniques the cost of operation can often be reduced or offset through the use of or sale of energy.

Incineration will not solve all waste problems. Some disadvantages include:

- The capital cost is high.
- Skilled operators are required.
- Not all materials are incinerable, e.g., high-aqueous wastes or noncombustible solids.
- Some materials require supplemental fuel to attain mandated efficiencies of destruction.

MATERIAL CLASSIFICATION

A substance will fall into one of three categories:

1. *Gas:* In this state of matter, the substance completely fills a container in volume and in shape.
2. *Liquid:* The substance will fill a container in shape but not in volume.
3. *Solid:* The substance will not fill a container in shape or in volume.

LIQUID OR SOLID

The difference between a solid and a liquid is not always clear. *Sludge, slurries, tars*, and *skimmings* are terms describing states of matter lying somewhere between a true solid and true liquid. One quantitative measure of this difference is viscosity, which is a measure of the resistance of a fluid to shear. The resistance to shear in a solid is extremely high. In a "pure" liquid such as water it is almost zero. Flow and burning characteristics of a liquid are a function of the nature of and value of its viscosity. Types of viscosity and resultant fluid properties are defined as follows:

Newtonian fluid. A "pure" liquid such as water or light-petroleum derivatives where the resistance to shear is initially zero and increases linearly with increase in shear, or velocity of flow.

Bingham plastic. A fluid which has an initial resistance to flow until a threshold is reached, at which point the fluid assumes newtonian flow. Examples of such liquids are sewage sludge and aqueous mixtures of grain.

Pseudoplastic fluids and dilatant flow. These materials have a viscosity that increases rapidly, exponentially, with increasing velocity. These fluids include paper pulp, quicksand, and beach sand.

Thixotropic fluids. Initially these materials have a high viscosity. Under uniform shear, the viscosity decreases with time. Typical fluids are catsup, paints, inks, mayonnaise, and drilling muds.

Rheopectic fluids. These fluids display a rapid increase in viscosity with time. They will "set up" upon shaking or tapping. Examples are aqueous suspensions of vanadium pentoxide or gypsum.

Viscoelastic fluids. Fluids which exhibit elastic recovery from deformation which occurs during flow. This property causes the *Weissenberg effect*, which is the tendency for the fluid to climb up a shaft rotating within it.

In general, a fluid can be pumped in a conventional manner if its viscosity is under 10,000 SSU. It can be atomized, i.e., burned in suspension, if its viscosity is below 750 SSU.

VISCOSITY

Figure 1.1 presents viscosities of a number of common fluids. Note the relative orders of magnitude, one substance to another, and the change in viscosity with respect to temperature.

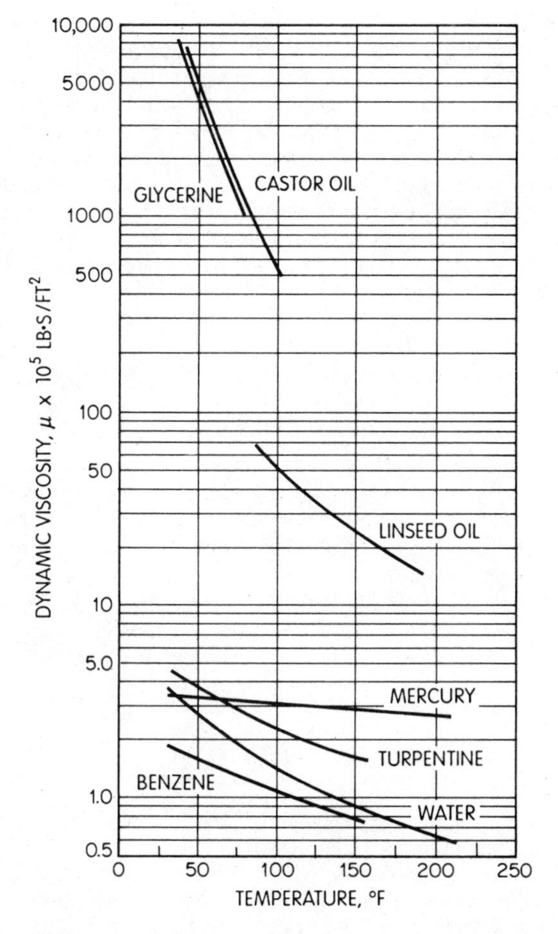

FIGURE 1.1 Viscosity values for typical liquids. *Source*: Ref. 1–1.

INCINERABLE WASTE

The Incinerator Institute of America was a national organization attempting to quantify and standardize incinerator design parameters. It went out of business over ten years ago; however, a number of their standards are still in use. One such standard is given in Table 1.1, which is used by manufacturers of small and packaged incinerators in rating their equipment. The classifications in the table represent incinerable wastes, wastes which are combustible and are viable candidates for incineration.

Incinerability can be defined more specifically by consideration of the following factors:

Waste moisture content. The greater the moisture content, the more fuel is required to destroy the waste. An aqueous waste with a moisture content greater than 95 percent or a sludge waste with less than 15 percent solids content would be considered poor candidates for incineration.

Heating value. Incineration is a thermal destruction process where the waste is degraded to nonputrescible form by the application and maintenance of a source of heat. With no significant heating value, incineration would not be a practical disposal method. Generally, a waste with a heating value less than 1000 Btu/lb as received, such as concrete blocks or stone, is not applicable for incineration. There are instances, however, where an essentially inert material has a relatively small content (or coating) of combustibles and incineration would be a viable option even with a small heating value. Two such cases are incineration of empty drums with a residual coating of organic material on their inner surfaces and incineration of grit from wastewater treatment plants. The grit adsorbs grease from within the wastewater flow which results in a slight heating value to the grit material, normally less than 500 Btu/lb.

Inorganic salts. Wastes rich in inorganic, alkaline salts are troublesome to dispose of in a conventional incineration system. A significant fraction of the salt will become airborne. It will collect on furnace surfaces, creating a slag, or cake, which severely reduces the ability of an incinerator to function properly.

High sulfur or halogen content. The presence of chlorides or sulfides in a waste will normally result in the generation of acid-forming compounds in the off-gas. The cost of protecting equipment from acid attack must be balanced against the cost of alternative disposal methods for the waste in question.

Radioactive waste. Incinerators have been developed specifically for the destruction of radioactive waste materials, as described in a later chapter. Unless designed specifically for radioactive waste disposal, however, an incinerator should not be used for the firing of a radioactive waste.

STANDARD INDUSTRIAL CLASSIFICATION

The Department of Commerce has established a classification of industries known as the standard industrial classification (SIC) code. A listing of the SIC codes of manufacturing industries which produce substantive quantities of incinerable wastes is presented in Table 1.2. Table 1.3 lists the number of such manufacturing establishments in the United States (as of 1978) related to SIC codes 20 through 39.

TABLE 1.1 Classification of Wastes to Be Incinerated

Classification of wastes								
Type	Description	Principal components	Approximate composition, % by weight	Moisture content, %	Incombustible solids, %	Refuse as fired, Btu/lb	Btu of auxiliary fuel per lb of waste to be included in combustion calculations	Recommended min Btu/h burner input per lb waste
[a]0	Trash	Highly combustible waste, paper, wood, cardboard cartons, including up to 10% treated papers, plastic or rubber scraps; commercial and industrial sources	Trash 100%	10	5	8500	0	0
[a]1	Rubbish	Combustible waste, paper, cartons, rags, wood scraps, combustible floor sweepings; domestic, commercial, and industrial sources	Rubbish 80% Garbage 20%	25	10	6500	0	0
[a]2	Refuse	Rubbish and garbage; residential sources	Rubbish 50% Garbage 50%	50	7	4300	0	1500
[a]3	Garbage	Animal and vegetable wastes, restaurants, hotels, markets; institutional, commercial, and club sources	Garbage 65% Rubbish 35%	70	5	2500	1500	3000
4	Animal solids and organic wastes	Carcasses, organs, solid organic wastes; hospital, laboratory, abattoirs, animal pounds, and similar sources	100% Animal and human tissue	85	5	1000	3000	8000 (5000 Primary) (3000 Secondary)
5	Gaseous, liquid or semi-liquid wastes	Industrial process wastes	Variable	Dependent on predominant components	Variable according to wastes survey	Variable according to wastes survey	Variable according to wastes survey	Variable according to wastes survey
6	Semi-solid and solid wastes	Combustibles requiring hearth, retort, or grate burning equipment	Variable	Dependent on predominant components	Variable according to wastes survey	Variable according to wastes survey	Variable according to wastes survey	Variable according to wastes survey

[a]The above figures on moisture content, ash, and Btu as fired have been determined by analysis of many samples. They are recommended for use in computing heat release, burning rate, velocity, and other details of incinerator designs. Any design based on these calculations can accommodate minor variations.

Source: Ref. 1-2.

TABLE 1.2 Sources and Types of Industrial Wastes

Code	SIC Group Classification	Waste Generating Processes	Expected Specific Wastes
17	Plumbing, heating, air conditioning Special trade contractors	Manufacturing and installation in homes, buildings, factories	Scrap metal from piping and duct work; rubber, paper, insulating materials, misc. construction, demolition debris
19	Ordnance and accessories	Manufacturing and assembling	Metals, plastic, rubber, paper, wood, cloth, chemical residues
20	Food and kindred products	Processing, packaging, shipping	Meats, fats, oils, bones, offal vegetables, fruits, nuts and shells, cereals
22	Textile mill products	Weaving, processing, dyeing and shipping	Cloth and fiber residues
23	Apparel and other finished products	Cutting, sewing, sizing, pressing	Cloth, fibers, metals, plastics, rubber
24	Lumber and wood products	Sawmills, mill work plants, wooden container, misc. wood products, manufacturing	Scrap wood, shavings, sawdust; in some instances metals, plastics, fibers, glues, sealers, paints solvents
25	Furniture, wood	Manufacture of household and office furniture, partitions, office and store fixtures, mattresses	Those listed under Code 24; in addition, cloth and padding residues
25	Furniture, metal	Manufacture of household and office furniture, lockers, bedsprings, frames	Metals, plastics, resins, glass, wood, rubber, adhesives, cloth, paper
26	Paper and allied products	Paper manufacture, conversion of paper and paperboard, manufacture of paperboard boxes and containers	Paper and fiber residues, chemicals, paper coatings and fillers, inks, glues, fasteners
27	Printing and publishing	Newspaper publishing, printing lithography, engraving, printing, and bookbinding	Paper, newsprint, cardboard, metals, chemicals, cloth, inks, glues
28	Chemicals and related products	Manufacture and preparation of inorganic chemicals (ranges from drugs and soups to paints and varnishes, and explosives)	Organic and inorganic chemicals, metals, plastics, rubber, glass, oils, pigments
29	Petroleum refining and related industries	Manufacture of paving and roofing materials	Asphalt and tars, felts, asbestos, paper, cloth, fiber
30	Rubber and miscellaneous plastic products	Manufacture of fabricated rubber and plastic products	Scrap rubber and plastics, lampblack, curing compounds, dyes

TABLE 1.2 Sources and Types of Industrial Wastes (Continued)

Code	SIC Group Classification	Waste Generating Processes	Expected Specific Wastes
31	Leather and leather products	Leather tanning and finishing; manufacture of leather belting and packing	Scrap leather, thread, dyes, oils, processing and curing compounds
32	Stone, clay, and glass products	Manufacture of flat glass, fabrication or forming of glass; manufacture of concrete, gypsum, and plaster products; forming and processing of stone and stone products, abrasives, asbestos, and misc. nonmineral products	Glass, cement, clay, ceramics, gypsum, asbestos, stome, paper, abrasives
33	Primary metal industries	Melting, casting, forging, drawing, rolling, forming, extruding operations	Ferrous and nonferrous metals scrap, slag, sand, cores, patterns, bonding agents
34	Fabricated metal products	Manufacture of metal cans, hand tools, general hardware, nonelectric heating apparatus, plumbing fixtures, fabricated structural products, wire, farm machinery and equipment, coating and engraving of metal	Metals, ceramics, sand, slag, scale, coatings, solvents, lubricants, pickling liquors
35	Machinery (except electrical)	Manufacture of equipment for construction, mining, elevators, moving stairways, conveyors, industrial trucks, trailers, stackers, machine tools, etc.	Slag, sand, cores, metal scrap, wood, plastics, resins, rubber, cloth, paints, solvents, petroleum products
36	Electrical	Manufacture of electric equipment, appliances, and communication apparatus, machining, drawing, forming, welding, stamping, winding, painting, plating, baking, firing, operations	Metal scrap, carbon, glass, exotic metals, rubber, plastics, resins, fibers, cloth residues
37	Transportation equipment	Manufacture of motor vehicles, truck and bus bodies, motor vehicle parts and accessories, aircraft and parts, ship and boat building and	Metal scrap, glass, fiber, wood, rubber, plastics, cloth, paints, solvents, petroleum products

TABLE 1.2 Sources and Types of Industrial Wastes (Continued)

Code	SIC Group Classification	Waste Generating Processes	Expected Specific Wastes
		repairing motorcycles and bicycles and parts, etc.	
38	Professional, scientific controlling instruments	Manufacture of engineering, laboratory, and research instruments and associated equipment	Metals, plastics, resins, glass, wood, rubber, fibers, abrasives
39	Miscellaneous manufacturing	Manufacture of jewelry, silverware, plated ware, toys, amusement, sport- and athletic goods, costume novelties, buttons, brooms, brushes, signs, advertising displays	Metals, glass, plastics, resins, leather, rubber, composition, bone, cloth, straw, adhesives, paints, solvents

Source: Ref. 1-3.

TABLE 1.3 Manufacturing Establishments in the United States by SIC Number

SIC Code		Number of Establishments in Nation (1978)
20	Food	18,195
21	Tobacco	167
22	Textile	5,990
23	Apparel	20,016
24	Lumber	28,881
25	Furniture	7,639
26	Paper	4,259
27	Printing	45,528
28	Chemical	8,407
29	Petroleum	1,494
30	Rubber, plastic	12,450
31	Leather	2,428
32	Stone, glass, and clay	13,385
33	Primary metal	5,477
34	Fabricated metal	27,453
35	Machinery	42,139
36	Electric machinery	13,144
37	Transportation equipment	9,078
38	Instruments	6,308
39	Miscellaneous manufacturing	16,052
	Total	288,490

DISPOSAL OPTIONS

There is no simple answer to the disposal problem. Proper waste management involves the examination of the entire breadth of available options for each type of material. Disposal options for waste materials include:

1. Physical treatment processes:
   - Gas cleaning
   - Liquid/solid separation
   - Removal of specific components
   - Blending of wastes
2. Chemical treatment processes:
   - Absorption
   - Chemical oxidation
   - Chemical reduction
3. Biological treatment processes
4. Ultimate disposal processes:
   - Deep well disposal
   - Sanitary landfill
   - Composting
   - Land burial
   - Encapsulation and solidification
   - Secure landfill
   - Ocean dumping
   - Dilute and disperse
   - Recycle/reuse
   - Incineration

Degradation options, as opposed to disposal, include the following:

- *Destruction* is the conversion of the waste materials to innocuous by-products (i.e., conversion of organic phosphate to CO_2 + H_2O + phosphoric acid).
- *Detoxification* is removal or destruction of the toxic component; this process renders the material nonhazardous (e.g., dechlorination of PCBs).

Degradation options include nonoxidative and oxidative techniques, as follows:

1. Nonoxidative techniques include:
   - Microbiological
   - Reductive dechlorination
   - High-energy radiation; microwaves
2. Oxidative techniques include:
   - Wet catalytic oxidation; aqueous phase reaction
   - Ozonation
   - Photochemical decomposition
   - Chemical oxidation (i.e., hydrogen peroxide, potassium permanganate, and chlorine dioxide oxidants)

- Electrochemical oxidation
- Flameless oxidation; low-temperature vapor combustion
- Incineration

3. *Refractory* is the inverse of destructible, i.e., it refers to the degree to which a material resists destruction. For instance, PCBs are highly refractory.

LOAD ESTIMATING

The quantity of solid waste generated in the United States, industrial and municipal, is approximately 300 million tons/year. Of this figure approximately 2000 lb of household refuse is produced per year per capita.

The estimation of incinerator loading, where the waste quantity is not known, usually requires a survey of the area in question including a study of past records, demographic trends, etc. Table 1.4 can be used as a guide in determining the solid waste produced from various sources.

Table 1.5 lists the average weight of various solid wastes, and Table 1.6 lists per capita waste generation in the United States.

Another major waste, sewage sludge, can be estimated to be generated at the rate of 0.2 lb/day of sludge solids per capita.

ESTIMATING SOLID WASTE QUALITY

While a general figure for waste generation can be obtained as noted in the previous sections, a more accurate means of determining the quality of a solid waste stream is by use of Table 1.7, Table 1.8, and/or Table 1.9. By a visual inspection of the waste, a percentage of each waste component as listed in these tables can be established. By multiplying the moisture percentage or heating value or density of each of these components by the indicated moisture, heating value, or density, a more accurate figure for the total waste quality can be estimated. (A more detailed analysis of heating value of wastes is included in a later chapter.)

As an example, to estimate the heating value of a particular municipal solid waste, with the waste components as listed below, using the heating value listed in Table 1.9, the total waste heating value is calculated as follows:

Component	Solid wastes, %	Inherent energy, Btu/lb	Total energy contribution, Btu/lb
Food wastes	15	2,000	300
Paper	40	7,200	2,880
Cardboard	5	7,000	350
Plastics	5	14,000	700
Wood	15	8,000	1,200
Glass	10	60	6
Tin cans	10	300	30
Total	100		5,466

Total energy content is therefore 5466 Btu/lb.

TABLE 1.4 Incinerator Capacity Chart

Classification	Building types	Quantities of waste produced
Industrial buildings	Factories	Survey must be made
	Warehouses	2 lb/(100 ft^2 · day)
Commercial buildings	Office buildings	1 lb/(100 ft^2 · day)
	Department stores	4 lb/(100 ft^2 · day)
	Shopping centers	Study of plans or survey required
	Supermarkets	9 lb/(100 ft^2 · day)
	Restaurants	2 lb per meal per day
	Drugstores	5 lb/(100 ft^2 · day)
	Banks	Study of plans or survey required
Residential	Private homes	5 lb basic & 1 lb per bedroom
	Apartment buildings	4 lb per sleeping room per day
Schools	Grade schools	10 lb per room & ½ lb per pupil per day
	High schools	8 lb per room & ½ lb per pupil per day
	Universities	Survey required
Institutions	Hospitals	15 lb per bed per day
	Nurses' or interns' homes	3 lb per person per day
	Homes for aged	3 lb per person per day
	Rest homes	3 lb per person per day
Hotels, etc.	Hotels—1st class	3 lb per room and 2 lb per meal per day
	Hotels—Medium class	1½ lb per room & 1 lb per meal per day
	Motels	2 lb per room per day
	Trailer camps	6 to 10 lb per trailer per day
Miscellaneous	Veterinary hospitals	Study of plans or survey required
	Industrial plants	
	Municipalities	

Do not estimate more than 7-h operation per shift of industrial installations.
Do not estimate more than 6-h operation per day for commercial buildings, institutions, and hotels.
Do not estimate more than 4-h operation per day for schools.
Do not estimate more than 3-h operation per day for apartment buildings.
Whenever possible an actual survey of the amount and nature of refuse to be burned should be carefully taken. The data herein are of value in estimating capacity of the incinerator where no survey is possible and also to double-check against an actual survey.
Source: Ref. 1-2.

TABLE 1.5 Average Weight of Solid Waste

Type	lb/ft^3
Type 0 waste	8 to 10
Type 1 waste	8 to 10
Type 2 waste	15 to 20
Type 3 waste	30 to 35
Type 4 waste	45 to 55
Garbage (70% H_2O)	40 to 45
Magazines and packaged paper	35 to 50
Loose paper	5 to 7
Scrap wood and sawdust	12 to 15
Wood shavings	6 to 8
Wood sawdust	10 to 12

Source: Ref. 1-2.

TABLE 1.6 Average Solid Waste Collected (lb per person per day)

Solid Wastes	Urban	Rural	National
Household	1.26	0.72	1.14
Commercial	0.46	0.11	0.38
Combined	2.63	2.60	2.63
Industrial	0.65	0.37	0.59
Demolition, construction	0.23	0.02	0.18
Street and alley	0.11	0.03	0.09
Miscellaneous	0.38	0.08	0.31
Totals	5.72	3.93	5.32

Source: Ref. 1-4.

TABLE 1.7 Industrial Solid Waste Density Data

Waste	lb/yd^3 as discarded
Department store waste	80
Hospital waste (not research)	100
School waste w/lunch program	110
Supermarket waste	100
Bakalite	600
Bitumen waste	1500
Brown paper	135
Cardboard	180
Cork	320
Corn cobs	300
Corrugated paper (loose)	100
Disposable hospital plastics	120
Grass, green	120
Hardboard	900
Latex	1200
Magazines	945
Meat scraps	400
Milk cartons, coated	80
Nylon	200
Paraffin—wax	1400
Plastic coated paper	135
Polyethylene film	20
Polystyrene	175
Polyurethane (foamed)	55
Resin bonded fiberglass	990
Rubber—synthetics	1200
Shoe leather	540
Tar paper	450
Textile waste (non-synthetic)	280
Textile waste (synthetic)	240
Vegetable food waste	375
Wax paper	150
Wood	300

Source: Ref. 1-5.

TABLE 1.8 Typical Moisture Content of Municipal Solid Waste (MSW) Components

Component	Moisture, percent	
	Range	Typical
Food wastes	50–80	70
Paper	4–10	6
Cardboard	4–8	5
Plastics	1–4	2
Textiles	6–15	10
Rubber	1–4	2
Leather	8–12	10
Garden trimmings	30–80	60
Wood	15–40	20
Glass	1–4	2
Tin cans	2–4	3
Nonferrous metals	2–4	2
Ferrous metals	2–6	3
Dirt, ashes, brick, etc.	6–12	8
Municipal solid waste	15–40	20

Source: Ref. 1-6.

TABLE 1.9 Typical Heating Value of MSW Components

Component	Energy, Btu/lb	
	Range	Typical
Food wastes	1500–3000	2000
Paper	5000–8000	7200
Cardboard	6000–7500	7000
Plastics	12000–16000	14000
Textiles	6500–8000	7500
Rubber	9000–12000	10000
Leather	6500–8500	7500
Garden trimmings	1000–8000	2800
Wood	7500–8500	8000
Glass	50–100	60
Tin cans	100–500	300
Nonferrous metals	—	—
Ferrous metals	100–500	300
Dirt, ashes, brick, etc.	1000–5000	3000
Municipal solid wastes	4000–6500	4500

Source: Ref. 1-5.

REFERENCES

1-1. Binder, R.: *Fluid Mechanics*, Prentice-Hall, Englewood Cliffs, N.J., 1973.

1-2. *Incinerator Standards*, Incinerator Institute of America, New York, 1968.

1-3. *Source Category Survey: Industrial Incinerators*, USEPA 450 3-80-013, May 1980.

1-4. Black, R., and Klee, A.: "The National Solid Wastes Survey: An Interim Report," presented at the 1968 Annual Meeting of the Institute of Solid Wastes of the American Public Works Association, Miami Beach, October 1968.

1-5. Brunner, C., and Schwarz, S.: *Energy and Resource Recovery from Waste*, Noyes, Park Ridge, N.J., 1983.

1-6. Tchobanoglous, G.: *Solid Wastes: Engineering-Principles and Management Issues*, McGraw-Hill, New York, 1977.

BIBLIOGRAPHY

Air Quality Control, National Association of Manufacturers, New York, 1975.

Arubuckle, J.: *Environmental Law Handbook*, Government Institutes, Washington, September 1979.

Bolton, R.: *Public Utilities Regulatory Policies Act* (*PURPA*), Malcolm Pirnie, Inc., White Plains, New York, May 1981.

Cross, F.: *Handbook on Environmental Monitoring*, Technomic, Westport, Connecticut, 1974.

The Federal Research and Development Plan for Air Pollution Control by Combustion Process Modification, USEPA Publication CPA/22-69-147, January 1977.

Municipal Incinerator Enforcement Manual, USEPA 340/1-76-013, January 1977.

CHAPTER 2
REGULATORY REQUIREMENTS

Each political entity in the United States shows concern for discharges from incineration by deferring to a higher authority or by establishing its own set of standards. Local zoning boards are often used to regulate the external effects of an incinerator facility. In this chapter the regulations of federal and state governments will be discussed.

FEDERAL REGULATIONS

The federal government has established an intensive program for regulating the incineration of hazardous wastes. Chapter 3 describes the federal hazardous waste incineration regulations. Nonhazardous waste incineration is regulated only if the waste is generated from municipal sources. Nonhazardous waste generated and incinerated by industry is not presently regulated by the federal government except for the provisions of the National Ambient Air Quality Standards, discussed later in this chapter.

Emissions from municipal solid waste incinerators are covered in the new-source performance standards, Title 40, part 60, subpart E. *Municipal solid waste* is defined as refuse, more than 50 percent of which is municipal-type waste consisting of a mixture of paper, wood, yard wastes, plastics, leather, rubber, and other combustibles, and noncombustibles such as glass and rock. This standard requires an air pollution emission rate of 0.08 gr of particulate per dry standard cubic foot of exhaust gas, corrected to 12 percent CO_2.

The second federal regulation established for control of emissions from municipal incinerators is subpart O of the above standard. This standard establishes the emissions from incinerators burning sewage sludge generated from municipal sewage treatment facilities. It limits particulate emissions from the facility to 1.3 lb per dry ton of sludge charged. It also limits visible emissions to 20 percent opacity.

Solid waste incinerator standards apply to incinerators constructed after August 17, 1971. The sewage sludge incinerator standards apply to incinerators constructed after June 11, 1973.

STATE PARTICULATE EMISSION STANDARDS

State emission standards are presented as grains per unit volume emitted or pounds per unit weight charged. Table 2.1 is a list of factors which allow conversion from one form of a standard to another. These conversions are a function of the constituents of the feed, and for this particular chart the waste feed chosen was 50 percent cellulose ($C_6H_{10}O_5$, the prime constituent of paper), 50 percent water, with a heating value of 4462 Btu/lb as charged.

Table 2.2 is a listing of the particulate emission regulations for each of the 50 states. The last column of this chart is an equivalent discharge in grains per cubic foot, in accordance with the conversion factors listed in Table 2.1 (as applicable) which provide a common basis of comparison for one state with another. These standards apply to all incineration facilities, e.g., municipal refuse, municipal sludge, industrial, and agricultural wastes.

STATE OPACITY STANDARDS

Opacity is a visual measure of particulate emission. The greater the particulate loading, the darker (or more opaque) the exhaust. In general, the standards governing particulate emissions have an effect on stack opacity. With opacity a visible quality of a plume, however, it was incumbent upon public officials to establish a standard that can be easily identified and verified by the layper-

TABLE 2.1 Emissions Standards Conversion Factors

	lb/ton Refuse (as received)	lb/1000 lb Flue gas at 50% excess air	lb/1000 lb Flue gas at 12% CO_2	gr/st ft^3 at 50% excess air	gr/st ft^3 at 12% CO_2	g/(N·m^3) at ntp, 7% CO_2
lb/ton refuse (as received)	1	0.089	0.10	0.047	0.053	0.067
lb/1000 lb flue gas at 50% excess air	11.27	1	1.12	0.52	0.585	0.74
lb/1000 lb flue gas at 12% CO_2	10.0	0.89	1	0.46	0.52	0.66
gr/st ft^3 at 50% excess air	21.31	1.93	2.16	1	1.12	1.42
gr/st ft^3 at 12% CO_2	18.85	1.71	1.92	0.89	1	1.26
g/(N·m^3) at ntp, 7% CO_2	15.0	1.36	1.53	0.704	0.79	1

Source: Ref. 2-1.

TABLE 2.2 Particulate Emission Limitations for New and Existing Incinerators

State	Regulation					Equivalent
	Value	Units	Corrected to	Process conditions	Validity	Common regulation (gr/dscf @ 12% CO_2)
1 Alabama	0.1	lb/100 lb charged		>50 TPD		0.12
	0.2	lb/100 lb charged		≤50 TPD		0.24
2 Alaska	0.3	gr/dscf	12% CO_2	≤200 lb/h		0.3
	0.2	gr/dscf	12% CO_2	200–1000 lb/h		0.2
	0.1	gr/dscf	12% CO_2	>1000 lb/h		0.1
3 Arizona	0.1	gr/dscf	12% CO_2			0.1
4 Arkansas	0.2	gr/dscf	12% CO_2	≥200 lb/h		0.2
	0.3	gr/dscf	12% CO_2	<200 lb/h		0.3
5 California	0.3	gr/dscf	12% CO_2	typical of the 43 APCDs		0.3
6 Colorado	0.1	gr/dscf	12% CO_2		designated control areas	0.1
	0.15	gr/dscf	12% CO_2		other areas	0.15
7 Connecticut	0.08	gr/dscf	12% CO_2		built after 6/1/72	0.08
	0.4	lb/1000 lb	50% excess air		built before 6/1/72	0.26
8 Delaware	0.2	lb/h		100 lb/h		0.24
	1.0	lb/h		500 lb/h		0.24
	2.0	lb/h		1000 lb/h		0.24
	5.0	lb/h		3000 lb/h		0.2
9 Florida	0.08	gr/dscf	50% excess air	≥50 TPD	built after 2/11/72	0.1
	0.1	gr/dscf	50% excess air	≥50 TPD	built before 2/11/72	0.12
10 Georgia	0.1	gr/dscf	12% CO_2	≤50 TPD–type 0, 1, 2 waste	new (built after 1/1/72)	0.1
	0.2	gr/dscf	12% CO_2	≤50 TPD–type 3, 4, 5, 6 waste	new (built after 1/1/72)	0.2
	0.2	gr/dscf	12% CO_2	type 0, 1, 2 waste	existing before 1/1/72	0.2
	0.3	gr/dscf	12% CO_2	type 3, 4, 5, 6 waste	existing before 1/1/72	0.3
	0.08	gr/dscf	12% CO_2	≥50 TPD	new (built after 1/1/72)	0.08
11 Hawaii	0.2	lb/100 lb charged				0.24
12 Idaho	0.2	lb/100 lb charged				0.24

TABLE 2.2 Particulate Emission Limitations for New and Existing Incinerators (Continued)

	Regulation					Equivalent
State	Value	Units	Corrected to	Process conditions	Validity	Common regulation (gr/dscf @ 12% CO_2)
13 Illinois	0.08	gr/dscf	12% CO_2	2000–60,000 lb/h		0.08
	0.2	gr/dscf	12% CO_2	≤2000 lb/h	built before 4/15/72	0.2
	0.1	gr/dscf	12% CO_2	≤2000 lb/h	built after 4/15/72	0.1
14 Indiana	0.3	lb/1000 lb gas	50% excess air	≥200 lb/h		0.19
	0.5	lb/1000 lb gas	50% excess air	<200 lb/h		0.32
15 Iowa	0.2	gr/dscf	12% CO_2	≥1000 lb/h		0.2
	0.35	gr/dscf	12% CO_2	<1000 lb/h		0.35
16 Kansas	0.3	gr/dscf	12% CO_2	<200 lb/h		0.3
	0.2	gr/dscf	12% CO_2	200–20,000 lb/h		0.2
	0.1	gr/dscf	12% CO_2	>20,000 lb/h		0.1
17 Kentucky	0.2	gr/dscf	12% CO_2	≤50 TPD		0.2
	0.08	gr/dscf	12% CO_2	>50 TPD		0.08
18 Louisiana	0.2	gr/dscf	12% CO_2			0.2
19 Maine	0.2	gr/dscf	12% CO_2			0.2
20 Maryland	0.1	gr/dscf	12% CO_2	<2000 lb/h	built after 1/17/72	0.1
	0.03	gr/dscf	12% CO_2	>2000 lb/h	built after 1/17/72	0.03
	0.3	gr/dscf	12% CO_2	<200 lb/h	built before 1/17/72	0.3
	0.2	gr/dscf	12% CO_2	>200 lb/h	built before 1/17/72	0.2
21 Massachusetts	0.1	gr/dscf	12% CO_2		existing	0.1
	0.05	gr/dscf	12% CO_2		new	0.05
22 Michigan	0.65	lb/1000 lb gas	50% excess air	0–100 lb/h		0.42
	0.3	lb/100 lb gas	50% excess air	>100 lb/h		0.19
23 Minnesota	0.3	gr/dscf	12% CO_2	<200 lb/h	existing before 8/17/71	0.3
	0.2	gr/dscf	12% CO_2	200–2000 lb/h	existing before 8/17/71	0.2
	0.1	gr/dscf	12% CO_2	>2000 lb/h	existing before 8/17/71	0.1
	0.2	gr/dscf	12% CO_2	<200 lb/h	new (built after 8/17/71)	0.2
	0.15	gr/dscf	12% CO_2	200–2000 lb/h	new (built after 8/17/71)	0.15
	0.1	gr/dscf	12% CO_2	>2000 lb/h	new (built after 8/17/71)	0.1
24 Mississippi	0.2	gr/dscf	12% CO_2	Design capacity		0.2
	0.1	gr/dscf	12% CO_2	New sources near residential areas		0.1
25 Missouri	0.2	gr/dscf	12% CO_2	≥200 lb/h		0.2
	0.3	gr/dscf	12% CO_2	<200 lb/h		0.3

26 Montana	0.2	gr/dscf	12% CO_2	>200 lb/h	existing before 9/5/75	0.2
	0.3	gr/dscf	12% CO_2	≤200 lb/h	existing before 9/5/75	0.3
	0.1	gr/dscf	12% CO_2		all others	0.1
27 Nebraska	0.2	gr/dscf	12% CO_2	<2000 lb/h		0.2
	0.1	gr/dscf	12% CO_2	≥2000 lb/h		0.1
28 Nevada	3.0	lb/ton charged		<2000 lb/h		0.18
	variable	$E = 40.7 \times 10^{-5}$ C	C, E = lb/h	>2000 lb/h		0.05
29 New Hampshire	0.3	gr/dscf	12% CO_2	≤200 lb/h		0.3
	0.2	gr/dscf	12% CO_2	>200 lb/h		0.2
	0.08	gr/dscf	12% CO_2	>50 TPD	built after 4/20/74	0.08
30 New Jersey	0.2	gr/dscf	12% CO_2	<2000 lb/h	type 0, 1, 2, 3 waste only	0.2
	0.2	gr/dscf	12% CO_2	all others		0.1
31 New Mexico	only opacity	regulations		≤50 TPD		—
	0.08	gr/dscf	12% CO_2	>50 TPD	new (built after 8/17/71)	0.08
32 New York	0.5	lb/100 lb charged		>2000 lb/h	built between 4/1/62 and 1/1/70	0.6
	0.5	lb/100 lb charged		≤2000 lb/h	built between 4/1/62 and 1/1/68	0.6
	variable (e.g., 0.3)	lb/h		≤100 lb/h	built after 1/1/68	0.36
	variable (e.g., 3.0)	lb/h		@1000 lb/h	built after 1/1/68	0.36
	variable (e.g., 7.5)	lb/h		@3000 lb/h	built after 1/1/70	0.3
33 North Carolina	0.2	lb/h		0–100 lb/h		0.24
	0.4	lb/h		@200 lb/h		0.24
	1.0	lb/h		@500 lb/h		0.24
	2.0	lb/h		@1000 lb/h		0.24
	4.0	lb/h		≥2000 lb/h		0.24
34 North Dakota	variable	lb/h		@100 lb/h		0.4
				@1000 lb/h		0.31
				@3000 lb/h		0.24
35 Ohio	0.1	lb/100 lb charged		≥100 lb/h		0.12
	0.2	lb/100 lb charged		<100 lb/h		0.24
36 Oklahoma	variable	lb/h		@100 lb/h		0.48
				@1000 lb/h		0.31
				@3000 lb/h		0.21

TABLE 2.2 Particulate Emission Limitations for New and Existing Incinerators (Continued)

State	Regulation					Equivalent
	Value	Units	Corrected to	Process conditions	Validity	Common regulation (gr/dscf @ 12% CO_2)
37 Oregon	0.3	gr/dscf		≥100 lb/h		0.3
	0.2	gr/dscf		>200 lb/h	built before 6/1/70	0.2
	0.1	gr/dscf		>200 lb/h	built after 6/1/70	0.1
38 Pennsylvania	0.1	gr/dscf	12% CO_2			0.1
39 Rhode Island	0.16	gr/dscf	12% CO_2	<2000 lb/h		0.16
	0.08	gr/dscf	12% CO_3	≥2000 lb/h		0.08
40 South Carolina	0.5	lb/10^6 Btu		@10 mm Btu/h		0.27
41 South Dakota	0.2	lb/100 lb charged				
42 Tennessee	0.2	% of charge		≤2000 lb/h		0.24
	0.1	% of charge		>2000 lb/h		0.12
43 Texas	variable	lb/h		@1000 lb/h		0.41
				@3000 lb/h		0.27
44 Utah	0.08	gr/dscf	12% CO_2	>50 TPD		0.08
45 Vermont	0.1	lb/100 lb charged				0.12
46 Virginia	0.14	gr/dscf	12% CO_2			0.14
47 Washington	0.1	gr/dscf	7% O_2			0.11
48 West Virginia	8.25	lb/ton		≤200 lb/h		0.5
	5.43	lb/ton		>200 lb/h		0.33
49 Wisconsin	0.2	lb/1000 lb exhaust gas	12% CO_2	500–4000 lb/h	built after 4/1/72	0.11
	0.3	lb/1000 lb exhaust gas	12% CO_2	≤500 lb/h	built after 4/1/72	0.17
	0.5	lb/1000 lb exhaust gas	12% CO_2	>500 lb/h	built before 4/1/72	0.28
	0.6	lb/1000 lb exhaust gas	12% CO_2	≤500 lb/h	built before 4/1/72	0.34
	0.15	lb/1000 lb exhaust gas	12% CO_2	≥4000 lb/h	built after 4/1/72	0.08
50 Wyoming	0.2	lb/100 lb charged				0.24

Source: Ref. 2-2.

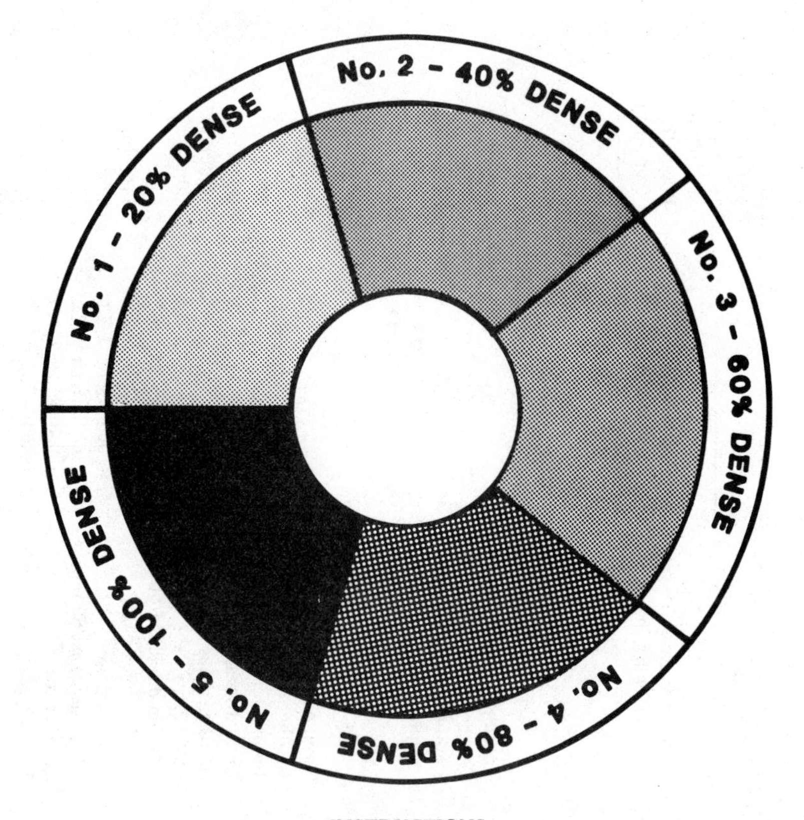

INSTRUCTIONS

1. Hold chart at arms length and view smoke through circle provided.
2. Observer should not be less than 100 ft, nor more than 1/4 mile from the stack.
3. Line of observation should be at right angles to the direction the smoke travels and viewed in the same light.
4. Do not try to observe smoke into direct sun or setting sun. Sun should always be overhead.
5. Match smoke with corresponding color on chart–note density number and time.

FIGURE 2.1 Opacity measurement. Alken smoke chart (Ringelmann type).

son. Figure 2.1 is a typical opacity chart. The chart is normally printed on a transparent surface, and the degree of opacity of the exhaust is matched to the chart reading to obtain a qualitative identification of degree of opacity.

Table 2.3 lists opacity requirements for each state. The percentage of opacity is that opacity noted in Fig. 2.1.

TABLE 2.3 Opacity Regulations for New and Existing Commercial and Industrial Incinerators

	Regulation				Equivalent
States	Value	Units	Process Conditions	Validity	Common Regulation (% opacity)
1 Alabama	60	% opacity	3 min discharge/60 min		60
	20	% opacity	all other times		20
2 Alaska	40	% opacity		installed before 7/1/72	40
	20	% opacity		installed after 7/1/72	20
3 Arizona	exempt		0.5 min discharge/60 min		exempt
	20	% opacity	all other times		20
4 Arkansas	No. 3	Ringelmann	5 min discharge/60 min		60
	No. 1	Ringelmann	all other times	built after 7/30/73	20
	No. 2	Ringelmann		built before 7/30/73	40
5 California					
6 Colorado	20	% opacity			20
7 Connecticut	40	% opacity	5 min discharge/60 min		40
	20	% opacity	all other times		20
8 Delaware	20	% opacity	3 min discharge/60 min		20
9 District of Columbia	20	% opacity	2 min discharge/60 min	existing	20
	prohibited		all other times	existing	prohibited
10 Florida	20	% opacity	⩽50 TPD, 3 min discharge/60 min		20
	prohibited		all other times		prohibited
11 Georgia	20	% opacity	all other times	installed after 1/1/72	20
	40	% opacity	6 min discharge/60 min	installed after 1/1/72	40
	40	% opacity	all other times	installed before 1/1/72	40
	60	% opacity	6 min discharge/60 min	installed before 1/1/72	60
12 Hawaii	40	% opacity			40
13 Idaho	No. 2	Ringelmann	3 min discharge/60 min	built before 4/1/72	40

	No. 1	Ringelmann	3 min discharge/60 min	built after 4/1/72	20
14 Illinois	30	% opacity	all other times		30
	30–40	% opacity	3 min discharge/60 min		30–60
15 Indiana	40	% opacity	15 min discharge/60 min		40
16 Iowa	40	% opacity			40
	60	% opacity	3 min discharge/60 min during breakdowns, etc.		60
17 Kansas	20	% opacity			20
18 Kentucky	20	% opacity			20
19 Louisiana	No. 1	Ringelmann	all other times		20
	>No. 1	Ringelmann	4 min discharge/60 min		>20
20 Maine	No. 1	Ringelmann			20
21 Maryland					
22 Massachusetts	No. 1	Ringelmann			20
23 Michigan	40	% opacity	3 min discharge/60 min		40
	20	% opacity	all other times		20
24 Minnesota	20	% opacity			20
25 Mississippi	40	% opacity			40
26 Missouri	No. 1	Ringelmann		built after 2/10/72	20
	No. 2	Ringelmann		built before 2/10/72	40
27 Montana	10	% opacity			10
28 Nebraska	20	% opacity			20
29 Neveda	20	% opacity	1 min discharge/60 min		20
30 New Hampshire	No. 1	Ringelmann	3 min discharge/60 min		20
31 New Jersey	No. 2	Ringelmann	3 consecutive minutes		40
	No. 1	Ringelmann	all other times		20
32 New Mexico	No. 1	Ringelmann	2 min discharge/60 min		20
33 New York (state)	40	% opacity		built before 1/26/67	40
(state)	20	% opacity		built after 1/26/67	20
(city)	No. 1	Ringelmann	3 min discharge/60 min		20
34 North Carolina					
35 North Dakota	No. 3	Ringelmann	4 min discharge/60 min		60
	No. 1	Ringelmann	all other times		20

TABLE 2.3 Opacity Regulations for New and Existing Commercial and Industrial Incinerators (Continued)

	Regulation				Equivalent
States	Value	Units	Process Conditions	Validity	Common Regulation (% opacity)
36 Ohio	60	% opacity	3 min discharge/60 min		60
	20	% opacity	all other times		20
37 Oklahoma	No. 1	Ringelmann	all other times		20
	No. 3	Ringelmann	5 min discharge/60 min		60
38 Oregon	40	% opacity	3 min discharge/60 min	built before 5/1/70	40
	20	% opacity	3 min discharge/60 min	built after 6/1/70	20
39 Pennsylvania	20	% opacity	3 min discharge/60 min		20
40 Puerto Rica	20	% opacity	all other times		20
	60	% opacity	6 min discharge/60 min		60
41 Rhode Island	20	% opacity	3 min discharge/60 min		20
42 South Carolina	No. 1	Ringelmann	3 min discharge/60 min		20
43 South Dakota	20	% opacity	all other times		20
	60	% opacity	3 min discharge/60 min		60
44 Tennessee	20	% opacity	5 min discharge/60 min		20
45 Texas	30	% opacity	5 min average	built before 1/31/72	30
	20	% opacity	5 min average	built after 1/31/72	20
46 Utah	No. 1	Ringelmann			20
47 Vermont	40	% opacity	6 min discharge/60 min	built before 4/30/70	40
	20	% opacity	6 min discharge/60 min	built after 4/30/70	20
48 Virginia	20	% opacity			20
49 Washington	20	% opacity	3 min discharge/60 min		20
	>20	% opacity	15 min/8 hr		>20
50 West Virginia	No. 1	Ringelmann			20
51 Wisconsin	20	% opacity		built after 4/1/72	20
52 Wyoming	20	% opacity			20

Source: Ref. 2-2.

THE NATIONAL AMBIENT AIR QUALITY STANDARDS

The federal government has established National Ambient Air Quality Standards (NAAQS) which attempt to define the desired, or permissible, maximum levels of pollutants in the air throughout the country. The NAAQS establishes clean air standards for minimal (class I), moderate (class II), and extensive (class III) growth areas. In conjunction with the NAAQS geographical locations are designated as either attainment or nonattainment areas:

Attainment area. Air in this geographical location is presently considered clean, i.e., within the definition of clean air in the NAAQS for the designated growth area classification.

Nonattainment area. Air quality in a nonattainment area is below the quality established in the NAAQS for the designated growth area classification.

PREVENTION OF SIGNIFICANT DETERIORATION

A set of regulations has been established by the federal government to help achieve the air quality standards designated in the NAAQS.

Implementation of NSR and BACT procedures is required, as noted below, for significant sources of air pollution. The PSD defines a significant source as either of the following:

- Any source which has the potential to emit over 250 tons/year of pollutants into the atmosphere. Pollutants are any of those pollutants listed in Table 2.4.
- Any one of the 28 source categories listed in Table 2.5 with the potential to discharge more than 100 tons/year of pollutants into the atmosphere.

The pollutants are measured at the stack. The term *potential* refers to the total amount of pollutants which may be discharged by the process in question. For instance, although a unit may be in operation only 30 weeks/year, it can be operated for 52 weeks/year. The potential emissions therefore refer to, in this illustration, 52 weeks/year operation, although the process may only operate a fraction of the year.

THE NEW-SOURCE REVIEW PROCESS

The chart designated Fig. 2.2 illustrates the *new-source review* (NSR) process. Following this chart:

If an installation does not constitute a major new or modified source, it is not subject to federal PSD review. It must, however, comply with applicable emissions regulations of the state in which it is located. Some states have adopted modified provisions of the National Ambient Air Quality Standards (NAAQS) and may require simplified source modeling. A public hearing for a nonmajor source may or may not be necessary, and a decision to hold a hearing is governed by state statute.

If a source is a major new or modified source, as defined previously, a deter-

TABLE 2.4 *De Minimis* Emission Values and Monitoring Exemption

Pollutant	Emission rate (tons/year)	Air quality impact ($\mu g/m^3$) and averaging time
Carbon monoxide	100.	575, 8-h
Nitrogen oxides	40.	14, 24-h
Sulfur dioxide	40.	13, 24-h
Total suspended particulates	25.	10, 24-h
Ozone (volatile organic compounds)	40.	—[a]
Lead	0.6	0.1, 24-h
Asbestos	0.007	—[b]
Beryllium	0.0004	0.0005, 24-h
Mercury	0.1	0.25, 24-h
Vinyl chloride	1.0	15, 24-h
Fluorides	3.	0.25, 24-h
Sulfuric acid mist	7.	—[b]
Total reduced sulfur (including H_2S)	10.	10, 1-h
Reduced sulfur (including H_2S)	10.	10, 1-h
Hydrogen sulfide	10.	0.023, 1-h

[a] All cases where VOC emissions are less than 100 tons/year.
[b] No satisfactory monitoring technique available at this time.
Source: Ref. 2-3.

TABLE 2.5 Major Stationary Sources of Air Pollution

Coal cleaning plants (with thermal dryers)
Kraft pulp mills
Portland cement plants
Primary zinc smelters
Iron and steel mills
Primary aluminum ore reduction plants
Primary copper smelters
Municipal incinerators capable of charging more than 250 tons of refuse per day
Hydrofluoric acid plants
Sulfuric acid plants
Nitric acid plants
Petroleum refineries
Lime plants
Phosphate rock processing plants
Coke oven batteries
Sulfur recovery plants
Carbon black plant (furnace process)
Primary lead smelters
Fuel conversion plants
Sintering plants
Secondary metal production facilities
Chemical process plants
Fossil-fuel boilers of more than 250,000,000 BTU/hr heat input
Petroleum storage and transfer facilities with a capacity exceeding 300,000 bbl
Taconite ore processing facilities
Glass fiber processing plants
Charcoal production facilities

Source: Ref. 2-3.

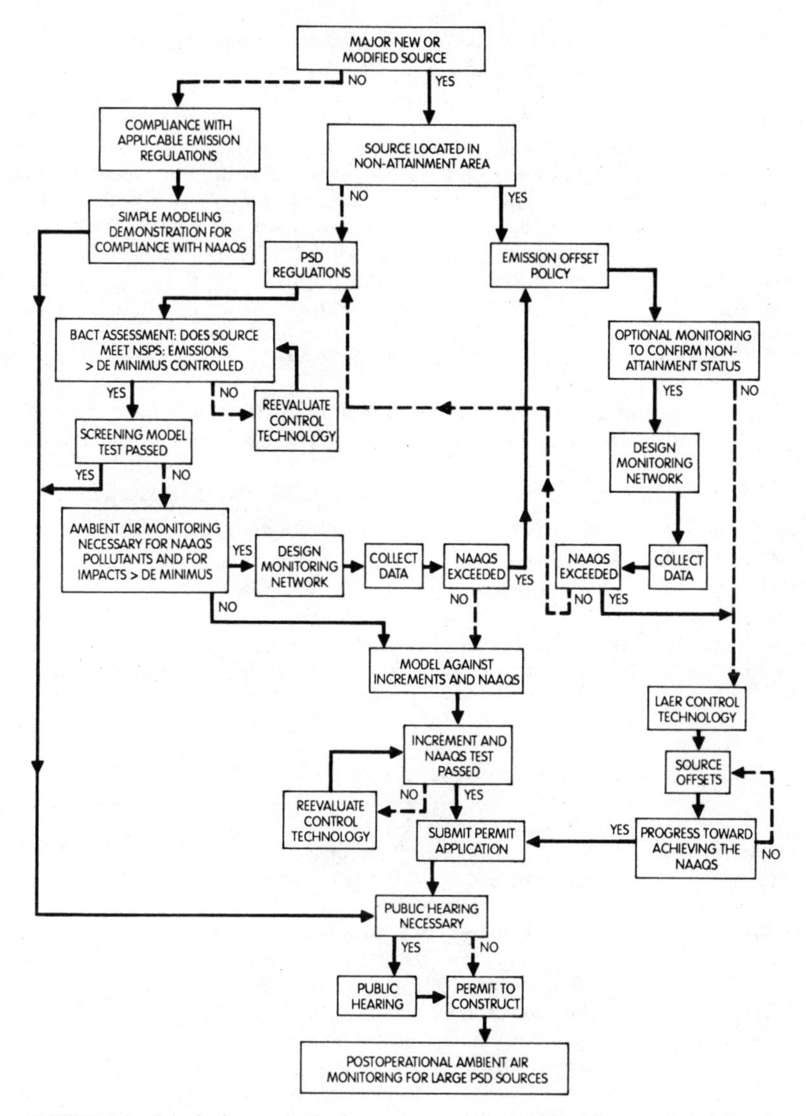

FIGURE 2.2 Principal steps in the new-source review process. *Source*: Ref. 2-3.

mination must first be made as to its location. If it is not located in a nonattainment area, the PSD requires that an assessment of the *best available control technology* (BACT) be applied to each pollutant emission in excess of the *de minimis* value. The BACT analysis evaluates the state of the art in air pollution control equipment for the application in question and considers the economic and energy implications of that equipment.

A screening model is applied. This is a relatively simple mathematical modeling technique to determine the order of magnitude of the resultant emissions on the NAAQS. If the result of the screening model is a discharge less than 50 percent of what is allowable under the NAAQS, further source study is not necessary and the permitting question can go to a public hearing. If the screening model indicates that a discharge equivalent to greater than 50 percent of the NAAQS allowable amount will occur, then further analysis must be performed. Ambient air monitoring is necessary, and a complete computer modeling procedure must be performed. This procedure utilizes inputs from the anticipated source (pollutants discharged, stack temperature, stack gas velocity, etc.), meteorological data (prevailing winds, latitude, storm frequency, mean temperature, precipitation, etc.), and receptor data (topography, i.e., presence of hills or mountains and their height, water surfaces, location of other structures, etc.). If this modeling indicates that the NAAQS will not be met, the *emissions offset policy* (EOP) procedures must be instituted. If the NAAQS will be satisfied, the way is open for a public hearing.

If the new (or modified) source is located in a nonattainment area, procedures of the emissions offset policy must be followed. The EOP requires the following:

- Source emissions must be controlled to the greatest degree possible.
- More than equivalent offsetting emission reductions must be secured for existing sources on the same site.
- Progress must be made toward achieving the NAAQS.

After the EOP is satisfied, the adequacy of the new source with respect to emissions can be demonstrated by a modeling procedure or by implementation of

TABLE 2.6 Air Emissions Glossary

AQCR	Air Quality Control Region
BACT	Best Available Control Technology
CAA	Clean Air Act
CAAA	Clean Air Act Amendments
CTG	Control Technology Guideline
DEC	(New York State) Department of Environmental Conservation
EOP	Emission Offset Policy
EPA	Environmental Protection Agency
GEP	Good Engineering Policy
LAER	Lowest Achievable Emission Rate
NA	Non-Attainment
NAAQS	National Ambient Air Quality Standards
NESHAPs	National Emission Standards for Hazardous Air Pollutants
NSPS	New Source Performance Standards
NSR	New Source Review
ORD	Office of Research and Development
PSD	Prevention of Significant Deterioration
RACT	Reasonable Available Control Technology
RCRA	Resource Conservation and Recovery Act
RDF	Resource-Derived Fuel
SIP	State Implementation Plan
TSCA	Toxic Substances Control Act
VOC	Volatile Organic Compound

the *lowest achievable emissions rate* (LAER) review. This review is similar to the BACT review except that economic or energy factors are not taken into account.

If the modeling procedure is utilized and it is found that the NAAQS is exceeded, the LAER must be instituted. If the NAAQS is not exceeded, the source can be considered as if it were located in an "attainment" area.

Application of the LAER procedure must demonstrate that there is progress toward meeting the NAAQS before a public hearing can be advertised.

An agency review of the above procedure must prove successful before a public hearing is called. A permit is granted after a public hearing and a designated comment period.

Table 2.6 is a glossary of the acronyms used in this section.

OTHER REQUIREMENTS

Regulatory requirements addressing central disposal systems (resource recovery or waste-to-energy plants) are discussed in Chapter 13. Specific requirements for incineration facilities for biomedical (hospital) waste streams are discussed in Chapter 14.

REFERENCES

2-1. Niessen, W., and Sarofin, A.: "Incinerator Air Pollution, Facts and Speculation," National Incinerator Conference, American Society of Mechanical Engineers, Orlando, Florida, 1970.

2-2. *Source Category Survey: Industrial Incinerators*, USEPA 450/3-85-916-013, July 1988.

2-3. Berkun, J., and Star, A.: *Implications of the New Source Review Process*, Malcolm Pirnie, Inc., White Plains, New York, 1980.

BIBLIOGRAPHY

Ambient Air Quality Standards, USEPA, 40CFR50, 1981.

Brunner, C. R.: *Hazardous Air Emissions from Incineration*, Chapman & Hall, New York, 1986.

Devitt, T.: *Inspection Manual for Enforcement of New Source Performance Standards: Municipal Incinerators*, USEPA 340/1-75-003, February 1975.

Devitt, T.: *Inspection Manual for Enforcement of New Source Performance Standards: Sewage Sludge Incinerators*, USEPA 340/1-75-004, February 1975.

Dravnieks, A.: "Odor Perception and Odorous Air Pollution," *Transactions of the Pulp and Paper Institute*, May 1972.

Gunther, C.: "Resource Recovery and the Clean Air Act," *Waste Age*, May 1981.

Matey, J.: "The Clean Air Act," *Pollution Engineering*, November 1981.

Regulatory Options for the Control of Odors, USEPA 450 5-80-003, February 1980.

A Review of Standards of Performance for New Stationary Sources: Incinerators, USEPA 60/4-79-009, March 1979.

A Review of Standards of Performance for New Stationary Sources: Sewage Sludge Incinerators, USEPA 60/4-79-010, March 1978.

"Swooping Down on Air Pollution," *Chemical Engineering*, June 1, 1981.

CHAPTER 3
HAZARDOUS WASTE

As a result of an increased public awareness of the effect of industrialization on the environment, the quality of our natural resources, and our national health, the federal government has established statutory control over discharges into the environment. In this chapter the nature of these requirements will be discussed with regard to incineration.

BACKGROUND

As population increases, the quantity of wastes generated will increase. With technological advances the quality of wastes will change. As energy costs increase, many so-called wastes are no longer discarded but are used as fuels. Not all wastes are hazardous to the environment or to the population. Most wastes generated, such as agricultural waste, domestic sewage, and household waste, are not hazardous, although they are objectionable and must be effectively disposed of. Of the thousands of millions of tons of wastes generated in this country annually, approximately 60 million tons is classified as hazardous. The federal government, as a first step toward control of hazardous waste disposal, has established definitions and classifications of hazardous waste.

HAZARDOUS WASTE CLASSIFICATION

The Environmental Protection Agency (EPA) through regulation 40 CFR part 261, dated May 19, 1980, and subsequent revisions has established a set of definitions of hazardous wastes:

1. Ignitable waste (ignitability), hazard code I, will have at least one of the following properties:
   a. A liquid having a flash point less than 140°F. An aqueous solution containing less than 20 percent alcohol by volume is excluded from this definition.
   b. A substance, other than a liquid, which can cause fire through friction, absorption of moisture, or spontaneous chemical changes, under standard temperature and pressure. When this substance burns, it does so vigorously and persistently.
   c. An ignitable compressed gas (see 49 CFR 173.300 for further definition).
   d. An oxidizer (see 40 CFR 173.151 for further definition).

An ignitable waste which is not listed elsewhere as a hazardous waste is given an EPA hazardous waste number of D001.

2. Corrosive waste (corrosivity), hazard code C, will have either or both of the following properties:
   a. An aqueous waste with pH equal to or less than 2.0 or a pH equal to or greater than 12.5.
   b. A liquid that corrodes carbon steel (grade SAE 1030) at a rate greater than 0.250 in/year.

A corrosive waste which is not listed elsewhere as a hazardous waste is given an EPA hazardous waste number of D002.

3. Reactive waste (reactivity), hazard code R, will have at least one of the following properties:
   a. A substance which is normally unstable and undergoes violent physical and/or chemical change without detonating
   b. A substance that reacts violently with water
   c. A waste which forms a potentially explosive mixture when wetted with water
   d. A substance which can generate harmful gases, vapors, or fumes when mixed with water
   e. A cyanide- or sulfide-bearing waste which can generate harmful gases, vapors, or fumes when exposed to pH conditions between 2 and 12.5
   f. A waste which, when subjected to a strong initiating source or when heated in confinement, will detonate and/or generate an explosive reaction
   g. A substance which is readily capable of detonation at standard temperature and pressure
   h. An explosive listed as class A, class B, or "forbidden" in accordance with 49 CFR 173

A reactive waste which is not listed elsewhere as a hazardous waste is given an EPA hazardous waste number of D003.

4. EP toxic waste (EP toxicity), hazard code E. If the extract from a representative sample of this waste (EP, extract procedure) contains contamination in excess of that allowed in Table 3.1, it is classified as a hazardous waste.

   An EP toxic waste which is not listed elsewhere as a hazardous waste is given an EPA hazardous waste number corresponding to that contaminant listed in Table 3.1 causing it to be hazardous.

5. Acute hazardous waste, hazard code H. A substance which has been found to be fatal to humans in low doses, or in the absence of data on human toxicity, has been found to be fatal in corresponding human concentrations in laboratory animals.
6. Toxic waste, hazard code T. Wastes that have been found, through laboratory studies, to have a carcinogenic, mutagenic, or teratogenic effect on human or other life forms. Definitions of these terms are as follows:

   - *Carcinogenic:* producing or tending to produce cancer
   - *Mutagenic:* capable of inducing mutations in future offspring
   - *Teratogenic:* producing abnormal growth in fetuses

TABLE 3.1 Allowable Contaminants

EPA hazardous waste number	Contaminant	Maximum concentration (mg/L)
D004	Arsenic	5.0
D005	Barium	100.0
D006	Cadmium	1.0
D007	Chromium	5.0
D008	Lead	5.0
D009	Mercury	0.2
D010	Selenium	1.0
D011	Silver	5.0
D012	Endrin (1,2,3,4,10,10-hexachloro-1,7-epoxy-1,4,4a,5,6, 7,8,8a-octahydro-1,4-endo, endo-5,8-dimethano naphthalene).	0.02
D013	Lindane (1,2,3,4,5,6-hexachlorocyclohexane, gamma isomer).	0.4
D014	Methoxychlor (1,1,1-Trichloro-2,2-bis [p-methoxyphenyl]ethane).	10.0
D015	Toxaphene ($C_{10}H_{10}Cl_8$, Technical chlorinated camphene, 67–69 percent chlorine).	0.5
D016	2,4-D (2,4-Dichlorophenoxyacetic acid).	10.0
D017	2,4,5-TP Silvex (2,4,5-Trichlorophenoxypropionic acid).	1.0

Source: Ref. 3-1.

HAZARDOUS WASTE LISTINGS

Based on the above definitions a set of lists has been included in the Resource Conservation and Recovery Act (RCRA) regulation identifying hazardous wastes.

1. *Nonspecific sources.* The listing in Table 3.2 identifies nonspecific sources the wastes of which are hazardous. The hazard code at the right of the listing refers to the quality of the waste creating its hazardous nature.
2. *Specific sources.* Table 3.3 lists specific processes where the wastes generated are classified as hazardous. The waste hazard code identifies the reason(s) for its hazard classification.
3. *Acute hazardous wastes.* The substances identified in Table 3.4 when discarded are classified as acute hazardous wastes.
4. *Toxic wastes.* The substances listed in Table 3.5 when discarded are classified as toxic hazardous wastes.
5. *Hazardous constituents.* A waste containing any of the substances listed in Table 3.6 is considered a hazardous waste.

ANCILLARY MATERIALS

The container or container liner in contact with a hazardous waste is itself hazardous. In addition, clothing, debris, soil, etc., which has become contaminated

TABLE 3.2 Hazardous Waste from Nonspecific Sources

Industry and EPA hazardous waste No.	Hazardous waste	Hazard code
Generic:		
F001	The spent halogenated solvents used in degreasing, tetrachloroethylene, trichloroethylene, methylene chloride, 1,1,1-trichloroethane, carbon tetrachloride, and the chlorinated fluorocarbons; and sludges from the recovery of these solvents in degreasing operations.	(T)
F002	The spent halogenated solvents, tetrachloroethylene, methylene chloride, trichloroethylene, 1,1,1-trichloroethane, chlorobenzene, 1,1,2-trichloro-1,2,2-trifluoroethane, o-dichlorobenzene, trichlorofluoromethane and the still bottoms from the recovery of these solvents.	(T)
F003	The spent non-halogenated solvents, xylene, acetone, ethyl acetate, ethyl benzene, ethyl ether, n-butyl alcohol, cyclohexanone, and the still bottoms from the recovery of these solvents.	(I)
F004	The spent non-halogenated solvents, cresols and cresylic acid, nitrobenzene, and the still bottoms from the recovery of these solvents	(T)
F005	The spent non-halogenated solvents, methanol, toluene, methyl ethyl ketone, methyl isobutyl ketone, carbon disulfide, isobutanol, pyridine and the still bottoms from the recovery of these solvents.	(I, T)
F006	Wastewater treatment sludges from electroplating operations	(T)
F007	Spent plating bath solutions from electroplating operations	(R, T)
F008	Plating bath sludges from the bottom of plating baths from electroplating operations	(R, T)
F009	Spent stripping and cleaning bath solutions from electroplating operations	(R, T)
F010	Quenching bath sludge from oil baths from metal heat treating operations	(R, T)
F011	Spent solutions from salt bath pot cleaning from metal heat treating operations	(R, T)
F012	Quenching wastewater treatment sludges from metal heat treating operations	(T)
F013	Flotation tailings from selective flotation from mineral metals recovery operations	(T)
F014	Cyanidation wastewater treatment tailing pond sediment from mineral metals recovery operations	(T)
F015	Spent cyanide bath solutions from mineral metals recovery operations	(R, T)
F016	Dewatered air pollution control scrubber sludges from coke ovens and blast furnaces	(T)

Source: Ref. 3-1.

TABLE 3.3 Hazardous Waste from Specific Sources

Industry and EPA hazardous waste No.	Hazardous waste	Hazard code
Wood Preservation: K001	Bottom sediment sludge from the treatment of wastewaters from wood preserving processes that use creosote and/or pentachlorophenol	(T)
Inorganic Pigments:		
K002	Wastewater treatment sludge from the production of chrome yellow and orange pigments	(T)
K003	Wastewater treatment sludge from the production of molybdate orange pigments	(T)
K004	Wastewater treatment sludge from the production of zinc yellow pigments	(T)
K005	Wastewater treatment sludge from the production of chrome green pigments	(T)
K006	Wastewater treatment sludge from the production of chrome oxide green pigments (anhydrous and hydrated)	(T)
K007	Wastewater treatment sludge from the production of iron blue pigments	(T)
K008	Oven residue from the production of chrome oxide green pigments	(T)
Organic Chemicals:		
K009	Distillation bottoms from the production of acetaldehyde from ethylene	(T)
K010	Distillation side cuts from the production of acetaldehyde from ethylene	(T)
K011	Bottom stream from the wastewater stripper in the production of acrylonitrile	(R, T)
K012	Still bottoms from the final purification of acrylonitrile in the production of acrylonitrile	(T)
K013	Bottom stream from the acetonitrile column in the production of acrylonitrile	(R, T)
K014	Bottoms from the acetronitrile purification column in the production of acrylonitrile	(T)
K015	Still bottoms from the distillation of benzyl chloride	(T)
K016	Heavy ends or distillation residues from the production of carbon tetrachloride	(T)
K017	Heavy ends (still bottoms) from the purification column in the production of epichlorohydrin	(T)
K018	Heavy ends from fractionation in ethyl chloride production	(T)
K019	Heavy ends from the distillation of ethylene dichloride in ethylene dichloride production	(T)
K020	Heavy ends from the distillation of vinyl chloride in vinyl chloride monomer production	(T)
K021	Aqueous spent antimony catalyst waste from fluoromethanes production	(T)
K022	Distillation bottom tars from the production of phenol/acetone from cumene	(T)
K023	Distillation light ends from the production of phthalic anhydride from naphthalene	(T)
K024	Distillation bottoms from the production of phthalic anhydride from naphthalene	(T)
K025	Distillation bottoms from the production of nitrobenzene by the nitration of benzene	(T)
K026	Stripping still tails from the production of methyl ethyl pyridines	(T)
K027	Centrifuge residue from toluene diisocyanate production	(R, T)
K028	Spent catalyst from the hydrochlorinator reactor in the production of 1,1,1-trichloroethane	(T)
K029	Waste from the product stream stripper in the production of 1,1,1-trichloroethane	(T)
K030	Column bottoms or heavy ends from the combined production of trichloroethylene and perchloroethylene	(T)
Pesticides:		
K031	By-products salts generated in the production of MSMA and cacodylic acid	(T)
K032	Wastewater treatment sludge from the production of chlordane	(T)
K033	Wastewater and scrub water from the chlorination of cyclopentadiene in the production of chlordane	(T)
K034	Filter solids from the filtration of hexachlorocyclopentadiene in the production of chlordane	(T)
K035	Wastewater treatment sludges generated in the production of creosote	(T)
K036	Still bottoms from toluene reclamation distillation in the production of disulfoton	(T)

TABLE 3.3 Hazardous Waste from Specific Sources (Continued)

Code	Description	Hazard
K037	Wastewater treatment sludges from the production of disulfoton	(T)
K038	Wastewater from the washing and stripping of phorate production	(T)
K039	Filter cake from the filtration of diethylphosphorodithoric acid in the production of phorate	(T)
K040	Wastewater treatment sludge from the production of phorate	(T)
K041	Wastewater treatment sludge from the production of toxaphene	(T)
K042	Heavy ends or distillation residues from the distillation of tetrachlorobenzene in the production of 2,4,5-T	(T)
K043	2,6-Dichlorophenol waste from the production of 2,4-D	(T)
Explosives:		
K044	Wastewater treatment sludges from the manufacturing and processing of explosives	(R)
K045	Spent carbon from the treatment of wastewater containing explosives	(R)
K046	Wastewater treatment sludges from the manufacturing, formulation and loading of lead-based initiating compounds	(T)
K047	Pink/red water from TNT operations	(R)
Petroleum Refining:		
K048	Dissolved air flotation (DAF) float from the petroleum refining industry	(T)
K049	Slop oil emulsion solids from the petroleum refining industry	(T)
K050	Heat exchanger bundle cleaning sludge from the petroleum refining industry	(T)
K051	API separator sludge from the petroleum refining industry	(T)
K052	Tank bottoms (leaded) from the petroleum refining industry	(T)
Leather Tanning Finishing:		
K053	Chrome (blue) trimmings generated by the following subcategories of the leather tanning and finishing industry: hair pulp/chrome tan/retan/wet finish; hair save/chrome tan/retan/wet finish; retan/wet finish; no beamhouse; through-the-blue; and shearling.	(T)
K054	Chrome (blue) shavings generated by the following subcategories of the leather tanning and finishing industry: hair pulp/chrome tan/retan/wet finish; hair save/chrome tan/retan/wet finish; retan/wet finish; no beamhouse; through-the-blue; and shearling.	(T)
K055	Buffing dust generated by the following subcategories of the leather tanning and finishing industry: hair pulp/chrome tan/retan/wet finish; hair save/chrome tan/retan/wet finish; retan/wet finish; no beamhouse; and through-the-blue.	(T)
K056	Sewer screenings generated by the following subcategories of the leather tanning and finishing industry: hair pulp/chrome tan/retan/wet finish; hair save/chrome tan/retan/wet finish; retan/wet finish; no beamhouse; through-the-blue; and shearling.	(T)
K057	Wastewater treatment sludges generated by the following subcategories of the leather tanning and finishing industry: hair pulp/chrome tan/retan/wet finish; hair save/chrome tan/retan/wet finish; retan/wet finish; no beamhouse; through-the-blue and shearling.	(T)
K058	Wastewater treatment sludges generated by the following subcategories of the leather tanning and finishing industry: hair pulp/chrome tan/retan/wet finish; hair save/chrome tan/retan/wet finish; and through-the-blue.	(R, T)
K059	Wastewater treatment sludges generated by the following subcategory of the leather tanning and finishing industry: hair save/non-chrome tan/retan/wet finish.	(R)
Iron and Steel:		
K060	Ammonia still lime sludge from coking operations	(T)
K061	Emission control dust/sludge from the electric furnace production of steel	(T)
K062	Spent pickle liquor from steel finishing operations	(C, T)
K063	Sludge from lime treatment of spent pickle liquor from steel finishing operations	(T)
Primary Copper: K064	Acid plant blowdown slurry/sludge resulting from the thickening of blowdown slurry from primary copper production	(T)
Primary Lead: K065	Surface impoundment solids contained in and dredged from surface impoundments at primary lead smelting facilities	(T)
Primary Zinc:		
K066	Sludge from treatment of process wastewater and/or acid plant blowdown from primary zinc production	(T)
K067	Electrolytic anode slimes/sludges from primary zinc production	(T)
K068	Cadmium plant leach residue (iron oxide) from primary zinc production	(T)
Secondary Lead: K069	Emission control dust/sludge from secondary lead smelting	(T)

Source: Ref. 3-1.

TABLE 3.4 Acute Hazardous Wastes

Hazardous waste No.	Substance
	1080 see P058
	1081 see P057
	(Acetato)phenylmercury see P092
	Acetone cyanohydrin see P069
P001	3-(alpha-Acetonylbenzyl)-4-hydroxycoumarin and salts
P002	1-Acetyl-2-thiourea
P003	Acrolein
	Agarin see P007
	Agrosan GN 5 see P092
	Aldicarb see P069
	Aldifen see P048
P004	Aldrin
	Algimycin see P092
P005	Allyl alcohol
P006	Aluminum phosphide (R)
	ALVIT see P037
	Aminoethylene see P054
P007	5-(Aminomethyl)-3-isoxazolol
P008	4-Aminopyridine
	Ammonium metavanadate see P119
P009	Ammonium picrate (R)
	ANTIMUCIN WDR see P092
	ANTURAT see P073
	AQUATHOL see P088
	ARETIT see P020
P010	Arsenic acid
P011	Arsenic pentoxide
P012	Arsenic trioxide
	Athrombin see P001
	AVITROL see P008
	Aziridene see P054
	AZOFOS see P061
	Azophos see P061
	BANTU see P072
P013	Barium cyanide
	BASENITE see P020
	BCME see P016
P014	Benzenethiol
	Benzoepin see P050
P015	Beryllium dust
P016	Bis(chloromethyl) ether
	BLADAN-M see P071
P017	Bromoacetone
P018	Brucine
P019	2-Butanone peroxide
	BUFEN see P092
	Butaphene see P020
P020	2-sec-Butyl-4,6-dinitrophenol
P021	Calcium cyanide
	CALDON see P020
P022	Carbon disulfide
	CERESAN see P092
	CERESAN UNIVERSAL see P092
	CHEMOX GENERAL see P020
	CHEMOX P.E. see P020
	CHEM-TOL see P090
P023	Chloroacetaldehyde
P024	p-Chloroaniline
P025	1-(p-Chlorobenzoyl)-5-methoxy-2-methylindole-3-acetic acid
P026	1-(o-Chlorophenyl)thiourea
P027	3-Chloropropionitrile
P028	alpha-Chlorotoluene
P029	Copper cyanide
	CRETOX see P108
	Coumadin see P001
	Coumafen see P001
P030	Cyanides
P031	Cyanogen
P032	Cyanogen bromide
P033	Cyanogen chloride
	Cyclodan see P050
P034	2-Cyclohexyl-4,6-dinitrophenol
	D-CON see P001
	DETHMOR see P001
	DETHNEL see P001
	DFP see P043
P035	2,4-Dichlorophenoxyacetic acid (2,4-D)
P036	Dichlorophenylarsine
	Dicyanogen see P031
P037	Dieldrin
	DIELDREX see P037
P038	Diethylarsine
P039	0,0-Diethyl-S-(2-(ethylthio)ethyl)ester of phosphorothioic acid
P040	0,0-Diethyl-0-(2-pyrazinyl)phosphorothioate
P041	0,0-Diethyl phosphoric acid, 0-p-nitrophenyl ester
P042	3,4-Dihydroxy-alpha-(methylamino)-methyl benzyl alcohol
P043	Di-isopropylfluorophosphate
	DIMETATE see P044
	1,4:5,8-Dimethanonaphthalene, 1,2,3,4,10,10-hexachloro-1,4,4a,5,8,8a-hexahydro endo, endo see P060
P044	Dimethoate
P045	3,3-Dimethyl-1-(methylthio)-2-butanone-O-[(methylamino)carbonyl] oxime
P046	alpha,alpha-Dimethylphenethylamine
	Dinitrocyclohexylphenol see P034
P047	4,6-Dinitro-o-cresol and salts
P048	2,4-Dinitrophenol
	DINOSEB see P020
	DINOSEBE see P020
	Disulfoton see P039
P049	2,4-Dithiobiuret
	DNBP see P020
	DOLCO MOUSE CEREAL see P108
	DOW GENERAL see P020
	DOW GENERAL WEED KILLER see P020
	DOW SELECTIVE WEED KILLER see P020
	DOWICIDE G see P090
	DYANACIDE see P092
	EASTERN STATES DUOCIDE see P001
	ELGETOL see P020
P050	Endosulfan
P051	Endrin
	Epinephrine see P042
P052	Ethylcyanide
P053	Ethylenediamine
P054	Ethyleneimine
	FASCO FASCRAT POWDER see P001
	FEMMA see P091
P055	Ferric cyanide
P056	Fluorine
P057	2-Fluoroacetamide
P058	Fluoroacetic acid, sodium salt
	FOLODOL-80 see P071
	FOLODOL M see P071
	FOSFERNO M 50 see P071
	FRATOL see P058
	Fulminate of mercury see P065
	FUNGITOX OR see P092
	FUSSOF see P057
	GALLOTOX see P092
	GEARPHOS see P071
	GERUTOX see P020
P059	Heptachlor

TABLE 3.4 Acute Hazardous Wastes (Continued)

Hazardous waste No.	Substance
P060	1,2,3,4,10,10-Hexachloro-1,4,4a,5,8,8a-hexahydro-1,4:5,8-endo, endo-dimethanonaphthalene
	1,4,5,6,7,7-Hexachloro-cyclic-5-norbornene-2,3-dimethanol sulfite see P050
P061	Hexachloropropene
P062	Hexaethyl tetraphosphate
	HOSTAQUICK see P092
	HOSTAQUIK see P092
	Hydrazomethane see P068
P063	Hydrocyanic acid
	ILLOXOL see P037
	INDOCI see P025
	Indomethacin see P025
	INSECTOPHENE see P050
	Isodrin see P060
P064	Isocyanic acid, methyl ester
	KILOSEB see P020
	KOP-THIODAN see P050
	KWIK-KIL see P108
	KWIKSAN see P092
	KUMADER see P001
	KYPFARIN see P001
	LEYTOSAN see P092
	LIQUIPHENE see P092
	MALIK see P050
	MAREVAN see P001
	MAR-FRIN see P001
	MARTIN'D MAR-FRIN see P001
	MAVERAN see P001
	MEGATOX see P005
P065	Mercury fulminate
	MERSOLITE see P092
	METACID 50 see P071
	METAFOS see P071
	METAPHOR see P071
	METAPHOS see P071
	METASOL 30 see P092
P066	Methomyl
P067	2-Methylaziridine
	METHYL-E 605 see P071
P068	Methyl hydrazine
	Methyl isocyanate see P064
P069	2-Methyllactonitrile
P070	2-Methyl-2-(methylthio)propionaldehyde-o-(methylcarbonyl) oxime
	METHYL NIRON see P042
P071	Methyl parathion
	METRON see P071
	MOLE DEATH see P108
	MOUSE-NOTS see P108
	MOUSE-RID see P108
	MOUSE-TOX see P108
	MUSCIMOL see P007
P072	1-Naphthyl-2-thiourea
P073	Nickel carbonyl
P074	Nickel cyanide
P075	Nicotine and salts
P076	Nitric oxide
P077	p-Nitroaniline
P078	Nitrogen dioxide
P079	Nitrogen peroxide
P080	Nitrogen tetroxide
P081	Nitroglycerine (R)
P082	N-Nitrosodimethylamine
P083	N-Nitrosodiphenylamine
P084	N-Nitrosomethylvinylamine
	NYLMERATE see P092
	OCTALOX see P037
P085	Octamethylpyrophosphoramide
	OCTAN see P092
P086	Oleyl alcohol condensed with 2 moles ethylene oxide
	OMPA see P085
	OMPACIDE see P085
	OMPAX see P085
P087	Osmium tetroxide
P088	7-Oxabicyclo[2.2.1]heptane-2,3-dicarboxylic acid
	PANIVARFIN see P001
	PANORAM D-31 see P037
	PANTHERINE see P007
	PANWARFIN see P001
P089	Parathion
	PCP see P090
	PENNCAP-M see P071
	PENOXYL CARBON N see P048
P090	Pentachlorophenol
	Pentachlorophenate see P090
	PENTA-KILL see P090
	PENTASOL see P090
	PENWAR see P090
	PERMICIDE see P090
	PERMAGUARD see P090
	PERMATOX see P090
	PERMITE see P090
	PERTOX see P090
	PESTOX III see P085
	PHENMAD see P092
	PHENOTAN see P020
P091	Phenyl dichloroarsine
	Phenyl mercaptan see P014
P092	Phenylmercury acetate
P093	N-Phenylthiourea
	PHILIPS 1861 see P008
	PHIX see P092
P094	Phorate
P095	Phosgene
P096	Phosphine
P097	Phosphorothioic acid, 0,0–dimethyl ester, 0–ester with N,N–dimethyl benzene sulfonamide
	Phosphorothioic acid 0,0–dimethyl–0–(p-nitrophenyl) ester see P071
	PIED PIPER MOUSE SEED see P108
P098	Potassium cyanide
P099	Potassium silver cyanide
	PREMERGE see P020
P100	1,2–Propanediol
	Propargyl alcohol see P102
P101	Propionitrile
P102	2–Propyn–1–o1
	PROTHROMADIN See P001
	QUICKSAM see P092
	QUINTOX see P037
	RAT AND MICE BAIT see P001
	RAT-A-WAY see P001
	RAT-B-GON see P001
	RAT-O-CIDE #2 see P001
	RAT-GUARD see P001
	RAT-KILL see P001
	RAT-MIX see P001
	RATS-NO-MORE see P001
	RAT-OLA see P001
	RATOREX see P001
	RATTUNAL see P001
	RAT-TROL see P001
	RO-DETH see P001
	RO-DEX see P108
	ROSEX see P001

TABLE 3.4 Acute Hazardous Wastes (Continued)

Hazardous waste No.	Substance
	ROUGH & READY MOUSE MIX see P001
	SANASEED see P108
	SANTOBRITE see P090
	SANTOPHEN see P090
	SANTOPHEN 20 see P090
	SCHRADAN see P085
P103	Selenourea
P104	Silver Cyanide
	SMITE see P105
	SPARIC see P020
	SPOR-KIL see P092
	SPRAY-TROL BRAND RODEN-TROL see P001
	SPURGE see P020
P105	Sodium azide
	Sodium coumadin see P001
P106	Sodium cyanide
	Sodium fluoroacetate see P056
	SODIUM WARFARIN see P001
	SOLFARIN see P001
	SOLFOBLACK BB see P048
	SOLFOBLACK SB see P048
P107	Strontium sulfide
P108	Strychnine and salts
	SUBTEX see P020
	SYSTAM see P085
	TAG FUNGICIDE see P092
	TEKWAISA see P071
	TEMIC see P070
	TEMIK see P070
	TERM-I-TROL see P090
P109	Tetraethyldithiopyrophosphate
P110	Tetraethyl lead
P111	Tetraethylpyrophosphate
P112	Tetranitromethane
	Tetraphosphoric acid, hexaethyl ester see P062
	TETROSULFUR BLACK PB see P048
	TETROSULPHUR PBR see P048
P113	Thallic oxide
	Thallium peroxide see P113
P114	Thallium selenite
P115	Thallium (I) sulfate
	THIFOR see P092
	THIMUL see P092
	THIODAN see P050
	THIOFOR see P050
	THIOMUL see P050
	THIONEX see P050
	THIOPHENIT see P071
P116	Thiosemicarbazide
	Thiosulfan tionel see P050
P117	Thiuram
	THOMPSON'S WOOD FIX see P090
	TIOVEL see P050
P118	Trichloromethanethiol
	TWIN LIGHT RAT AWAY see P001
	USAF RH-8 see P069
	USAF EK-4890 see P002
P119	Vanadic acid, ammonium salt
P120	Vanadium pentoxide
	VOFATOX see P071
	WANADU see P120
	WARCOUMIN see P001
	WARFARIN SODIUM see P001
	WARFICIDE see P001
	WOFOTOX see P072
	YANOCK see P057
	YASOKNOCK see P058
	ZIARNIK see P092
P121	Zinc cyanide
P122	Zinc phosphide (R,T)
	ZOOCOUMARIN see P001

Note: The Agency included those trade names of which it was aware; an omission of a trade name does not imply that the omitted material is not hazardous. The material is hazardous if it is listed under its generic name.

Source: Ref. 3-1.

with a hazardous waste is considered hazardous. These hazardous waste definitions are subject to the small-quantity exclusion noted later in this chapter.

NONHAZARDOUS WASTES

A number of wastes are specifically excluded from classification as hazardous wastes under RCRA, including the following:

1. Domestic sewage
2. Irrigation return flows
3. Nuclear waste
4. Household waste

TABLE 3.5 Toxic Wastes

Hazardous Waste No.	Substance
	AAF see U005
U001	Acetaldehyde
U002	Acetone (I)
U003	Acetonitrile (I,T)
U004	Acetophenone
U005	2-Acetylaminoflourene
U006	Acetyl chloride (C,T)
U007	Acrylamide
	Acetylene tetrachloride see U209
	Acetylene trichloride see U228
U008	Acrylic acid (I)
U009	Acrylonitrile
	AEROTHENE TT see U226
	3-Amino-5-(p-acetamidophenyl)-1H-1,2,4-triazole, hydrate see U011
U010	6-Amino-1,1a,2,8,8a,8b-hexahydro-8-(hydroxymethyl)8-methoxy-5-methylcarbamate azirino(2',3':3,4) pyrrolo(1,2-a) indole-4, 7-dione (ester)
U011	Amitrole
U012	Aniline (I)
U013	Asbestos
U014	Auramine
U015	Azaserine
U016	Benz[c]acridine
U017	Benzal chloride
U018	Benz[a]anthracene
U019	Benzene
U020	Benzenesulfonyl chloride (C,R)
U021	Benzidine
	1,2-Benzisothiazolin-3-one, 1,1-dioxide see U202
	Benzo[a]anthracene see U018
U022	Benzo[a]pyrene
U023	Benzotrichloride (C,R,T)
U024	Bis(2-chloroethoxy)methane
U025	Bis(2-chloroethyl) ether
U026	N,N-Bis(2-chloroethyl)-2-naphthylamine
U027	Bis(2-chloroisopropyl) ether
U028	Bis(2-ethylhexyl) phthalate
U029	Bromomethane
U030	4-Bromophenyl phenyl ether
U031	n-Butyl alcohol (I)
U032	Calcium chromate
	Carbolic acid see U188
	Carbon tetrachloride see U211
U033	Carbonyl fluoride
U034	Chloral
U035	Chlorambucil
U036	Chlordane
U037	Chlorobenzene
U038	Chlorobenzilate
U039	p-Chloro-m-cresol
U040	Chlorodibromomethane
U041	1-Chloro-2,3-epoxypropane
	CHLOROETHENE NU see U226
U042	Chloroethyl vinyl ether
U043	Chloroethene
U044	Chloroform (I,T)
U045	Chloromethane (I,T)
U046	Chloromethyl methyl ether
U047	2-Chloronaphthalene
U048	2-Chlorophenol
U049	4-Chloro-o-toluidine hydrochloride
U050	Chrysene
	C.I. 23060 see U073
U051	Cresote
U052	Cresols
U053	Crotonaldehyde
U054	Cresylic acid
U055	Cumene
	Cyanomethane see U003
U056	Cyclohexane (I)
U057	Cyclohexanone (I)
U058	Cyclophosphamide
U059	Daunomycin
U060	DDD
U061	DDT
U062	Diallate
U063	Dibenz[a,h]anthracene
	Dibenzo[a,h]anthracene see U063
U064	Dibenzo[a,i]pyrene
U065	Dibromochloromethane
U066	1,2-Dibromo-3-chloropropane
U067	1,2-Dibromoethane
U068	Dibromomethane
U069	Di-n-butyl phthalate
U070	1,2-Dichlorobenzene
U071	1,3-Dichlorobenzene
U072	1,4-Dichlorobenzene
U073	3,3'-Dichlorobenzidine
U074	1,4-Dichloro-2-butene
	3,3'-Dichloro-4,4'-diaminobiphenyl see U073
U075	Dichlorodifluoromethane
U076	1,1-Dichloroethane
U077	1,2-Dichloroethane
U078	1,1-Dichloroethylene
U079	1,2-trans-dichloroethylene
U080	Dichloromethane
	Dichloromethylbenzene see U017
U081	2,4-Dichlorophenol
U082	2,6-Dichlorophenol
U083	1,2-Dichloropropane
U084	1,3-Dichloropropene
U085	Diepoxybutane (I,T)
U086	1,2-Diethylhydrazine
U087	0,0-Diethyl-S-methyl ester of phosphorodithioic acid
U088	Diethyl phthalate
U089	Diethylstilbestrol
U090	Dihydrosafrole
U091	3,3'-Dimethoxybenzidine
U092	Dimethylamine (I)
U093	p-Dimethylaminoazobenzene
U094	7,12-Dimethylbenz[a]anthracene
U095	3,3'-Dimethylbenzidine
U096	alpha,alpha-Dimethylbenzylhydroperoxide (R)
U097	Dimethylcarbamoyl chloride
U098	1,1-Dimethylhydrazine
U099	1,2-Dimethylhydrazine
U100	Dimethylnitrosoamine
U101	2,4-Dimethylphenol
U102	Dimethyl phthalate
U103	Dimethyl sulfate
U104	2,4-Dinitrophenol
U105	2,4-Dinitrotoluene
U106	2,6-Dinitrotoluene
U107	Di-n-octyl phthalate
U108	1,4-Dioxane
U109	1,2-Diphenylhydrazine
U110	Dipropylamine (I)
U111	Di-n-propylnitrosamine
	EBDC see U114
	1,4-Epoxybutane see U213
U112	Ethyl acetate (I)
U113	Ethyl acrylate (I)
U114	Ethylenebisdithiocarbamate
U115	Ethylene oxide (I,T)

TABLE 3.5 Toxic Wastes (Continued)

Hazardous Waste No.	Substance
U116	Ethylene thiourea
U117	Ethyl ether (I,T)
U118	Ethylmethacrylate
U119	Ethyl methanesulfonate
	Ethylnitrile see U003
	Firemaster T23P see U235
U120	Fluoranthene
U121	Fluorotrichloromethane
U122	Formaldehyde
U123	Formic acid (C,T)
U124	Furan (I)
U125	Furfural (I)
U126	Glycidylaldehyde
U127	Hexachlorobenzene
U128	Hexachlorobutadiene
U129	Hexachlorocyclohexane
U130	Hexachlorocyclopentadiene
U131	Hexachloroethane
U132	Hexachlorophene
U133	Hydrazine (R,T)
U134	Hydrofluoric acid (C,T)
U135	Hydrogen sulfide
	Hydroxybenzene see U188
U136	Hydroxydimethyl arsine oxide
	4,4'-(Imidocarbonyl)bis(N,N-dimethyl)aniline see U014
U137	Indeno(1,2,3-cd)pyrene
U138	Iodomethane
U139	Iron Dextran
U140	Isobutyl alcohol
U141	Isosafrole
U142	Kepone
U143	Lasiocarpine
U144	Lead acetate
U145	Lead phosphate
U146	Lead subacetate
U147	Maleic anhydride
U148	Maleic hydrazide
U149	Malononitrile
	MEK Peroxide see U160
U150	Melphalan
U151	Mercury
U152	Methacrylonitrile
U153	Methanethiol
U154	Methanol
U155	Methapyrilene
	Methyl alcohol see U154
U156	Methyl chlorocarbonate
	Methyl chloroform see U226
U157	3-Methylcholanthrene
	Methyl chloroformate see U156
U158	4,4'-Methylene-bis-(2-chloroaniline)
U159	Methyl ethyl ketone (MEK) (I,T)
U160	Methyl ethyl ketone peroxide (R)
	Methyl iodide see U138
U161	Methyl isobutyl ketone
U162	Methyl methacrylate (R,T)
U163	N-Methyl-N'-nitro-N-nitrosoguanidine
U164	Methylthiouracil
	Mitomycin C see U010
U165	Naphthalene
U166	1,4-Naphthoquinone
U167	1-Naphthylamine
U168	2-Naphthylamine
U169	Nitrobenzene (I,T)
	Nitrobenzol see U169
U170	4-Nitrophenol
U171	2-Nitropropane (I)
U172	N-Nitrosodi-n-butylamine
U173	N-Nitrosodiethanolamine
U174	N-Nitrosodiethylamine
U175	N-Nitrosodi-n-propylamine
U176	N-Nitroso-n-ethylurea
U177	N-Nitroso-n-methylurea
U178	N-Nitroso-n-methylurethane
U179	N-Nitrosopiperidine
U180	N-Nitrosopyrrolidine
U181	5-Nitro-o-toluidine
U182	Paraldehyde
	PCNB see U185
U183	Pentachlorobenzene
U184	Pentachloroethane
U185	Pentachloronitrobenzene
U186	1,3-Pentadiene (I)
	Perc see U210
	Perchlorethylene see U210
U187	Phenacetin
U188	Phenol
U189	Phosphorous sulfide (R)
U190	Phthalic anhydride
U191	2-Picoline
U192	Pronamide
U193	1,3-Propane sultone
U194	n-Propylamine (I)
U196	Pyridine
U197	Quinones
U200	Reserpine
U201	Resorcinol
U202	Saccharin
U203	Safrole
U204	Selenious acid
U205	Selenium sulfide (R,T)
	Silvex see U233
U206	Streptozotocin
	2,4,5-T see U232
U207	1,2,4,5-Tetrachlorobenzene
U208	1,1,1,2-Tetrachloroethane
U209	1,1,2,2-Tetrachloroethane
U210	Tetrachloroethene
	Tetrachloroethylene see U210
U211	Tetrachloromethane
U212	2,3,4,6-Tetrachlorophenol
U213	Tetrahydrofuran (I)
U214	Thallium (I) acetate
U215	Thallium (I) carbonate
U216	Thallium (I) chloride
U217	Thallium (I) nitrate
U218	Thioacetamide
U219	Thiourea
U220	Toluene
U221	Toluenediamine
U222	o-Toluidine hydrochloride
U223	Toluene diisocyanate
U224	Toxaphene
	2,4,5-TP see U233
U225	Tribromomethane
U226	1,1,1-Trichloroethane
U227	1,1,2-Trichloroethane
U228	Trichloroethene
	Trichloroethylene see U228

TABLE 3.5 Toxic Wastes (Continued)

Hazardous Waste No.	Substance	Hazardous Waste No.	Substance
U229	Trichlorofluoromethane	U237	Uracil mustard
U230	2,4,5-Trichlorophenol	U238	Urethane
U231	2,4,6-Trichlorophenol		Vinyl chloride see U043
U232	2,4,5-Trichlorophenoxyacetic acid		Vinylidene chloride see U078
U233	2,4,5-Trichlorophenoxypropionic acid alpha, alpha, alpha- Trichlorotoluene see U023	U239	Xylene
	TRI-CLENE see U228		
U234	Trinitrobenzene (R,T)		
U235	Tris(2,3-dibromopropyl) phosphate		
U236	Trypan blue		

Note: The Agency included those trade names of which it was aware; an omission of a trade name does not imply that it is not hazardous. The material is hazardous if it is listed under its generic name.
Source: Ref. 3-1.

TABLE 3.6 Hazardous Constituents

Acetaldehyde
(Acetato)phenylmercury
Acetonitrile
3-(alpha-Acetonylbenzyl)-4-hydroxycoumarin and salts
2-Acetylaminofluorene
Acetyl chloride
1-Acetyl-2-thiourea
Acrolein
Acrylamide
Acrylonitrile
Aflatoxins
Aldrin
Allyl alcohol
Aluminum phosphide
4-Aminobiphenyl
6-Amino-1,1a,2,8,8a,8b-hexahydro-8-(hydroxymethyl)-8a-methoxy-5-methylcarbamate azirino(2',3':3,4) pyrrolo(1,2-a)indole-4,7-dione (ester) (Mitomycin C)
5-(Aminomethyl)-3-isoxazolol
4-Aminopyridine
Amitrole
Antimony and compounds, N.O.S.[1]
Aramite
Arsenic and compounds, N.O.S.
Arsenic acid
Arsenic pentoxide
Arsenic trioxide
Auramine
Azaserine
Barium and compounds, N.O.S.
Barium cyanide
Benz[c]acridine
Benz[a]anthracene
Benzene
Benzenearsonic acid
Benzenethiol
Benzidine
Benzo[a]anthracene
Benzo[b]fluoranthene
Benzo[j]fluoranthene
Benzo[a]pyrene
Benzotrichloride
Benzyl chloride
Beryllium and compounds, N.O.S.
Bis(2-chloroethoxy)methane
Bis(2-chloroethyl) ether
N,N-Bis(2-chloroethyl)-2-naphthylamine
Bis(2-chloroisopropyl) ether
Bis(chloromethyl) ether
Bis(2-ethylhexyl) phthalate
Bromoacetone
Bromomethane
4-Bromophenyl phenyl ether
Brucine
2-Butanone peroxide
Butyl benzyl phthalate
2-sec-Butyl-4,6-dinitrophenol (DNBP)
Cadmium and compounds, N.O.S.
Calcium chromate
Calcium cyanide
Carbon disulfide
Chlorambucil
Chlordane (alpha and gamma isomers)

[1] The abbreviation N.O.S. signifies those members of the general class "not otherwise specified" by name in this listing.

TABLE 3.6 Hazardous Constituents (Continued)

Chlorinated benzenes, N.O.S.
Chlorinated ethane, N.O.S.
Chlorinated naphthalene, N.O.S.
Chlorinated phenol, N.O.S.
Chloroacetaldehyde
Chloroalkyl ethers
p-Chloroaniline
Chlorobenzene
Chlorobenzilate
1-(p-Chlorobenzoyl)-5-methoxy-2-methylindole-3-acetic acid
p-Chloro-m-cresol
1-Chloro-2,3-epoxybutane
2-Chloroethyl vinyl ether
Chloroform
Chloromethane
Chloromethyl methyl ether
2-Chloronaphthalene
2-Chlorophenol
1-(o-Chlorophenyl)thiourea
3-Chloropropionitrile
alpha-Chlorotoluene
Chlorotoluene, N.O.S.
Chromium and compounds, N.O.S.
Chrysene
Citrus red No. 2
Copper cyanide
Creosote
Crotonaldehyde
Cyanides (soluble salts and complexes), N.O.S.
Cyanogen
Cyanogen bromide
Cyanogen chloride
Cycasin
2-Cyclohexyl-4,6-dinitrophenol
Cyclophosphamide
Daunomycin
DDD
DDE
DDT
Diallate
Dibenz[a,h]acridine
Dibenz[a,j]acridine
Dibenz[a,h]anthracene(Dibenzo[a,h]anthracene)
7H-Dibenzo[c,g]carbazole
Dibenzo[a,e]pyrene
Dibenzo[a,h]pyrene
Dibenzo[a,i]pyrene
1,2-Dibromo-3-chloropropane
1,2-Dibromoethane
Dibromomethane
Di-n-butyl phthalate
Dichlorobenzene, N.O.S.
3,3′-Dichlorobenzidine
1,1-Dichloroethane
1,2-Dichloroethane
trans-1,2-Dichloroethane
Dichloroethylene, N.O.S.
1,1-Dichloroethylene
Dichloromethane
2,4-Dichlorophenol
2,6-Dichlorophenol
2,4-Dichlorophenoxyacetic acid (2,4-D)
Dichloropropane
Dichlorophenylarsine
1,2-Dichloropropane
Dichloropropanol, N.O.S.
Dichloropropene, N.O.S.
1,3-Dichloropropene
Dieldrin
Diepoxybutane
Diethylarsine
0,0-Diethyl-S-(2-ethylthio)ethyl ester of phosphorothioic acid
1,2-Diethylhydrazine
0,0-Diethyl-S-methylester phosphorodithioic acid
0,0-Diethylphosphoric acid, 0-p-nitrophenyl ester
Diethyl phthalate
0,0-Diethyl-0-(2-pyrazinyl)phosphorothioate
Diethylstilbestrol
Dihydrosafrole
3,4-Dihydroxy-alpha-(methylamino)-methyl benzyl alcohol
Di-isopropylfluorophosphate (DFP)
Dimethoate
3,3′-Dimethoxybenzidine
p-Dimethylaminoazobenzene
7,12-Dimethylbenz[a]anthracene
3,3′-Dimethylbenzidine
Dimethylcarbamoyl chloride
1,1-Dimethylhydrazine
1,2-Dimethylhydrazine
3,3-Dimethyl-1-(methylthio)-2-butanone-0-((methylamino) carbonyl)oxime
Dimethylnitrosoamine
alpha,alpha-Dimethylphenethylamine
2,4-Dimethylphenol
Dimethyl phthalate
Dimethyl sulfate
Dinitrobenzene, N.O.S.
4,6-Dinitro-o-cresol and salts
2,4-Dinitrophenol
2,4-Dinitrotoluene
2,6-Dinitrotoluene Di-n-octyl phthalate
1,4-Dioxane
1,2-Diphenylhydrazine

TABLE 3.6 Hazardous Constituents (Continued)

Di-n-propylnitrosamine
Disulfoton
2,4-Dithiobiuret
Endosulfan
Endrin and metabolites
Epichlorohydrin
Ethyl cyanide
Ethylene diamine
Ethylenebisdithiocarbamate (EBDC)
Ethyleneimine
Ethylene oxide
Ethylenethiourea
Ethyl methanesulfonate
Fluoranthene
Fluorine
2-Fluoroacetamide
Fluoroacetic acid, sodium salt
Formaldehyde
Glycidylaldehyde
Halomethane, N.O.S.
Heptachlor
Heptachlor epoxide (alpha, beta, and gamma isomers)
Hexachlorobenzene
Hexachlorobutadiene
Hexachlorocyclohexane (all isomers)
Hexachlorocyclopentadiene
Hexachloroethane
1,2,3,4,10,10-Hexachloro-1,4,4a,5,8,8a-hexahydro-1,4:5,8-endo,endo-dimethanonaphthalene
Hexachlorophene
Hexachloropropene
Hexaethyl tetraphosphate
Hydrazine
Hydrocyanic acid
Hydrogen sulfide
Indeno(1,2,3-c,d)pyrene
Iodomethane
Isocyanic acid, methyl ester
Isosafrole
Kepone
Lasiocarpine
Lead and compounds, N.O.S.
Lead acetate
Lead phosphate
Lead subacetate
Maleic anhydride
Malononitrile
Melphalan
Mercury and compounds, N.O.S.
Methapyrilene
Methomyl
2-Methylaziridine
3-Methylcholanthrene
4,4'-Methylene-bis-(2-chloroaniline)
Methyl ethyl ketone (MEK)
Methyl hydrazine
2-Methyllactonitrile
Methyl methacrylate
Methyl methanesulfonate
2-Methyl-2-(methylthio)propionaldehyde-o-(methylcarbonyl) oxime
N-Methyl-N'-nitro-N-nitrosoguanidine
Methyl parathion
Methylthiouracil
Mustard gas
Naphthalene
1,4-Naphthoquinone
1-Naphthylamine
2-Naphthylamine
1-Naphthyl-2-thiourea
Nickel and compounds, N.O.S.
Nickel carbonyl
Nickel cyanide
Nicotine and salts
Nitric oxide
p-Nitroaniline
Nitrobenzene
Nitrogen dioxide
Nitrogen mustard and hydrochloride salt
Nitrogen mustard N-oxide and hydrochloride salt
Nitrogen peroxide
Nitrogen tetroxide
Nitroglycerine
4-Nitrophenol
4-Nitroquinoline-1-oxide
Nitrosamine, N.O.S.
N-Nitrosodi-N-butylamine
N-Nitrosodiethanolamine
N-Nitrosodiethylamine
N-Nitrosodimethylamine
N-Nitrosodiphenylamine
N-Nitrosodi-N-propylamine
N-Nitroso-N-ethylurea
N-Nitrosomethylethylamine
N-Nitroso-N-methylurea
N-Nitroso-N-methylurethane
N-Nitrosomethylvinylamine
N-Nitrosomorpholine
N-Nitrosonornicotine
N-Nitrosopiperidine
N-Nitrosopyrrolidine
N-Nitrososarcosine
5-Nitro-o-toluidine
Octamethylpyrophosphoramide
Oleyl alcohol condensed with 2 moles ethylene oxide
Osmium tetroxide

TABLE 3.6 Hazardous Constituents (Continued)

7-Oxabicyclo[2.2.1]heptane-2,3-dicarboxylic acid
Parathion
Pentachlorobenzene
Pentachloroethane
Pentachloronitrobenzene (PCNB)
Pentacholorophenol
Phenacetin
Phenol
Phenyl dichloroarsine
Phenylmercury acetate
N-Phenylthiourea
Phosgene
Phosphine
Phosphorothioic acid, O,O-dimethyl ester, O-ester with N,N-dimethyl benzene sulfonamide
Phthalic acid esters, N.O.S.
Phthalic anhydride
Polychlorinated biphenyl, N.O.S.
Potassium cyanide
Potassium silver cyanide
Pronamide
1,2-Propanediol
1,3-Propane sultone
Propionitrile
Propylthiouracil
2-Propyn-1-ol
Pryidine
Reserpine
Saccharin
Safrole
Selenious acid
Selenium and compounds, N.O.S.
Selenium sulfide
Selenourea
Silver and compounds, N.O.S.
Silver cyanide
Sodium cyanide
Streptozotocin
Strontium sulfide
Strychnine and salts
1,2,4,5-Tetrachlorobenzene
2,3,7,8-Tetrachlorodibenzo-p-dioxin (TCDD)
Tetrachloroethane, N.O.S.
1,1,1,2-Tetrachloroethane
1,1,2,2-Tetrachloroethane
Tetrachloroethene (Tetrachloroethylene)
Tetrachloromethane
2,3,4,6-Tetrachlorophenol
Tetraethyldithiopyrophosphate
Tetraethyl lead
Tetraethylpyrophosphate
Thallium and compounds, N.O.S.
Thallic oxide
Thallium (I) acetate
Thallium (I) carbonate
Thallium (I) chloride
Thallium (I) nitrate
Thallium selenite
Thallium (I) sulfate
Thioacetamide
Thiosemicarbazide
Thiourea
Thiuram
Toluene
Toluene diamine
o-Toluidine hydrochloride
Tolylene diisocyanate
Toxaphene
Tribromomethane
1,2,4-Trichlorobenzene
1,1,1-Trichloroethane
1,1,2-Trichloroethane
Trichloroethene (Trichloroethylene)
Trichloromethanethiol
2,4,5-Trichlorophenol
2,4,6-Trichlorophenol
2,4,5-Trichlorophenoxyacetic acid (2,4,5-T)
2,4,5-Trichlorophenoxypropionic acid (2,4,5-TP) (Silvex)
Trichloropropane, N.O.S.
1,2,3-Trichloropropane
0,0,0-Triethyl phosphorothioate
Trinitrobenzene
Tris(1-azridinyl)phosphine sulfide
Tris(2,3-dibromopropyl) phosphate
Trypan blue
Uracil mustard
Urethane
Vanadic acid, ammonium salt
Vanadium pentoxide (dust)
Vinyl chloride
Vinylidene chloride
Zinc cyanide
Zinc phosphide

Note: The abbreviation N.O.S. signifies those members of the general class "not otherwise specified" by name in this listing.
Source: Ref. 3-1.

5. Wastes generated from the growing of crops and the raising of animals (manure) and which are returned to the soil as fertilizers
6. Mining overburden returned to the mine site
7. Fly ash, bottom ash, slag waste, and waste from flue gas emissions control systems when generated from the burning of coal or other fossil fuels
8. Wastes associated with the exploration, development, or production of crude oil, natural gas, or geothermal energy such as drilling fluids and oil-laden waters

USABLE HAZARDOUS WASTE

Regulations applicable to disposal of hazardous wastes are not currently applicable to a hazardous waste that meets any of the following criteria:

1. The waste is being recycled or reclaimed.
2. The waste is in storage or is being treated prior to its reclamation.

Reclamation includes the use of a waste for the generation of heat energy. Of note, however, is that the waste must have significant heating value. The regulations state specifically that a waste cannot be fired to avoid regulation under this provision unless it can legitimately be used for heat generation and recovery.

QUANTITY EXCLUSION

Provisions are included in the RCRA regulations to exempt small generators of hazardous waste from the rigorous procedures necessary for compliance with the disposal statutes. The small-generator exclusion is as follows:

1. Generation of less than 1000 kg (2200 lb) per month of hazardous wastes, unless otherwise specified
2. Accumulation of less than 1000 kg (2200 lb) of hazardous waste in any one calendar month, unless otherwise specified
3. A waste containing no more than 1 kg (2.2 lb) of any product or chemical listed in Table 3.4 (acute hazardous waste)
4. Containers not greater than 20 L (5.3 gal) or liners no greater than 10 kg (22 lb) in weight, holding materials listed in Table 3.4 (acute hazardous waste)
5. No greater than 100 kg (220 lb) of residue, soil contaminated from a spill, etc. from a material listed in Table 3.4 (acute hazardous waste)

The regulations allow that a hazardous waste subject to the small-quantity exclusion may be mixed with nonhazardous waste and remain subject to this exclusion even though the resultant mixture exceeds the small-quantity limitations.

GENERAL CONSIDERATIONS

It is the responsibility of the waste generator to determine if the waste is hazardous and then to classify it with an appropriate hazardous waste code. The generator is then obligated to report the existence of this waste, its transport, and method of disposal, using a manifest which is described in 40 CFR part 260.

These regulations will undoubtedly be subject to intense review as they are applied. Litigation and subsequent revisions are to be expected, and these regulations should be monitored with regularity because of these expected changes.

HAZARDOUS WASTE INCINERATION REGULATIONS

If a hazardous waste is hazardous only because it has characteristics of ignitability or of corrosivity or if the hazardous waste will be reused, reclaimed, or used for heat recovery, its incineration does not have to comply with hazardous waste incineration criteria. In addition, certain wastes which are hazardous solely because of reactivity may be incinerated without compliance with these criteria. These criteria require that a test burn be run on the *principal organic hazardous constituent* (POHC). The POHC is that compound listed in Table 3.6 that is present in the waste. If more than one compound from Table 3.6 can be identified, the selection of the POHC will be a matter of judgment based on the difficulty of incineration, quantity present, and toxicity of that compound. More than one of the compounds present may be designated as the POHC.

The test burn is one phase of a four-phase permit process. Permitting provides for a one-month shake-down phase prior to performance of a test burn. The test burn is the second phase of the permitting process. The third phase comprises limited operation while results of the test burn are evaluated. The fourth phase is the final or permanent operating phase. The permit requirements, promulgated in federal regulations 40 CFR parts 122, 264, and 265, include the following provisions:

1. Identification, by the generator, of the POHCs.
2. Operation of incineration equipment to achieve a *destruction and removal efficiency* (DRE) of at least 99.99 percent. The DRE is defined, with W_{in} the POHC mass rate into the system and W_{out} the POHC mass rate leaving in the incinerator exhaust gas stream, as follows:

$$\text{DRE} = \frac{(W_{in} - W_{out}) \times 100\%}{W_{in}}$$

3. If the hydrogen chloride exiting the stack is less than 4 lb/h, no HCl removal is necessary. If the stack emission contains in excess of 4 lb/h of hydrogen chloride, 99 percent of the hydrogen chloride produced must be removed from the exhaust gas stream.
4. Particulate emissions into the atmosphere must not exceed 0.08 gr per dry standard cubic foot when corrected to 50 percent excess air. A correction to 50 percent excess air is provided by the following formula:

$$P_c = P_m \times \frac{14}{21 - Y}$$

where P_m is the measured particulate, Y is the percent oxygen by volume, and P_c is the corrected particulate concentration.

5. Continuous monitoring is required for combustion temperature, waste feed rate, combustion gas flow rate, and carbon monoxide (CO) in the exhaust gas stream.

In lieu of a test burn, which requires EPA approval prior to its implementation, operating criteria can be established on the basis of published data on disposal of the POHCs in question. The published data, however, must have been generated from incinerator equipment similar in type to the proposed incinerator.

WHEN CRITERIA ARE NOT AVAILABLE

For purposes of initial design, where test data are not immediately available for a particular waste, a reasonable estimate of combustion criteria can be assumed, as follows:

1. For a nonhalogenated hazardous waste provide 1000°C (1832°F) in the combustion chamber, with a retention time of 2 s and 2 percent oxygen in the exhaust.
2. For halogenated hazardous waste (one containing at least 0.5 percent chlorine) provide a temperature of 1200°C (2192°F) in the combustion chamber, with a retention time of 2 s, and 3 percent oxygen in the exhaust.

Note that these criteria are not stated within RCRA or other federal statutory requirements. They are only suggested here as criteria to be used as a guide to equipment design where no other data are available and are not presented in lieu of a test burn.

PCB INCINERATION

Incineration of substances containing PCBs (polychlorinated biphenyls) is covered by the Toxic Substances Control Act (TSCA) 44FR106 paragraph 761.41. The incineration of PCBs is subject to the following criteria, excerpted from TSCA.

1. Combustion at 1200°C (2192°F), with a 2-s retention time at 3 percent oxygen in the exhaust gas or 1600°C (2912°F), with a 1½-s retention time at 2 percent oxygen in the exhaust.
2. Combustion efficiency (CE) of 99.9 percent. With C_{CO_2} and C_{CO} the concentration of CO_2 and CO in the exhaust gas, respectively, the CE is calculated as follows:

$$CE = \frac{C_{CO_2}}{C_{CO_2} + C_{CO}} \times 100\%$$

3. The PCB charging rate and total feed must be monitored at least every 15 min.
4. The combustion temperature must be monitored on a continuous basis.
5. Upon a drop in temperature below 1200°C (or 1600°C) the flow of PCBs shall automatically cease. PCB flow shall also cease if there is a failure in the monitoring operations or if exhaust oxygen falls below that required.
6. When an incinerator is initially used for disposal of PCBs, the following stack emissions must be monitored:
   a. Oxygen (O_2)
   b. Carbon monoxide (CO)
   c. Carbon dioxide (CO_2)
   d. Oxides of nitrogen (NO_x)
   e. Hydrogen chloride (HCl)
   f. Total chlorinated organic component (RCl)
   g. PCBs
   h. Total particulate matter
7. During normal operation of the incinerator the CO_2 concentration in the exhaust gas shall be monitored on a periodic basis. On a continuous basis the O_2 and CO components of the exhaust shall be monitored.
8. Water scrubbers or equivalent gas cleaning equipment shall be used to control the HCl emission in the exhaust gas. Spent scrubber water effluent must be monitored and shall be in compliance with applicable effluent standards.
9. If the PCBs to be incinerated are nonliquid, in addition to the above requirements the mass air emissions shall not be greater than 1 lb PCB per 10^6 lb of PCBs charged into the furnace.

COMMENTARY ON PCB INCINERATOR REGULATIONS

The EPA intends to include PCB regulations in the RCRA statutes at a later date, removing these regulations from TSCA. In this transferral of authority changes may be made to these regulations.

Other considerations regarding the present PCB incineration regulations include the following:

1. Instrumentation for incinerator monitoring of low levels of CO (for combustion efficiency calculations) for PCB and HCl detection from incinerator systems is not readily available. It is being developed on an application by application basis, advancing the present state of the art.
2. A requirement of 15-min monitoring is impractical for manual recordkeeping. It would require a full-time operator who does nothing but record data. This requirement mandates use of continuous automatic monitoring equipment.

REFERENCE

3-1. "Hazardous Waste and Consolidated Permit Regulations (RCRA)," *The Federal Register*, 45(98), May 19, 1980.

BIBLIOGRAPHY

Beck, E., and Plehn, S.: "EPA's Cradle to Grave Approach to Hazardous Waste Management." *Consulting Engineer*, September 1980.

Frick, G.: "What Wastes Are Hazardous under RCRA?," *Consulting Engineer*, September 1980.

Hart, F., and Tusa, W.: "Organizing to Deal with RCRA," *Waste Age*, April 1981.

Hazardous Waste Facilities: Standards for Incinerators, USEPA SW-908, January 1981.

Hazardous Waste Management: A Guide to the Regulations, USEPA, 1980.

"Incinerator Standards for Owners and Operators of Hazardous Waste Management Facilities," *Federal Register*, January 23, 1981.

Ross, R.: "The Burning Issue: Incineration of Hazardous Wastes," *Pollution Engineering*, August 1979.

Sittig, M.: *Incineration of Industrial Hazardous Wastes and Sludges*, Noyes, Park Ridge, N.J., 1979.

CHAPTER 4
BASIC THERMODYNAMICS

Thermodynamics is defined as the science concerned with the conversion of heat to energy. Many texts deal with heat conversion, thermodynamic properties of materials, and the production of energy. This chapter is not intended as a substitution for these texts but a concise explanation of basic thermodynamic concepts relative to the incineration process including enthalpy, energy levels, and psychometric analysis.

ELEMENTAL LAWS OF THERMODYNAMICS

The two basic laws of thermodynamics can be simply stated as follows:

1. In any reaction the energy output must equal the energy input of the system.
2. In any process the flow of heat will always be from a region of higher to a region of lower temperature.

UNITS OF ENERGY

The unit of energy in common use in the United States is the British thermal unit (Btu). One Btu is that amount of heat required to raise the temperature of one pound of water one degree Fahrenheit. The calorie is an equivalent unit, that amount of heat which will raise the temperature of one gram of water one degree centigrade.

In accordance with the first law of thermodynamics the energy input to a system must be accounted for within that system. A corollary of this law is the equivalence of heat and work. For instance, a fixed quantity of water stirred by a mixer (mechanical work) will experience an equivalent rise in temperature (heat energy). The constants of conversion of mechanical and heat energy, as well as other thermodynamic conversions, are as follows:

Heat Conversions

1 Btu	≡ 778 foot-pounds
1 Btu	≡ 1055 joules
1 Btu	≡ 252 calories
1 Btu	≡ 0.252 calorie
1 Btu	≡ 0.0002931 kilowatthour
1 horsepower-hour	≡ 2544 Btu
1 kilowatthour	≡ 3412 Btu
1 calorie	≡ 3.968 Btu
1 calorie	≡ 1 kilocalorie (1000 calories)

ENTHALPY

To facilitate quantification of these laws, a number of material properties have been defined. Of interest to an understanding of combustion and related processes, the total energy or enthalpy is of particular concern. Enthalpy h, Btu/lb, is the total energy of a substance at one temperature relative to the energy of the same substance at another temperature. It is a relative property, and whenever it is used, it must be related to a base or datum point.

Table 4.1 lists the enthalpy of water vapor. The enthalpy of air is listed relative to the enthalpy of air at 60 and at 80°F. The enthalpy of water vapor (steam) is listed relative to liquid water at both 60 and 80°F.

Table 4.2 lists the enthalpy of gases other than air and moisture (steam) normally found in flue gas, i.e., CO_2, N_2, and O_2.

REACTION TEMPERATURES

The temperature of a reaction can be calculated given the heat produced by the reaction and the quantities of products generated in the reaction. For example,

> If 15 lb of water vapor (steam) is produced in a reaction generating 20,000 Btu and all this heat is absorbed by the water vapor, and assuming that the water vapor entered the reaction as liquid water at 60°F, what is the temperature of the water vapor (steam) exiting the reaction?
>
> In this case the steam carries 20,000 Btu ÷ 15 lb or 1333 Btu/lb, a definition of its enthalpy. Consulting Table 4.1, an enthalpy of 1333 Btu/lb corresponds to a temperature of approximately 650°F, based on a 60°F datum.

This simple example illustrates that the temperature of a substance is that temperature at which its enthalpy is equal to the total heat contained within that substance.

For a more complex example,

TABLE 4.1 Enthalpy, Air, and Moisture

Relative to 60°F			Relative to 80°F	
H_{Air}, Btu/lb	H_{H_2O}, Btu/lb	Temp., °F	H_{Air}, Btu/lb	H_{H_2O}, Btu/lb
21.61	1091.92	150	16.82	1071.91
33.65	1116.62	200	28.86	1096.61
45.71	1140.72	250	40.92	1120.71
57.81	1164.52	300	53.02	1144.51
69.98	1188.22	350	65.19	1168.21
82.19	1211.82	400	77.40	1191.81
94.45	1235.47	450	89.66	1215.46
106.79	1259.22	500	102.00	1239.21
119.21	1283.07	550	114.42	1263.06
131.69	1307.12	600	126.90	1287.11
144.25	1331.27	650	139.46	1311.26
156.87	1355.72	700	152.08	1335.71
169.59	1380.27	750	164.80	1360.26
187.38	1405.02	800	177.59	1385.01
195.26	1430.02	850	190.47	1410.01
208.21	1455.32	900	203.42	1435.31
221.25	1480.72	950	216.46	1460.71
234.36	1506.42	1000	229.57	1486.41
247.55	1532.40	1050	242.76	1512.40
260.81	1558.32	1100	256.02	1538.31
274.15	1584.80	1150	264.36	1564.80
287.55	1611.22	1200	282.76	1591.21
301.02	1638.26	1250	296.23	1618.20
314.56	1665.12	1300	309.77	1645.11
328.17	1692.15	1350	323.38	1672.15
341.85	1719.82	1400	337.06	1699.81
355.58	1747.70	1450	350.82	1727.70
369.37	1775.52	1500	364.58	1755.51
397.17	1832.12	1600	392.33	1812.11
425.08	1890.11	1700	420.29	1870.10
453.24	1948.02	1800	448.45	1928.01
481.57	2007.17	1900	476.78	1987.70
510.07	2067.42	2000	505.28	2047.41
538.72	2128.70	2100	533.93	2108.70
567.52	2189.92	2200	562.73	2169.91
596.45	2252.60	2300	591.66	2232.60
625.52	2315.32	2400	620.73	2295.31
654.70	2377.80	2500	649.91	2357.80
684.01	2443.30	2600	679.22	2423.30
713.42	2511.88	2700	708.63	2491.80

Source: Refs. 4-1 and 4-2.

A liquid at 60°F is combusted and releases 700,000 Btu. The products of combustion are 1000 lb of dry gas, which has properties identical to those of air, and 250 lb of moisture. Using Table 4.1, assuming a reaction temperature of 1200°F:

TABLE 4.2 Enthalpies of Various Gases Expressed in Btu/lb of Gas

Temp, °F	CO_2	N_2	O_2
100	5.8	6.4	8.8
150	17.6	20.6	19.8
200	29.3	34.8	30.9
250	40.3	47.7	42.1
300	51.3	59.8	53.4
350	63.1	73.3	64.8
400	74.9	84.9	76.2
450	87.0	97.5	87.8
500	99.1	110.1	99.5
550	111.8	122.9	111.3
600	124.5	135.6	123.2
700	150.2	161.4	147.2
800	176.8	187.4	171.7
900	204.1	213.8	196.5
1000	231.9	240.5	221.6
1100	260.2	267.5	247.0
1200	289.0	294.9	272.7
1300	318.0	326.1	298.5
1400	347.6	350.5	324.6
1500	377.6	378.7	350.8
1600	407.8	407.3	377.3
1700	438.2	435.9	403.7
1800	469.1	464.8	430.4
1900	500.1	493.7	457.3
2000	531.4	523.0	484.5
2100	562.8	552.7	511.4
2200	594.3	582.0	538.6
2300	626.2	612.3	566.1
2400	658.2	642.3	593.5
2500	690.2	672.3	621.0
3000	852.3	823.8	760.1
3500	1017.4	978.0	901.7

Source: Ref. 4-3.

Air: 287.55 Btu/lb × 1000 lb = 287,550 Btu

H_2O: 1611.22 Btu/lb × 250 lb = 402,805 Btu

Total heat content = 690,355 Btu

The heat release, 700,000 Btu, is greater than the heat contained by the gas mixture at 1200°F. Therefore, the mixture temperature must be greater than 1200°F. Try 1250°F:

Air: 301.02 Btu/lb × 1000 lb = 301,020 Btu

H_2O: 1638.26 Btu/lb × 250 lb = 409,565 Btu

Total heat content = 710,585 Btu

The heat content at 1250°F is greater than that of the heat released, 700,000 Btu. The mixture temperature x must lie between 1200 and 1250°F, as follows:

Temperature, °F	Heat content, Btu
1200	690,355
x	700,000
1250	710,585

By interpolation:

$$x = 1200 + (1250 - 1200)\frac{700{,}000 - 690{,}355}{710{,}585 - 690{,}355} = 1224°\text{F}$$

The mixture temperature is 1224°F.

The determination of temperature is a trial-and-error calculation, based on component enthalpies and heat release.

Note that the moisture enthalpy values in Table 4.2 are based on moisture in the liquid phase at 60 or 80°F. If moisture enters a reaction in the vapor phase, its heat of vaporization, that amount of heat required to change water from a liquid to a gas (steam) without a change in temperature, must be subtracted from the moisture value in Table 4.1. At 60°F the heat of vaporization is 1059.6 Btu/lb, and at 80°F the heat of vaporization is 1048.3 Btu/lb.

In the previous example, with moisture entering the reaction at 60°F, in the vapor phase, the temperature of the mixture y is calculated as follows:

At 1800°F:

Air: 435.24 Btu/lb × 1000 lb = 453,240 Btu

H_2O: (1948.02 − 1059.6) Btu/lb × 250 lb = 222,104 Btu

Total heat content = 675,344 Btu

At 1900°F:

Air: 481.57 Btu/lb × 1000 lb = 481,570 Btu

H_2O: (2007.17 − 1059.6) Btu/lb × 250 lb = 236,893 Btu

Total heat content = 718,463 Btu

To summarize:

Temperature, °F	Heat content, Btu
1800	675,344
y	700,000
1900	718,463

By interpolation:

$$y = 1800 + (1900 - 1800)\frac{700{,}000 - 675{,}345}{718{,}463 - 675{,}345} = 1857°\text{F}$$

The mixture temperature is 1857°F. Note that the heat of vaporization is a significant factor in the calculated heat level. When present as liquid water, the resultant temperature is 1224°F. Without inclusion of the heat of vaporization, i.e., if moisture enters the reaction as steam or vapor, the temperature is 1857°F.

THE PERFECT GAS LAW

In general, gases at low (atmospheric) pressures will behave in accordance with the perfect gas law, as follows:

$$144P = WRT$$

where P = pressure, lb/in^2 absolute
W = specific weight, lb/ft^3
T = absolute temperature, degrees Rankine (the absolute temperature is equal to $t + 460$, where t is in degrees Fahrenheit)
R = universal gas constant divided by molecular weight of gas, ft/°F

The universal gas constant, 1545, is unvarying for any gas. The molecular weight and the R value of some common gases are listed in Table 4.3.

TABLE 4.3

	Gas Molecular Weight	R Value (ft/°R)
Air	28.9	53.3
Oxygen	32.00	48.3
Hydrogen	2.02	764.9
Nitrogen	28.02	55.1
Carbon monoxide	28.01	55.2
Carbon dioxide	44.01	35.1
Sulfur dioxide	64.06	24.01
Water vapor	18.02	85.83

The gas law is convenient for calculating W, the specific weight of a gas at a particular temperature and pressure. For example,

What is the specific weight of nitrogen at 14.7 lb/in^2 absolute and 100°F? Using the perfect gas law, with $R = 55.1$ from Table 4.3, $P = 14.7$, and $T = 100°F + 460° = 560°R$,

$$W = \frac{144P}{RT} = \frac{144 \times 14.7}{55.1 \times 560} = 0.0686 \text{ lb/ft}^3$$

Table 4.4 lists specific volumes, ft^3/lb, the reciprocal of specific weight, for air and water vapor at atmospheric pressure, 14.7 lb/in^2 absolute.

TABLE 4.4 Specific Volume

T, °F	Air, ft^3/lb	H_2O, ft^3/lb	T, °F	Air, ft^3/lb	H_2O, ft^3/lb
70	13.3	21.5	2000	61.9	99.7
100	14.1	22.7	2100	64.5	103.8
200	16.6	26.8	2200	67.0	107.9
300	19.1	30.8	2300	69.5	111.9
400	21.7	34.9	2400	72.0	116.0
500	24.2	38.9	2500	74.5	120.0
600	26.7	43.0	2600	77.0	124.1
700	29.2	47.0	2700	79.6	128.1
800	31.7	51.1	2800	82.1	132.2
900	34.2	55.1	2900	84.6	136.2
1000	36.8	59.2	3000	87.1	140.3
1100	39.3	63.3	3100	89.6	144.3
1200	41.8	67.3	3200	92.2	148.4
1300	44.3	71.4	3300	94.7	152.5
1400	46.8	75.4	3400	97.2	156.5
1500	49.4	79.5	3500	99.7	160.6
1600	51.9	83.5	3600	102.2	164.6
1700	54.4	87.6	3700	104.7	168.7
1800	56.9	91.6	3800	107.3	172.7
1900	59.4	95.7	3900	109.8	176.8

A MIXTURE OF GASES

A mixture of two or more gases has the following properties:

- The temperature of each gas is the same as that of the mixture.
- The weight of the mixture is equal to the sum of the weights of each of the component gases.
- The volume of the mixture is equal to the sum of the volumes of each gas each calculated at the mixture pressure.
- The enthalpy of the mixture is equal to the sum of the total enthalpies of each of the component gases.

An illustration of calculations involving these mixture properties is as follows:

A mixture of 1000 lb of air plus 200 lb of steam is contained at a pressure of 14.7 lb/in^2 absolute at 1200°F. Determine the total enthalpy of the mixture, the weight, and volume percentage of each component.

The enthalpy of the mixture is equal to the sum of the component enthalpies (from Table 4.1, 60°F datum):

Air at 1200°F:	287.55 Btu/lb × 1000 lb =	287,550 Btu
H_2O at 1200°F:	1611.22 Btu/lb × 200 lb =	322,244 Btu
	Total enthalpy =	609,794 Btu

The total weight of the mixture is equal to the sum of the component weights:

Weight of air	1000 lb
Weight of H_2O	200 lb
Total weight	1200 lb

Therefore, the enthalpy is

$$609{,}794 \div 1200 \text{ lb} = 508 \text{ Btu/lb mixture}$$

The weight percentage:

Air: $1000 \text{ lb}/1200 \text{ lb} = 83\%$

H_2O: $200 \text{ lb}/1200 \text{ lb} = 17\%$

From Table 4.4, the mixture volume is

Air at 1200°F: $41.8 \text{ ft}^3/\text{lb} \times 1000 \text{ lb} = 41{,}800 \text{ ft}^3$

H_2O at 1200°F: $67.3 \text{ ft}^3/\text{lb} \times 200 \text{ lb} = 13{,}460 \text{ ft}^3$

Total $= 55{,}260 \text{ ft}^3$

Volume percentage is

Air: $41{,}800 \div 55{,}260 = 76\%$

H_2O: $13{,}460 \div 55{,}260 = 24\%$

AIR-MOISTURE MIXTURES

At atmospheric pressure, 14.7 lb/in^2 absolute, the air can contain a maximum amount of moisture, moisture of saturation, as shown in Fig. 4.1. The saturation time is the 100 percent curve, to the left of the series of vertical curves. The moisture of saturation, the humidity, is listed on the vertical axis of the chart as pounds of water vapor per pound of dry air. The horizontal axis lists the temperature of the mixture. The saturation moisture can be found as follows:

> At 100°F the moisture of saturation, that maximum quantity of moisture that the air can contain, by following the 100 percent curve, is 0.043 lb H_2O/lb dry air.

For this example if the temperature of the mixture were increased to 150°F, the humidity of the air would drop from 100 to 20 percent, following the horizontal line representing 0.043 lb H_2O/lb dry air horizontally to the right. If the mixture were to cool to 80°F, the saturation moisture would be 0.022 lb H_2O/lb dry air, reading the 100 percent saturation line. The difference between the 0.043 lb H_2O/lb dry air contained in the mixture at 150°F and the 0.022 lb H_2O/lb dry air would condense, i.e., would leave the mixture as water droplets.

The properties of the air-moisture mixture including saturation humidity, volume, and enthalpy are listed in Table 4.5.

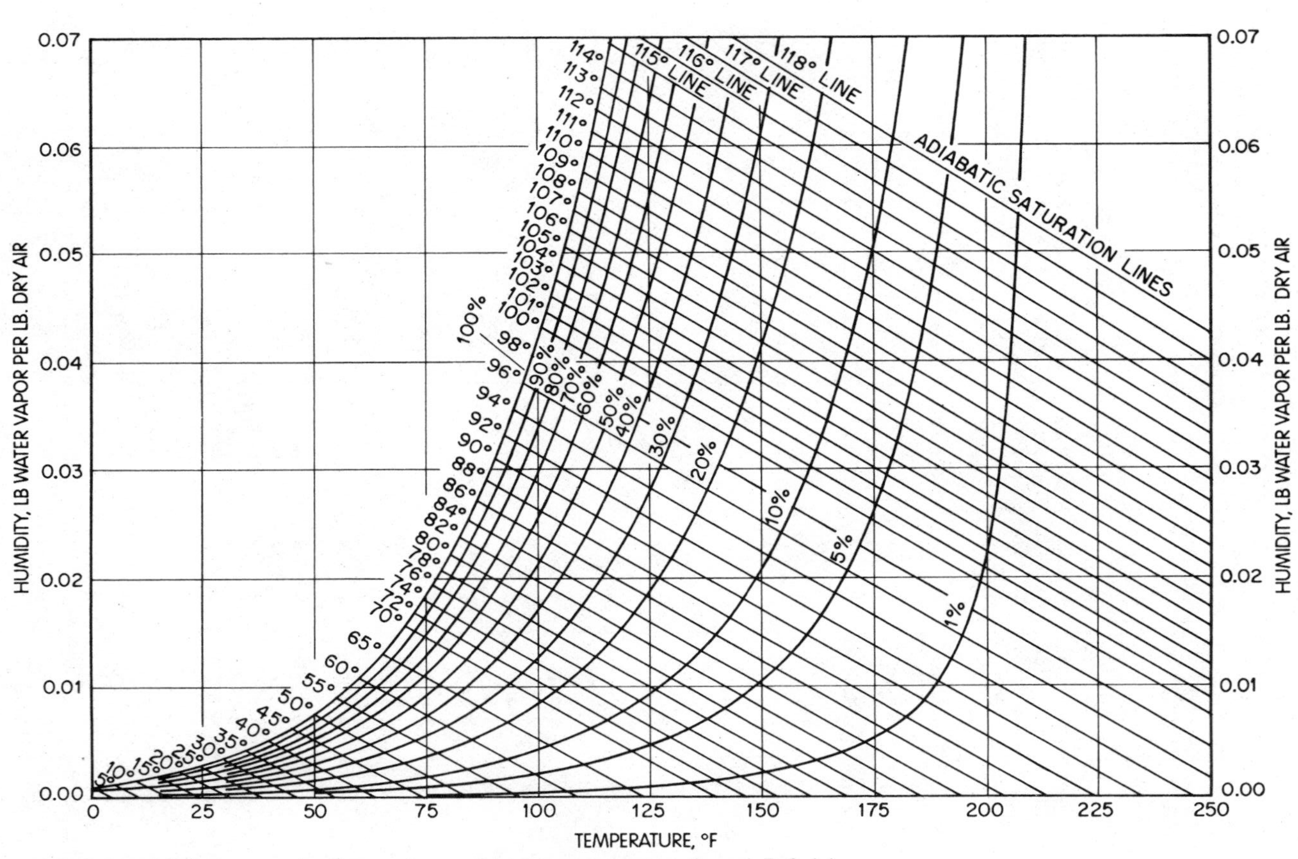

FIGURE 4.1 Adiabatic saturation lines and saturation percentage curves. *Source*: Ref. 4-4.

TABLE 4.5 Properties of Dry Air, Water Vapor, and Saturated Air–Water Vapor Mixtures

Temperature Range, −100 to +211°F; Pressure, 29.921 inHg

Temp., °F	Saturation pressure		Saturation humidity, wt. water vapor/lb. dry air		Saturation moisture content, wt. water vapor/ft³ sat. mixture	
	lb/in² × 10⁴	inHg × 10⁴	lb × 10⁵	gr	lb × 10⁷	gr
−100	0.47315	0.9633	0.2002	0.01402	2.213	0.001549
−99	.51235	1.0431	.2168	.01518	2.390	.001673
−98	.55446	1.1289	.2347	.01643	2.579	.001806
−97	.59979	1.2212	.2538	.01777	2.783	.001948
−96	.64852	1.3204	.2745	.01922	3.000	.002100
−95	0.70090	1.4270	0.2966	0.02077	3.234	0.002264
−94	.75721	1.5417	.3205	.02243	3.484	.002439
−93	.81760	1.6646	.3460	.02422	3.751	.002626
−92	.88249	1.7968	.3735	.02614	4.038	.002827
−91	.95209	1.9385	.4029	.02821	4.345	.003041
−90	1.0267	2.0904	0.4345	0.03042	4.672	0.003271
−89	1.1068	2.2534	.4684	.03279	5.023	.003516
−88	1.1925	2.4279	.5047	.03533	5.397	.003778
−87	1.2844	2.6150	.5436	.03805	5.797	.004058
−86	1.3828	2.8153	.5852	.04095	6.225	.004357
−85	1.4880	3.0296	0.6298	0.04408	6.680	0.004676
−84	1.6007	3.2590	.6774	.04742	7.167	.005017
−83	1.7212	3.5043	.7284	.05099	7.686	.005380
−82	1.8500	3.7665	.7829	.05481	8.239	.005767
−81	1.9876	4.0468	.8412	.05888	8.828	.006180
−80	2.1346	4.3460	0.9034	0.06324	9.456	0.006619
−79	2.2914	4.6653	0.9698	.06788	10.124	.007087
−78	2.4592	5.0070	1.0408	.07286	10.837	.007586
−77	2.6379	5.3709	1.1164	.07815	11.593	.008115
−76	2.8266	5.7590	1.1971	.08380	12.399	.008679
−75	3.0320	6.1732	1.283	0.08983	13.26	0.009279
−74	3.2486	6.6142	1.375	.09624	14.17	.009916
−73	3.4793	7.0838	1.473	.10308	15.13	.010592
−72	3.7254	7.5850	1.577	.11037	16.16	.011312
−71	3.9869	8.1173	1.687	.11811	17.25	.012074
−70	4.2653	8.6842	1.805	0.1264	18.41	0.01289
−69	4.5619	9.2882	1.931	.1352	19.64	.01375
−68	4.8769	9.9295	2.064	.1445	20.94	.01466
−67	5.2118	10.6114	2.206	.1544	22.32	.01562
−66	5.5678	11.3362	2.357	.1650	23.78	.01595
−65	5.9460	12.106	2.517	0.1762	25.33	0.01773
−64	6.3477	12.924	2.687	.1881	26.98	.01888
−63	6.7741	13.792	2.867	.2007	28.71	.02010
−62	7.2262	14.713	3.058	.2141	30.55	.02139
−61	7.7061	15.690	3.262	.2283	32.50	.02275
−60	8.2152	16.726	3.477	0.2434	34.56	0.02419
−59	8.7545	17.824	3.705	.2594	36.74	.02572
−58	9.3265	18.989	3.947	.2763	39.04	.02733
−57	9.9325	20.223	4.204	.2943	41.47	.02903
−56	10.5745	21.530	4.476	.3133	44.04	.03083

TABLE 4.5 Properties of Dry Air, Water Vapor, and Saturated Air–Water Vapor Mixtures (Continued)

Temperature Range, −100 to +211°F; Pressure, 29.921 in Hg

Saturation density, wt. air plus water vapor/ft³ sat. mixture		Volume			Enthalpy			Temp., °F
lb	gr	Dry air, ft³/lb	Water vapor, ft³/lb[a]	Saturated mixture, ft³/lb dry air	Dry air, Btu/lb	Water vapor, Btu/lb	Saturated mixture, Btu/lb dry air	
0.1105	773.8	9.047	14.580	9.047	−31.691	1016.6	−31.689	−100
.1102	771.6	9.072	14.620	9.072	−31.451	1017.1	−31.449	−99
.1099	769.4	9.098	14.661	9.098	−31.211	1017.5	−31.209	−98
.1096	767.3	9.123	14.702	9.123	−30.971	1018.0	−30.968	−97
.1093	765.2	9.148	14.742	9.148	−30.731	1018.4	−30.728	−96
0.1090	763.1	9.173	14.783	9.173	−30.491	1018.8	−30.488	−95
.1087	761.0	9.199	14.823	9.199	−30.251	1019.3	−30.248	−94
.1084	758.9	9.224	14.864	9.224	−30.011	1019.7	−30.007	−93
.1081	756.8	9.249	14.904	9.249	−29.771	1020.2	−29.767	−92
.1078	754.7	9.275	14.944	9.275	−29.531	1020.6	−29.527	−91
0.1075	752.7	9.300	14.985	9.300	−29.291	1021.1	−29.287	−90
.1072	750.6	9.325	15.026	9.326	−29.051	1021.5	−29.046	−89
.1069	748.6	9.351	15.066	9.351	−28.811	1021.9	−28.806	−88
.1067	746.6	9.376	15.107	9.376	−28.571	1022.4	−28.565	−87
.1064	744.6	9.402	15.147	9.402	−28.331	1022.8	−28.325	−86
0.1061	742.6	9.427	15.188	9.427	−28.091	1023.3	−28.085	−85
.1058	740.6	9.452	15.229	9.452	−27.851	1023.7	−27.844	−84
.1055	738.6	9.478	15.269	9.478	−27.611	1024.2	−27.604	−83
.1052	736.6	9.503	15.310	9.503	−27.371	1024.6	−27.362	−82
.1050	734.7	9.528	15.350	9.528	−27.130	1025.0	−27.122	−81
0.1047	732.7	9.554	15.390	9.554	−26.890	1025.5	−26.881	−80
.1044	730.7	9.579	15.431	9.579	−26.650	1025.9	−26.640	−79
.1041	728.8	9.604	15.472	9.604	−26.410	1026.4	−26.399	−78
.1039	727.0	9.630	15.512	9.630	−26.170	1026.8	−26.159	−77
.1036	725.0	9.655	15.553	9.655	−25.930	1027.3	−25.918	−76
0.1033	723.1	9.680	15.593	9.680	−25.690	1027.7	−25.677	−75
.1030	721.2	9.706	15.634	9.706	−25.450	1028.1	−25.436	−74
.1028	719.3	9.731	15.675	9.731	−25.210	1028.6	−25.195	−73
.1025	717.5	9.756	15.715	9.757	−24.970	1029.0	−24.954	−72
.1022	715.6	9.782	15.756	9.782	−24.730	1029.5	−24.712	−71
0.1020	713.8	9.807	15.796	9.807	−24.490	1029.9	−24.471	−70
.1017	711.9	9.832	15.837	9.833	−24.250	1030.4	−24.230	−69
.1014	710.1	9.858	15.877	9.858	−24.010	1030.8	−23.989	−68
.1012	708.3	9.883	15.917	9.883	−23.770	1031.2	−23.747	−67
.1009	706.4	9.908	15.958	9.909	−23.530	1031.7	−23.506	−66
0.10067	704.7	9.934	15.999	9.934	−23.290	1032.1	−23.264	−65
.10041	702.9	9.959	16.039	9.959	−23.050	1032.6	−23.022	−64
.10016	701.1	9.984	16.080	9.985	−22.810	1033.0	−22.780	−63
.09990	699.3	10.010	16.120	10.010	−22.570	1033.5	−22.538	−62
.09965	697.6	10.035	16.160	10.035	−22.330	1033.9	−22.296	−61
0.09940	695.8	10.060	16.201	10.061	−22.090	1034.3	−22.054	−60
.09915	694.0	10.086	16.242	10.086	−21.850	1034.8	−21.812	−59
.09890	692.3	10.111	16.283	10.112	−21.610	1035.2	−21.569	−58
.09865	690.6	10.136	16.323	10.137	−21.370	1035.7	−21.326	−57
.09841	688.9	10.162	16.364	10.162	−21.130	1036.1	−21.084	−56

TABLE 4.5 Properties of Dry Air, Water Vapor, and Saturated Air–Water Vapor Mixtures (Continued)

Temperature Range, −100 to +211°F; Pressure, 29.921 inHg

Temp., °F	Saturation pressure		Saturation humidity, wt. water vapor/lb. dry air		Saturation moisture content, wt. water vapor/ft³ sat. mixture	
	lb/in² × 10²	inHg × 10²	lb × 10⁴	gr	lb × 10⁶	gr
−55	0.11254	0.22913	0.4763	0.3334	4.675	0.03273
−54	.11973	.24377	.5067	.3547	4.962	.03473
−53	.12733	.25926	.5389	.3773	5.264	.03685
−52	.13538	.27564	.5730	.4011	5.583	.03908
−51	.14389	.29296	.6090	.4263	5.919	.04143
−50	0.15288	0.31126	0.6471	0.4529	6.273	0.04391
−49	.16238	.33061	.6873	.4811	6.647	.04653
−48	.17242	.35106	.7298	.5109	7.041	.04929
−47	.18305	.37269	.7748	.5424	7.456	.05220
−46	.19421	.39543	.8221	.5754	7.892	.05525
−45	0.20603	0.41949	0.8721	0.6105	8.352	0.05846
−44	.21850	.44488	0.9249	.6474	8.836	.06185
−43	.23164	.47161	0.9805	.6863	9.345	.06541
−42	.24552	.49988	1.0392	.7275	9.830	.06916
−41	.26014	.52964	1.1011	.7708	10.444	.07311
−40	0.27555	0.56102	1.166	0.8165	11.04	0.07725
−39	.29177	.59405	1.235	.8645	11.66	.08161
−38	.30889	.62890	1.308	.9153	12.31	.08618
−37	.32687	.66551	1.384	.9686	13.00	.09099
−36	.34582	.70409	1.464	1.0247	13.72	.09604
−35	0.36577	0.74472	1.548	1.084	14.48	0.1013
−34	.38676	.78744	1.637	1.146	15.27	.1069
−33	.40881	.83236	1.730	1.211	16.10	.1127
−32	.43204	.87964	1.829	1.280	16.98	.1189
−31	.45643	.92929	1.932	1.353	17.89	.1253
−30	0.48207	0.9815	2.041	1.429	18.86	0.1320
−29	.50900	1.0363	2.155	1.508	19.86	.1390
−28	.53727	1.0939	2.275	1.592	20.92	.1464
−27	.56697	1.1544	2.400	1.680	22.02	.1542
−26	.59816	1.2179	2.533	1.773	23.18	.1623
−25	0.63090	1.2845	2.671	1.870	24.39	0.1707
−24	.66522	1.3544	2.817	1.972	25.66	.1796
−23	.70129	1.4278	2.969	2.079	26.99	.1889
−22	.73899	1.5046	3.129	2.190	28.38	.1986
−21	.77856	1.5852	3.297	2.308	29.83	.2088
−20	0.82009	1.6697	3.473	2.431	31.35	0.2194
−19	.86362	1.7583	3.657	2.560	32.93	.2305
−18	.90924	1.8512	3.850	2.695	34.59	.2422
−17	.95698	1.9484	4.053	2.837	36.33	.2543
−16	1.00696	2.0502	4.265	2.985	38.14	.2670
−15	1.0592	2.1567	4.486	3.140	40.03	0.2802
−14	1.1141	2.2683	4.718	3.303	42.00	.2940
−13	1.1714	2.3849	4.961	3.473	44.07	.3085
−12	1.2313	2.5069	5.215	3.651	46.22	.3235
−11	1.2940	2.6346	5.481	3.837	48.46	.3392

TABLE 4.5 Properties of Dry Air, Water Vapor, and Saturated Air–Water Vapor Mixtures (Continued)

Temperature Range, −100 to +211°F; Pressure, 29.921 inHg

Saturation density, wt. air plus water vapor/ft³ sat. mixture		Volume			Enthalpy			Temp., °F
lb	gr	Dry air, ft³/lb	Water vapor, ft³/lb[a]	Saturated mixture, ft³/lb dry air	Dry air, Btu/lb	Water vapor, Btu/lb	Saturated mixture, Btu/lb dry air	
0.09816	687.1	10.187	16.404	10.188	−20.890	1036.6	−20.841	−55
.09792	685.4	10.212	16.445	10.213	−20.650	1037.0	−20.597	−54
.09768	683.7	10.238	16.485	10.238	−20.410	1037.5	−20.354	−53
.09744	682.1	10.263	16.526	10.264	−20.170	1037.9	−20.111	−52
.09720	680.4	10.289	16.566	10.289	−19.930	1038.3	−19.867	−51
0.09696	678.7	10.313	16.607	10.315	−19.690	1038.8	−19.623	−50
.09672	677.0	10.339	16.647	10.340	−19.450	1039.2	−19.379	−49
.09648	675.4	10.364	16.688	10.365	−19.210	1039.7	−19.134	−48
.09625	673.8	10.389	16.728	10.391	−18.970	1040.1	−18.889	−47
.09601	672.1	10.415	16.787	10.416	−18.730	1040.6	−18.644	−46
0.09578	670.5	10.440	16.809	10.441	−18.490	1041.0	−18.399	−45
.09555	668.8	10.465	16.850	10.467	−18.250	1041.4	−18.154	−44
.09532	667.2	10.491	16.891	10.492	−18.010	1041.9	−17.908	−43
.09509	665.6	10.516	16.931	10.518	−17.769	1042.3	−17.661	−42
.09486	664.0	10.541	16.971	10.543	−17.529	1042.8	−17.414	−41
0.09463	662.4	10.567	17.012	10.569	−17.289	1043.2	−17.167	−40
.09441	660.8	10.592	17.053	10.594	−17.049	1043.7	−16.920	−39
.09418	659.3	10.617	17.093	10.619	−16.809	1044.1	−16.672	−38
.09396	657.7	10.642	17.134	10.645	−16.569	1044.5	−16.424	−37
.09373	656.1	10.668	17.174	10.670	−16.329	1045.0	−16.176	−36
0.09351	654.6	10.693	17.215	10.696	−16.089	1045.4	−15.927	−35
.09329	653.0	10.718	17.255	10.721	−15.849	1045.9	−15.678	−34
.09307	651.5	10.744	17.296	10.747	−15.609	1046.3	−15.428	−33
.09285	649.9	10.769	17.336	10.772	−15.369	1046.8	−15.178	−32
.09263	648.4	10.794	17.376	10.798	−15.129	1047.2	−14.927	−31
0.09241	646.9	10.820	17.418	10.823	−14.888	1047.6	−14.674	−30
.09220	645.4	10.845	17.458	10.849	−14.648	1048.1	−14.422	−29
.09198	643.9	10.870	17.498	10.874	−14.408	1048.5	−14.170	−28
.09177	642.4	10.896	17.539	10.900	−14.168	1049.0	−13.916	−27
.09155	640.9	10.921	17.580	10.925	−13.928	1049.4	−13.662	−26
0.09134	639.4	10.946	17.620	10.951	−13.688	1049.9	−13.408	−25
.09113	637.9	10.971	17.661	10.977	−13.448	1050.3	−13.152	−24
.09092	636.4	10.997	17.701	11.002	−13.208	1050.7	−12.895	−23
.09071	635.0	11.022	17.742	11.028	−12.968	1051.2	−12.639	−22
.09050	633.5	11.047	17.782	11.053	−12.727	1051.6	−12.380	−21
0.09029	632.0	11.073	17.823	11.079	−12.487	1052.1	−12.122	−20
.09009	630.6	11.098	17.863	11.105	−12.247	1052.5	−11.862	−19
.08988	629.2	11.123	17.904	11.130	−12.007	1053.0	−11.602	−18
.08967	627.7	11.149	17.944	11.156	−11.767	1053.4	−11.340	−17
.08947	626.3	11.174	17.985	11.182	−11.527	1053.8	−11.078	−16
0.08926	624.8	11.199	18.026	11.207	−11.287	1054.3	−10.814	−15
.08906	623.5	11.225	18.066	11.233	−11.047	1054.7	−10.549	−14
.08886	622.0	11.250	18.106	11.259	−10.807	1055.2	−10.284	−13
.08866	620.6	11.275	18.147	11.285	−10.567	1055.6	−10.016	−12
.08846	619.2	11.301	18.188	11.310	−10.326	1056.1	−9.748	−11

TABLE 4.5 Properties of Dry Air, Water Vapor, and Saturated Air–Water Vapor Mixtures (Continued)

Temperature Range, −100 to +211°F; Pressure, 29.921 inHg

Temp., °F	Saturation pressure		Saturation humidity, wt. water vapor/lb. dry air		Saturation moisture content, wt. water vapor/ft³ sat. mixture	
	lb/in²	inHg	lb × 10³	gr	lb × 10⁵	gr
−10	0.013595	0.027680	0.5759	4.031	5.080	0.3556
−9	.014281	.029076	.6050	4.235	5.324	.3727
−8	.014997	.030534	.6353	4.447	5.579	.3905
−7	.015745	.032056	.6670	4.669	5.844	.4091
−6	.016526	.033648	.7002	4.901	6.121	.4284
−5	0.017343	0.035310	0.7348	5.144	6.409	0.4486
−4	.018196	.037047	.7710	5.397	6.709	.4697
−3	.019085	.038858	.8088	5.661	7.022	.4915
−2	.020014	.040748	.8482	5.937	7.347	.5143
−1	.020982	.042720	.8893	6.225	7.686	.5380
0	0.021994	0.044779	0.9322	6.525	8.039	0.5627
1	.023048	.046925	0.9769	6.839	8.406	.5884
2	.024148	.049165	1.0236	7.166	8.788	.6151
3	.025294	.051500	1.0723	7.506	9.185	.6430
4	.026488	.053929	1.1230	7.861	9.598	.6718
5	0.027733	0.056464	1.176	8.231	10.03	0.7019
6	.029039	.059106	1.231	8.617	10.47	.7331
7	.030384	.061861	1.289	9.020	10.94	.7657
8	.031790	.064724	1.348	9.438	11.42	.7994
9	.033253	.067705	1.411	9.874	11.92	.8344
10	0.034779	0.070810	1.475	10.33	12.44	0.8708
11	.036367	.074043	1.543	10.80	12.98	.9086
12	.038026	.077409	1.613	11.29	13.54	.9479
13	.039735	.080901	1.686	11.80	14.12	.9886
14	.041522	.084539	1.762	12.34	14.73	1.0308
15	0.043384	0.088331	1.842	12.89	15.37	1.076
16	.045316	.092264	1.924	13.47	16.00	1.120
17	.047325	.096355	2.009	14.07	16.68	1.168
18	.049411	.100602	2.098	14.69	17.38	1.216
19	.051583	.105024	2.191	15.34	18.10	1.267
20	0.053838	0.10961	2.287	16.01	18.85	1.320
21	.056179	.11438	2.387	16.71	19.63	1.376
22	.058606	.11932	2.490	17.43	20.44	1.431
23	.061133	.12447	2.598	18.19	21.28	1.489
24	.063751	.12980	2.710	18.97	22.14	1.550
25	0.066474	0.13534	2.826	19.78	23.04	1.613
26	.069295	.14109	2.947	20.63	23.97	1.678
27	.072225	.14705	3.072	21.50	24.93	1.745
28	.075263	.15324	3.202	22.41	25.92	1.815
29	.078417	.15966	3.337	23.36	26.95	1.887
30	0.081684	0.16631	3.476	24.33	28.02	1.962
31	.085072	.17321	3.621	25.35	29.12	2.039
32	.088579	.18035	3.787	26.51	30.38	2.127
33	.092218	.18776	3.943	27.60	31.57	2.210
34	.096000	.19546	4.106	28.74	32.80	2.296

TABLE 4.5 Properties of Dry Air, Water Vapor, and Saturated Air–Water Vapor Mixtures (Continued)

Temperature Range, −100 to +211°F; Pressure, 29.921 inHg

Saturation density, wt. air plus water vapor/ft³ sat. mixture		Volume			Enthalpy			Temp., °F
lb	gr	Dry air, ft³/lb	Water vapor, ft³/lb[a]	Saturated mixture, ft³/lb dry air	Dry air, Btu/lb	Water vapor, Btu/lb	Saturated mixture, Btu/lb dry air	
0.08826	617.8	11.326	18.228	11.336	−10.086	1056.5	−9.478	−10
.08806	616.5	11.351	18.269	11.362	−9.846	1056.9	−9.207	−9
.08787	615.1	11.376	18.309	11.388	−9.606	1057.4	−8.934	−8
.08767	613.7	11.402	18.350	11.414	−9.366	1057.8	−8.660	−7
.08747	612.3	11.427	18.390	11.440	−9.126	1058.3	−8.385	−6
0.08727	610.9	11.452	18.431	11.466	−8.886	1058.7	−8.108	−5
.08709	609.6	11.478	18.471	11.492	−8.646	1059.2	−7.829	−4
.08689	608.2	11.503	18.512	11.518	−8.406	1059.6	−7.549	−3
.08670	606.9	11.528	18.552	11.544	−8.165	1060.0	−7.266	−2
.08651	605.6	11.553	18.593	11.570	−7.925	1060.5	−6.982	−1
0.08632	604.2	11.579	18.634	11.596	−7.685	1060.9	−6.696	0
.08613	602.9	11.604	18.674	11.622	−7.445	1061.4	−6.408	1
.08594	601.6	11.629	18.715	11.649	−7.205	1061.8	−6.118	2
.08575	600.2	11.655	18.755	11.675	−6.965	1062.3	−5.826	3
.08556	598.9	11.680	18.796	11.701	−6.725	1062.7	−5.531	4
0.08537	597.6	11.705	18.836	11.727	−6.485	1063.1	−5.234	5
.08518	596.3	11.731	18.877	11.754	−6.244	1063.6	−4.935	6
.08500	595.0	11.756	18.917	11.780	−6.004	1064.0	−4.633	7
.08481	593.7	11.781	18.958	11.807	−5.764	1064.5	−4.329	8
.08463	592.4	11.806	18.998	11.833	−5.524	1064.9	−4.022	9
0.08444	591.1	11.832	19.039	11.860	−5.284	1065.4	−3.712	10
.08426	589.8	11.857	19.079	11.886	−5.044	1065.8	−3.399	11
.08405	588.4	11.882	19.120	11.913	−4.804	1066.2	−3.083	12
.08389	587.3	11.908	19.161	11.940	−4.563	1066.7	−2.765	13
.08371	586.0	11.933	19.201	11.967	−4.323	1067.3	−2.443	14
0.08360	585.2	11.958	19.242	11.984	−4.083	1067.6	−2.1171	15
.08335	583.5	11.983	19.282	12.021	−3.843	1068.0	−1.7882	16
.08317	582.2	12.009	19.322	12.048	−3.603	1068.5	−1.4558	17
.08299	580.9	12.034	19.363	12.075	−3.363	1068.9	−1.1197	18
.08281	579.7	12.059	19.403	12.102	−3.122	1069.3	−0.7797	19
0.08264	578.5	12.085	19.444	12.129	−2.882	1069.8	−0.43572	20
.08246	577.2	12.110	19.485	12.156	−2.642	1070.2	−0.08768	21
.08228	576.0	12.135	19.525	12.184	−2.402	1070.7	+0.26436	22
.08210	574.7	12.160	19.566	12.211	−2.162	1071.1	0.62115	23
.08194	573.6	12.186	19.606	12.239	−1.922	1071.6	0.98221	24
0.08175	572.3	12.211	19.647	12.267	−1.6813	1072.0	1.348	25
.08158	571.1	12.236	19.687	12.294	−1.4411	1072.5	1.719	26
.08140	569.8	12.262	19.728	12.322	−1.2010	1072.9	2.095	27
.08123	568.6	12.287	19.769	12.350	−0.9608	1073.3	2.476	28
.08106	567.4	12.312	19.809	12.378	−0.7206	1073.8	2.862	29
0.08089	566.3	12.337	19.850	12.405	−0.4804	1074.2	3.254	30
.08071	565.0	12.363	19.890	12.435	−0.2402	1074.7	3.651	31
.08055	563.8	12.388	19.918	12.463	0.0000	1075.1	4.071	32
.08037	562.6	12.413	19.959	12.492	0.2402	1075.5	4.481	33
.08020	561.4	12.439	20.000	12.520	0.4804	1076.0	4.898	34

TABLE 4.5 Properties of Dry Air, Water Vapor, and Saturated Air–Water Vapor Mixtures (Continued)

Temperature Range, −100 to +211°F; Pressure, 29.921 inHg

Temp., °F	Saturation pressure		Saturation humidity, wt. water vapor/lb. dry air		Saturation moisture content, wt. water vapor/ft³ sat. mixture	
	lb/in²	inHg	lb	gr	lb × 10⁴	gr
35	0.09991	0.20342	0.004274	29.92	3.406	2.384
36	.10396	.21166	.004449	31.14	3.537	2.476
37	.10815	.22019	.004629	32.41	3.672	2.570
38	.11250	.22905	.004817	33.72	3.812	2.668
39	.11699	.23819	.005011	35.08	3.956	2.769
40	0.12164	0.24766	0.005212	36.48	4.105	2.874
41	.12646	.25747	.005420	37.94	4.259	2.981
42	.13144	.26761	.005635	39.45	4.418	3.093
43	.13660	.27812	.005859	41.01	4.582	3.208
44	.14194	.28899	.006090	42.63	4.752	3.327
45	0.14745	0.30021	0.006329	44.30	4.927	3.449
46	.15316	.31183	.006576	46.03	5.107	3.575
47	.15906	.32385	.006833	47.83	5.294	3.706
48	.16516	.33627	.007098	49.68	5.486	3.840
49	.17146	.34909	.007369	51.59	5.682	3.977
50	0.17798	0.36237	0.007655	53.59	5.888	4.121
51	.18471	.37607	.007948	55.64	6.099	4.269
52	.19167	.39024	.008252	57.76	6.316	4.421
53	.19885	.40486	.008565	59.96	6.540	4.578
54	.20627	.41997	.008889	62.23	6.771	4.739
55	0.21394	0.43558	0.009225	64.57	7.009	4.906
56	.22185	.45169	.009571	67.00	7.254	5.078
57	.23002	.46832	.009929	69.50	7.506	5.254
58	.23845	.48548	.010299	72.09	7.766	5.436
59	.24716	.50321	.010681	74.77	8.034	5.624
60	0.25614	0.52150	0.01108	77.53	8.310	5.817
61	.26541	.54038	.01149	80.40	8.595	6.016
62	.27497	.55984	.01191	83.34	8.886	6.220
63	.28483	.57991	.01234	86.39	9.188	6.431
64	.29500	.60062	.01279	89.54	9.498	6.649
65	0.30549	0.62198	0.01326	92.79	9.816	6.871
66	.31630	.64399	.01374	96.15	10.144	7.101
67	.32744	.66667	.01423	99.61	10.479	7.335
68	.33893	.69006	.01474	103.19	10.829	7.580
69	.35077	.71417	.01527	106.88	11.186	7.830
70	0.36297	0.73901	0.01581	110.7	11.55	8.086
71	.37554	.76460	.01638	114.6	11.93	8.352
72	.38848	.79095	.01700	118.7	12.32	8.623
73	.40182	.81811	.01755	122.9	12.72	8.902
74	.41556	.84608	.01817	127.2	13.13	9.189
75	0.42969	0.87485	0.01881	131.7	13.55	9.484
76	.44425	.90449	.01946	136.2	13.98	9.787
77	.45923	.93499	.02014	141.0	14.43	10.098
78	.47467	.96643	.02084	145.9	14.88	10.418
79	.49055	.99876	.02156	150.9	15.35	10.746

TABLE 4.5 Properties of Dry Air, Water Vapor, and Saturated Air–Water Vapor Mixtures (Continued)

Temperature Range, −100 to +211°F; Pressure, 29.921 inHg

Saturation density, wt. air plus water vapor/ft³ sat. mixture		Volume			Enthalpy			Temp., °F
lb	gr	Dry air, ft³/lb	Water vapor, ft³/lb[a]	Saturated mixture, ft³/lb dry air	Dry air, Btu/lb	Water vapor, Btu/lb	Saturated mixture, Btu/lb dry air	
0.08003	560.2	12.464	20.040	12.549	0.7206	1076.4	5.321	35
.07986	559.0	12.489	20.080	12.578	0.9608	1076.9	5.751	36
.07969	557.8	12.514	20.121	12.607	1.2010	1077.3	6.188	37
.07951	556.6	12.540	20.162	12.637	1.4412	1077.8	6.633	38
.07935	555.4	12.565	20.203	12.666	1.6814	1078.2	7.084	39
0.07918	554.3	12.590	20.243	12.695	1.9216	1078.7	7.544	40
.07901	553.1	12.615	20.284	12.725	2.1618	1079.1	8.010	41
.07884	551.9	12.641	20.324	12.755	2.4020	1079.5	8.485	42
.07867	550.7	12.666	20.365	12.785	2.6422	1080.0	8.969	43
.07851	549.6	12.691	20.404	12.815	2.8824	1080.4	9.462	44
0.07834	548.4	12.717	20.445	12.846	3.1226	1080.9	9.963	45
.07817	547.2	12.742	20.485	12.876	3.3628	1081.3	10.474	46
.07801	546.0	12.767	20.525	12.907	3.6030	1081.8	10.995	47
.07784	544.9	12.792	20.565	12.938	3.8432	1082.2	11.505	48
.07767	543.7	12.818	20.605	12.970	4.0834	1082.7	12.062	49
0.07750	542.5	12.843	20.646	13.002	4.3238	1083.1	12.615	50
.07734	541.4	12.868	20.686	13.033	4.5639	1083.5	13.176	51
.07717	540.2	12.894	20.726	13.065	4.8041	1084.0	13.749	52
.07701	539.0	12.919	20.766	13.097	5.0444	1084.4	14.332	53
.07684	537.9	12.944	20.807	13.129	5.2846	1084.9	14.929	54
0.07668	536.7	12.969	20.847	13.162	5.5248	1085.3	15.536	55
.07651	535.6	12.995	20.887	13.195	5.7650	1085.8	16.157	56
.07635	534.4	13.020	20.927	13.228	6.0053	1086.2	16.790	57
.07618	533.3	13.045	20.967	13.262	6.2456	1086.7	17.438	58
.07602	532.1	13.070	21.006	13.295	6.4858	1087.1	18.097	59
0.07586	531.0	13.096	21.046	13.329	6.7260	1087.5	18.771	60
.07569	529.9	13.121	21.087	13.363	6.9664	1087.9	19.461	61
.07553	528.7	13.146	21.126	13.398	7.2067	1088.3	20.164	62
.07536	527.5	13.172	21.166	13.433	7.4469	1088.8	20.885	63
.07520	526.4	13.197	21.208	13.468	7.6872	1089.2	21.620	64
0.07503	525.2	13.222	21.247	13.504	7.9275	1089.7	22.373	65
.07487	524.1	13.247	21.287	13.540	8.1679	1090.1	23.140	66
.07469	522.8	13.273	21.327	13.580	8.4081	1090.5	23.926	67
.07454	521.8	13.298	21.367	13.613	8.6484	1090.9	24.729	68
.07438	520.7	13.323	21.407	13.650	8.8888	1091.3	25.552	69
0.07421	519.5	13.348	21.447	13.688	9.1290	1091.7	26.392	70
.07405	518.3	13.374	21.486	13.726	9.3694	1092.2	27.255	71
.07389	517.2	13.399	21.527	13.764	9.6098	1092.7	28.137	72
.07372	516.0	13.424	21.566	13.803	9.8500	1093.1	29.037	73
.07356	514.9	13.450	21.606	13.842	10.0904	1093.6	29.962	74
0.07339	513.7	13.475	21.646	13.882	10.331	1094.0	30.907	75
.07323	512.6	13.500	21.685	13.922	10.571	1094.4	31.865	76
.07306	511.4	13.525	21.725	13.963	10.812	1094.9	32.865	77
.07290	510.3	13.551	21.765	14.004	11.052	1095.3	33.880	78
.07273	509.1	13.576	21.804	14.046	11.292	1095.7	34.920	79

TABLE 4.5 Properties of Dry Air, Water Vapor, and Saturated Air–Water Vapor Mixtures (Continued)

Temperature Range, −100 to +211°F; Pressure, 29.921 inHg

Temp., °F	Saturation pressure		Saturation humidity, wt. water vapor/lb. dry air		Saturation moisture content, wt. water vapor/ft³ sat. mixture	
	lb/in²	inHg	lb	gr	lb × 10³	gr
80	0.50689	1.0320	0.02231	156.1	1.583	11.08
81	.52370	1.0663	.02308	161.5	1.645	11.51
82	.54099	1.1015	.02387	167.1	1.684	11.79
83	.55878	1.1377	.02468	172.8	1.736	12.15
84	.57707	1.1749	.02552	178.7	1.789	12.53
85	0.59588	1.2132	0.02639	184.7	1.844	12.91
86	.61522	1.2526	.02728	191.0	1.901	13.30
87	.63510	1.2931	.02821	197.4	1.958	13.71
88	.65555	1.3347	.02916	204.1	2.018	14.12
89	.67656	1.3775	.03014	210.9	2.079	14.55
90	0.69816	1.4215	0.03115	218.0	2.141	14.99
91	.72036	1.4667	.03219	225.3	2.205	15.44
92	.74316	1.5131	.03326	232.8	2.271	15.90
93	.76659	1.5608	.03437	240.6	2.338	16.37
94	.79065	1.6098	.03551	248.5	2.407	16.85
95	0.81537	1.6601	0.03668	256.8	2.478	17.34
96	.84074	1.7118	.03789	265.3	2.550	17.85
97	.86681	1.7648	.03914	274.0	2.624	18.37
98	.89358	1.8193	.04043	283.0	2.701	18.91
99	.92105	1.8753	.04175	292.3	2.779	19.45
100	0.94926	1.9327	0.04312	301.8	2.859	20.01
101	0.97821	1.9916	.04453	311.7	2.941	20.58
102	1.00792	2.0521	.04498	321.9	3.025	21.17
103	1.03842	2.1142	.04748	332.3	3.111	21.77
104	1.06965	2.1788	.04902	343.1	3.198	22.39
105	1.1018	2.2432	0.05061	354.3	3.289	23.02
106	1.1347	2.3103	.05225	365.7	3.381	23.67
107	1.1685	2.3790	.05394	377.6	3.476	24.33
108	1.2031	2.4495	.05568	389.7	3.572	25.00
109	1.2386	2.5218	.05747	402.3	3.671	25.70
110	1.2750	2.5959	0.05932	415.3	3.772	26.41
111	1.3123	2.6719	.06123	428.6	3.876	27.13
112	1.3506	2.7497	.06319	442.4	3.982	27.88
113	1.3897	2.8295	.06522	456.5	4.090	28.63
114	1.4300	2.9114	.06731	471.2	4.202	29.41
115	1.4711	2.9952	0.06946	486.2	4.315	30.20
116	1.5133	3.0811	.07168	501.8	4.431	31.02
117	1.5566	3.1691	.07397	517.8	4.550	31.85
118	1.6008	3.2593	.07633	534.3	4.671	32.69
119	1.6462	3.3517	.07877	551.4	4.795	33.57
120	1.6927	3.4463	0.08128	569.0	4.922	34.45
121	1.7403	3.5432	.08388	587.1	5.052	35.36
122	1.7890	3.6424	.08655	605.9	5.184	36.29
123	1.8389	3.7440	.08931	625.2	5.320	37.24
124	1.8900	3.8480	.09216	645.1	5.458	38.21

TABLE 4.5 Properties of Dry Air, Water Vapor, and Saturated Air–Water Vapor Mixtures (Continued)

Temperature Range, −100 to +211°F; Pressure, 29.921 inHg

Saturation density, wt. air plus water vapor/ft³ sat. mixture		Volume			Enthalpy			Temp., °F
lb	gr	Dry air, ft³/lb	Water vapor, ft³/lb[a]	Saturated mixture, ft³/lb dry air	Dry air, Btu/lb	Water vapor, Btu/lb	Saturated mixture, Btu/lb dry air	
0.07257	508.0	13.601	21.844	14.088	11.533	1096.2	35.985	80
.07240	506.8	13.626	21.884	14.131	11.773	1096.6	37.077	81
.07223	505.6	13.652	21.923	14.175	12.013	1097.1	38.197	82
.07206	504.4	13.677	21.964	14.219	12.254	1097.5	39.342	83
.07190	503.3	13.702	22.003	14.264	12.494	1097.9	40.515	84
0.07173	502.1	13.728	22.043	14.309	12.735	1098.3	41.718	85
.07156	500.9	13.753	22.082	14.355	12.975	1098.7	42.952	86
.07139	499.8	13.778	22.122	14.402	13.217	1099.1	44.216	87
.07123	498.6	13.803	22.161	14.449	13.457	1099.6	45.515	88
.07106	497.4	13.829	22.200	14.497	13.696	1100.0	46.845	89
0.07088	496.2	13.854	22.240	14.547	13.938	1100.5	48.212	90
.07071	495.0	13.879	22.279	14.597	14.177	1100.9	49.612	91
.07054	493.8	13.904	22.319	14.647	14.418	1101.4	51.050	92
.07037	492.6	13.930	22.358	14.699	14.658	1101.8	52.522	93
.07020	491.4	13.955	22.398	14.751	14.899	1102.3	54.037	94
0.07003	490.2	13.980	22.437	14.804	15.139	1102.7	55.586	95
.06985	489.0	14.006	22.476	14.854	15.380	1103.1	57.179	96
.06968	487.8	14.031	22.515	14.913	15.620	1103.5	58.810	97
.06951	486.6	14.056	22.555	14.968	15.860	1103.9	60.486	98
.06933	485.3	14.081	22 594	15.025	16.101	1104.3	62.209	99
0.06916	484.1	14.107	22.633	15.083	16.341	1104.8	63.980	100
.06898	482.9	14.132	22.673	15.142	16.582	1105.2	65.794	101
.06880	481.6	14.157	22.712	15.202	16.822	1105.6	67.657	102
.06863	480.4	14.182	22.751	15.263	17.063	1106.1	69.577	103
.06845	479.2	14.208	22.790	15.325	17.304	1106.5	71.541	104
0.06827	477.9	14.233	22.829	15.389	17.544	1106.9	73.563	105
.06809	476.7	14.258	22.868	15.453	17.785	1107.3	75.639	106
.06791	475.4	14.283	22.907	15.519	18.025	1107.7	77.771	107
.06773	474.1	14.309	22.946	15.587	18.266	1108.1	79.961	108
.06755	472.8	14.334	22.985	15.655	18.506	1108.5	82.215	109
0.06737	470.6	14.359	23.024	15.725	18.747	1109.0	84.535	110
.06718	470.3	14.384	23.063	15.796	18.987	1109.4	86.915	111
.06700	469.0	14.410	23.101	15.869	19.228	1109.8	89.360	112
.06681	467.7	14.435	23.140	15.944	19.469	1110.2	91.873	113
.06662	466.4	14.460	23.178	16.020	19.709	1110.6	94.461	114
0.06643	465.0	14.486	23.218	16.098	19.950	1111.1	97.128	115
.06624	463.7	14.516	23.256	16.178	20.190	1111.5	99.866	116
.06605	462.4	14.536	23.295	16.259	20.431	1112.0	102.688	117
.06586	461.0	14.561	23.333	16.343	20.672	1112.4	105.586	118
.06567	459.7	14.587	23.372	16.428	20.912	1112.8	108.571	119
0.06547	458.3	14.612	23.408	16.515	21.153	1113.3	111.65	120
.06528	457.0	14.637	23.448	16.603	21.394	1113.7	114.81	121
.06508	455.6	14.662	23.487	16.695	21.634	1114.2	118.07	122
.06488	454.2	14.688	23.526	16.789	21.875	1114.6	121.42	123
.06468	452.8	14.713	23.565	16.885	22.116	1115.0	124.88	124

TABLE 4.5 Properties of Dry Air, Water Vapor, and Saturated Air–Water Vapor Mixtures (Continued)

Temperature Range, −100 to +211°F; Pressure, 29.921 inHg

Temp., °F	Saturation pressure		Saturation humidity, wt. water vapor/lb. dry air		Saturation moisture content, wt. water vapor/ft³ sat. mixture	
	lb/in²	inHg	lb	gr	lb	gr
125	1.9423	3.9544	0.09511	665.8	0.005600	39.20
126	1.9958	4.0634	.09815	687.1	.005745	40.22
127	2.0506	4.1749	.10129	709.0	.005893	41.25
128	2.1066	4.2891	.10453	731.7	.006045	42.33
129	2.1640	4.4059	.10788	755.2	.006199	43.40
130	2.2227	4.5255	0.1113	779.4	0.006357	44.50
131	2.2828	4.6479	.1149	804.4	.006519	45.63
132	2.3442	4.7729	.1186	830.2	.006683	46.78
133	2.4072	4.9010	.1224	856.9	.006851	47.96
134	2.4715	5.0320	.1264	884.5	.007024	49.16
135	2.5373	5.1659	0.1304	913.0	0.007199	50.39
136	2.6045	5.3028	.1346	942.5	.007377	51.64
137	2.6733	5.4429	.1390	973.1	.007561	52.92
138	2.7436	5.5861	.1435	1,004.5	.007747	54.23
139	2.8155	5.7324	.1482	1,037.3	.007937	55.56
140	2.8890	5.8821	0.1530	1,071	0.008131	56.92
141	2.9641	6.0349	.1580	1,106	.008329	58.31
142	3.0409	6.1912	.1632	1,142	.008532	59.72
143	3.1193	6.3509	.1685	1,180	.008738	61.17
144	3.1915	6.5141	.1741	1,219	.008949	62.64
145	3.2814	6.6809	0.1798	1,259	0.009163	64.14
146	3.3651	6.8512	.1858	1,301	.009383	65.68
147	3.4506	7.0253	.1920	1,344	.009607	67.25
148	3.5379	7.2032	.1984	1,389	.009835	68.84
149	3.6271	7.3847	.2051	1,436	.010066	70.46
150	3.7182	7.5703	0.2120	1,484	0.01030	72.13
151	3.8113	7.7597	.2192	1,534	.01055	73.82
152	3.9063	7.9531	.2267	1,587	.01079	75.54
153	4.0033	8.1506	.2344	1,641	.01104	77.30
154	4.1023	8.3523	.2425	1,698	.01130	79.09
155	4.2034	8.5581	0.2509	1,756	0.01156	80.92
156	4.3066	8.7682	.2596	1,817	.01183	82.78
157	4.4120	8.9828	.2688	1,881	.01210	84.69
158	4.5195	9.2016	.2782	1,948	.01237	86.61
159	4.6292	9.4251	.2881	2,005	.01265	88.58
160	4.7412	9.6531	0.2985	2,089	0.01294	90.59
161	4.8554	9.8856	.3092	2,165	.01323	92.63
162	4.9720	10.1231	.3205	2,244	.01353	94.72
163	5.0909	10.3652	.3323	2,326	.01384	96.85
164	5.2122	10.6162	.3446	2,412	.01414	99.00
165	5.3358	10.864	0.3575	2,502	0.01446	101.2
166	5.4621	11.121	.3710	2,597	.01478	103.5
167	5.5908	11.383	.3851	2,696	.01511	105.8
168	5.7220	11.650	.4000	2,800	.01543	108.1
169	5.8558	11.922	.4156	2,909	.01577	110.4

TABLE 4.5 Properties of Dry Air, Water Vapor, and Saturated Air–Water Vapor Mixtures (Continued)

Temperature Range, −100 to +211°F; Pressure, 29.921 inHg

Saturation density, wt. air plus water vapor/ft³ sat. mixture		Volume			Enthalpy			Temp., °F
lb	gr	Dry air, ft³/lb	Water vapor, ft³/lb[a]	Saturated mixture, ft³/lb dry air	Dry air, Btu/lb	Water vapor, Btu/lb	Saturated mixture, Btu/lb dry air	
0.06448	451.4	14.738	23.602	16.983	22.356	1115.4	128.44	125
.06428	450.0	14.763	23.641	17.084	22.597	1115.8	132.11	126
.06408	448.5	14.789	23.679	17.187	22.838	1116.3	135.91	127
.06387	447.1	14.814	23.718	17.293	23.079	1116.7	139.81	128
.06366	445.6	14.839	23.756	17.402	23.319	1117.1	143.83	129
0.06345	441.2	14.864	23.794	17.514	23.560	1117.6	147.99	130
.06325	442.7	14.890	23.833	17.628	23.801	1118.0	152.27	131
.06303	441.2	14.915	23.871	17.746	24.041	1118.4	156.64	132
.06282	439.7	14.940	23.910	17.867	24.282	1118.8	161.23	133
.06261	438.2	14.965	23.948	17.991	24.523	1119.2	165.95	134
0.06239	436.7	14.991	23.987	18.119	24.764	1119.6	170.79	135
.06217	435.2	15.016	24.025	18.251	25.005	1120.1	175.81	136
.06195	433.7	15.041	24.063	18.386	25.245	1120.5	181.01	137
.06173	432.1	15.067	24.103	18.525	25.486	1120.9	186.35	138
.06150	430.5	15.092	24.139	18.669	25.727	1121.3	191.88	139
0.06128	428.9	15.117	24.178	18.816	25.968	1121.7	197.59	140
.06105	427.3	15.142	24.215	18.969	26.209	1122.1	203.50	141
.06082	425.7	15.168	24.253	19.126	26.449	1122.5	209.62	142
.06058	424.1	15.193	24.290	19.288	26.690	1122.9	215.94	143
.06035	422.5	15.218	24.329	19.454	26.931	1123.3	222.49	144
0.06012	420.8	15.243	24.367	19.626	27.172	1123.7	229.26	145
.05988	419.1	15.269	24.405	19.804	27.413	1124.1	236.29	146
.05964	417.5	15.294	24.443	19.987	27.654	1124.6	243.59	147
.05940	415.8	15.320	24.481	20.176	27.895	1125.0	251.12	148
.05915	414.0	15.344	24.518	20.374	28.135	1125.4	258.94	149
0.05890	412.3	15.370	24.555	20.576	28.376	1125.8	267.06	150
.05866	410.6	15.395	24.593	20.786	28.617	1126.2	275.48	151
.05840	408.8	15.420	24.630	21.004	28.858	1126.6	284.22	152
.05815	407.0	15.445	24.668	21.229	29.099	1127.0	293.30	153
.05789	405.3	15.471	24.705	21.462	29.340	1127.4	302.73	154
0.05763	403.4	15.496	24.742	21.704	29.581	1127.8	312.55	155
.05737	401.6	15.521	24.780	21.955	29.822	1128.2	322.75	156
.05711	399.8	15.546	24.818	22.216	30.063	1128.7	333.41	157
.05684	397.9	15.572	24.855	22.487	30.304	1129.1	344.46	158
.05657	396.0	15.597	24.892	22.769	30.545	1129.5	356.00	159
0.05630	394.1	15.622	24.929	23.063	30.786	1129.9	368.13	160
.05603	392.2	15.647	24.966	23.368	31.027	1130.3	380.56	161
.05575	390.3	15.673	25.003	23.685	31.268	1130.7	393.67	162
.05547	388.3	15.698	25.040	24.017	31.509	1131.1	407.35	163
.05519	386.3	15.723	25.077	24.365	31.750	1131.5	421.66	164
0.05490	384.3	15.748	25.114	24.725	31.991	1131.9	436.61	165
.05462	382.3	15.774	25.151	25.102	32.232	1132.3	452.30	166
.05434	380.3	15.799	25.188	25.492	32.473	1132.7	468.72	167
.05402	378.2	15.824	25.225	25.914	32.714	1133.1	485.95	168
.05373	376.1	15.849	25.261	26.347	32.955	1133.5	504.05	169

TABLE 4.5 Properties of Dry Air, Water Vapor, and Saturated Air–Water Vapor Mixtures (Continued)

Temperature Range, −100 to +211°F; Pressure, 29.921 inHg

Temp., °F	Saturation pressure		Saturation humidity, wt. water vapor/lb. dry air		Saturation moisture content, wt. water vapor/ft³ sat. mixture	
	lb/in²	inHg	lb	gr	lb	gr
170	5.9923	12.200	0.4320	3,024	0.01612	112.8
171	6.1314	12.484	.4493	3,145	.01647	115.3
172	6.2733	12.772	.4675	3,273	.01682	117.8
173	6.4179	13.067	.4867	3,407	.01719	120.3
174	6.5653	13.367	.5070	3,549	.01756	122.9
175	6.7156	13.673	0.5284	3,699	0.01793	125.5
176	6.8687	13.985	.5511	3,858	.01832	128.2
177	7.0247	14.302	.5752	4,026	.01871	130.9
178	7.1838	14.626	.6008	4,205	.01910	133.7
179	7.3458	14.956	.6279	4,395	.01950	136.5
180	7.5109	15.292	0.6569	4,598	0.01992	139.4
181	7.6791	15.635	.6878	4,815	.02033	142.3
182	7.8504	15.983	.7209	5,046	.02076	145.3
183	8.0247	16.339	.7563	5,294	.02119	148.3
184	8.2027	16.701	.7943	5,560	.02163	151.4
185	8.3836	17.069	0.8352	5,847	0.02207	154.5
186	8.5678	17.444	.8794	6,156	.02253	157.7
187	8.7554	17.826	.9271	6,490	.02299	160.9
188	8.9465	18.215	.9790	6,853	.02346	164.2
189	9.1411	18.611	1.0355	7,249	.02393	167.5
190	9.3392	19.015	1.097	7,681	0.02442	170.9
191	9.5409	19.425	1.165	8,155	.02491	174.4
192	9.7463	19.844	1.240	8,679	.02542	177.9
193	9.9553	20.269	1.322	9,257	.02592	181.5
194	10.1684	20.703	1.414	9,900	.02644	185.1
195	10.385	21.143	1.517	10,629	0.02697	188.8
196	10.605	21.591	1.633	11,432	.02750	192.5
197	10.829	22.048	1.765	12,352	.02805	196.3
198	11.057	22.513	1.915	13,405	.02860	200.2
199	11.289	22.985	2.089	14,620	.02916	204.1
200	11.526	23.466	2.292	16,046	0.02973	208.1
201	11.766	23.955	2.532	17,725	.03031	212.2
202	12.010	24.453	2.820	19,739	.03090	216.3
203	12.259	24.960	3.173	22,212	.03149	220.5
204	12.512	25.474	3.614	25,301	.03210	224.7
205	12.769	25.998	4.181	29,269	0.03272	229.0
206	13.031	26.531	4.939	34,576	.03346	234.2
207	13.297	27.073	6.000	42,000	.03398	237.9
208	13.568	27.624	7.594	53,161	.03463	242.4
209	13.843	28.184	10.248	71,736	.03528	247.0
210	14.122	28.753	15.54	108,773	0.03595	251.6
211	14.407	29.332	31.49	220,451	.03667	256.7

TABLE 4.5 Properties of Dry Air, Water Vapor, and Saturated Air–Water Vapor Mixtures (Continued)

Temperature Range, −100 to +211°F; Pressure, 29.921 inHg

Saturation density, wt. air plus water vapor/ ft³ sat. mixture		Volume			Enthalpy			Temp., °F
lb	gr	Dry air, ft³/lb	Water vapor, ft³/lb[a]	Saturated mixture, ft³/lb dry air	Dry air, Btu/lb	Water vapor, Btu/lb	Saturated mixture, Btu/ lb dry air	
0.05343	374.0	15.875	25.298	26.804	33.196	1133.9	523.06	170
.05314	372.0	15.900	25.335	27.272	33.437	1134.3	543.08	171
.05281	369.7	15.925	25.372	27.787	33.678	1134.7	564.15	172
.05251	367.5	15.950	25.408	28.315	33.919	1135.1	586.38	173
.05219	365.3	15.976	25.445	28.876	34.161	1135.5	609.84	174
0.05187	363.1	16.001	25.481	29.465	34.402	1135.9	634.63	175
.05155	360.9	16.026	25.518	30.089	34.643	1136.3	660.88	176
.05123	358.6	16.051	25.554	30.749	34.884	1136.7	688.69	177
.05090	356.3	16.077	25.591	31.449	35.125	1137.1	718.25	178
.05057	356.1	16.102	25.627	32.193	35.366	1137.5	749.63	179
0.05023	351.6	16.127	25.664	32.984	35.607	1137.9	783.08	180
.04989	349.2	16.152	25.700	33.829	35.848	1138.3	818.78	181
.04955	346.8	16.178	25.736	34.731	36.090	1138.7	856.97	182
.04920	344.4	16.203	25.772	35.694	36.331	1139.1	897.79	183
.04885	342.0	16.228	25.809	36.728	36.572	1139.5	941.68	184
0.04850	339.5	16.253	25.844	37.839	36.813	1139.9	988.88	185
.04814	337.0	16.279	25.881	39.037	37.054	1140.3	1,039.79	186
.04778	334.5	16.304	25.916	40.332	37.296	1140.7	1,094.88	187
.04742	331.9	16.329	25.953	41.737	37.537	1141.1	1,154.70	188
.04705	329.3	16.354	25.988	43.265	37.778	1141.5	1,219.80	189
0.04667	326.7	16.380	26.024	44.935	38.019	1141.9	1,291.0	190
.04630	324.1	16.405	26.060	46.764	38.261	1142.3	1,369.5	191
.04592	321.4	16.430	26.093	48.780	38.502	1142.7	1,455.1	192
.04553	318.7	16.455	26.131	51.011	38.743	1143.1	1,550.4	193
.04514	316.0	16.481	26.167	53.488	38.985	1143.5	1,656.2	194
0.04474	313.2	16.506	26.202	56.265	39.226	1143.9	1,775.0	195
.04434	310.4	16.530	26.238	59.381	39.467	1144.3	1,908.3	196
.04394	307.6	16.556	26.273	62.918	39.709	1144.7	2,059.6	197
.04353	304.7	16.582	26.309	66.963	39.950	1145.1	2,232.8	198
.04312	301.8	16.607	26.344	71.630	40.191	1145.5	2,432.7	199
0.04270	298.9	16.632	26.380	77.102	40.433	1145.9	2,667.2	200
.04228	296.0	16.657	26.415	83.543	40.674	1146.2	2,943.0	201
.04185	293.0	16.683	26.450	91.270	40.916	1146.6	3,274.2	202
.04142	290.0	16.708	26.486	100.750	41.158	1146.0	3,679.6	203
.04098	286.9	16.733	26.521	112.590	41.399	1147.4	4,188.6	204
0.04054	283.8	16.758	26.556	127.80	41.640	1147.7	4,840.5	205
.04024	281.7	16.784	26.591	147.60	41.882	1148.1	5,712.8	206
.03965	277.5	16.809	26.626	176.56	42.124	1148.4	6,932.3	207
.03919	274.3	16.834	26.661	219.30	42.366	1148.8	8,766.8	208
.03873	271.1	16.859	26.696	290.44	42.608	1149.2	11,820.3	209
0.03826	267.8	16.885	26.731	432.25	42.849	1149.6	17,906	210
.03779	264.5	16.910	26.765	859.82	43.090	1150.0	36,260	211

[a]Theoretical volume calculated from the relationship: Vol. at 1 atm = Vol. at saturation press. × $\frac{\text{saturation press., inHg}}{29.921}$

Source: Ref. 4-4.

REFERENCES

4-1. Keenan, J., and Keyes, F.: *Thermodynamic Properties of Steam*, John Wiley & Sons, New York, 1957.

4-2. Keenan, J., and Kaye, J.: *Gas Tables*, John Wiley & Sons, New York, 1957.

4-3. Danielson, J.: *Air Pollution Engineering Manual*, County of Los Angeles, Air Pollution Control District, AP-40, May 1973.

4-4. Zimmerman, O., and Lavine, I.: *Psychrometric Tables and Charts*, Industrial Research Service, Dover, N.H., 1964.

BIBLIOGRAPHY

Baumeister and Marks: *Standard Handbook for Mechanical Engineers*, 7th ed., McGraw-Hill, New York, 1982.

Combustion Fundamentals for Waste Incineration, American Society of Mechanical Engineers, New York, 1974.

Costello, F.: "The Second Law of Thermodynamics," American Society of Mechanical Engineers, Winter Annual Meeting, New York, 1979.

Guidebook for Industrial and Commercial Gas Fired Incineration, American Gas Association, New York, 1963.

Kennan, J., and Keyes, F.: *Theoretical Steam Rate Tables*, American Society of Mechanical Engineers, New York, 1938.

Lee, J., and Sears, F.: *Thermodynamics*, 2d ed., Addison Wesley, Reading, Mass., 1963.

Norris, E., Therkelson, E., and Pratt, L.: *Applied Thermodynamics*, McGraw-Hill, New York, 1955.

Obert, M.: *Thermodynamics*, McGraw-Hill, New York, 1948.

Perry and Chilton: *Chemical Engineering Handbook*, 5th ed., McGraw-Hill, New York, 1969.

Shapiro, A.: *The Dynamics and Thermodynamics of Compressible Fluid Flow*, Ronald Press, New York, 1958.

CHAPTER 5
HEAT-TRANSFER SYSTEMS

Incineration is a thermal process, and the control of the flow of heat is basic to this process. In this chapter heat-transfer calculations will be explained, and thermal properties of materials will be presented.

THE NATURE OF HEAT TRANSFER

There are three modes of heat transfer:

- *Conduction.* Heat transfer through a medium by progressive heating of adjacent elements.
- *Convection.* Transfer of heat by physical motion of the heated, or cooled, medium, as motion of a fluid.
- *Radiation.* Heat transfer by electromagnetic radiation between two boundaries not in physical contact with each other and at different temperatures or between one body and another medium, such as air, at different temperatures.

Normally these three heat-transfer modes are all present at the same time and must be taken into account in combination. The calculations required to determine heat-transfer parameters are complex, but methods have been developed to expedite this calculation process.

CONDUCTION

Referring to Fig. 5.1, the mechanism of conduction transfers heat from a hot surface, T_i, to a cooler surface, T_0, both expressed in degrees Fahrenheit. The conducting material, with width X, in inches, has a property of conduction, or conductivity, K, which has the units of Btu · in/(ft^2 · h · °F). Good insulators, such as fiberglass or perlite, have K values in the range of 0.2 to 3.0. Moderate insulators, such as lightweight castable refractory or lightweight firebrick, have K values in the range of 3.0 to 6.0. Poor insulators, such as standard firebrick, have K values above 6.0, whereas conductors have K values over 100.0.

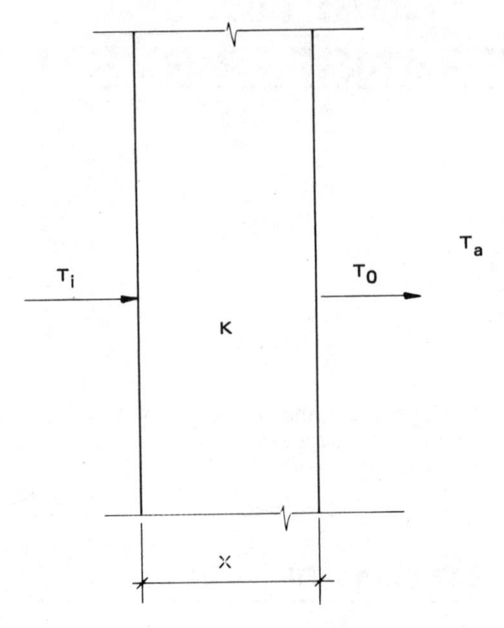

Fig. 5-1. Single wall.

The basic law of heat transfer by conduction is

$$Q' = \frac{KA(T_i - T_0)}{X} \tag{5.1}$$

where Q' is the flow of heat in Btu per hour and A is the area, in square feet, perpendicular to the direction of heat flow.

For convenience the quantity resistance R is defined as follows:

$$R = \frac{X}{K} \tag{5.2}$$

For a wall composed of more than one material, the heat transfer (see Fig. 5.2) with

$$Q = \frac{Q'}{A} \tag{5.3}$$

becomes

$$Q = \frac{K_1}{X_1}(T_i - T_1) = \frac{K_2}{X_2}(T_1 - T_0) \tag{5.4}$$

or

$$Q = \frac{1}{R_1}(T_i - T_1) = \frac{1}{R_2}(T_1 - T_0) \tag{5.5}$$

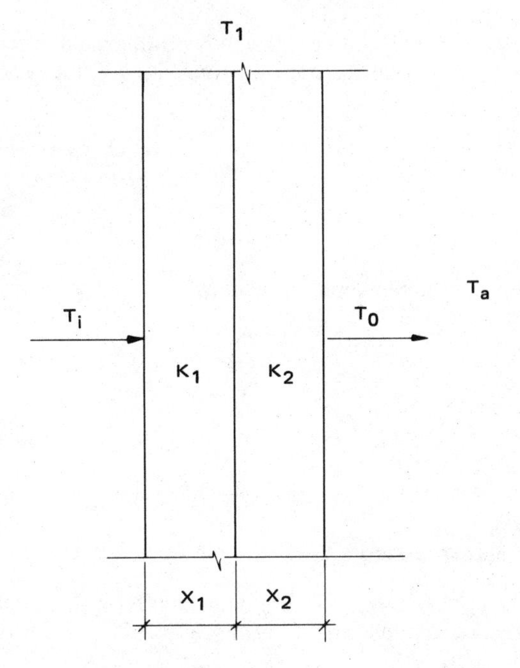

Fig. 5-2. Dual wall.

Solving for the interface temperature T_1 gives

$$T_1 = T_i - Q\frac{X_1}{K_1} = T_0 + Q\frac{X_2}{K_2} \tag{5.6}$$

and solving for Q yields

$$Q = \frac{T_i - T_0}{X_1/K_1 + X_2/K_2} \tag{5.7}$$

or by substituting resistance

$$Q = \frac{1}{R}(T_i - T_0) \tag{5.8}$$

where

$$R = R_1 + R_2 = \frac{X_1}{K_1} + \frac{X_2}{K_2} \tag{5.9}$$

With multiple-wall systems the resistance is equal to the sum of the resistance of each wall:

$$R = R_1 + R_2 + \cdots + R_n = \frac{X_1}{K_1} + \frac{X_2}{K_2} + \cdots + \frac{X_n}{K_n} \tag{5.9a}$$

The temperature drop across any wall is proportional to the resistance of that wall compared to the total resistance. From Fig. 5.2, therefore,

$$T_i - T_1 = \frac{R_1}{R}(T_i - T_0) \tag{5.10}$$

$$T_1 - T_0 = \frac{R_2}{R}(T_i - T_0) \tag{5.11}$$

For multiple walls the temperature drop ΔT across a wall with resistance R_x is

$$\Delta T = \frac{R_x}{R}(T_i - T_0) \tag{5.12}$$

EXAMPLES

Conduction across a Single Wall

What is the cold-face temperature with 1200°F on the hot face of 3 in of insulation with a conductivity of 1.33 Btu · in/(h · ft^2 · °F), allowing a heat loss of 430 Btu/(h · ft^2)?

From (5.2), $R = X/K$ = 3 in/[1.33 Btu · in/(h · ft^2 · °F)] = 2.256 h · ft^2 · °F/Btu.

From (5.8), $T_0 = T_i - RQ$ = 1200°F − 2.256 h · ft^2 · °F/Btu × 430 Btu/(h · ft^2) = 230°F, the cold-face temperature.

Conduction across a Composite Wall

What are the interface temperatures across a composite wall composed of 4 in of firebrick [K = 9.2 Btu · in/(h · ft^2 · °F)], 2 in of insulating block [K = 0.8 Btu · in/(h · ft^2 · °F)], and 1 in of steel [K = 380 Btu · in/(h · ft^2 · °F)]? Determine the interface temperatures and heat flow with 1600°F hot-face (T_i) and 150°F (T_0) cold-face temperatures.

From (5.2):

$$R_1 = \frac{X_1}{K_1} = \frac{4.0 \text{ in}}{9.2 \text{ Btu} \cdot \text{in/(h} \cdot \text{ft}^2 \cdot °\text{F)}} = 0.435 \text{ h} \cdot \text{ft}^2 \cdot °\text{F/Btu (firebrick)}$$

$$R_2 = \frac{X_2}{K_2} = \frac{2.0 \text{ in}}{0.8 \text{ Btu} \cdot \text{in/(h} \cdot \text{ft}^2 \cdot °\text{F)}} = 2.500 \text{ h} \cdot \text{ft}^2 \cdot °\text{F/Btu (insulation)}$$

$$R_3 = \frac{X_3}{K_3} = \frac{1.0 \text{ in}}{380 \text{ Btu} \cdot \text{in/(h} \cdot \text{ft}^2 \cdot °\text{F)}} = 0.003 \text{ h} \cdot \text{ft}^2 \cdot °\text{F/Btu (steel)}$$

$$R = \Sigma R = 0.435 + 2.500 + 0.003 = 2.938 \text{ h} \cdot \text{ft}^2 \cdot °\text{F/Btu}$$

From (5.10):

$$T_i - T_1 = \frac{R_1}{R}(T_i - T_0) = \frac{0.435}{2.938}(1600 - 150) = 215°\text{F}$$

$$T_1 = T_i - 215 = 1600 - 215 = 1385°\text{F}$$

Also

$$T_1 - T_2 = \frac{R_2}{R}(T_i - T_0) = \frac{2.500}{2.938}(1600 - 150) = 1234°\text{F}$$

$$T_2 = T_1 - 1234 = 1385 - 1234 = 151°\text{F}$$

and

$$T_2 - T_0 = \frac{R_3}{R}(T_i - T_0) = \frac{0.003}{2.938}(1600 - 150) = 1°\text{F}$$

$$T_2 = T_0 + 1 = 150 + 1 = 151°\text{F}$$

Therefore, the interface temperature between the brick and the insulation is 1385°F and between the steel and the insulation 151°F.

From (5.8):

$$Q = \frac{1}{R}(T_i - T_0) = \frac{1}{2.938} \times (1600 - 150) = 494 \text{ Btu/(h} \cdot \text{ft}^2)$$

HEAT CONVECTION

Heat convection is a function of the properties of the convected medium, a fluid, and the geometry of the heated surface. Of specific interest is the flow of heat by convection of air. The basic equation for convection of heat is

$$Q_c' = hA(T_0 - T_a) \tag{5.13}$$

or

$$Q_c = h(T_0 - T_a) \tag{5.14}$$

where Q is the flow of heat by conduction in Btu/(h · ft^2), h is the film coefficient, Btu/(h · ft^2 · °F), and T_0 and T_a are the interface temperature and the ambient temperature, respectively, as shown in Fig. 5.1. The film coefficient is difficult to determine, and it is generally calculated from empirical data. One set of such data is as follows:

$$h = 0.25(T_0 - T_a)^{0.25} \quad \text{for vertical plates} \tag{5.15}$$

$$h = 0.38(T_0 - T_a)^{0.25} \quad \text{for plates facing up} \tag{5.16}$$

$$h = 0.20(T_0 - T_a)^{0.25} \quad \text{for plates facing down} \tag{5.17}$$

These values are used for convection in still air. For air at a velocity V, ft/s:

$$h = 1.0 + 0.225V \tag{5.18}$$

RADIATION

Radiation from a body is governed by the texture and color of that body, expressed as a dimensionless constant, emissivity ϵ. The equation describing radiation heat loss Q_R, Btu/(h · ft² · °F), is

$$Q_R = Q_R A \tag{5.19}$$

$$Q_R = 0.174\epsilon\left[\left(\frac{T_0 + 460}{100}\right)^4 - \left(\frac{T_a + 460}{100}\right)^4\right] \tag{5.20}$$

This equation is derived from the Stephan-Boltzmann equations. Typical values for emissivity ϵ are listed in Table 5.1.

TABLE 5.1 Typical Values of Emissivity

Material	ϵ
Polished aluminum	0.10
Aluminum paint	0.50
White paper	0.70
Rubber, soft gray	0.85
Brick	0.85
Steel surface, oxidized	0.90

Source: Ref. 5-1.

COMBINED HEAT TRANSFER

There will always be flow of heat from a heated to a cooler medium. In practically all situations this flow of heat involves the three modes of heat transfer. Heat is conducted through a medium and is dispersed at the medium boundary through convection and radiation. Expressing this mathematically, we get

$$Q = Q_R + Q_C \tag{5.21}$$

and from Eqs. (5.8), (5.14), and (5.20),

$$Q = \frac{1}{R}(T_i - T_0) = 0.174\epsilon\left[\left(\frac{T_0 + 460}{100}\right)^4 - \left(\frac{T_a + 460}{100}\right)^4\right] + h(T_0 - T_a) \tag{5.22}$$

The values for ϵ are as listed in Table 5.1 and can also be found in the heat-transfer literature. Values for h are given in Eqs. (5.15), (5.16), (5.17), and (5.18).

Typical values of conduction are presented in Tables 5.2 through 5.5. Note that conduction varies as a function of temperature whereas the values for ϵ and

TABLE 5.2 Dense Castables, Typical Materials

Manufacturer: / Material:	Johns Manville Std. Firecrete	Quigley Q-Cast 30-50	JH France Hydricon 186
Maximum temp., °F	2500	3000	2500
Weight, as placed, pcf	130	132	115
Conductivity, BTU-in./ft^2-°F-hr			
800°F	4.2	6.3	4.9
1000°F	4.3	6.2	5.0
1200°F	4.6	6.1	5.2
1400°F	4.4	6.1	5.4
1600°F	5.2	6.2	5.5
1800°F	5.5	6.4	5.6

TABLE 5.3 Insulating Castables, Typical Materials

Manufacturer: / Material:	Harbison Walker Lightweight Castable 26	Quigley Insulcrete 22	JH France Light Weight Hydricon 2400
Maximum temp., °F	2600	2200	2400
Weight, as placed, pcf	50	43	75
Conductivity, BTU-in./ft^2-°F-hr			
800°F	1.6	1.2	1.6
1000°F	1.7	1.3	1.7
1200°F	1.8	1.4	1.7
1400°F	1.9	1.7	1.8
1600°F	2.0	2.0	1.8
1800°F	2.2	2.4	1.9

TABLE 5.4 Firebrick, Typical Materials

Manufacturer: / Material:	Harbison Walker Standard Firebrick VARNON	Johns Manville Insulating Firebrick JM-20	JH France Insulating Firebrick FR-20
Maximum temp., °F	3100	2000	2600
Weight, pcf	146	29	33
Conductivity, BTU-in./ft^2-°F-hr			
800°F	9.8	1.1	1.9
1000°F	9.8	1.1	2.0
1200°F	9.9	1.2	2.1
1400°F	10.0	1.2	2.2
1600°F	10.2	1.2	2.3
1800°F	10.5	1.3	2.4

TABLE 5.5 Industrial Insulation, Typical Materials

Manufacturer: / Material:	Celotex Insulation board IMF-50	Eagle Picher Insulation board Epitherm 1200	Johns Manville Pipe and block insulation Superex-M	Owens Corning Pipe insulation Fiberglass 25
Maximum temp., °F	1050	1200	1600	450
Density, lb/ft^3	8	11	21	4
Conductivity, Btu·in/(ft^2·°F·h)				
200°F	0.34	0.33	0.58	0.30
400°F	0.50	0.45	0.62	0.45
600°F	0.68	0.59	0.66	—
800°F	1.10	0.78	0.69	—
1000°F	—	1.07	0.73	—

Note: Aluminum covering has a thermal conductivity of 1370 Btu·in/(ft^2·°F·h), carbon steel is 380 Btu·in/(ft^2·°F·h), and stainless steel is 130 Btu·in/(ft^2·°F·h), all evaluated at approximately 300°F. These values do not change significantly for metals with changes in temperature.

h are not, as presented here, temperature-dependent. Normally, the internal, external, and ambient temperatures are known. The value R must be determined. Once R is known ($R = X/k$), the interface temperatures must be approximated to determine appropriate K values. This is a cumbersome trial-and-error procedure. The first requirement in heat-transfer calculations is obtaining the value of R from Eq. (5.22), a complex, unwieldy expression.

THE RESISTANCE GRAPH

A method has been developed to present a complete heat-transfer profile on one graph. Referring to Fig. 5.3, the hot-face temperature is on the vertical axis. By choosing a hot-face temperature and drawing a horizontal line, then choosing a cold-face temperature, from the bottom axis, and drawing a vertical line, the R value required is found at the intersection of these two lines. The hot-face and cold-face temperatures define R. The R, or resistance, value is that resistance required to obtain the desired cold-face temperature with a given hot-face temperature.

Referring again to Fig. 5.3, as an example, for a hot-face temperature of 2100°F, a resistance value of 6.0 is required to obtain a cold-face temperature of 200°F. A resistance value of 6.0 defines the type and thickness of refractory wall required. From Eq. (5.2),

$$R = \frac{X}{K} = 6 \text{ ft}^2 \cdot \text{°F} \cdot \text{h/Btu}$$

Using insulating firebrick with $K = 1.2$ Btu · in/(ft^2 · °F · h),

$$X = RK = 6 \text{ ft}^2 \cdot \text{°F} \cdot \text{h/Btu} \times 1.2 \text{ Btu} \cdot \text{in/(ft}^2 \cdot \text{°F} \cdot \text{h)}$$
$$= 7.2 \text{ in}$$

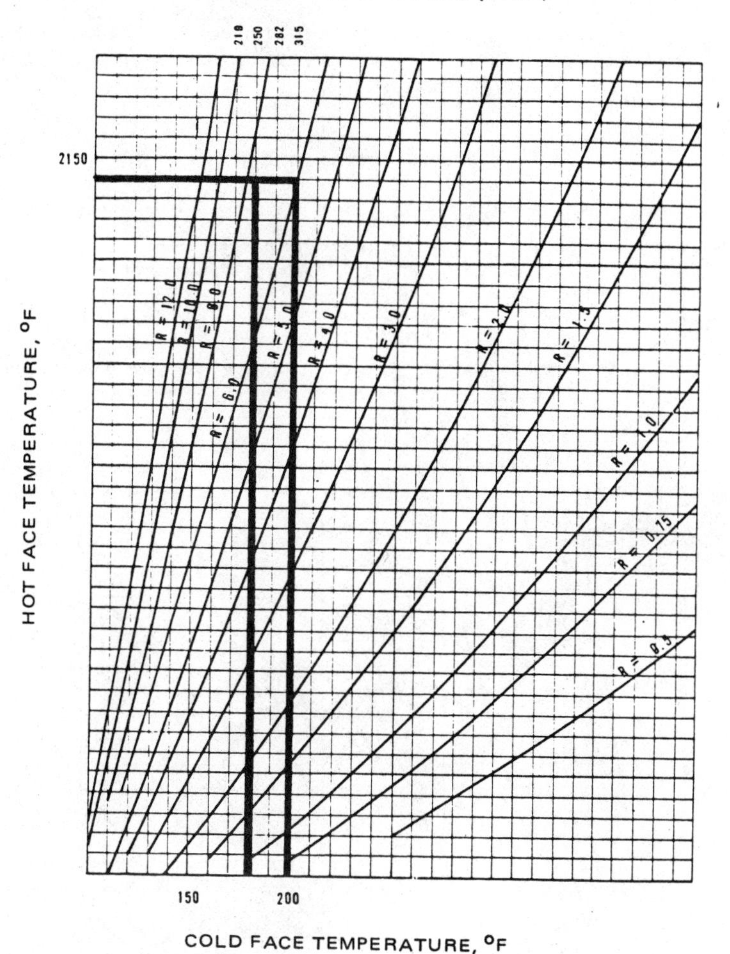

Fig. 5-3. Resistance graph design example.

The calculated thickness, 7.2 in, is not a standard size. For the next thickest standard size of brick, 9.0 in, the actual R value is

$$R = \frac{X}{K}$$

$$= \frac{9.0 \text{ in}}{1.2 \text{ Btu} \cdot \text{in}/(\text{ft}^2 \cdot {}^\circ\text{F} \cdot \text{h})}$$

$$= 7.5 \text{ ft}^2 \cdot {}^\circ\text{F} \cdot \text{h/Btu}$$

The cold-face temperature, again from Fig. 5.3, is read from the line $R = 7.5$, where intersected by the horizontal line representing the 2100°F hot-face temperature. The cold face is approximately 180°F, less than the idealized 200°F by 20°F.

The heat leaving the cold face is found on the top horizontal line and is a function of the cold-face temperature. For the face in this example it is 250 Btu/(h · ft²), and for 200°F it is 315 Btu/(h · ft²).

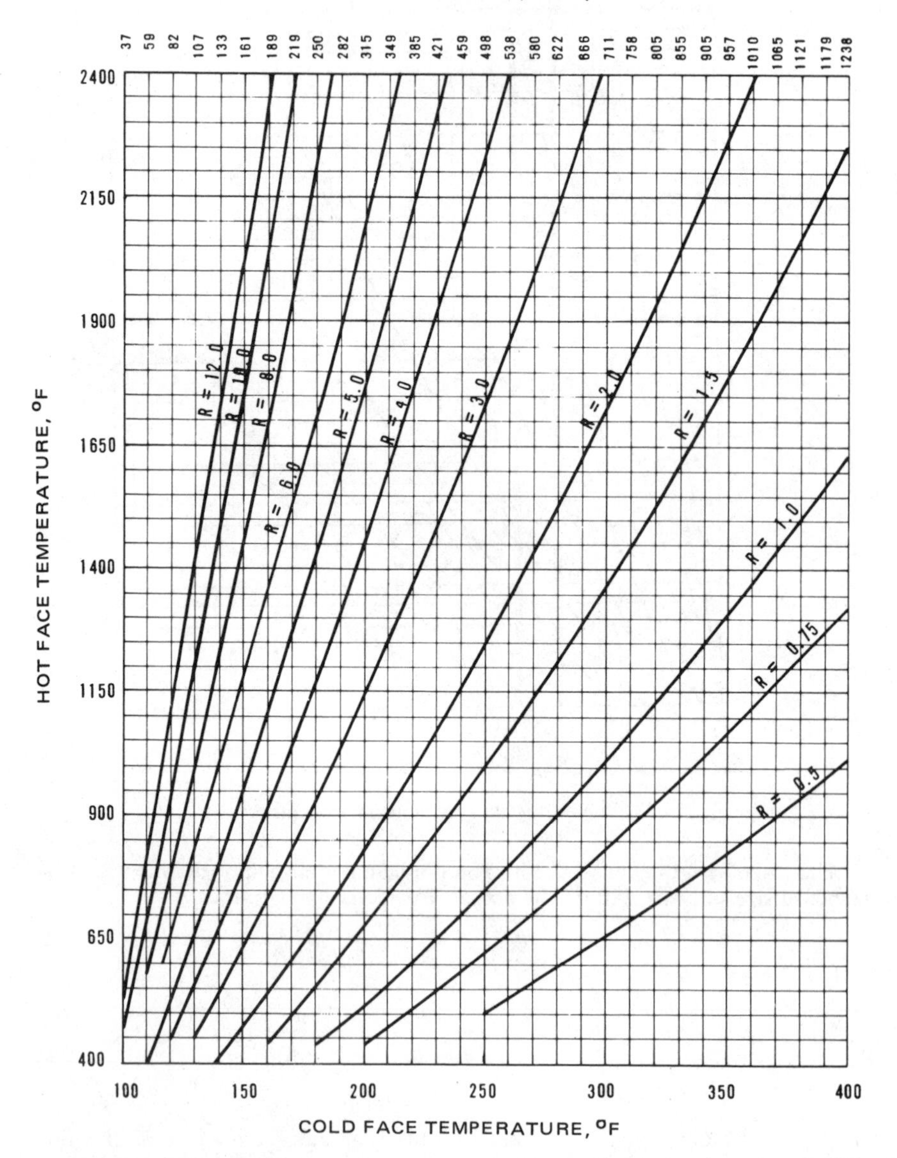

Fig. 5-4. Resistance graph, $E = 0.9$, facing up.

Four resistance graphs have been included in this chapter. They are each based on an ambient temperature of 80°F. Figure 5.4 is for an emissivity of 0.9 with the cold face horizontal, facing up. The resistance graph in Fig. 5.5 is for an emissivity of 0.9 with the cold face vertical. In Fig. 5.6 the cold face is horizontal, looking down. Figure 5.7 is for an emissivity of 0.1 with the cold face vertical.

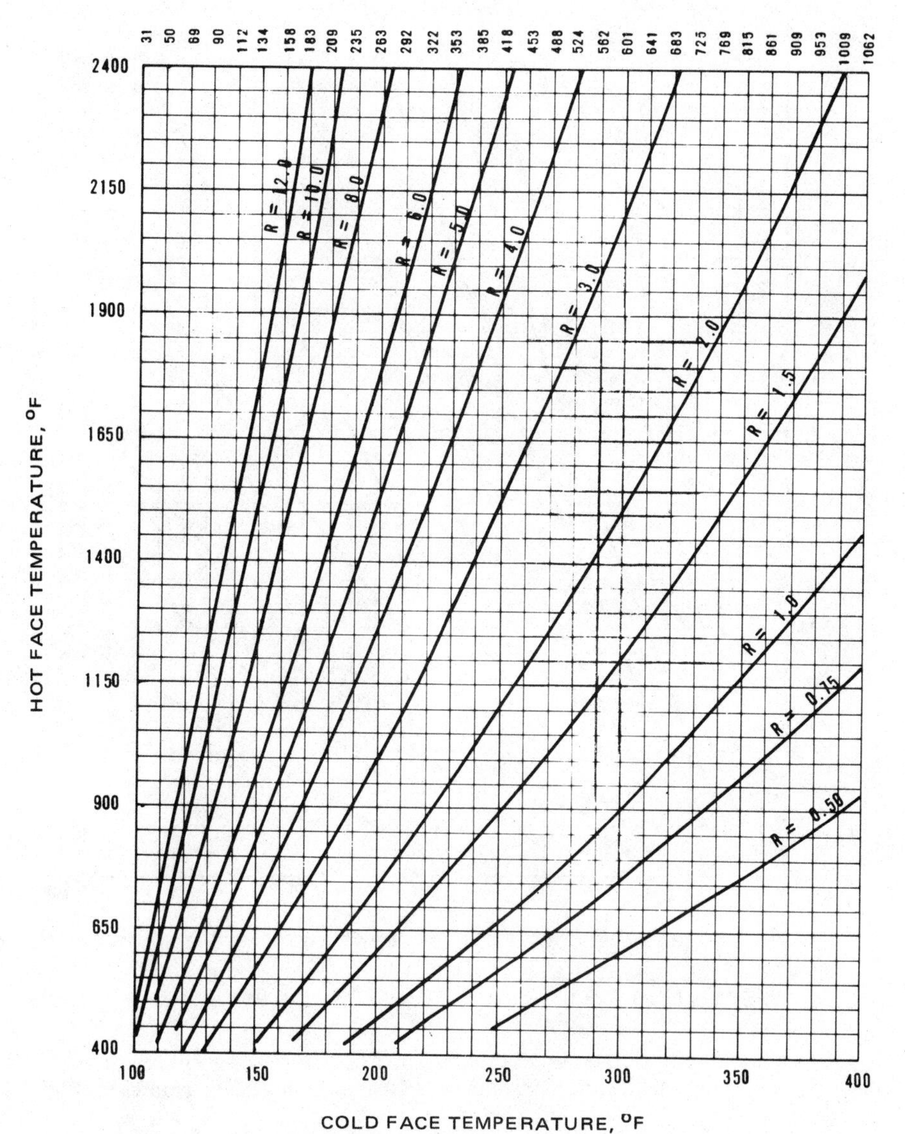

Fig. 5-5. Resistance graph, $E = 0.9$, vertical wall.

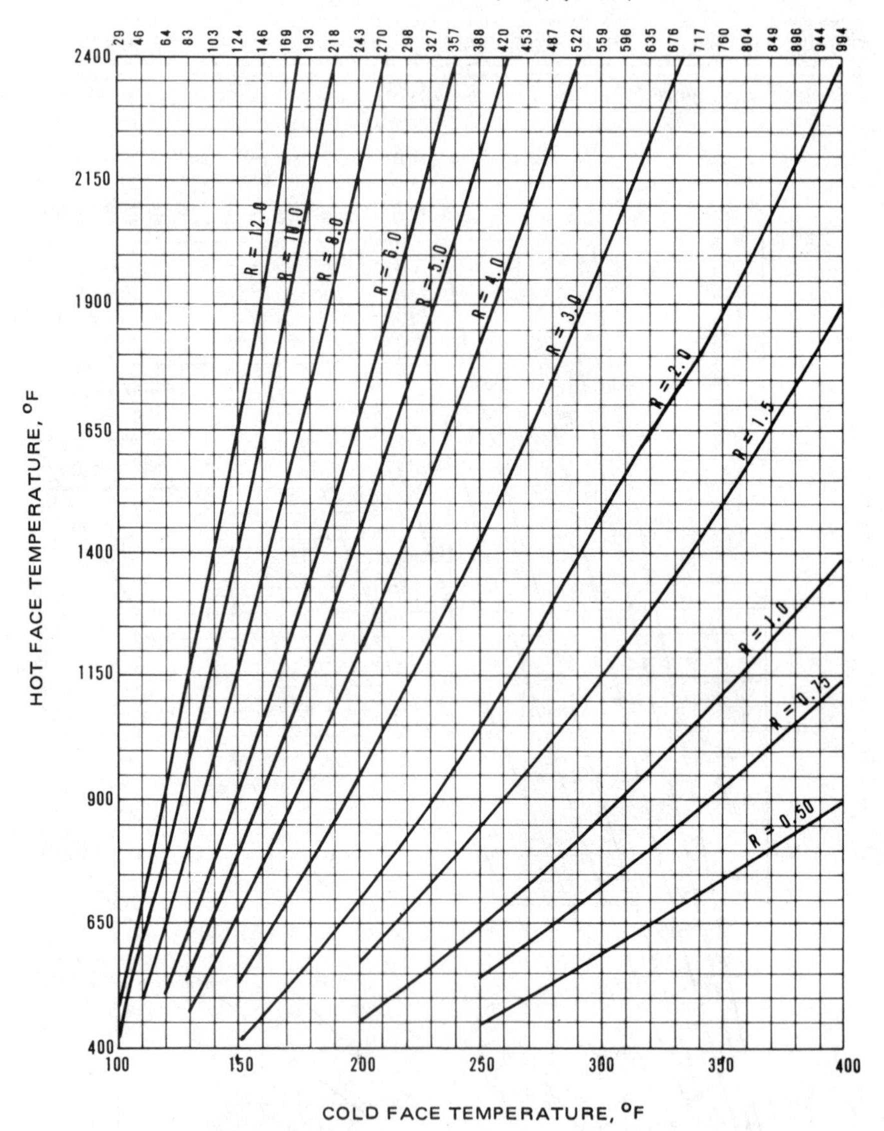

Fig. 5-6. Resistance graph, $E = 0.9$, facing down.

The example used in the above discussion, Fig. 5.3, was a representation of the resistance graph with emissivity 0.9, cold face horizontal, facing upward (Fig. 5.4).

The resistance graph is constructed as follows: Values for h, ϵ, and T_a are inserted into Eq. (5.22). Heat transfer Q is calculated for a range of T_0. In Figs. 5.4

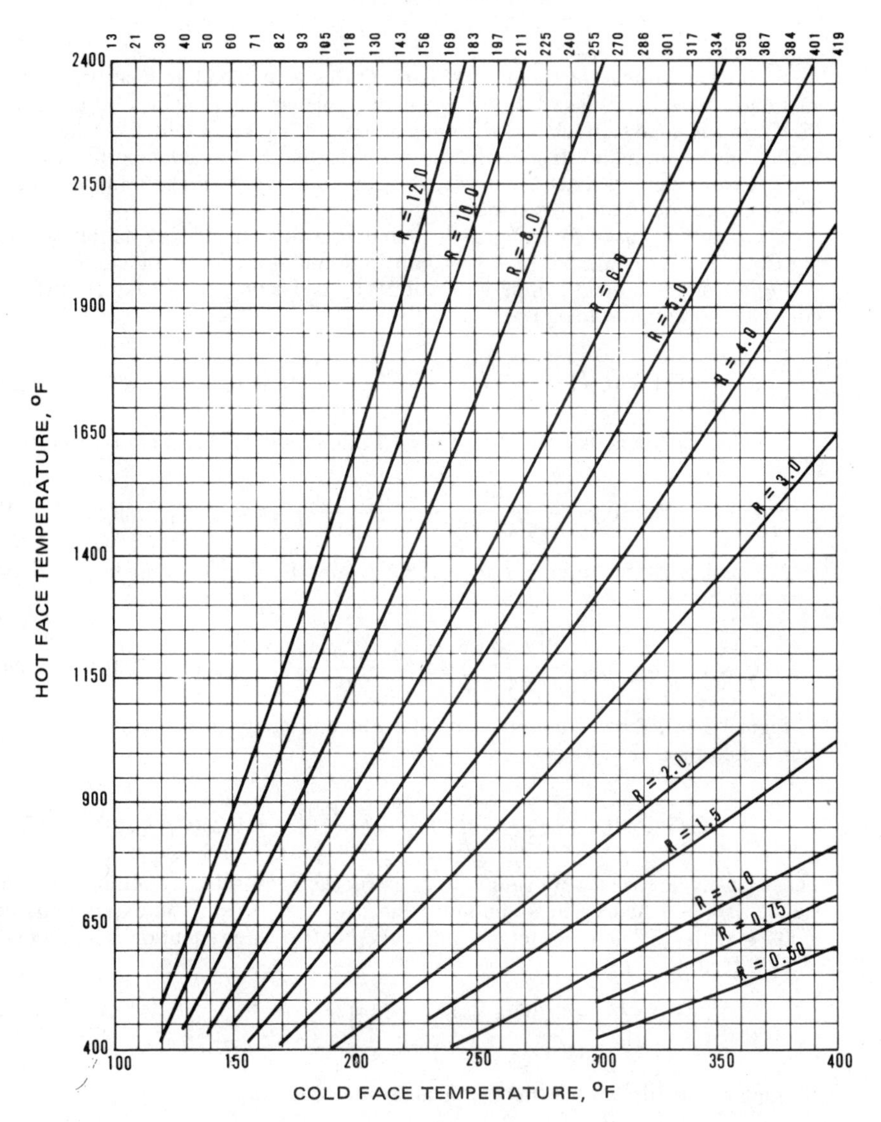

Fig. 5-7. Resistance graph, $E = 0.1$, vertical wall.

through 5.7, values of T_0 from 100 to 400°F in 10°F increments were calculated. The cold-face temperature T_0 is inserted on the bottom horizontal axis, and the heat transfer Q is inserted opposite T_0, on the top horizontal axis.

The hot-face temperature T_i and the resistance R must be integrated into this graph.

From Eq. (5.8), $Q = (1/R)(T_i - T_0)$ or

$$T_i = T_0 + QR \tag{5.23}$$

With the cold-face temperature T_0 and Q the corresponding heat transfer, as calculated above, values of T_i are calculated for various R values by using Eq. (5.23). In Figs. 5.4 through 5.7, R values of 0.5 to 12.0 are used in this calculation.

Plotting resulting hot-face temperature T_i against cold-face temperature T_0 for each R value defines a set of curves, the curves drawn in Figs. 5.4 through 5.7. These curves define the resistance graph.

The equations used for the graphs included in this chapter are as follows, using Eqs. (5.15), (5.16), and (5.17) for convection and (5.19) for radiation.

Figure 5.4: Ambient temperature of 80°F, emissivity of 0.9 (steel surface), cold face horizontal facing up:

$$Q = 0.174 \times 0.9\left[\left(\frac{T_0 + 460}{100}\right)^4 + \left(\frac{80 + 460}{100}\right)^4\right] + 0.38(T_0 - 80)^{1.25} \tag{5.24}$$

Figure 5.5: Ambient temperature of 80°F, emissivity of 0.9, cold face vertical:

$$Q = 0.174 \times 0.9\left[\left(\frac{T_0 + 460}{100}\right)^4 + \left(\frac{80 + 460}{100}\right)^4\right] + 0.25(T_0 - 80)^{1.25} \tag{5.25}$$

Figure 5.6: Ambient temperature of 80°F, emissivity of 0.9, cold face horizontal facing down:

$$Q = 0.174 \times 0.9\left[\left(\frac{T_0 + 460}{100}\right)^4 + \left(\frac{80 + 460}{100}\right)^4\right] + 0.20(T_0 - 80)^{1.25} \tag{5.26}$$

Figure 5.7: Ambient temperature of 80°F, emissivity of 0.1 (polished aluminum), cold face vertical:

$$Q = 0.174 \times 0.1\left[\left(\frac{T_0 + 460}{100}\right)^4 + \left(\frac{80 + 460}{100}\right)^4\right] + 0.25(T_0 - 80)^{1.25} \tag{5.27}$$

These graphs are relatively simple to construct for any set of conditions. For instance, if a polished vertical aluminum surface is exposed to an external air velocity V with 40°F ambient temperature, the heat-transfer equation is as follows, using h from (5.18):

$$Q = 0.174 \times 0.1\left[\left(\frac{T_0 + 460}{100}\right)^4 + \left(\frac{40 + 460}{100}\right)^4\right] + (1.0 + 0.225V)(T_0 - 40) \tag{5.28}$$

A graph can be drawn for this or for any other condition.

REFRACTORY AND INSULATION

There are a wide variety of refractory and insulation products available. Their significant properties include:

- Maintaining strength under high-temperature conditions.

- Cold crushing strength, the ability to withstand handling and shipping without damage, and impact strength at low-temperature operations.
- Insulating value, the ability to provide resistance to the flow of heat.
- Porosity, the susceptibility to penetration by slags or gases.
- Reheat (ASTM method C113), a measure of changes in linear dimensions under repeated heating and cooling.
- Modulus of rupture, a standard measure of structural strength provided by load testing (ASTM methods C16 and C216).
- Erosion, the washing away or physical destruction of material, under forces due to physical contact with the gaseous, liquid, or solid materials on the hot face.
- Corrosion, the chemical process that destroys the refractory bond or the chemical integrity of the insulation.
- Thermal expansion, the reversible change in linear dimensions under heating and cooling. Where refractory is used as a liner in a flue, for instance, the expansion of the refractory must be evaluated with respect to thermal expansion of the flue. If the expansions do not match, provision for material expansion must be considered.
- Operating environment. Of particular interest is the operation of refractory in an oxidizing or reducing atmosphere. A reducing atmosphere, one deficient in oxygen, will tend to degrade refractory material containing iron or silicon components.

MATERIALS DESCRIPTION

Castables (Refractory Concrete)

These materials are supplied dry, to be mixed with water before installation. They are installed by pouring (casting in place), troweling, pneumatic gunning (as with gunned fireproofing), or ramming. A castable refractory provides a smooth, continuous, monolithic mass. Castable materials are normally placed in an area prepared with pins, mesh, or other anchor devices to hold the refractory in place during placement and curing. Mesh, grid, studs, or needles may also be used to enhance the strength of the refractory installation.

Castable refractories are classified as dense or lightweight (insulating). Dense castables have excellent mechanical strength and low permeability. Their insulating properties, however, are relatively poor. As dense materials, over 100 lb/ft^3 specific weight, they offer good resistance in wet service such as quencher linings.

Lightweight castables are excellent insulators. All castable materials have the advantage of placement in irregular areas such as furnace transitions, burner openings, etc. They are cast in place. Poured castables require a form; however, plastic castable does not. It is rammed in place. Gunned castable is blown in place, and formwork is not required. The use of a gunned material, however, is often dictated by the size of the job in question. Gunning requires installation equipment including air compressors and gunning applicators that would be uneconomical to provide when compared to the relatively small amount of equipment needed for poured or plastic castable installation. Normally, gunning is not economical for jobs less than a week in duration.

The quality of a castable installation is a function of mixing and curing as well as application. Mixing is performed at the job site with water added by the installation contractor in accordance with the refractory manufacturer's instructions. If the amount of water added is not within the product specifications, the refractory installation will be deficient. This is related to placement, as well as to the strength developed by the refractory in curing and in service. Too dry a mix will not flow in a form, and voids will appear in certain areas. Thickness requirements may not be satisfied in other areas. Too much water in a mix will promote segregation of the mix as it solidifies and as it cures, providing inconsistency in refractory quality as well as poor refractory quality. Note that poured refractory is often vibrated to aid its flow, similar to placement of concrete.

Curing refractory is necessary to set a ceramic bond, creating the strength of the refractory matrix. Without proper curing the refractory will not develop mechanical strength or chemical resistance (resistance to corrosion). Refractory is usually either air-set or heat-cured. Air setting requires only a quiescent period where the refractory is not subject to mechanical changes (motion) or heating effects. Heat curing requires the refractory be heated at a controlled rate to establish the ceramic bond. This bond may require that temperatures of 1000 to over 2000°F be reached and maintained. Often incinerator temperatures are not high enough to cure heat-setting refractories, and special operating procedures are necessary to bring the equipment to the curing temperature.

Firebrick

Conventional firebrick is kiln-baked to uniform, controlled consistency and quality. The term *firebrick* refers to dense brick, over 100 lb/ft^3, normally placed in direct contact with the hot gas stream. It has relatively poor insulating quality. Insulating firebrick (IFB) is a lightweight, porous brick, normally less than 50 lb/ft^3, which can be placed in direct contact with the gas stream and which provides good insulating characteristics. IFB is machined to its final shape, providing excellent dimensional control as compared to firebrick which is used as cast.

IFB is lower in strength than firebrick and because of its porosity is a soft material not effective with erosive gas streams, i.e., gas streams with high particulate components. This low abrasion resistance limits the maximum velocities allowed adjacent to it.

Where refractory brick is required and high abrasive resistance is necessary, firebrick will often be provided with insulation block as back-up between the firebrick and the furnace-flue wall.

Anchors are normally provided to hold brick in place. Firebrick and IFB are both self-supporting. However, because of the relative structural strengths of these materials IFB requires a more elaborate anchoring system than firebrick. Brick manufacturers each have their own unique anchor and anchor systems designs.

Table 5.6 lists properties of typical refractory brick materials.

TABLE 5.6 Properties of Refractory Brick

Type	Class	Weight, %			Temperature limit, °F	Density, lb/ft³	Modulus of rupture, lb/in²	Cold crushing strength, lb/in²	Thermal expansion ambient to 1800°F, %
		Silica (SiO_2)	Alumina (Al_2O_3)	Titania (TiO_2)					
Fireclay	super duty	40–56	40–44	1–3	3185	140/145	730/1975	1830/7325	0.6
	high duty	51–61	40–44	1–3	3175	120/145	610/3660	1830/8550	0.6
	medium duty	57–60	25–38	1–2	3040	120/145	975/3050	2075/7325	0.5–0.6
	low duty	60–70	22–33	1–2	2815	120/145	1220/3050	2440/7325	0.5–0.6
	semi-silica	72–80	18–26	1–2	3040	115/125	365/1100	1220/3660	0.7
High aluminum	45–48%	44–51	45–48	2–3	3245	135/160	1220/1950	3050/7325	0.4–0.5
	60%	31–37	58–62	2–3	3295	140/160	730/2200	2200/8550	0.6
	80%	11–15	78–82	3–4	3390	155/180	1340/3660	4880/10975	0.7
	90%	8–9	89–91	0.4–1	3500	165/190	1465/4275	4880/10975	0.7
	mullite	18–34	60–78	0.5–3	3360	145/165	1220/4275	4275/10975	0.7
	corundum	0.2–1	98–99	trace	3660	170/200	2200/3660	6100/10975	0.8
Silica	super duty	95–97	0.1–0.3 +	lime	—	105/118	610/1220	1830/4275	1.3
	conventional	94–97	0.4–1.4 +	lime	—	105/118	730/1465	2200/4880	1.2
	lightweight	94–97	0.4–1.4 +	lime	—	60	—	—	
Silicon carbide	bonded	—	—	—	3360	145/165	2440/4880	3050/18300	0.5
Insulating	1600°F	—	—	—	—	36	85/125	110/135	low
	2300°F	—	—	—	—	46	120/200	135/230	low
	2800°F	—	—	—	—	59	200/365	205/370	low
	3000°F	—	—	—	—	60	490/730	975/1220	low
Chrome	fired	Cr_2O_3 (28–38), MgO (14–49), Al_2O_3 (15–34), Fe_2O_3 (11–17)			—	185/205	855/1590	2440/4880	0.8

Source: Selected manufacturers' data.

REFERENCE

5-1. Jakob, M.: *Elements of Heat Transfer*, 3d ed., John Wiley & Sons, New York, 1957.

BIBLIOGRAPHY

Brunner, C.: "Program Calculates Heat Transfer through Composite Walls," *Chemical Engineering*, June 16, 1980.

Darroudi, T.: "Thermal Shock Damage Resistance Parameters of Refractory Castables," *Industrial Heating*, May 1980.

Harrison, M., and Pelanne, C.: "Cost Effective Thermal Insulation," *Chemical Engineering*, December 17, 1977.

"Hazardous Waste Laws Put New Strains on Refractories Used in Incineration," *Refractory Magazine*, December 1981.

Hess, L.: *Insulation Guide for Buildings and Industrial Processes*, Noyes, Park Ridge, N.J., 1979.

Marino, J.: "Thermal Design of Refractory and Insulation Systems," *Industrial Heating*, May 1980.

Miner, R : "Evaluation of Improved Silicon Refractories," *Industrial Heating*, May 1980.

O'Keefe, W.: "Thermal Insulation," *Power Magazine*, August 1974.

"Saving Heat Energy in Refractory Lined Equipment," *Chemical Engineering*, May 4, 1981.

Trinklein, R.: "Development of a Lightweight, Thermal Insulating, Corrosion Resistant Refractory Monolith," *Industrial Heating*, April 1979.

Weber, J.: "Predict Thermal Conductivity of Pure Gases," *Chemical Engineering*, January 12, 1981.

CHAPTER 6
COMBUSTION CALCULATIONS

Incineration is a burning process and as such involves the reaction of combustible components with air. In this chapter the relationship among combustion, air, and products of combustion will be determined.

WASTE CHARACTERISTICS FOR COMBUSTION

Almost by its very definition waste is not able to be classified into a single, or even multiple, known chemical structure. Often the chemical nature of a particular waste is unknown. Calculations involving the combustion of a waste material, therefore, in most cases must necessarily be based on a number of assumptions. These assumptions include the following:

- All hydrogen present converts to water vapor, H_2O, unless otherwise noted below.
- All chloride (or fluoride) converts to hydrogen chloride, HCl (or hydrogen fluoride, HF).
- All carbon converts to carbon dioxide, CO_2.
- All sulfur converts to sulfur dioxide, SO_2.
- Alkali metals convert to hydroxides: sodium to sodium hydroxide ($2Na + O_2 + H_2 \rightarrow 2NaOH$) and potassium to potassium hydroxide ($2K + O_2 + H_2 \rightarrow 2KOH$).
- Nonalkali metals convert to oxides: copper to copper oxide ($2Cu + O_2 \rightarrow 2CuO$), iron to iron oxide ($4Fe + 3O_2 \rightarrow 2Fe_2O_3$).
- All nitrogen from the waste, the fuel, or air will take the form of a diatomic molecule; i.e., nitrogen is present as N_2.

EQUILIBRIUM EQUATIONS

Chemical equilibrium equations are developed to indicate conservation of matter; i.e., the molecular weight on the left side of the equation equals the molecular weight on the right. For carbon, for instance,

$$C + O_2 \rightarrow CO_2 \tag{6.1}$$

From Table 6.1, the atomic weight of carbon is 12.01, and of oxygen, 16.00. For this simple example, therefore, the atomic weight on the left-hand side is $1 \times 12.01 + 2 \times 16.00 = 44.01$, which is equal to the weight of CO_2 on the right-hand side of the equation.

In the wastes and fuels normally encountered, the major constituents include carbon and hydrogen. For hydrogen:

$$2H_2 + O_2 \rightarrow 2H_2O \tag{6.2}$$

Note that oxygen or hydrogen exists as diatomic molecules (O_2 and H_2), as does nitrogen (N_2).

The equilibrium equation for methane, CH_4, is

$$CH_4 + 2O_2 \rightarrow CO_2 + 2H_2O \tag{6.3}$$

TABLE 6.1 Chemical Elements

Element	Symbol	Atomic No.	Atomic Weight
Actinium	Ac	89	227.
Aluminum	Al	13	26.98
Americium	Am	95	243.
Antimony	Sb	51	121.75
Argon	Ar	18	39.95
Arsenic	As	33	74.92
Astatine	At	85	210.
Barium	Ba	56	137.34
Berkelium	Bk	97	247.
Beryllium	Be	4	9.01
Bismuth	Bi	83	208.98
Boron	B	5	10.81
Bromine	Br	35	79.90
Cadmium	Cd	48	112.40
Calcium	Ca	20	40.08
Californium	Cf	98	251.
Carbon	C	6	12.01
Cerium	Ce	58	140.12
Cesium	Cs	55	132.91
Chlorine	Cl	17	35.45
Chromium	Cr	24	52.00
Cobalt	Co	27	58.93
Columbium (*see* Niobium)			
Copper	Cu	29	63.55
Curium	Cm	96	247.
Dysprosium	Dy	66	162.50
Einsteinium	Es	99	254.
Erbium	Er	68	167.26
Europium	Eu	63	151.96
Fermium	Fm	100	253.
Fluorine	F	9	19.00

TABLE 6.1 Chemical Elements (Continued)

Element	Symbol	Atomic No.	Atomic Weight
Francium	Fr	87	223.
Gadolinium	Gd	64	157.25
Gallium	Ga	31	69.72
Germanium	Ge	32	72.59
Gold	Au	79	196.97
Hafnium	Hf	72	178.49
Helium	He	2	4.00
Holmium	Ho	67	164.93
Hydrogen	H	1	1.01
Indium	In	49	114.82
Iodine	I	53	126.90
Iridium	Ir	77	192.2
Iron	Fe	26	55.85
Krypton	Kr	36	83.80
Lanthanum	La	57	138.91
Lead	Pb	82	207.19
Lithium	Li	3	6.94
Lutetium	Lu	71	174.97
Magnesium	Mg	12	24.31
Manganese	Mn	25	54.94
Mendelevium	Md	101	256.
Mercury	Hg	80	200.59
Molybdenum	Mo	42	95.94
Neodymium	Nd	60	144.24
Neon	Ne	10	20.18
Neptunium	Np	93	237.
Nickel	Ni	28	58.71
Niobium	Nb	41	92.91
Nitrogen	N	7	14.01
Nobelium	No	102	254.
Osmium	Os	76	190.2
Oxygen	O	8	16.00
Palladium	Pd	46	106.4
Phosphorus	P	15	30.97
Platinum	Pt	78	195.09
Plutonium	Pu	94	242.
Polonium	Po	84	210.
Potassium	K	19	39.10
Praseodymium	Pr	59	140.91
Promethium	Pm	61	147.
Protactinium	Pa	91	231.
Radium	Ra	88	226.
Radon	Rn	86	222.
Rhenium	Re	75	186.2
Rhodium	Rh	45	102.91
Rubidium	Rb	37	85.47

TABLE 6.1 Chemical Elements (Continued)

Element	Symbol	Atomic No.	Atomic Weight
Ruthenium	Ru	44	101.07
Samarium	Sm	62	150.53
Scandium	Sc	21	44.96
Selenium	Se	34	78.96
Silicon	Si	14	28.09
Silver	Ag	47	107.87
Sodium	Na	11	22.99
Strontium	Sr	38	87.62
Sulphur	S	16	32.06
Tantalum	Ta	73	180.95
Technetium	Tc	43	99.
Tellurium	Te	52	127.60
Terbium	Tb	65	158.92
Thallium	Tl	81	204.37
Thorium	Th	90	232.04
Thulium	Tm	69	168.93
Tin	Sn	50	118.69
Titanium	Ti	22	47.90
Tungsten	W	74	183.85
Uranium	U	92	238.03
Vanadium	V	23	50.94
Xenon	Xe	54	131.30
Ytterbium	Yb	70	173.04
Yttrium	Y	39	88.91
Zinc	Zn	30	65.37
Zirconium	Zr	40	91.22

Source: Refs. 6-1 and 6-2.

Note that the total molecules of each element on the left are equal to the total molecules of each element on the right (one C, four H, four O).

AIR PROPERTIES

Air is a mixture of nitrogen, oxygen, water vapor, helium, carbon dioxide, and other gases. Its composition varies in different parts of the world, at different elevations, and at different times of the year. For purposes of combustion calculations, however, air is considered to be a mixture of only nitrogen and oxygen, as noted in Table 6.2.

BASIC COMBUSTION CALCULATIONS

In most situations waste, or fuel, is combusted with air. Air, as shown in Table 6.2, is composed of nitrogen and oxygen. For a burning reaction to proceed, ox-

TABLE 6.2 Air Composition

	By Weight	By Volume
Oxygen in air	0.2315	0.21
Nitrogen in air	0.7685	0.79
Air to oxygen	4.3197	4.7619
Air to nitrogen	1.3012	1.2658
Oxygen to nitrogen	0.3012	0.2658
Nitrogen to oxygen	3.3197	3.7619
Molecular weight, average	28.9414	

ygen must be introduced, and nitrogen will also be present along with the oxygen, 3.76 molecules N_2 per molecule O_2 from Table 6.2. For carbon, from Eq. (6.1),

$$C + O_2 + 3.76N_2 \rightarrow CO_2 + 3.76N_2 \tag{6.4}$$

The number of molecules of each substance present (moles) is proportional to the volume of that substance.

The weight of each constituent is proportional to its atomic weight. Inserting the atomic weights as listed in Table 6.1 into Eq. (6.3) gives

$$\begin{array}{ccccccccc} 12.01 & & 32.00 & & 28.02 & & 44.01 & & 28.02 \\ C & + & O_2 & + & 3.76N_2 & \rightarrow & CO_2 & + & 3.76N_2 \\ 12.01 & & 32.00 & & 105.36 & & 44.01 & & 105.36 \\ 1.00 & & 2.66 & & 8.77 & & 3.66 & & 8.77 \end{array} \tag{6.5}$$

The first line beneath (6.5), the equilibrium equation, represents the total weights of each element present as calculated from its molecular weight above the equation. The second line relates each element to 1 lb of carbon and is obtained by dividing the constituent weights by the weight of carbon, 12.01.

Note first that the sum of the constituent weights on one side, 12.01 + 32.00 + 105.36 = 149.37, equals that on the other side of the equation, 44.01 + 105.36 = 149.37. Likewise, normalized to 1 lb of carbon, the constituent weights are equal, one side to the other: 1.00 + 2.66 + 6.77 = 12.43, 3.66 + 8.77 = 12.43.

By normalizing this reaction to 1 lb of carbon the following determination is apparent:

For burning 1 lb of carbon, 2.66 lb O_2 is required, 2.66 + 8.77 = 11.43 lb of air is required, 3.66 lb of CO_2 is generated, and 3.66 + 8.77 = 12.43 lb of combustion products is produced.

Applying similar calculations to the combustion of hydrogen, Eq. (6.2), gives

$$\begin{array}{ccccccccc} 2.02 & & 32.00 & & 28.02 & & 18.02 & & 28.02 \\ 2H_2 & + & O_2 & + & 3.76N_2 & \rightarrow & 2H_2O & + & 3.76N_2 \\ 4.04 & & 32.00 & & 105.36 & & 35.04 & & 105.36 \\ 1.00 & & 7.92 & & 26.08 & & 8.92 & & 26.08 \end{array} \tag{6.6}$$

Therefore, for burning 1 lb of hydrogen 7.92 lb O_2 is required, 7.92 + 26.08 = 34.00 lb of air is required, 8.92 lb of H_2O is generated, 8.92 + 26.08 = 35.00 lb of combustion products is produced.

STOICHIOMETRIC BURNING

The examples of combustion illustrated in Eqs. (6.5) and (6.6) utilized only that amount of air required for complete combustion to CO_2 or to H_2O. All the oxygen provided was used up in the generation of CO_2 and H_2O, and no free oxygen was left in the products of combustion. This condition is known as *complete combustion*, provision of 100 percent of total air, zero excess air, and defines the *stoichiometric* condition. When stoichiometric air or stoichiometric oxygen is referred to, it defines the condition where only that amount of air, or oxygen, is provided to ensure complete combustion in accordance with the chemical equilibrium equation. For carbon the stoichiometric air requirement is 11.43 lb per pound carbon, and for hydrogen, 34.00 lb stoichiometric air per pound of hydrogen, as previously calculated.

EXCESS AIR

Combustion calculations must relate to actual burning equipment requirements if they are to have any validity. In practice the air provided for complete burning must be greater than the stoichiometric requirement. The stoichiometric air value implies a burning process with 100 percent efficiency, where air is in contact with 100 percent of the waste, or fuel, surface, no air is wasted, and the burning reaction occurs instantaneously. In actuality, to achieve complete burning, an amount of air greater than the stoichiometric requirement must be provided. For burning gaseous fuel a total air requirement of 5 percent above stoichiometric may be sufficient, but for burning solid wastes or sludges in a multiple-chamber furnace excess air of 100 to 200 percent of stoichiometric may be required.

EXCESS AIR CALCULATIONS

Burning a simple fuel, benzene, C_6H_6, at stoichiometric conditions

$$\begin{array}{cccccc} 78.12 & 32.00 & 28.02 & 44.01 & 18.02 & 28.02 \\ C_6H_6 + & 7.5O_2 + & (7.5 \times 3.76)N_2 \rightarrow & 6CO_2 + & 3H_2O + & 28.20N_2 \\ 78.12 & 240.00 & 790.16 & 264.06 & 54.06 & 790.16 \\ 1.00 & 3.07 & 10.11 & 3.38 & 0.69 & 10.11 \end{array} \tag{6.7}$$

The stoichiometric oxygen is 3.07 lb/lb C_6H_6, 7.5 mol O_2/mol C_6H_6. For 20 percent excess air an additional amount of oxygen, 20 percent of 7.5 or 1.5 mol of oxygen, must be added to this reaction. With 7.5 mol of oxygen required for combustion of fuel, therefore, the additional 1.5 mol of oxygen added in the air will appear in the flue gas, as follows:

$$\begin{array}{ccccccc} 78.12 & 32.00 & 28.02 & 44.01 & 18.02 & 32.00 & 28.02 \\ C_6H_6 + & (7.5 + 1.5)O_2 + & (9 \times 3.76)N_2 \rightarrow & 6CO_2 + & 3H_2O + & 1.5O_2 + & 33.64N_2 \\ 78.12 & 288.00 & 948.20 & 264.06 & 54.06 & 48.00 & 948.20 \\ 1.00 & 3.69 & 12.14 & 3.38 & 0.69 & 0.62 & 12.14 \end{array} \tag{6.8}$$

When a waste or fuel containing oxygen, such as cellulose ($C_6H_{10}O_5$), the main constituent of paper, is burned, calculations of excess air must consider the oxygen component of the fuel.

For stoichiometric conditions:

$$\begin{array}{cccccc} 162.16 & 32.00 & 28.02 & 44.01 & 18.02 & 28.02 \\ C_6H_{10}O_5 + & 6O_2 + & (6 \times 3.76)N_2 \rightarrow & 6CO_2 + & 5H_2O + & 22.56N_2 \\ 162.16 & 192.00 & 632.13 & 264.06 & 90.10 & 632.13 \\ 1.00 & 1.19 & 3.90 & 1.63 & 0.56 & 3.90 \end{array} \quad (6.9)$$

When calculating excess air requirements for burning, consider that 6 mol of O_2 defines the stoichiometric requirements. (This is not true when calculating heating value or NO_x and CO emissions which will be covered in a later chapter.)

Therefore, for 150 percent excess air, an additional $6 \times 1.5 = 9.0$ mol of O_2 must be introduced, as follows:

$$\begin{array}{ccccccc} 162.16 & 32.00 & 28.02 & 44.01 & 18.02 & 32.00 & 28.02 \\ C_6H_{10}O_5 + & (6 + 9)O_2 + & (15 \times 3.76)N_2 \rightarrow & 6CO_2 + & 5H_2O + & 9O_2 + & 56.40N_2 \\ 162.16 & 480.00 & 1580.33 & 264.06 & 90.10 & 288.00 & 1580.33 \\ 1.00 & 2.69 & 9.75 & 1.63 & 0.56 & 1.77 & 9.75 \end{array} \quad (6.10)$$

MEASUREMENT OF EXCESS AIR

Burning equipment is usually rated to be utilized within a specific range of excess air. If air below this range is provided, the waste or fuel will not combust completely. Above this range the excess air is greater than that required to ensure complete combustion, and this airflow will reduce the combustion temperature, increasing the supplemental fuel requirement. A measurement of excess air, therefore, is an important parameter of operation of burning equipment or systems.

Normally an oxygen or carbon dioxide gas analyzer is installed in the furnace exhaust stream, and the oxygen or carbon dioxide present is related to excess air. From the burning of benzene and cellulose, as previously discussed, this relationship can be illustrated. Of note is that four modes of measurement are used:

1. Fraction O_2, or CO_2, in total flue gas, by weight
2. Fraction O_2, or CO_2, in dry flue gas, by weight
3. Fraction O_2, or CO_2, in total flue gas, by volume
4. Fraction O_2, or CO_2, in dry flue gas, by volume

Flue gas is the gaseous product of combustion exiting a furnace.

For benzene, at stoichiometric conditions (0 percent excess air), from Eq. (6.7):

$$\text{Total flue gas weight} = 3.38 \text{ lb } CO_2 + 0.69 \text{ lb } H_2O + 10.11 \text{ lb } N_2 = 14.18 \text{ lb}$$

$$\text{Dry flue gas weight} = 3.38 \text{ lb } CO_2 + 10.11 \text{ lb } N_2 = 13.49 \text{ lb}$$

$$\text{Total flue gas volume} = 6CO_2 + 3H_2O + 28.20N_2 = 37.20 \text{ mol}$$

$$\text{Dry flue gas volume} = 6CO_2 + 28.20N_2 = 34.20 \text{ mol}$$

For oxygen in the flue gas, at stoichiometric conditions, the O_2 content in the flue gas is zero.

For CO_2 in the flue gas (3.38 lb or 6 mol):

$$CO_2 \text{ by weight, total gas:} \quad \frac{3.38}{14.18} = 23.84\%$$

$$\text{Dry gas:} \quad \frac{3.38}{13.49} = 25.06\%$$

$$CO_2 \text{ by volume, total gas:} \quad \frac{6.00}{37.20} = 16.13\%$$

$$\text{Dry gas:} \quad \frac{6.00}{34.20} = 17.54\%$$

For 20 percent excess air, Eq. (6.8):

$$\text{Total flue gas, by weight} = 3.38 \text{ lb } CO_2 + 0.69 \text{ lb } H_2O + 0.62 \text{ lb } O_2 + 12.14 \text{ lb } N_2$$

$$= 16.83 \text{ lb}$$

$$\text{Dry flue gas, by weight} = 3.36 \text{ lb } CO_2 + 0.62 \text{ lb } O_2 + 12.14 \text{ lb } N_2$$

$$= 16.14 \text{ lb}$$

$$\text{Total flue gas, by volume} = 6CO_2 + 3H_2O + 1.5O_2 + 33.64N_2$$

$$= 44.34 \text{ mol}$$

$$\text{Dry flue gas, by volume} = 6CO_2 + 1.5O_2 + 33.84N_2 = 41.34 \text{ mol}$$

Oxygen fraction (0.62 lb, 1.5 mol in flue gas):

$$\text{Oxygen, by weight, total gas:} \quad \frac{0.62}{16.83} = 3.68\%$$

$$\text{Dry gas:} \quad \frac{0.62}{16.14} = 3.84\%$$

$$\text{Oxygen, by volume, total gas:} \quad \frac{1.5}{44.34} = 3.38\%$$

$$\text{Dry gas:} \quad \frac{1.5}{41.34} = 3.63\%$$

CO_2 fraction (3.38 lb, 6 mol in flue gas)

$$CO_2\text{, by weight, total gas:} \quad \frac{3.38}{16.83} = 20.08\%$$

$$\text{Dry gas:} \quad \frac{3.38}{16.14} = 20.94\%$$

CO_2, by volume, total gas: $\frac{6}{44.34} = 13.53\%$

Dry gas: $\frac{6}{41.34} = 14.51\%$

Similarly, for cellulose, at stoichiometric or 100 percent total air, Eq. (6.9):

O_2, by weight or volume: zero

CO_2, by weight, total gas: $\frac{1.63}{6.9} = 26.77\%$

Dry gas: $\frac{1.63}{5.53} = 29.48\%$

CO_2, by volume, total gas: $\frac{6}{33.56} = 17.88\%$

Dry gas: $\frac{6}{28.56} = 21.01\%$

At 150 percent excess air, or 250 percent total air, Eq. (6.10):

O_2, by weight, total gas: $\frac{1.77}{13.71} = 12.91\%$

Dry gas: $\frac{1.77}{13.15} = 13.46\%$

O_2, by volume, total gas: $\frac{9}{76.40} = 11.78\%$

Dry gas: $\frac{9}{71.40} = 12.61\%$

O_2, by weight, total gas: $\frac{1.63}{13.71} = 11.89\%$

Dry gas: $\frac{1.63}{13.15} = 12.40\%$

O_2, by volume, total gas: $\frac{6}{76.40} = 7.85\%$

Dry gas: $\frac{6}{71.40} = 8.40\%$

A more desirable method of measurement is based on volumetric percentage, with moisture eliminated from the calculations, i.e., based on dry flue gas by volume. Equipment is available to dry a sample prior to measurement. This eliminates concern for the moisture content of the fuel, as will be discussed in a later chapter. Summarizing the flue gas constituents, dry flue gas by volume in Table 6.3, note the following:

- As excess air increases, the oxygen component of the flue gas increases.
- As excess air increases, the CO_2 component of the flue gas decreases.

TABLE 6.3 O_2 and CO_2 Variation with Excess Air

	Benzene		Cellulose	
	0% excess air	20% excess air	0% excess air	150% excess air
O_2	0	3.63%	0	12.61%
CO_2	17.54%	14.51%	21.01%	8.4%

The oxygen fraction will increase, by definition, as the excess-air quantity increases. The excess, that which is not used in the reaction, must necessarily appear in the products of combustion. The reason for a decrease in CO_2 level with an increase in excess air is that a fixed amount of carbon produces a fixed amount of CO_2 regardless of the amount of excess air introduced. As the total gas flow increases, however, and excess air is increased, the proportion of CO_2 will decrease; i.e., the amount of CO_2 is fixed in an increased volume of flue gas.

PYROLYSIS

Up to now in this chapter burning processes have been discussed. Another thermal process of interest is *pyrolysis*, the degradation of carbonaceous material (a material that contains carbon) in the absence of oxygen, or air, upon application of heat. For instance, pyrolysis of cellulose is

$$C_6H_{10}O_5 \rightarrow 2CO + CH_4 + 3H_2O + 3C \qquad (6.11)$$

Note that in the absence of oxygen the cellulose will degrade, or fractionate, to carbon monoxide (CO), methane (CH_4), and water vapor (H_2O), which will be in a gaseous phase, and carbon (C), a char, which will be in a solid or liquid phase. The off-gas has a heating value, since two of its constituents, CO and CH_4, will burn in oxygen (or air), providing a heat release.

In practice true pyrolysis does not occur because it is difficult to reduce the oxygen infiltration to zero. A condition between the pyrolysis and stoichiometric burning is defined as a *starved air* mode. There is equipment designed to operate in a starved air mode, as will be discussed in a subsequent chapter. Often the terms *pyrolysis* and *starved air* are incorrectly used interchangeably.

FLUE GAS PROPERTIES

A good approximation of the properties of flue gas can be made by using the properties of air as the dry flue gas component. Flue gas will contain carbon dioxide, nitrogen, oxygen, and water vapor. The proportions of these elements will vary with changes in excess air and other conditions of operation. If calculations of carbon dioxide, oxygen, and nitrogen were performed for each variation in operation, analysis of flue gas properties would be exceedingly complex and time-consuming. Using air as dry gas is a convenient simplification which allows quick calculations, accurate calculations, use of air and humidity tables, etc. Incineration normally involves large amounts of excess air for proper combustion, and this excess-air supply passes to the exiting flue gas. A large component of the flue gas is, therefore, air from the excess-air supply.

To illustrate the properties of air with regard to the individual gas components of the flue gas, consider the flue gas produced when burning cellulose with 150 percent excess air [Eq. (6.10)], assuming 8500 Btu produced. The reaction temperature will be determined.

Flue gas components:

1.63 lb CO_2
1.77 lb O_2
9.75 lb N_2
0.56 lb H_2O

Enthalpy, from Chap. 4:

	CO_2	O_2	N_2	H_2O	Total
	1.63 lb	1.77 lb	9.75 lb	0.56 lb	
h at t = 2100°F, Btu/lb	562.8	511.4	552.7	2128.70	8403
h at t = 2200°F, Btu/lb	594.3	538.6	582.0	2189.92	8823

The reaction temperature for generation of 8500 Btu is interpolated from the above data as follows:

$$t = 2100 + (2200 - 2100)\,\frac{8500 - 8403}{8823 - 8403} = 2123°\text{F}$$

Considering the dry flue gas as air:

$$1.63 + 1.77 + 9.75 = 13.15 \text{ lb air}$$

$$0.56 \text{ lb } H_2O$$

	Air	H_2O	Total
h at t = 2100, Btu/lb	538.72	2128.70	8276
h at t = 2200, Btu/lb	567.52	2189.92	8689

The reaction temperature for generation of 8500 Btu is interpolated as follows:

$$t = 2100 + (2200 - 2100)\,\frac{8500 - 8276}{8689 - 8276} = 2154°\text{F}$$

Compare the calculated reaction temperature, using the actual gas components so that temperature is calculated with air as the dry gas component:

$$100\% \times \frac{2154 - 2123}{2123} = 1.46\%$$

This error will vary with the amount of excess air introduced and with the nature of the combusted waste. But an error on the order of 1½ percent is truly insignificant considering the reliability of other data used in incinerator design, particularly heating value. Heating value is never known with such a degree of accuracy. At best, the determination of heating value is rarely repeatable to an accuracy better than 5 percent even with the best of instruments. The variability of waste quality from one lot to another will often exceed 10 or 15 percent in heating value and other parameters.

REFERENCES

6-1. *Combustion Fundamentals for Waste Incineration*, The American Society of Mechanical Engineers, New York, 1974.

6-2. Stewart, E.: "Determine Excess Air from Wet Stack Gas Analysis," *Pollution Engineering*, April 1979.

BIBLIOGRAPHY

Cone, C.: *Energy Management for Industrial Furnaces*, John Wiley & Sons, New York, 1980.

Eshbach, O.: *Handbook of Engineering Fundamentals*, 2d ed., John Wiley & Sons, New York, 1961.

Fundamentals of Gas Combustion, American Gas Association, New York, 1973.

Guidebook for Industrial and Commercial Gas Fired Incineration, American Gas Association, New York, 1963.

CHAPTER 7
AIR EMISSIONS CALCULATIONS

As more attention is focused upon incineration as a method of waste disposal, air pollution control is becoming increasingly rigorous. The identification and control of emissions from incinerators are often the subject of intense public scrutiny. In this chapter methods of predicting the generation of air emissions will be presented.

THE AIR

Besides oxygen, nitrogen, and water vapor, clean air at sea level will typically contain the components listed in Table 7.1. These compounds are the result of natural processes occurring on the earth, from plant photosynthesis to forest fires, lightning, and the eruption of volcanoes.

The concentration of human life in centralized locations around the globe necessarily results in discharges into the earth, water, and air environments. Of interest in the evaluation of incineration and other burning processes are discharges into the air environment. Such emissions include inorganic gas and organic gas discharges and the release of particulate matter. Some of these discharges appear harmless; however, many of them have been found to be injurious to life: plant, animal, and human. The map in Fig. 7.1 is an indication of the visual effects of air pollution within the United States.

INORGANIC GAS DISCHARGES

The burning process can produce carbon dioxide, carbon monoxide, oxides of nitrogen and, where sulfur is present, oxides of sulfur. Carbon dioxide is not considered a pollutant; however, there is concern that excessive quantities of this gas within the atmosphere might produce a "greenhouse effect." This is the mechanism whereby carbon dioxide molecules retain heat energy as does glass in a greenhouse, preventing the normal radiation of heat from the earth. An indication of this effect can be that the first half of this century saw a rise in the mean temperature of the earth of approximately 1°F. At the same time the level of carbon dioxide in the atmosphere rose over 10 percent.

Carbon monoxide is a danger to human health. It has the ability to pass

TABLE 7.1 Minor Components of Clean, Dry Air at Sea Level

	% by Volume	ppm
Argon	0.93	9300
Carbon dioxide	0.0318	318
Neon	0.0018	18
Helium	0.00052	512
Krypton	0.0001	1
Xenon	0.000008	0.08
Nitrous oxide	0.000025	0.25
Hydrogen	0.00005	0.5
Methane	0.00015	1.5
Nitrogen dioxide	0.0000001	0.001
Ozone	0.000002	2.02
Sulfur dioxide	0.00000002	0.0002
Ammonia	0.000001	0.01
Carbon monoxide	0.00001	0.1

Source: Ref. 7-1.

through the lungs, directly into the bloodstream of an organism, where it reduces the ability of red blood cells to carry oxygen. At an exposure of only 0.10 percent carbon monoxide in air by volume (1000 ppm), a human being will be comatose in less than 2 h. The federal government has established maximum exposure standards for carbon monoxide of 9.0 ppm for an 8-h average and 13.0 ppm for any 1 h.

Nitrogen is an extremely active substance forming a wide range of compounds with oxygen, as listed in Table 7.2. As seen in this listing, nitrogen dioxide is most significant as a pollutant of all the nitrogen oxides. Federal standards for NO_x, which normally include NO and NO_2, expressed as NO_2, are 0.05 ppm maximum on a yearly average and 0.13 ppm maximum for any 24-h average.

Note that where NO and NO_2 are present, the calculation for NO_x is as follows:

> NO_x is expressed as NO_2. The molecular weight of NO_2 is 46.01 whereas the molecular weight of NO is 30.01. If 10 lb of NO_2 and 150 lb of NO are present, the quantity of NO_x expressed as NO_2 is the sum of NO present × ratio of weight of NO_2 to NO, or

$$150 \text{ lb NO} \times \frac{46.01 \text{ lb NO}_2}{30.01 \text{ lb NO}} = 299.97 \text{ lb NO}_2 \text{ equivalent}$$

$$\begin{array}{rl} \text{plus NO}_2 \text{ present} & \underline{10.00 \text{ lb}} \\ \text{Total NO}_x & 309.97 \text{ lb expressed as NO}_2 \end{array}$$

It is important to note that given only the quantity of NO_x present it is not possible to estimate the NO and/or the NO_2 components.

The concept of NO_x is a convenient way of describing the magnitude of nitrogen oxide pollutants existing in a gas stream.

If the NO and NO_2 components of a gas stream are required, and the NO_x

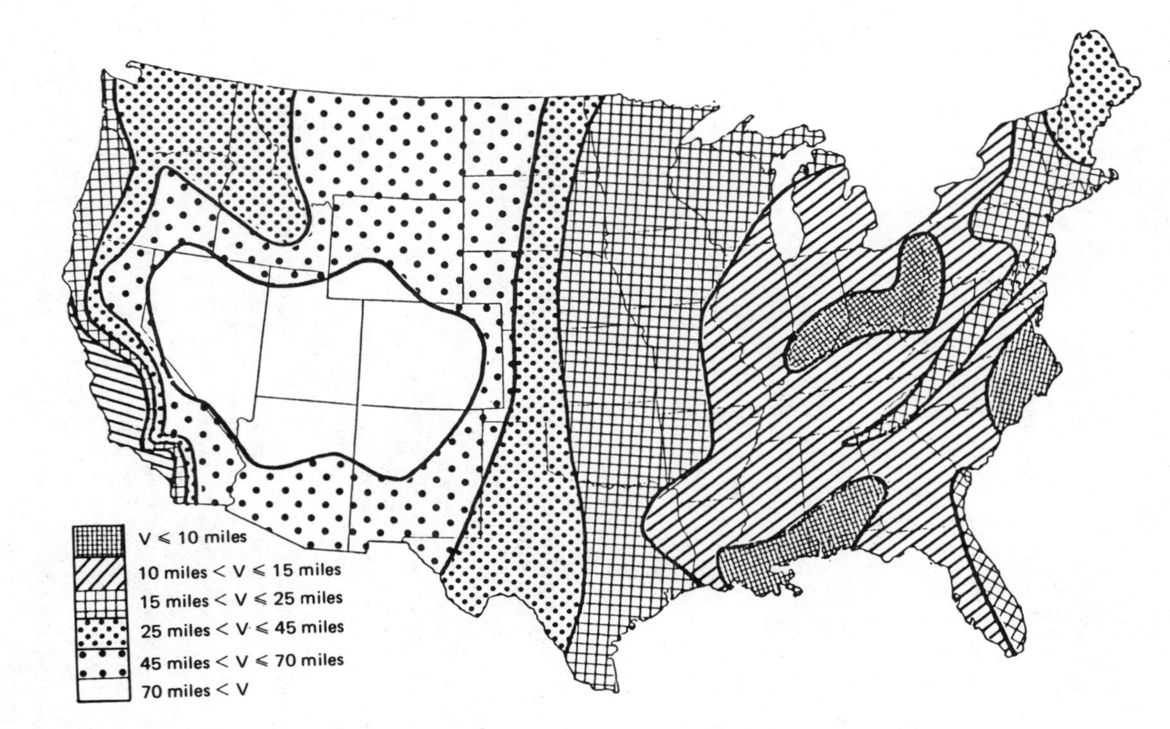

FIGURE 7.1 Visual range in suburban and nonurban airports of the United States, 1974–1976. *Source:* Ref. 7-2.

TABLE 7.2 Oxides of Nitrogen

Formula	Name	Effects
N_2O	nitrous oxide	Inert, not a pollutant (laughing gas)
NO	nitric oxide	Main product of combustion, considered harmless by itself. Converts to NO_2. Some indication that it may disintegrate the ability of red blood cells to carry oxygen.
N_2O_3	dinitrogen trioxide	Unstable, rare, not a significant pollutant
NO_2	nitrogen dioxide	Causes significant effects in the atmosphere, e.g., smog, yellows white fabric, creates plant leaf injury, reduces plant yields.
N_2O_5	dinitrogen pentoxide	Unstable, rare, not a significant pollutant.

quantity is known, by assuming a ratio of NO to NO_2, these components can be estimated. For example:

> A gas stream contains 1000 lb of NO_x, expressed as the NO_2 with a ratio of NO to NO_2 of 100:1. Determine the NO and NO_2 present.
>
> With the NO_2 quantity as X and the NO quantity as $100X$, equivalent to $46.01/30.01 \times 100X = 153.32X$ NO_2, the total NO_2 equivalent is $154.32X$ lb. For a total flow of $1000 \text{ lb} = 154.32X$,

$$X = 6.48 \text{ lb } NO_2$$

$$100X = 648.00 \text{ lb NO}$$

Sulfur is released into the atmosphere from burning processes in the form of sulfur dioxide, SO_2, and sulfur trioxide, SO_3. More than 95 percent of sulfur oxides generated are SO_2, which will slowly oxidize to SO_3. Sulfur trioxide is highly soluble in water and forms sulfuric acid, H_2SO_4. Sulfur dioxide is less soluble and forms sulfurous acid, H_2SO_3, in water.

Sulfur oxides are considered significant pollutants. They will spot and bleach leaves of plants and trees and have been found, in concentrations as low as 0.5 ppm, to cause damage to fruit trees such as apple and pear. Alfalfa, barley, and various species of pine and other conifers are also sensitive to the presence of sulfur oxides.

Rain will wash sulfur oxides from the atmosphere; however, the rain will turn acidic. Acid rain is increasingly of concern because of its detrimental effect on plant life, fabrics, metals, and other structural materials. It is a major cause of deterioration of statuary, building façades, and other structures throughout the industrial (and semi-industrial) world.

In human life, SO_2 is an eye irritant and causes and aggravates respiratory diseases such as emphysema and bronchitis, and studies have found a link between sulfur oxides and the occurrence of lung cancer.

Federal standards have been established for SO_x, measured as SO_2 (see the previous discussion of NO_x to relate SO_2 and SO_3 emissions to SO_x), as 0.14 ppm for a 24-h period and 0.03 ppm on a yearly basis.

Besides specific direct pollutant effects of sulfur oxides, they are contributors to gross atmospheric effects such as smog and haze.

ORGANIC GAS DISCHARGES

The majority of gaseous organic discharges into the atmosphere occur from natural sources. Of the many organic discharges from industrial sources, the more significant ones are as follows:

- Oxygenated hydrocarbons such as aldehydes, ketones, alcohols, and acids. In sufficient quantity they will produce eye irritation, reduce visibility in the atmosphere, and react with other components of the atmosphere to form additional pollutants.
- Halogenated hydrocarbons, such as carbon tetrachloride, perchloroethylene, etc. These will act as the oxygenated compounds, above, and may also generate odor.
- Hydrocarbons, such as paraffins, olefins, and aromatics. These compounds may cause plant damage as well as eye irritation, odor, and a reduction in atmospheric visibility. Such compounds may also be highly reactive, readily combining with other elements of the atmospheric environment to create additional significant danger to life and comfort.

In general, of the above organic gases the incineration process will produce reactive hydrocarbons. These compounds are significant in that they will form complex and harmful components in a series of reactions with nitrogen oxides and sulfur oxides.

PARTICULATE DISCHARGE

The most obvious particulate discharge from a burning process is smoke, which will be discussed in a later section of this chapter.

Of note is that in 1980 there were five major eruptions of Mount St. Helens, the Washington state volcano. Just one of those eruptions discharged over 325,000,000 tons of particulate into the atmosphere. If all the municipal refuse and sewage sludge generated in the United States were incinerated (assuming 5 lb per capita per day for municipal refuse and 0.2 lb per capita per day for sewage solids), at current emissions standards, it would take over 8000 years to equal a single day's eruption of the volcano.

This calculation was made as follows:

$$220{,}000{,}000 \text{ people} \times 5 \text{ lb} \times 365 \text{ days} = 200{,}000{,}000 \text{ tons/year refuse}$$

Assuming 7 lb/ton emissions × 0.05 discharge (95 percent efficiency),

$$200{,}000{,}000 \text{ tons} \times 0.35 \text{ lb/ton} = 35{,}000 \text{ tons/year particulate from refuse}$$

$$220{,}000{,}000 \text{ people} \times 0.2 \text{ lb} \times 365 \text{ days} = 8{,}000{,}000 \text{ tons/year sewage solids}$$

With 1.3 lb/dry ton allowable emissions,

$$8{,}000{,}000 \text{ tons/year} \times 1.3 \text{ lb/ton} = 5200 \text{ tons/year particulate from sludge}$$

$$\text{Total sludge and refuse} = 40{,}200 \text{ tons/year particulate}$$

$$\text{Volcano} = 325{,}000{,}000 \text{ tons}$$

$$\text{Allowable years} = \frac{325{,}000{,}000 \text{ tons}}{40{,}200 \text{ tons/year}} = 8085 \text{ years}$$

Metallic particulate discharges into the atmosphere account for less than 1 percent of the total particulate discharge. However, metals can have severe toxic effects. It has been found that inhalation of metals has a greater effect on the human body than receiving the same metals damage by digestion.

There are four metals which have been shown to have a significant negative effect on human health:

1. *Lead.* This metal accumulates within the human body. It causes symptoms such as anemia, headaches, sterility, miscarriages, or the birth of handicapped children. Such handicaps, known as lead encephalopathy, include convulsions, coma, blindness, mental retardation, and/or death.
2. *Nickel.* Nickel carbonyl, NiCO, is considered to be a form of nickel which is hazardous to human health. It causes changes in the lung structure, which causes respiratory diseases, including lung cancer. It has been found to be present in tobacco smoke.
3. *Cadmium.* This element is associated with the incineration of cadmium-containing products such as automobile tires and certain plastics. In the human body it has been found to interfere with the natural processes of zinc and copper metabolism. Cadmium has also been found to cause cardiovascular disease and hypertension.
4. *Mercury.* Mercury has long been recognized as a severe pollutant in the water environment. Because of its low boiling point it is also released into the atmosphere when it is present within a combustible, or heated, material. Mercurial poisoning in humans is characterized by blindness, progressive weakening of the muscles, numbness, paralysis, coma, and death. It also leads to severe birth deformities.

SMOKE

Smoke is a suspension of solid or liquid particulate matter within a gaseous discharge. The particles range from fractions of a micrometer (micron) to over 50 μm. The visibility of smoke is related to the quantity of particles present, rather than the weight of the particulate matter. The weight of particulate emissions is therefore not necessarily indicative of the density of a smoke discharge. The color of smoke is not necessarily indicative of smoke density either.

White Smoke

The formation of white or other opaque, nonblack smoke is usually due to insufficient furnace temperatures when burning carbonaceous materials. Hydrocarbons will be heated to a level where evaporation and/or cracking will occur within the furnace when white smoke is produced. The temperatures will not be high enough to produce complete combustion of these hydrocarbons. With a stack temperature in the range of 300 to 500°F, many of these hydrocarbons will condense to liquid particulate, and with the solid hydrocarbons these will appear as nonblack smoke. A method of control of white smoke is, therefore, an increase in the furnace or stack temperatures and increased turbulence, to help ensure uniformity of this higher temperature within the off-gas flow. Excessive airflow may provide excessive cooling, and an evaluation of reducing white smoke discharges would include investigating the air quantity introduced into the furnace. Inorganics in the exit gas may also produce a nonblack smoke discharge. For instance, sulfur and sulfur compounds will appear yellow in a discharge; calcium and silicon oxides in the discharge will appear light to dark brown.

Black Smoke

When burned in an oxygen-deficient atmosphere, hydrocarbons will not be completely destroyed, and carbon particles will be found in the off-gas. Related to oxygen deficiency is poor atomization, inadequate turbulence (or mixing), and poor air distribution within a furnace chamber. These factors will all produce carbon particles which produce dark, black smoke in the off-gas. Black smoke is also present where hydrocarbons are heated with insufficient oxygen and undergo a pyrolysis reaction. This generates stable, less complex hydrocarbon compounds that form a dark, minute particulate, generating black smoke.

One common method of reducing, or eliminating, black smoke is steam injection into the furnace. The carbon present is converted to methane and carbon monoxide as follows:

$$3C \text{ (smoke)} + 2H_2O \rightarrow CH_4 + 2CO$$

Similar reactions occur with other hydrocarbons present. The methane and carbon monoxide produced burn clean in the heat of the furnace, eliminating the black carbonaceous smoke that would have been produced without steam injection:

$$CH_4 + 2O_2 \rightarrow CO_2 + 2H_2O \qquad \text{(smokeless)}$$

$$2CO + O_2 \rightarrow 2CO_2 \qquad \text{(smokeless)}$$

Steam injection normally requires from 20 to 80 lb steam/100 lb flue gas.

GENERATION OF CARBON MONOXIDE

Carbon monoxide, CO, is produced where insufficient oxygen is provided to completely combust a fuel. It is, therefore, indicative of burning, or combustion efficiency. The greater the amount of air present and the greater the degree of turbulence, the less carbon monoxide will be formed.

Turbulence as a combustion parameter cannot be easily quantified. The amount of air present and the combustion temperature affect the equilibrium constant, the relationship of CO to CO_2 produced for a given reaction. Table 7.3 is derived from equilibrium analyses and lists the formation of CO as a function of excess air, temperatures, and the ratio of carbon to hydrogen in the fuel.

Note that the generation of CO decreases with an increase in the air supply and increases with an increase in temperature. As expected, with increased carbon in the fuel, there will be an increased rate of CO formed.

TABLE 7.3 Generation of Carbon Monoxide (CO), lb CO/lb Stoichiometric Air

	Excess Air							
Temp. °F	0	10%	20%	30%	50%	100%	150%	200%
For C_1H_0 (carbon)								
1000	–	–	–	–	–	–	–	–
1500	2.060E-07	1.407E-09	1.039E-09	8.827E-10	7.344E-10	5.997E-10	5.474E-10	5.193E-00
1832	9.544E-06	1.480E-07	1.093E-07	9.287E-08	7.725E-08	6.303E-08	5.758E-08	5.474E-00
2192	1.170E-04	6.110E-06	4.506E-06	3.827E-06	3.183E-06	2.597E-06	2.370E-06	2.249E-00
2500	4.830E-04	5.128E-05	3.778E-05	3.207E-05	2.666E-05	2.176E-05	1.985E-05	1.883E-00
3000	3.418E-03	9.796E-04	7.248E-04	6.159E-04	5.118E-04	4.174E-04	3.807E-04	3.610E-00
For C_3H_4 ($C_1H_{1.33}$)								
1000	–	–	–	–	–	–	–	–
1500	2.591E-08	1.080E-09	7.968E-10	6.755E-10	5.606E-10	4.554E-10	4.149E-10	3.928E-00
1832	8.131E-06	1.137E-07	8.375E-08	7.105E-08	5.896E-08	4.791E-08	4.362E-08	4.133E-00
2192	8.858E-05	4.692E-06	3.454E-06	2.930E-06	2.429E-06	1.975E-06	1.796E-06	1.702E-00
2500	3.870E-04	5.468E-05	2.897E-05	2.456E-05	2.036E-05	1.653E-05	1.505E-05	1.425E-00
3000	2.792E-03	7.567E-04	5.574E-04	4.723E-04	3.912E-04	3.175E-04	2.889E-04	2.735E-00
For CH_4								
1000	2.919E-10	–	–	–	–	–	–	–
1500	8.122E-08	7.970E-10	5.418E-10	4.589E-10	3.799E-10	3.075E-10	2.794E-10	2.642E-00
1832	5.572E-06	7.749E-08	5.699E-08	4.828E-08	3.997E-08	3.236E-10	2.940E-08	2.778E-00
2192	6.161E-05	3.199E-06	2.350E-06	1.991E-06	1.647E-06	1.333E-06	1.204E-06	1.144E-00
2500	2.743E-04	2.686E-05	1.972E-05	1.669E-05	1.380E-05	1.116E-05	1.014E-05	9.581E-00
3000	2.022E-03	5.174E-04	3.801E-04	3.213E-04	2.654E-04	2.146E-04	1.946E-04	1.839E-00

Note: 1.5E-03 equals .0015.
Source: Derived from Ref. 7-3.

In burning methane, CH_4, the quantity of carbon monoxide formed can be estimated as follows:

16.05	32.00	44.01	18.02	28.01
CH_4 +	$2O_2$ →	CO_2 +	$2H_2O$ +	CO (trace)
16.05	64.00	44.01	36.04	
1.00	3.99	2.74	2.25	

The equilibrium equation defines the stoichiometric oxygen required for burning methane, 3.99 lb oxygen/lb methane. Assuming that this reaction will take place at 2500°F, with 20 percent excess air supplied, the amount of CO formed per pound of stoichiometric air is 1.972×10^{-5}, from Table 7.3. For 1 lb of methane, therefore:

$$3.99 \text{ lb } O_2 \text{ required}$$

$$3.99 \text{ lb } O_2 \times 4.3197 \text{ lb air/lb } O_2 = 17.24 \text{ lb air}$$

$$17.24 \text{ lb air} \times 1.972 \times 10^{-5} \text{ lb CO/lb air} = 0.00034 \text{ lb CO}$$

Therefore 0.00034 lb CO is produced when 1 lb of methane is burned at 2500°F with 20 percent excess air.

To determine the volume of CO produced in parts per million (ppm), relative to the volume of methane, the volumes of each component must be calculated. Using the perfect gas law (Chap. 4) gives

$$W = \frac{144P}{RT}$$

where

$$W = \frac{m}{V} \qquad \text{lb/ft}^3$$

where M is weight, lb, and V is volume, ft^3. Therefore

$$V = \frac{mRT}{144P}$$

Substituting $R = 1545/M$, where M is the molecular weight, gives

$$V = \frac{1545mT}{144PM}$$

For the case in question, the volume of CO vs. the volume of methane, both components are at the same temperature and pressure. Therefore the volume ratio is simplified to the following:

$$\frac{V_{CO}}{V_{CH_4}} = \frac{m_{CO}}{m_{CH_4}} \times \frac{M_{CH_4}}{M_{CO}}$$

$$= \frac{0.00034 \text{ lb}}{1.0 \text{ lb}} \times \frac{16.05}{28.01} \times 1{,}000{,}000$$

$$= 195 \text{ ppm by volume, CO to CH}_4$$

To relate the CO present to the total flue gas flow, at 20 percent excess air, the flue gas quantity must be determined, noting 20 percent excess air corresponds to 1.2 × 2 = 2.4 molecules of oxygen, which carries with it 2.4 × 3.7619 = 9.03 molecules nitrogen.

16.05	(2 × 16.00)	(2 × 14.01)		44.01	18.02	(2 × 16.00)	(2 × 14.01)
CH_4 +	$2.4O_2$ +	$9.03N_2$	→	CO_2 +	$2H_2O$ +	$0.4O_2$ +	$9.03N_2$
16.05	76.80	253.02		44.01	36.04	12.80	253.02
1.00	4.79	15.76		2.74	2.25	0.80	15.76

The outlet flue gas volume is

$$V_{FG} = \frac{1545}{144} \frac{T}{P} \left(\frac{m_{CO_2}}{M_{CO_2}} + \frac{m_{H_2O}}{M_{H_2O}} + \frac{m_{O_2}}{M_{O_2}} + \frac{m_{N_2}}{M_{N_2}} \right)$$

The ratio of CO to flue gas volume is therefore

$$\frac{V_{CO}}{V_{FG}} = \frac{m_{CO}/M_{CO}}{m_{FG}/M_{FG}}$$

$$= \frac{\dfrac{0.00034 \text{ lb}}{28.01 \text{ lb}}}{\dfrac{2.74 \text{ lb}}{24.01 \text{ lb}} + \dfrac{2.25 \text{ lb}}{18.02 \text{ lb}} + \dfrac{0.80 \text{ lb}}{16.00 \text{ lb}} + \dfrac{15.76 \text{ lb}}{14.01 \text{ lb}}} \times 1{,}000{,}000$$

$$= 8.91 \text{ ppm by volume CO in exiting flue gas}$$

GENERATION OF NITROGEN OXIDES

The amount of nitrogen oxides generated is a function of temperature, excess air, and fuel composition. The greater the amount of excess air, the greater the amount of nitrogen present, and higher quantities of NO_x will be expected. The series of graphs in Fig. 7.2 shows this relationship. The scale lb NO_x per 10^6 Btu relates to fuel type or composition and is directly related to mols of oxygen required for stoichiometric combustion, which is the second of the vertical scales.

Tables 7.4 and 7.5 list generation rates for NO and NO_2, respectively. To illustrate the use of these charts, from the example for the burning of methane, it was found that 17.24 lb air was the stoichiometric demand for burning 1 lb of methane, CH_4. Also methane has the formula CH_4, which is between the values for C_3H_4 and C_0H_2 listed in Tables 7.3 and 7.4. A closer examination of the tables, however, indicates that NO and NO_2 production is not significantly affected by fuel composition at the 20 percent excess air and 2500°F parameters used in

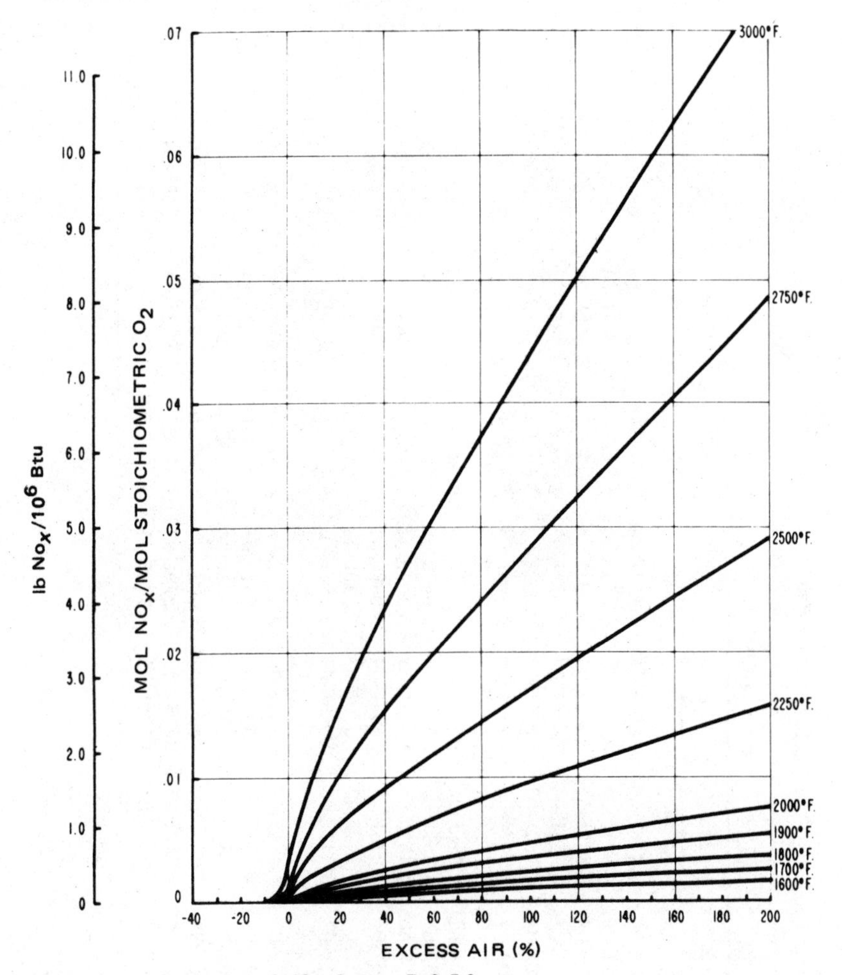

FIGURE 7.2 Generation of NO_x. *Source:* Ref. 7-3.

this example. Therefore, for NO use 1.258×10^{-3}, and for NO_2 use 2.5×10^{-6} lb generated per pound of stoichiometric air.

For NO (atomic weight of 30.01),

$$17.24 \text{ lb air} \times \frac{1.258 \times 10^{-3} \text{ lb NO}}{\text{lb air}} = 0.022 \text{ lb NO}$$

$$\frac{V_{NO}}{V_{CH_4}} = \frac{m_{NO}}{m_{CH_4}} \times \frac{M_{CH_4}}{M_{NO}} \times 1{,}000{,}000 = \frac{0.022 \text{ lb}}{1 \text{ lb}} \times \frac{16.05}{30.01} \times 1{,}000{,}000$$

$$= 11{,}766 \text{ ppm by volume, NO to } CH_4$$

TABLE 7.4 Generation of Nitrogen Oxide (NO), lb NO/lb Stoichiometric Air

	Excess Air							
Temp. °F	0	10%	20%	30%	50%	100%	150%	200%
For C_1H_0 (carbon)								
1000	–	9.587E-07	1.416E-06	1.805E-06	2.503E-06	4.088E-06	5.599E-06	7.082E-06
1500	–	2.931E-05	4.331E-05	5.523E-05	7.657E-05	1.251E-04	1.712E-04	2.166E-04
1832	1.712E-06	1.260E-04	1.862E-04	2.375E-04	3.293E-04	5.380E-04	7.366F-04	9.316E-04
2192	2.022E-05	4.081E-04	6.033E-04	7.696E-04	1.074E-03	1.744E-03	2.388E-03	3.022E-03
2500	8.198E-05	8.510E-04	1.260E-03	1.607E-03	2.232E-03	3.645E-03	4.993E-03	6.318E-03
3000	5.677E-04	2.204E-03	3.242E-03	4.194E-03	5.760E-03	9.420E-03	1.291E-02	1.633E-02
For C_3H_4 ($C_1H_{1.33}$)								
1000	–	9.587E-07	1.416E-06	1.805E-06	2.503E-06	4.088E-06	5.599E-06	7.082E-06
1500	2.575E-07	2.933E-05	4.331E-05	5.523E-05	7.657E-05	1.251E-04	1.712E-04	2.166E-04
1832	1.932E-06	1.260E-04	1.862E-04	2.375E-04	3.293E-04	5.380E-04	7.366E-04	9.316E-04
2192	1.899E-05	4.079E-04	6.033E-04	7.694E-04	1.068E-03	1.744E-03	2.388E-03	3.022E-03
2500	7.870E-05	8.502E-04	1.259E-03	1.670E-03	2.230E-03	3.645E-03	4.993E-03	6.318E-03
3000	5.341E-04	2.191E-03	3.241E-03	4.138E-03	5.751E-03	9.411E-03	1.290E-02	1.632E-02
For C_0H_2 (hydrogen)								
1000	–	9.587E-07	1.416E-06	1.805E-06	3.372E-06	4.088E-06	5.599E-06	7.082E-00
1500	2.835E-07	2.933E-05	4.331E-05	5.523E-05	7.657E-05	1.251E-04	1.712E-04	2.166E-00
1832	1.849E-06	1.260E-04	1.862E-04	2.375E-04	3.293E-04	5.380E-04	7.366E-04	9.316E-00
2192	1.792E-05	4.079E-04	6.031E-04	7.694E-04	1.068E-03	1.744E-03	2.388E-03	3.018E-00
2500	6.949E-05	8.493E-04	1.258E-03	1.606E-03	2.230E-03	3.645E-03	4.993E-03	6.315E-00
3000	4.281E-04	2.169E-03	3.222E-03	4.125E-03	5.738E-03	9.398E-03	1.288E-02	1.632E-00

Note: 1.5E-03 equals .0015.
Source: Derived from Ref. 7-3.

TABLE 7.5 Generation of Nitrogen Dioxide (NO_2), lb NO_2/lb Stoichiometric Air

Temp. °F	Excess Air 0	10%	20%	30%	50%	100%	150%	200%
For C_1H_0 (carbon)								
1000	–	1.208E-07	2.416E-07	3.622E-07	6.039E-07	1.208E-06	1.812E-06	2.416E-00
1500	–	4.045E-07	8.086E-07	1.213E-06	2.022E-06	4.045E-06	6.065E-06	8.089E-00
1832	–	6.528E-07	1.352E-06	2.029E-06	3.386E-06	6.768E-06	1.015E-05	1.354E-00
2192	6.538E-10	1.018E-06	2.008E-06	3.066E-06	5.113E-06	1.024E-05	1.536E-05	2.049E-00
2500	1.357E-08	1.330E-06	2.671E-06	4.015E-06	6.701E-06	1.343E-05	2.015E-05	2.668E-00
3000	1.359E-07	1.863E-06	3.718E-06	5.589E-06	9.341E-06	1.875E-05	2.816E-05	3.758E-00
For C_3H_4 ($C_1H_{1.33}$)								
1000	–	1.180E-07	2.364E-07	3.552E-07	5.936E-07	1.192E-06	1.793E-06	2.395E-00
1500	–	3.948E-07	7.916E-07	1.189E-06	1.988E-06	3.991E-06	6.006E-06	8.020E-00
1832	–	6.598E-07	1.324E-06	1.989E-06	3.326E-06	6.681E-06	1.005E-05	1.342E-00
2192	6.961E-10	9.944E-07	1.998E-06	3.004E-06	5.027E-06	1.010E-05	1.262E-05	2.031E-00
2500	1.219E-08	1.296E-06	2.611E-06	3.932E-06	6.585E-06	1.325E-05	1.994E-05	2.664E-00
3000	1.172E-07	1.798E-06	3.615E-06	5.453E-06	9.151E-06	1.846E-05	2.782E-05	3.718E-00
For C_0H_2 (hydrogen)								
1000	–	1.107E-07	2.228E-07	3.362E-07	5.656E-07	1.149E-06	1.740E-06	2.335E-00
1500	–	3.705E-07	7.460E-07	1.126E-06	1.894E-06	3.848E-06	5.826E-06	7.820E-00
1832	–	6.189E-07	1.247E-06	1.883E-06	3.168E-06	6.438E-06	9.751E-06	1.309E-00
2192	–	9.318E-07	1.882E-06	2.843E-06	4.787E-06	9.734E-06	1.475E-05	1.980E-00
2500	1.471E-08	1.213E-06	2.457E-06	3.718E-06	6.269E-06	1.276E-05	1.934E-05	2.597E-00
3000	7.021E-08	1.654E-06	3.376E-06	5.123E-06	8.679E-06	1.775E-05	2.694E-05	3.655E-00

Note: 1.5E-03 equals .0015.
Source: Derived from Ref. 7-3.

For NO_2 (atomic weight of 46.01):

$$17.24 \text{ lb air} \times \frac{2.5 \times 10^{-6} \text{ lb } NO_2}{1 \text{ lb air}} = 0.0000431 \text{ lb } NO_2$$

$$\frac{V_{NO_4}}{V_{CH_4}} = \frac{m_{NO_2}}{m_{CH_4}} \times \frac{M_{CH_4}}{M_{NO_2}} \times 1{,}000{,}000 = \frac{0.000431 \text{ lb}}{1 \text{ lb}} \times \frac{16.05}{46.01} \times 1{,}000{,}000$$

$$= 15 \text{ ppm by volume, } NO_2 \text{ to } CH_4$$

To obtain the total weight of NO_x expressed as NO_2:

NO: $$0.022 \text{ lb NO} \times \frac{46.01 \text{ lb } NO_2}{30.01 \text{ lb NO}} = 0.034 \text{ lb } NO_2 \text{ equivalent}$$

NO_2: Total NO_x $$\frac{0.0000431 \text{ lb}}{0.034 \text{ lb/lb } CH_4}$$

Similarly, the volumetric presence of NO_x would be calculated as

$$11{,}766 \text{ ppm NO} \times \frac{46.01 \text{ lb } NO_2}{30.01 \text{ lb NO}} = 18{,}034 \text{ ppm NO}$$

NO_2: 15 ppm NO_2

Total: 18,054 ppm NO_x by volume related to CH_4

To relate the NO, NO_2, and NO_x generated to the exiting flue gas flue, use the methods in the previous discussion of CO generation.

SULFUR OXIDE EMISSIONS

It is difficult to predict the quantities of SO_2 and SO_3 generated in incineration. It has been found, however, that many wastes, municipal refuse, and sewage sludge, for instance, have low sulfur content, and the release of sulfur oxides is not significant.

After determining the sulfur component in the waste, assume that all the sulfur exits the stack and that this airborne sulfur is in the form of SO_2.

Table 7.6 lists expected emissions from refuse-burning incinerators which can be used as a guide when no other data are available.

PARTICULATE GENERATION

The generation of particulate matter is a function of the waste burned, furnace design, charging method, combustion air supply, and other furnace parameters. Table 7.6 lists particulate emission which can be expected when burning refuse in a number of different types of incinerators. There are additional tables in other chapters of this book which provide additional estimates of particulate emission.

TABLE 7.6 Emission Factors for Refuse Incinerators without Controls

Incinerator Type	Particulates lb/ton	Sulfur Oxides[a] lb/ton	Carbon Monoxide lb/ton	Hydrocarbons[b] lb/ton	Nitrogen Oxides[c] lb/ton
Industrial/commercial					
Multiple chamber	7	2.5	10	3	3
Single chamber	15	2.5	20	15	2
Conical (wood waste)	7	0.1	130	11	1

[a]Expressed as sulfur dioxide.
[b]Expressed as methane.
[c]Expressed as nitrogen dioxide.
Source: Ref. 7-4.

METALS DISCHARGES

Few data are available on the release of heavy metals into the gas stream from the incineration process. Of the four metals of interest, lead, nickel, cadmium, and mercury, a worst-case situation is that they are released to the stack as their oxide and that no portion of the elements remains in the ash. Table 7.7 lists a typical heavy-metals mass balance for the burning of sewage sludge in a fluid bed incinerator. Note that the ash is removed from the gas stream with the scrubber water.

TABLE 7.7 Heavy-Metal Mass Balance Fluid Bed Incinerator Burning Sewage Sludge

Metal	Ash	Scrubber Water Percent by Weight	Stack
Cadmium	80	20	0
Chromium	95	4	1
Copper	78	21	1
Lead	87	12	1
Mercury	0.4	2	97.6
Nickel	80	20	0
Zinc	79	20	1

Source: Ref. 7-5.

ORGANIC GAS GENERATION

It has not been possible to reliably predict the quantities or even the nature of organic components discharged into the atmosphere. These discharges should be determined by measurement at the incinerator or at the stack discharge of an operating unit.

REFUSE INCINERATOR EMISSIONS

In the absence of other data, the emissions listed in Table 7.8, Emissions from Municipal Waste Incinerators, Mass Burn, can be used.

TABLE 7.8 Emissions from Municipal Waste Incinerators, Mass Burn

Emission	lb/ton[a]	lb/ton[b]	ppm or (gr/DSCF)
Particulate	0.371	0.34	(0.02)[c]
Sulfur dioxide	2.88	2.4	80
Nitrogen oxides	7.24	1.6	75
Carbon monoxide	1.259	1.9	150
Hydrocarbons	0.472	0.12	16
Hydrogen chloride	0.820	3.4	200
Hydrogen fluoride	0.0105	0.06	6.5
Sulfuric acid	0.0095	0.04	(0.0023)[c]
Arsenic	1.28E-04	—	—
Beryllium	4.56E-08	5.1E-08	(3.0E-09)[c]
Cadmium	9.93E-04	—	—
Chromium	3.71E-04	—	—
Lead	0.024	0.012	(6.8E-04)[c]
Mercury	0.00889	0.0064	(2.4E-04)[c]
Nickel	2.52E-04	—	—
PCDD, total	1.46E-06	—	—
PCDF, total	3.42E-06	—	—
Polynuclear aromatics	—	1.0E-05	—
Polychlorinated biphenyls	—	1.3E-04	—
Tetrachlorodibenzo-*p*-dioxin	1.406E-8	1.0E-08	—

[a]Emissions downstream of a baghouse/dry scrubber. From: R. Getter, "Emission Estimates for Modern Resource Recovery Facilities," *Proceedings of the Thirteenth Biennial Conference of the Solid Waste Processing Division of the American Society of Mechanical Engineers*, Philadelphia.

[b]Emissions downstream of an electrostatic precipitator. From: W. O'Connell, G. Stotler, and R. Clark, "Emissions and Emission Control in Modern Municipal Incinerators," *Proceedings of the Tenth Biennial Conference of the Solid Waste Processing Division of the American Society of Mechanical Engineers*, New York.

[c]Corrected to 12 percent CO_2.

REFERENCES

7-1. Pryde, L.: *Environmental Chemistry*, Cummings, Menlo Park, Calif., 1973.

7-2. Trijonis, J., and Shapland, D.: *Existing Visibility Levels in the US*, USEPA 450/5-79-010, 1979.

7-3. *Combustion Fundamentals for Waste Incineration*, The American Society of Mechanical Engineers, 1974.

7-4. *Source Category Survey: Industrial Incinerators*, USEPA 450/3-80-013, May 1980.

7-5. Dewling, R., Manganelli, R., and Baer, G.: "Fate and Behavior of Selected Heavy Metals in Incinerated Sludge," *Journal of the Water Pollution Control Federation*, October 1980.

BIBLIOGRAPHY

Air Pollution Aspects of Sludge Incineration, USEPA 625/4-75-009, June 1975.

Air Pollution Manual, American Industrial Hygiene Association, 1972.

"Air Quality Criteria for Particulate Matter," North Atlantic Treaty Organization Committee on the Environment, PB-240-570, Government Printing Office, Washington, November 1971.

Community Air Quality Guides, American Industrial Hygiene Association, undated.

Compilation of Air Pollutant Emission Factors, USEPA AP-42, April 1973.

Copeland, B.: "A Study of Heavy Metal Emissions from Fluidized Bed Incinerators," Industrial Waste Conference, May 1975.

Dioxin From Combustion Sources, American Society of Mechanical Engineers, 1981.

"Dioxin Issue Resolved," *Waste Age*, November 1981.

Greenberg, R., Zoller, W., and Gordon, J.: "Atmospheric Emissions of Elements from the Parkway Sewage Sludge Incinerator," *Environmental Science and Technology*, January 1981.

Industrial Noise Manual, American Industrial Hygiene Association, undated.

Introduction to Manual Methods for Measuring Air Quality, California State Air Resources Board, July 1974.

Jahnke, J.: "A Research Study of Gaseous Emissions from a Municipal Incinerator," *Journal of the Air Pollution Control Association*, August 1971.

Peterson, M., and Stutzenberger, F.: "Microbiological Evaluation of Incinerator Operations," *Applied Microbiological Journal*, July 1967.

Rinaldi, G.: *An Evaluation of Emission Factors for Waste to Energy Systems*, USEPA, 1979.

Shen, T.: "Air Pollutants from Sewage Sludge Incineration," *Journal of the American Society of Civil Engineers*, February 1979.

Test Methods to Determine the Mercury Emissions from Sludge Incineration Plants, USEPA 60/4-79-058, September 1979.

CHAPTER 8
WASTE CHARACTERISTICS

Although a waste may be heterogeneous, defying rigorous analysis, assumptions about waste composition and quality must be made in order to determine incinerator parameters. In this chapter waste characteristics with respect to combustion will be presented.

WASTE CHARACTERIZATION

In Chapter 1 general characteristics of waste streams were presented. Design of incineration systems and equipment requires identification of many additional parameters of waste generation and composition, as follows:

1. *General parameters.* Compositional weight fractions including component percentages of paper, metals, glass, special wastes, etc.

   Process weight fractions such as percentages of combustibles, noncombustibles, salvageable materials, etc.

2. *Physical parameters.* The total waste, its unit size and unit shape, weight, density, age, odor, void fraction, generation and storage temperatures, compactibility, angle of repose, physical state, etc.

   For a solid waste, its solubility, combustibility, specific heat, heat conduction, volatile and ash fractions, shape, melting point, etc.

   For a liquid waste, temperature-viscosity relationship, turbidity, specific gravity, specific heat, vapor pressure, total, settleable and suspended solids, aqueous fraction, etc.

   For a gas waste, its temperature and pressure quality, viscosity, density, odor, color, etc.

3. *Chemical parameters.* In general, its chemical constituents, heating value, products of combustion, organic and inorganic component fractions, etc.

4. *Caveats.* Is the waste a hazardous waste in accordance with the definition of a hazardous waste as it is generated, stored, transported, or disposed of? Can it be hazardous under any condition of handling and/or operation? Is the waste dangerous in any other way with regard to storage or disposal?

PROXIMATE ANALYSIS

The proximate analysis is a relatively quick and inexpensive laboratory determination of the percentages of moisture, volatile matter, fixed carbon, and ash. The analytic procedure is as follows:

1. Heat a sample for 1 h at 105 to 110°C (221 to 230°F). Report the weight loss fraction as moisture percentage.
2. Raise the temperature of the dried sample, in a covered crucible, to 725°C (1337°F) and hold it at this temperature for 7 min. Report the sample weight loss fraction as volatile matter percentage.
3. Ignite the remaining sample in an open crucible at 950°C (1742°F) and allow it to burn to a constant weight. Report the sample weight loss as percentage of fixed carbon.
4. The sample residual is to be reported as percentage of ash. The sum of moisture, volatiles, fixed carbon, and ash should equal 100 percent.

ULTIMATE ANALYSIS

An ultimate analysis is a standard procedure used for a determination of the quantities of elemental components present in a sample. It is required to determine the products of combustion of a material, its combustion air requirement, and the nature of the off-gas or combustion products.

In this procedure the following element percentages are normally determined:

- Carbon
- Hydrogen
- Sulfur
- Oxygen
- Nitrogen
- Halogens (chlorine, fluorine, etc.)
- Heavy metals (mercury, lead, etc.)
- Other elements that can effect the combustion process

In addition to these components, analyses may be performed under the heading of ultimate analysis for the presence of certain compounds that may be in the waste such as benzene, PCBs (polychlorinated biphenyls), dioxins, etc.

Ultimate analyses are performed by specialty laboratories with specialty equipment developed specifically for elemental analyses such as gas chromatographs, infrared scanners, or mass spectrometers.

HEATING VALUE MEASUREMENT

The most reliable method of determining heating value quantities is by testing. The most common equipment for testing for heating value is the oxygen bomb calorimeter. In this instrument a measured sample, usually 1 g, is ignited in an atmosphere of pure oxygen by an electric wire. The sample heat of combustion heats a water bath surrounding the bomb. The temperature rise of the water is measured, and the heat of combustion is calculated from this temperature increase.

When one is dealing with waste, this method of analysis is often impractical. A sample as small as 1 g will not be an accurate representation of a heterogeneous waste such as refuse or wastewater skimmings. Attempts are being made by the Environmental Protection Agency (EPA) to design and construct a calorimeter for analysis of a 100-g to 1-kg sample. Even when it is operational, this calorimeter may still not be able to handle a representative sample of a material as diffuse as common waste materials encountered.

With heterogeneous waste streams, heating value measurement is made on a case-by-case procedure.

TABULAR VALUES

Heating values and other combustion characteristics of common chemical substances are listed in Table 8.1. Note that the high (or gross) heating value assumes that the moisture of combustion condenses within the combustion system; i.e., the exit temperature of the flue gas is below 212°F at atmospheric pressure. The lower heating value (net) is based on the water present in the vapor phase, i.e., if the exit temperature from the combustion system is greater than 212°F.

Table 8.2 lists heating values of common wastes, their density, and typical ash and moisture fractions.

HEAT OF FORMATION

The *heat of formation* of a compound is that amount of heat absorbed when it is formed from its prime elements. If the formation reaction generates heat, the heat of formation ΔH_f has a negative sign and the reaction is determined as exothermic. If the reaction absorbs heat, it is termed *endothermic* and ΔH_f is positive.

Given values for the heat of formation, the heat of combustion can be calculated. The *heat of combustion* of a substance is equal to that amount of heat released when the substance is completely burned.

Table 8.3 lists the heat of formation of selected organic compounds, Table 8.4 lists that of the solid phase of inorganic oxides, and Table 8.5 lists the heat of formation of miscellaneous materials. The units used are those in which this parameter is normally stated, kilocalorie per mole and gram per mole. The conversion to English units is as follows:

$$\left(\frac{\text{kcal}}{\text{mol}} \times \frac{\text{Btu}}{0.252\text{ kcal}}\right) \div \left(\frac{\text{g}}{\text{mol}} \times \frac{\text{lb}}{454\text{ g}}\right)$$

$$1\text{ kcal/g} = 1802\text{ Btu/lb}$$

TABLE 8.1 Combustion Constants

							Heat of Combustion			
							Btu/ft^3		Btu/lb	
No.	Substance	Formula	Molecular weight	lb/ft^3	ft^3/lb	Sp gr air 1.0000	Gross (high)	Net (low)	Gross (high)	Net (low)
1	Carbon[a]	C	12.01	—	—	—	—	—	14,093	14,093
2	Hydrogen	H_2	2.016	0.0053	187.723	0.0696	325	275	61,100	51,623
3	Oxygen	O_2	32.000	0.0846	11.819	1.1053	—	—	—	—
4	Nitrogen (atm)	N_2	28.016	0.0744	13.443	0.9718	—	—	—	—
5	Carbon monoxide	CO	28.01	0.0740	13.506	0.9672	322	322	4,347	4,347
6	Carbon dioxide	CO_2	44.01	0.1170	8.548	1.5282	—	—	—	—
	Paraffin series									
7	Methane	CH_4	16.041	0.0424	23.565	0.5543	1013	913	23,879	21,520
8	Ethane	C_2H_6	30.067	0.0803	12.455	1.0488	1792	1641	22,320	20,432
9	Propane	C_3H_8	44.092	0.1196	8.365	1.5617	2590	2385	21,661	19,944
10	*n*-Butane	C_4H_{10}	58.118	0.1582	6.321	2.0665	3370	3113	21,308	19,680
11	Isobutane	C_4H_{10}	58.118	0.1582	6.321	2.0665	3363	3105	21,257	19,629
12	*n*-Pentane	C_5H_{12}	72.144	0.1904	5.252	2.4872	4016	3709	21,091	19,517
13	Isopentane	C_5H_{12}	72.144	0.1904	5.252	2.4872	4008	3716	21,052	19,478
14	Neopentane	C_5H_{12}	72.144	0.1904	5.252	2.4872	3993	3693	20,970	19,396
15	*n*-Hexane	C_6H_{14}	86.169	0.2274	4.398	2.9704	4762	4412	20,940	19,403
	Olefin series									
16	Ethylene	C_2H_4	28.051	0.0746	13.412	0.9740	1614	1513	21,644	20,295
17	Propylene	C_2H_6	42.077	0.1110	9.007	1.4504	2336	2186	21,041	19,691
18	*n*-Butene	C_4H_8	56.102	0.1480	6.756	1.9336	3084	2885	20,840	19,496
19	Isobutene	C_4H_8	56.102	0.1480	6.756	1.9336	3068	2869	20,730	19,382
20	*n*-Pentene	C_5H_{10}	70.128	0.1852	5.400	2.4190	3836	3586	20,712	19,363
	Aromatic series									
21	Benzene	C_6H_6	78.107	0.2060	4.852	2.6920	3751	3601	18,210	17,480
22	Toluene	C_7H_8	92.132	0.2431	4.113	3.1760	4484	4284	18,440	17,620
23	Xylene	C_8H_{10}	106.158	0.2803	3.567	3.6618	5230	4980	18,650	17,760
	Miscellaneous gases									
24	Acetylene	C_2H_2	26.036	0.0697	14.344	0.9107	1499	1448	21,500	20,776
25	Naphthalene	$C_{10}H_8$	128.162	0.3384	2.955	4.4208	5854	5654	17,298	16,708
26	Methyl alcohol	CH_3OH	32.041	0.0846	11.820	1.1052	868	768	10,259	9,078
27	Ethyl alcohol	C_2H_5OH	46.067	0.1216	8.221	1.5890	1600	1451	13,161	11,929
28	Ammonia	NH_3	17.031	0.0456	21.914	0.5961	441	365	9,668	8,001
29	Sulfur[a]	S	32.06	—	—	—	—	—	3,983	3,983
30	Hydrogen sulfide	H_2S	34.076	0.0911	10.979	.1898	647	596	7,100	6,545
31	Sulfur dioxide	SO_2	64.06	0.1733	5.770	2.2640	—	—	—	—
32	Water vapor	H_2O	18.016	0.0476	21.017	0.6215	—	—	—	—
33	Air	—	28.9	0.0766	13.063	1.0000	—	—	—	—

[a]Carbon and sulfur are considered as gases for molal calculations only.

Note: This table is reprinted from *Fuel Flue Gases*, 1941 edition, courtesy of American Gas Association. All gas volumes corrected to 60°F and 30 inHg dry.

For 100% total air, mol/mol or ft^3/ft^3 combustible						For 100% total air, lb/lb combustible					
Required for combustion			Flue products			Required for combustion			Flue products		
O_2	N_2	Air	CO_2	H_2O	N_2	O_2	N_2	Air	CO_2	H_2O	N_2
1.0	3.76	4.76	1.0	—	3.76	2.66	8.86	11.53	3.66	—	8.86
0.5	1.88	2.38	—	1.0	1.88	7.94	26.41	34.34	—	8.94	26.41
—	—	—	—	—	—	—	—	—	—	—	—
—	—	—	—	—	—	—	—	—	—	—	—
0.5	1.88	2.38	1.0	—	1.88	0.57	1.90	2.47	1.57	—	1.90
—	—	—	—	—	—	—	—	—	—	—	—
2.0	7.53	9.53	1.0	2.0	7.53	3.99	13.28	17.27	2.74	2.25	13.28
3.5	13.18	16.68	2.0	3.0	13.18	3.73	12.39	16.12	2.93	1.80	12.39
5.0	18.82	23.82	3.0	4.0	18.82	3.63	12.07	15.70	2.99	1.63	12.07
6.5	24.47	30.97	4.0	5.0	24.47	3.58	11.91	15.49	3.03	1.55	11.91
6.5	24.47	30.97	4.0	5.0	24.47	3.58	11.91	15.49	3.03	1.55	11.91
8.0	30.11	38.11	5.0	6.0	30.11	3.55	11.81	15.35	3.05	1.50	11.81
8.0	30.11	38.11	5.0	6.0	30.11	3.55	11.81	15.35	3.05	1.50	11.81
8.0	30.11	38.11	5.0	6.0	30.11	3.55	11.81	15.35	3.05	1.50	11.81
9.5	35.76	45.26	6.0	7.0	35.76	3.53	11.74	15.27	3.06	1.46	11.74
3.0	11.29	14.29	2.0	2.0	11.29	3.42	11.39	14.81	3.14	1.29	11.39
4.5	16.94	21.44	3.0	3.0	16.94	3.42	11.39	14.81	3.14	1.29	11.39
6.0	22.59	28.59	4.0	4.0	22.59	3.42	11.39	14.81	3.14	1.29	11.39
6.0	22.59	28.59	4.0	4.0	22.59	3.42	11.39	14.81	3.14	1.29	11.39
7.5	28.23	35.73	5.0	5.0	28.23	3.42	11.39	14.81	3.14	1.29	11.39
7.5	28.23	35.73	6.0	3.0	28.23	3.07	10.22	13.30	3.38	0.69	10.22
9.0	33.88	42.88	7.0	4.0	33.88	3.13	10.40	13.53	3.34	0.78	10.40
10.5	39.52	50.02	8.0	5.0	39.52	3.17	10.53	13.70	3.32	0.85	10.53
2.5	9.41	11.91	2.0	1.0	9.41	3.07	10.22	13.30	3.38	0.69	10.22
12.0	45.17	57.17	10.0	4.0	45.17	3.00	9.97	12.96	3.43	0.56	9.97
1.5	5.65	7.15	1.0	2.0	5.65	1.50	4.98	6.48	1.37	1.13	4.98
3.0	11.29	14.29	2.0	3.0	11.29	2.08	6.93	9.02	1.92	1.17	6.93
0.75	2.82	3.57	—	1.5	3.32	1.41	4.69	6.10	—	1.59	5.51
			SO_2						SO_2		
1.0	3.76	4.76	1.0	—	3.76	1.00	3.29	4.29	2.00	—	3.29
1.5	5.65	7.15	1.0	1.0	5.65	1.41	4.69	6.10	1.88	0.53	4.69
—	—	—	—	—	—	—	—	—	—	—	—
—	—	—	—	—	—	—	—	—	—	—	—
—	—	—	—	—	—	—	—	—	—	—	—

TABLE 8.2 Characteristics of Selected Materials

Waste	Btu/lb as fired	Weight, lb/ft^3 (loose)	Weight, lb/ft^3	Content by weight in percentage	
				Ash	Moisture
Type 0 waste	8,500	10		5	10
Type 1 waste	6,500	10		10	25
Type 2 waste	4,300	20		7	50
Type 3 waste	2,500	35		5	70
Type 4 waste	1,000	55		5	85
Kerosene	18,900		50	0.5	0
Benzene	18,210		55	0.5	0
Toluene	18,440		52	0.5	0
Hydrogen	61,000		0.0053	0	0
Acetic acid	6,280		65.8	0.5	0
Methyl alcohol	10,250		49.6	0	0
Ethyl alcohol	13,325		49.3	0	0
Turpentine	17,000		53.6	0	0
Naphtha	15,000		41.6	0	0
Newspaper	7,975	7		1.5	6
Brown paper	7,250	7		1.0	6
Magazines	5,250	35		22.5	5
Corrugated paper	7,040	7		5.0	5
Plastic coated paper	7,340	7		2.6	5
Coated milk cartons	11,330	5		1.0	3.5
Citrus rinds	1,700	40		0.75	75
Shoe leather	7,240	20		21.0	7.5
Butyl sole composition	10,900	25		30.0	1
Polyethylene	20,000	40–60	60	0	0
Polyurethane (foamed)	13,000	2	2	0	0
Latex	10,000	45	45	0	0
Rubber waste	9,000–11,000	62–125		20–30	
Carbon	14,093		138	0	0
Wax paraffin	18,621		54–57	0	0
⅓ wax-⅔ paper	11,500	7–10		3	1
Tar or asphalt	17,000	60		1	0
⅓ tar-⅔ paper	11,000	10–20		2	1
Wood sawdust (pine)	9,600	10–12		3	10
Wood sawdust	7,800–8,500	10–12		3	10
Wood bark (fir)	9,500	12–20		3	10
Wood bark	8,000–9,000	12–20		3	10
Corn cobs	8,000	10–15		3	5
Rags (silk or wool)	8,400–8,900	10–15		2	5
Rags (linen or cotton)	7,200	10–15		2	5
Animal fats	17,000	50–60			0
Cotton seed hulls	8,600	25–30		2	10
Coffee grounds	10,000	25–30		2	20
Linoleum scrap	11,000	70–100		20–30	1

The above chart shows the various Btu values of materials commonly encountered in incinerator designs. The values given are approximate and may vary based on their exact characteristics or moisture content. The Btu value is the higher heating value.

Source: Ref. 8-1.

TABLE 8.3 Heat of Formation of Selected Organic Compounds

Formula	Weight, g/mol	Name	ΔH_f, kcal/mol
CBr_3	251.71	Carbon tribromide (g)	+ 42.0
CBr_4	331.61	Carbon tetrabromide (g)	+ 19.0
CBr_4	331.61	Carbon tetrabromide (c)	+ 4.5
CCl_3	118.36	Carbon trichloride (g)	+ 14.0
CCl_4	153.81	Carbon tetrachloride (g)	− 24.6
CCl_4	153.81	Carbon tetrachloride (liq)	− 32.4
CF_2O	66.01	Carbonyl difluoride (g)	− 151.7
CF_3	69.01	Carbon trifluoride (g)	− 114.0
CF_4	88.01	Carbon tetrafluoride (g)	− 223.0
CO_3Fe	115.86	Iron carbonate (c)	− 177.0
$CHBr_3$	252.72	Bromoform (g)	+ 4.0
$CHBr_3$	252.72	Bromoform (liq)	− 6.8
$CHCl_3$	119.37	Chloroform (g)	− 24.7
$CHCl_3$	119.37	Chloroform (liq)	32.1
CHF_3	70.02	Fluoroform (g)	− 164.5
CHI_3	393.72	Iodoform (c)	33.7
CH_2N_4	70.07	Tetrazole (c)	+ 56.7
CH_2N_4O	86.07	5-hydroxytetrazole (c)	+ 1.5
CH_2O	30.03	Formaldehyde (g)	− 26.0
CH_2O_2	46.03	Formic acid (liq)	− 101.5
CH_3Br	94.94	Methyl bromide (g)	− 8.4
CH_3Cl	50.49	Methyl chloride (g)	− 19.3
CH_3Hg	215.63	Methyl mercury (g)	+ 40.0
CH_3I	141.94	Methyl iodide (g)	+ 3.1
CH_3I	141.94	Methyl iodide (liq)	− 3.7
CH_4	16.05	Methane (g)	− 17.9
CH_4N_2O	60.07	Urea (c)	− 79.7
CH_4O	32.05	Methanol (g)	− 48.0
CH_4O	32.05	Methanol (liq)	− 57.0
$CMnO_3$	114.95	Manganese carbonate (c)	− 213.7
CO_3Zn	125.38	Zinc carbonate (c)	− 194.3
C_2F_4	100.02	Tetrafluoroethylene (g)	− 155.5
C_2F_6	138.02	Hexaflouroethane (g)	− 310.0
$C_2H_2Cl_4$	167.84	1,1,2,2-tetrachloroethane (g)	− 35.7
$C_2H_2CoO_4$	148.97	Cobaltous formate (c)	− 208.7
$C_2H_2CuO_4$	153.59	Copper formate (c)	− 186.7
$C_2H_2MnO_4$	144.98	Manganese formate (c)	− 249.7
$C_2H_2NiO_4$	148.75	Nickel formate (c)	− 208.4
$C_2H_2O_2$	58.04	Glyoxal (g)	− 50.7
$C_2H_2O_4$	90.04	Oxalic acid (c)	− 197.7
$C_2H_2O_4Pb$	297.23	Lead formate (c)	− 210.0
$C_2H_2O_4Zn$	155.41	Zinc formate (c)	− 235.8
$C_2H_3AgO_2$	166.92	Silver acetate (c)	− 95.3
C_2H_3Br	106.95	Vinyl bromide (g)	+ 18.7
C_2H_3BrO	122.95	Acetyl bromide (liq)	− 53.4
$C_2H_3Br_3O_2$	298.75	Bromal hydrate (c)	− 112.0

TABLE 8.3 Heat of Formation of Selected Organic Compounds (Continued)

Formula	Weight, g/mol	Name	ΔH_f, kcal/mol
C_2H_3Cl	62.50	Vinyl chloride (g)	+ 8.5
C_2H_3Cl	62.50	Vinyl chloride (liq)	+ 3.5
C_2H_3ClO	78.50	Acetyl chloride (g)	- 58.2
C_2H_3ClO	78.50	Acetyl chloride (liq)	- 65.4
$C_2H_3ClO_2$	94.50	Chloral hydrate (c)	- 137.7
$C_2H_3ClO_2$	94.50	Chloral hydrate (g)	- 107.2
C_2H_4	28.06	Ethylene (g)	+ 12.5
$C_2H_4O_2$	60.06	Acetic acid (liq)	- 115.7
$C_2H_4O_2$	60.06	Methyl formate (g)	- 83.7
$C_2H_4O_2$	60.06	Methyl formate (liq)	- 90.6
$C_2H_4O_3$	76.06	Gylcolic acid (c)	- 158.6
$C_2H_4O_4$	92.06	Glyoxylic acid (c)	- 199.7
C_2H_5Br	108.97	Ethyl bromide (g)	- 15.4
C_2H_5Br	108.97	Ethyl bromide (liq)	- 22.0
C_2H_5Cl	64.52	Ethyl chloride (g)	- 26.8
C_2H_5Cl	64.52	Ethyl chloride (liq)	- 32.6
C_2H_5ClO	80.52	Ethylene chlorohydrin (liq)	- 70.6
C_2H_5I	143.89	Ethyl iodide (g)	- 1.8
C_2H_5I	143.89	Ethyl iodide (liq)	- 9.6
C_2H_5N	43.08	Ethylenimine (liq)	+ 21.9
C_2H_6	30.08	Ethane (g)	- 20.2
C_2H_6O	46.08	Ethanol (g)	- 56.2
C_2H_6O	46.08	Ethanol (liq)	- 66.4
$C_3H_4O_3$	78.07	Ethylene carbonate (c)	- 138.9
C_3H_6	42.09	Propylene (g)	+ 4.9
C_3H_6O	58.09	Propanone (g)	- 51.8
C_3H_6O	58.09	Propanone (liq)	- 59.2
$C_3H_6O_2$	74.09	Methane acetate (liq)	- 106.4
C_3H_8	44.11	Propane (g)	- 24.8
$C_3H_9N_5O_4$	179.17	Acetamideguanidine nitrate (c)	- 119.1
C_3H_9P	76.09	Trimethylphomphine (liq)	- 30.0
$C_3H_{10}N_2$	74.15	1,2-propanediamine (liq)	- 23.4
$C_3H_{10}N_2O_3$	122.15	Trimethylamine nitrate (c)	- 74.1
$C_3H_{10}O_3Si$	122.22	Trimethexymilane (liq)	- 199.0
$C_3H_{10}Sn$	164.82	Trimethyl tin (g)	+ 5.0
$C_3H_{10}Sn$	164.82	Trimethyl tin (liq)	- 2.1
$C_3H_{12}N_6O_3$	180.21	Guanidine carbonate (c)	- 232.1
$C_4H_2N_2S$	110.14	4-cyanothiazole (c)	+ 52.6
$C_4H_4N_2$	80.10	Pyraxine (c)	+ 33.4
$C_4H_4N_2$	80.10	Pyraxine (g)	+ 46.9
C_4H_4NS	99.16	4-Methylthiazole (liq)	+ 16.3
C_4H_6	54.1	Butadiene (g)	+ 38.8
C_4H_6	54.0	1-butyne (g)	+ 39.5
$C_4H_6O_4Pb$	325.29	Lead acetate (c)	- 230.4
$C_4H_6O_4Zn$	183.47	Zinc acetate (c)	- 257.8
$C_4H_6O_6$	150.10	Tartaric acid (c)	- 308.5

TABLE 8.3 Heat of Formation of Selected Organic Compounds (Continued)

Formula	Weight, g/mol	Name	ΔH_f, kcal/mol
C_4H_8	56.12	Cyclobutane (liq)	+ 0.8
C_4H_8	56.12	Butene (g)	− 2.7
C_4H_8O	72.12	Butanone (g)	− 59.3
C_4H_8O	72.12	Butanone (liq)	− 65.3
C_4H_9N	71.14	Pyrrolidine (g)	− 0.9
C_4H_9N	71.14	Pyrrolidine (liq)	− 9.9
C_4H_{10}	58.14	Butane (g)	− 30.1
$C_4H_{10}O$	74.14	Butanol (liq)	− 78.2
$C_4H_{10}O$	74.14	Diethyl ether (g)	− 60.3
$C_4H_{11}N$	73.16	Butylamine (liq)	− 30.5
$C_5H_4N_4O$	136.13	Hypoxanthine (c)	− 26.2
$C_5H_4N_4O_2$	152.13	Xanthine (c)	− 90.5
$C_5H_4N_4O_3$	168.13	Uric acid (c)	− 147.7
$C_5H_4O_2$	96.09	Furfural (liq)	− 47.8
C_5H_5N	79.11	Pyridine (g)	+ 33.6
C_5H_5N	79.11	Pyridine (liq)	+ 24.0
$C_5H_6N_4O_4$	186.15	Pseudouric acid (c)	− 221.7
$C_5H_6O_2$	98.11	Furfuryl alcohol (liq)	− 66.0
C_5H_8	68.13	Pentadine (g)	+ 33.6
$C_5H_9NO_2$	115.15	DL-proline (c)	− 125.7
$C_5H_9NO_3$	131.15	L-hydroxyproline (c)	− 158.1
C_5H_{10}	70.15	Cyclopentane (liq)	− 25.3
C_5H_{10}	70.15	Pentene (g)	− 5.0
$C_5H_{10}O$	86.15	2-pentanone (liq)	− 71.1
C_5H_{12}	72.17	Pentane (g)	− 35.0
C_5H_{12}	88.17	Pentane (liq)	− 41.4
$C_5H_{12}O$	88.17	Pentanol (liq)	− 64.3
$C_5H_{12}O_5$	152.17	Xylitol (c)	− 267.3
C_6H_6	78.12	Benzene (g)	+ 19.6
C_6H_6	78.12	Benzene (liq)	+ 11.7
C_6H_6O	94.12	Phenol (c)	− 39.5
C_6H_7N	93.14	Aniline (liq)	+ 7.6
C_6H_7N	93.14	Methylpyridine (g)	+ 24.1
C_6H_7N	93.14	Methylpyridine (liq)	+ 13.8
$C_6H_8O_2$	112.14	Sorbic acid (c)	− 93.4
$C_6H_8O_6$	176.14	Ascorbic acid (c)	− 278.3
$C_6H_8O_7$	192.14	Citric acid, anhydrous (c)	− 269.0
C_6H_{10}	82.16	Hexyne (g)	+ 29.6
$C_6H_{10}O_2$	114.16	Hydrosorbic acid (liq)	− 110.2
$C_6H_{10}O_5$	162.16	Saccharinic acid lactone (c)	− 249.6
$C_6H_{10}O_8$	210.16	Citric acid monohydrate (c)	− 439.4
$C_6H_{10}O_8$	210.16	Mucic acid (c)	− 423.0
C_6H_{12}	84.18	Cyclohexane (liq)	− 37.3
C_6H_{12}	84.18	Hexene (g)	− 9.9
$C_6H_{12}O_2$	116.18	Caproic acid (liq)	− 139.7
$C_6H_{12}O_3$	132.18	Acetone glycerol (liq)	− 163.0

TABLE 8.3 Heat of Formation of Selected Organic Compounds (Continued)

Formula	Weight, g/mol	Name	ΔH_f, kcal/mol
$C_6H_{12}O_5$	164.18	Fucose (c)	− 262.7
$C_6H_{12}O_6$	180.18	Fructose (c)	− 302.2
$C_6H_{12}O_6$	180.18	Glucose (c)	− 203.8
$C_6H_{12}O_7$	196.18	Gluconic acid (c)	− 379.3
C_6H_{14}	86.20	Dimethylbutane (liq)	− 51.0
C_6H_{14}	86.20	Hexane (liq)	− 47.5
$C_6H_{14}O$	102.20	Hexanol (liq)	− 90.7
$C_6H_{14}O_6$	182.20	Dulcitol (c)	− 321.8
$C_6H_{14}O_7$	198.20	Glucose hydrate (c)	− 375.0
$C_6H_{15}N$	101.22	Triethylamine (liq)	− 32.1
C_7H_6O	106.13	Benzaldehyde (liq)	− 20.1
$C_7H_6O_2$	122.13	Benzoic acid (c)	− 92.0
$C_7H_6O_3$	138.13	Salicylic acid (c)	− 140.9
C_7H_7N	105.15	Vinylpyridine (liq)	+ 37.2
C_7H_7NO	121.15	Benzamide (c)	− 48.4
C_7H_8	92.15	Toluene (liq)	+ 2.9
C_7H_8O	108.15	Cresol (c)	− 48.9
C_7H_8S	124.21	Benzyl mercaptan (liq)	+ 10.5
C_7H_9N	107.17	Ethylpyridine (liq)	− 1.2
C_7H_9N	107.17	Methylaniline (liq)	+ 7.7
C_7H_{12}	96.19	Heptyne (g)	+ 24.6
C_7H_{14}	98.21	Cycloheptane (liq)	− 37.5
C_7H_{14}	98.21	Heptene (g)	− 14.9
C_7H_{14}	98.21	Methylcyclohexane (liq)	− 45.4
C_7H_{16}	100.23	Dimethylpentane (liq)	− 57.0
C_7H_{16}	100.23	Heptane (liq)	− 53.6
$C_8H_6O_4$	166.14	Phthalic acid (c)	− 191.9
C_8H_7N	117.16	Indole (c)	+ 29.8
C_8H_7NO	133.16	Oxindole (c)	− 41.2
$C_8H_7NO_2$	149.16	Dioxindole (c)	− 76.9
C_8H_8	104.16	Ethenylbenzene (liq)	+ 24.8
$C_8H_8N_2O_2$	164.18	Phthalamide (c)	− 104.4
$C_8H_8O_2$	136.16	Methyl benzoate (liq)	− 79.8
C_8H_9NO	135.18	Acetanilide (c)	− 50.3
C_8H_{10}	106.18	Ethylbenzene (liq)	− 3.0
$C_8H_{10}N_4O_2$	194.22	Caffeine (c)	− 76.2
$C_8H_{10}O$	122.18	Ethylphenol (c)	− 49.9
$C_8H_{11}N$	121.20	Dimethylaniline (liq)	+ 8.2
C_8H_{14}	110.22	Octyne (g)	+ 19.7
C_8H_{16}	112.24	Cyclooctane (liq)	− 40.6
C_8H_{16}	112.24	Dimethylcyclohexane (liq)	− 53.3
C_8H_{16}	112.24	Dimethylcyclohexane (liq)	− 51.5
C_8H_{16}	112.24	Dimethylcyclohexane (liq)	− 51.5
C_8H_{16}	112.24	Dimethylcyclohexane (liq)	− 53.1
C_8H_{16}	112.24	Ethylcyclohexane (liq)	− 50.7
C_8H_{16}	112.24	Propylcyclopentane (liq)	− 45.2

TABLE 8.3 Heat of Formation of Selected Organic Compounds (Continued)

Formula	Weight, g/mol	Name	ΔH_f, kcal/mol
$C_8H_{16}N_2O_3$	188.26	DL-leucylglycine (c)	− 205.1
$C_8H_{16}O_2$	144.24	Caprylic acid (liq)	− 151.9
$C_8H_{17}N$	127.26	Couline (liq)	− 57.6
C_8H_{18}	114.26	Dimethylhexane (liq)	− 62.6
C_8H_{18}	114.26	Ethylhexane (liq)	− 59.9
C_8H_{18}	114.26	Octane (liq)	− 59.7
$C_8H_{20}Pb$	323.47	Tetraethyl lead (g)	− 26.2
$C_8H_{20}Pb$	323.47	Tetraethyl lead (liq)	+ 12.6
C_9H_9N	131.19	Skatole (c)	+ 16.3
$C_9H_{10}N_2$	146.21	Dipyrrylmethane (c)	+ 31.4
C_9H_{12}	120.21	Isopropylbenzene (liq)	− 9.8
$C_9H_{14}O_6$	218.23	Glyceryl triacetate (liq)	− 318.3
$C_9H_{14}O_6$	218.23	Mannitol triformal (c)	− 242.0
$C_9H_{15}N$	137.25	Phyllopyrrole (c)	− 20.4
C_9H_{16}	124.25	Nonyne (g)	+ 14.8
C_9H_{18}	126.27	Nonene (g)	− 24.7
C_9H_{18}	126.27	Propylcyclohexane (liq)	− 57.0
$C_9H_{18}O_2$	158.27	Methyl caprylate (liq)	− 141.1
C_9H_{20}	128.29	Nonane (liq)	− 65.8
$C_9H_{21}N$	143.31	Tri-*n*-propylamine (liq)	− 49.5
$C_{10}H_8$	128.18	Naphthalene (c)	+ 18.0
$C_{10}H_8$	128.18	Naphthalene (g)	+ 35.6
$C_{10}H_9N$	143.20	Phenylpyrrole (c)	+ 34.5
$C_{10}H_9N$	143.20	Quinaldene (c)	+ 39.3
$C_{10}H_{10}O_4$	194.20	Dimethyl phthalate (c)	− 171.0
$C_{10}H_{11}NO_3$	193.22	Benzoyl sarcosine (c)	− 135.7
$C_{10}H_{11}NO_4$	209.22	Animoyl glycine (c)	− 180.9
$C_{10}H_{12}O_4$	196.22	Glyceryl benzoate (c)	− 185.8
$C_{10}H_{13}NO_2$	179.24	Phenacetin (c)	− 101.1
$C_{10}H_{14}N_2$	162.26	Nicotine (liq)	+ 9.4
$C_{10}H_{16}O_2$	168.26	Dehydrocampholenolactone (c)	− 130.0
$C_{10}H_{16}O_3$	184.26	Methyl ethyl heptane lactone (c)	− 183.5
$C_{10}H_{18}$	136.26	Decyne (g)	+ 9.9
$C_{10}H_{20}$	140.30	Decene (g)	− 29.6
$C_{10}H_{20}O_2$	172.30	Capric acid (c)	− 170.6
$C_{10}H_{20}O_2$	172.30	Capric acid (liq)	− 163.6
$C_{10}H_{20}O_2$	172.30	Methyl pelargonate (liq)	− 147.3
$C_{10}H_{22}$	142.32	Decane (liq)	− 71.9
$C_{10}H_{22}O_7$	254.32	Dipentaerythritot (c)	− 376.9
$C_{11}H_{12}N_2O_2$	204.25	Tryptophane (c)	− 99.8
$C_{11}H_{14}N_2O_3$	222.27	Glycylphenylalanine (c)	− 163.9
$C_{11}H_{20}$	152.31	1-undecyne (g)	+ 5.0
$C_{11}H_{20}N_2O_2$	212.33	Valylleucyl anhydride (c)	− 150.1
$C_{11}H_{22}$	154.33	1-undecene (g)	− 34.6
$C_{11}H_{22}O_2$	186.33	Methyl caprate (liq)	− 153.1
$C_{11}H_{24}$	156.35	Undecane (liq)	− 78.0

TABLE 8.3 Heat of Formation of Selected Organic Compounds (Continued)

Formula	Weight, g/mol	Name	ΔH_f, kcal/mol
$C_{12}H_8N_2$	180.22	Phenazine (c)	+ 56.4
$C_{12}H_9N$	167.22	Carbazole (c)	+ 30.3
$C_{12}H_{10}S_2$	218.34	Diphenyl disulfide (c)	+ 35.8
$C_{12}H_{14}N_2O_2$	218.28	Alanylphenylalanyl anhydride (c)	- 89.3
$C_{12}H_{14}H_4O_6$	310.30	Desoxyamalic acid (c)	- 285.7
$C_{12}H_{14}N_4O_8$	342.30	Amalic acid (c)	- 367.0
$C_{12}H_{15}NO_4$	237.28	DL-phenylalanine-*N*-carboxylic acid dimethyl ester (c)	- 184.3
$C_{12}H_{16}N_2O_3$	236.30	Alanylphenylalanine (c)	- 170.2
$C_{12}H_{16}O_8$	288.28	Levoglucosan triacetate (c)	- 371.3
$C_{12}H_{22}$	166.34	Dodecyne (g)	+ 0.1
$C_{12}H_{22}N_2O_2$	226.36	Leucine anhydride (c)	- 160.0
$C_{12}H_{22}O_6$	262.34	Diacetonemannitol (c)	- 350.0
$C_{12}H_{22}O_{11}$	342.34	Cellobiose (c)	- 532.5
$C_{12}H_{22}O_{11}$	342.34	Lactose (c)	- 530.1
$C_{12}H_{22}O_{11}$	342.34	Maltose (c)	- 530.8
$C_{12}H_{22}O_{11}$	342.34	Sucrose (c)	- 531.9
$C_{12}H_{24}$	168.36	Dodecene (g)	- 39.5
$C_{12}H_{24}O_2$	200.36	Lauric acid (c)	- 185.1
$C_{12}H_{24}O_2$	200.36	Lauric acid (liq)	- 176.4
$C_{12}H_{24}O_2$	200.36	Methyl undecylate (liq)	- 159.0
$C_{12}H_{24}O_{12}$	360.36	Lactose monohydrate (c)	- 602.0
$C_{12}H_{26}$	170.38	Dodecane (liq)	- 84.1
$C_{12}H_{26}O_{13}$	378.38	Trehalose dihydrate (c)	- 676.1
$C_{13}H_9N$	179.23	Acridine (c)	+ 44.8
$C_{13}H_{10}O$	182.23	Benzophenone (c)	- 8.0
$C_{13}H_{11}NO$	197.25	Benzanilide (c)	- 22.3
$C_{13}H_{24}$	180.37	Tridecyne (g)	- 4.9
$C_{13}H_{26}$	182.39	Tridecene (g)	- 44.4
$C_{13}H_{26}O_2$	214.39	Methyl laurate (liq)	- 165.6
$C_{13}H_{28}$	184.41	Tridecane (liq)	- 90.2
$C_{14}H_{10}$	178.24	Anthracene (c)	+ 29.0
$C_{14}H_{10}$	178.24	Anthracene (g)	+ 53.7
$C_{14}H_{10}$	178.24	Phenanthrene (c)	+ 27.3
$C_{14}H_{10}$	178.24	Phenanthrene (g)	+ 48.4
$C_{14}H_{18}N_2O_2$	246.34	Valylphenylalanyl anhydride (c)	- 94.3
$C_{14}H_{19}N_3O_4$	293.36	Glycylalanyl phenylalanine (c)	- 222.0
$C_{14}H_{10}N_2O_3$	254.26	Valylphenylalanine (c)	- 183.5
$C_{14}H_{20}O_9$	332.34	Rhamnose triacetate (c)	- 455.4
$C_{14}H_{23}N_3O_{10}$	393.40	Diethylenetriaminepentaacetic acid (c)	- 531.8
$C_{14}H_{26}$	194.40	Tetradecyne (g)	- 9.8
$C_{14}H_{28}$	196.42	Tetradecene (g)	- 49.3
$C_{14}H_{28}O_2$	228.42	Myristic acid (c)	- 199.2
$C_{14}H_{28}O_2$	228.42	Myristic acid (liq)	- 188.5
$C_{14}H_{30}$	198.44	Tetradecane (liq)	- 96.3
$C_{15}H_{20}O_6$	296.35	Tricyclobutyrin (liq)	- 247.0

TABLE 8.3 Heat of Formation of Selected Organic Compounds (Continued)

Formula	Weight, g/mol	Name	ΔH_f, kcal/mol
$C_{15}H_{28}$	208.43	Pentadecyne (g)	− 14.7
$C_{15}H_{30}$	210.45	Pentadecene (g)	− 54.3
$C_{15}H_{30}O_2$	242.45	Methyl myristate (liq)	− 177.8
$C_{15}H_{32}$	212.47	Pentadecane (liq)	− 102.4
$C_{16}H_{10}$	202.26	Fluoranthene (c)	+ 45.8
$C_{16}H_{10}$	202.26	Pyrene (c)	+ 27.4
$C_{16}H_{10}N_2O_2$	262.28	Indigotin (c)	− 32.0
$C_{16}H_{12}N_2O_4$	296.30	Insatide (c)	− 139.0
$C_{16}H_{15}NO_3$	269.32	Benzoylphenylalanine (c)	− 129.6
$C_{16}H_{22}O_{11}$	390.38	Galactose pentaacetate (c)	− 532.8
$C_{16}H_{22}O_{11}$	390.38	Glucose pentaacetate (c)	− 532.1
$C_{16}H_{30}$	222.46	Hexadecyne (g)	− 19.6
$C_{16}H_{32}$	224.48	Hexadecene (g)	− 59.1
$C_{16}H_{32}O_2$	256.48	Palmitic acid (c)	− 213.1
$C_{16}H_{32}O_2$	256.48	Palmitic acid (liq)	− 200.4
$C_{16}H_{34}$	226.50	Hexadecene (liq)	− 108.5
$C_{17}H_{21}NO_4$	303.39	Morphine monohydrate (c)	− 170.1
$C_{17}H_{32}$	236.49	Heptadecyne (g)	− 24.6
$C_{17}H_{34}$	238.51	Heptadecene (g)	− 64.1
$C_{17}H_{36}$	240.53	Heptadecane (liq)	− 114.6
$C_{18}H_{12}$	228.30	Naphthacene (c)	+ 38.3
$C_{18}H_{12}$	228.30	Triphenylene (c)	+ 33.7
$C_{18}H_{18}N_2O_2$	294.38	Phenylalanyl anhydride (c)	− 69.3
$C_{18}H_{23}NO_4$	317.42	Codeine monohydrate (c)	− 151.2
$C_{18}H_{26}O_6$	338.44	Tricyclovalerin (liq)	− 270.0
$C_{18}H_{32}O_{16}$	504.50	Melezitose (c)	− 815.0
$C_{18}H_{32}O_{16}$	504.50	Raffinose (c)	− 761.0
$C_{18}H_{34}$	250.52	Octadecyne (g)	− 29.5
$C_{18}H_{34}O_2$	282.52	Oleic acid (c)	− 187.2
$C_{18}H_{34}O_2$	282.52	Oleic acid (liq)	− 178.9
$C_{18}H_{36}$	252.54	Octadecene (g)	− 69.0
$C_{18}H_{36}O_2$	284.54	Stearic acid (c)	− 226.5
$C_{18}H_{36}O_2$	284.54	Stearic acid (liq)	− 212.5
$C_{18}H_{38}$	254.56	*n*-octadeacne (liq)	− 120.7
$C_{18}H_{42}O_{21}$	594.60	Raffinose penthydrate (c)	−1122.0
$C_{19}H_{21}NO_3$	311.41	Thebaine (c)	+ 63.0
$C_{19}H_{22}N_2O$	294.43	Cinchonine (c)	+ 7.4
$C_{19}H_{24}N_2O$	296.45	Cinchonamine (c)	− 10.4
$C_{19}H_{25}N_3O_4$	359.47	Cinchonamine nitrate (c)	− 79.9
$C_{19}H_{32}$	260.51	Androstane (c)	− 75.0
$C_{19}H_{36}$	264.55	Nonadecyne (g)	− 34.4
$C_{19}H_{36}O_2$	296.55	Methyl elaidate (liq)	− 175.8
$C_{19}H_{36}O_2$	296.55	Methyl oleate (liq)	− 174.2
$C_{19}H_{38}$	266.57	1-nonadecene (g)	− 73.9
$C_{19}H_{40}$	268.59	Nonadecane (liq)	− 126.8
$C_{20}H_{12}$	252.32	Perylene (c)	+ 43.7

TABLE 8.3 Heat of Formation of Selected Organic Compounds (Continued)

Formula	Weight, g/mol	Name	ΔH_f, kcal/mol
$C_{20}H_{21}NO_4$	339.42	Papaverine (c)	- 120.2
$C_{20}H_{24}N_2O_2$	324.26	Quinidine (c)	- 38.3
$C_{20}H_{24}N_2O_2$	324.26	Quinine (c)	- 37.1
$C_{20}H_{27}NO_{11}$	457.48	Amygdalin (c)	- 455.0
$C_{20}H_{38}$	278.58	Elcosyne (g)	- 39.4
$C_{20}H_{40}$	280.60	Elcosene (g)	- 78.9
$C_{20}H_{40}O_2$	312.60	Arachidic acid (c)	- 241.8
$C_{20}H_{40}O_2$	312.60	Arachidic acid (liq)	- 224.6
$C_{20}H_{42}$	282.62	Eicosane (liq)	- 132.9
$C_{21}H_{16}N_2$	296.39	Lophine (c)	+ 65.0
$C_{21}H_{18}N_2$	298.41	Amarine (c)	+ 63.0
$C_{21}H_{19}N_2O_5$	379.42	Amarine heminhydrate (c)	+ 29.0
$C_{21}H_{22}N_2O_2$	334.45	Strychnine (c)	- 41.0
$C_{21}H_{42}O_4$	358.63	Glyceryl ntearate (c)	- 315.8
$C_{22}H_{23}NO_7$	413.46	Narcotine (c)	- 210.9
$C_{22}H_{42}O_2$	338.64	Erucic acid (c)	- 207.0
$C_{22}H_{44}O_2$	340.66	Behenic acid (c)	- 235.0
$C_{22}H_{44}O_4$	372.66	Dihydroxybehenic acid (c)	- 337.0
$C_{23}H_{26}N_2O_4$	394.51	Brucine (c)	- 118.6
$C_{23}H_{31}NO_{10}$	481.55	Narceine dihydrate (c)	- 421.2
$C_{24}H_{20}O_6$	404.44	Glyceryl tribenzoate (c)	- 214.0
$C_{24}H_{24}N_2O_3$	388.50	Anisine (c)	- 51.0
$C_{24}H_{40}O_{20}$	648.64	Diamylose (c)	- 850.0
$C_{24}H_{42}O_{21}$	666.66	Stachyose (c)	- 987.0
$C_{32}H_{36}N_4O_2$	508.72	Pyrroporphyrin monomethyl ester (c)	+ 88.6
$C_{32}N_{38}N_4$	478.74	Aetioporphyrin (c)	- 1.8
$C_{34}H_{34}N_4O_4$	562.72	Protoporphyrin (c)	- 120.6
$C_{34}H_{36}N_4O_3$	548.74	Phylloerythrin monomethyl ester (c)	- 83.2
$C_{36}H_{36}N_4O_6$	620.76	Methyl pheophorbide b (c)	- 200.7
$C_{36}H_{38}N_4O_4$	590.78	Protoporphyrin dimethyl ester (c)	- 122.1
$C_{36}H_{38}N_4O_5$	606.78	Methyl pheophorbide a (c)	- 156.1
$C_{36}H_{38}N_4O_5$	606.78	Pheoporphyrin a_5 dimethyl ester (c)	- 164.3
$C_{36}H_{40}N_4O_6$	624.80	Chlorin p_6 trimethyl ester (c)	- 292.0
$C_{36}H_{42}N_4O_4$	594.82	Mesoporphyrin (1x) dimethyl ester (c)	- 196.4
$C_{36}H_{46}N_4$	534.86	Octaethylporphyrin (c)	- 39.9
$C_{36}H_{60}O_{30}$	972.96	Tetamylose (c)	-1360.0
$C_{37}H_{40}N_4O_7$	652.81	Dimethyl pheopurpurin 7 (c)	- 245.8
$C_{39}H_{74}O_6$	639.13	Glyceryl trilaurate (c)	- 489.0
$C_{40}H_{46}N_4O_8$	710.90	Coproporphyrin (I) tetramethyl ester (c)	- 348.3
$C_{45}H_{86}O_6$	723.31	Glyceryl trimyristate (c)	- 520.3
$C_{47}H_{88}O_5$	733.35	Glyceryl dibrassidate (c)	- 472.0
$C_{47}H_{88}O_5$	733.35	Glyceryl dierucate (c)	- 447.0
$C_{48}H_{54}N_4O_{16}$	943.06	Isouroporphyrian octamethyl ester (c)	- 620.1
$C_{48}H_{80}O_{40}$	1297.28	Hexamylose (c)	-1853.0
$C_{69}N_{128}O_6$	1053.97	Glyceryl tribrassidate (c)	- 625.0
$C_{69}H_{128}O_6$	1053.97	Glyceryl trierucate (c)	- 596.0

(g) Gaseous.
(c) Crystalline.
(liq) Liquid.
Source: Compiled from Refs. 8-2, 8-3, and 8-4.

TABLE 8.4 Heat of Formation of Inorganic Oxides, Solid State

Formula	Weight, g/mol	Name	ΔH_f, kcal/mol
Al_2O_3	101.96	Aluminum oxide	−390.0
B_2O_3	69.62	Boric oxide	−299.8
BaO	153.34	Barium oxide	−132.3
BeO	25.01	Beryllium oxide	−145.7
BrO_2	111.90	Bromine dioxide	11.6
CaO	56.08	Calcium oxide	−151.8
CdO	128.40	Cadmium oxide	−61.7
Co_3O_4	240.79	Cobalt oxide	−213.0
CrO_2	84.00	Chromium dioxide	−143.0
CrO_3	100.00	Chromium trioxide	−140.9
Cr_3O_4	220.00	Chromium oxide	−366.0
CuO	79.55	Cupric oxide	−37.6
Cu_2O	143.10	Cuprous oxide	−40.3
FeO	71.85	Ferrous oxide	−65.0
Fe_2O_3	159.70	Ferric oxide	−197.0
HgO	216.59	Mercuric oxide	−21.4
K_2O	94.20	Potassium oxide	−86.8
Li_2O	29.88	Lithium oxide	−143.1
MgO	40.31	Magnesium oxide	−142.9
MnO	70.94	Manganese oxide	−92.1
MnO_2	86.94	Manganese dioxide	−124.3
MoO_2	127.94	Molybdenum dioxide	−140.8
Na_2O	61.98	Sodium oxide	−99.9
NiO	74.71	Nickel monoxide	−57.3
P_4O_6	219.88	Phosphorus trioxide	−392.0
PbO	223.19	Lead monoxide	−52.3
PbO_2	239.19	Lead dioxide	−66.3
Pt_3O_4	649.27	Platinum oxide	−39.0
PuO_2	274.00	Plutonium dioxide	−252.9
RaO	242.00	Radium oxide	−125.0
RhO	118.91	Rhodium monoxide	92.0
RhO_2	134.91	Rhodium dioxide	44.0
SO_3	80.06	Sulfur trioxide	−108.6
SiO_2	60.09	Silicon dioxide	−217.3
SnO	134.69	Tin monoxide	−68.3
SnO_2	150.69	Tin dioxide	−138.8
SrO	103.62	Strontium monoxide	−141.5
SrO_2	119.62	Strontium peroxide	−151.4
TeO_2	159.60	Tellurium dioxide	−77.1
ThO_2	264.04	Thorium dioxide	−293.2
TiO_2	79.90	Titanium dioxide	−224.6
UO_2	270.03	Uranium dioxide	−259.2
UO_3	286.03	Uranium trioxide	−292.0
VO	66.94	Vanadium monoxide	−103.2
V_2O_5	181.88	Vanadium pentoxide	−370.6
WO_2	215.85	Tungsten dioxide	−140.9
ZnO	81.37	Zinc oxide	−83.2
ZrO_2	123.22	Zirconium dioxide	−263.0

Source: Compiled from Refs. 8-2, 8-3, and 8-4.

TABLE 8.5 Heat of Formation of Miscellaneous Materials

Formula	Weight, g/mol	Name	ΔH_f, kcal/mol
CO	28.01	Carbon monoxide, gas	- 26.4
CO_2	44.01	Carbon dioxide, gas	- 94.1
$CaCO_3$	100.09	Calcium carbonate, solid	-289.5
$Ca(OH)_2$	74.10	Calcium hydroxide, solid	-235.6
$Ca(OH)_2$	74.10	Calcium hydroxide, liquid	-239.7
ClO	51.45	Chlorine monoxide, gas	24.3
ClO_2	67.45	Chlorine dioxide, gas	24.5
ClO_3	83.45	Chlorine trioxide, gas	37.0
Cl_2O	86.90	Dichlorine monoxide, gas	19.2
FO	35.00	Fluorine monoxide, gas	- 5.2
F_2O	54.00	Oxygen difluoride, gas	4.3
HCl	36.46	Hydrogen chloride, gas	- 22.1
HCl	36.46	Hydrogen chloride, liquid	- 40.0
H_2O	18.02	Water, liquid	- 68.3
H_2O	18.02	Water, steam	- 57.8
H_2O_2	34.02	Hydrogen peroxide, liquid	- 44.9
H_2O_2	34.02	Hydrogen peroxide, gas	- 32.6
H_2SO_3	82.02	Sulfurous acid, liquid	-146.8
H_2SO_4	98.08	Sulfuric acid, liquid	-193.9
IO	142.90	Iodine monoxide, gas	41.8
MgS	56.37	Magnesium sulfide, solid	- 84.2
$Mg(OH)_2$	58.33	Magnesium hydroxide, solid	-221.9
$MgCO_3$	84.32	Magnesium carbonate, solid	-261.7
$MgSO_4$	120.37	Magnesium sulfate, solid	-304.9
NO	30.01	Nitric oxide, gas	21.6
NO_2	46.01	Nitrogen dioxide, gas	7.9
N_2O	44.02	Nitrous oxide, gas	19.6
Na_2CO_3	105.99	Sodium carbonate, solid	-269.5
Na_2CO_3	105.99	Sodium carbonate, liquid	-275.1
N_2O_5	108.02	Nitrogen pentoxide, gas	2.7
$NaCl$	58.44	Sodium chloride, solid	- 98.2
$NaCl$	58.44	Sodium chloride, aqueous	- 97.3
$NaHCO_3$	84.01	Sodium bicarbonate, solid	-226.0
$NaHCO_3$	84.01	Sodium bicarbonate, liquid	-222.1
NaF	41.99	Sodium flouride, solid	-135.9
NaI	149.89	Sodium iodide, solid	- 69.3
$NaOH$	40.00	Sodium hydroxide, liquid	-112.2
O_3	48.00	Ozone, gas	34.1
SO	48.06	Sulfur monoxide, gas	1.5
SO_2	64.06	Sulfur dioxide, gas	- 70.9
SO_3	80.06	Sulfur trioxide, gas	- 94.6
SbO	137.75	Antimony monoxide, gas	47.7
VO	66.94	Vanadium monoxide, gas	25.0
VO_2	82.94	Vanadium dioxide, gas	- 57.1

Source: Compiled from Refs. 8-2, 8-3, and 8-4.

The calculation for heating value using the heat of combustion can be illustrated by the burning of benzene, C_6H_6, as follows:

$$\begin{array}{llll} +11.7 & & -94.1 & -57.8\ (g) \\ & & & -68.3\ (l) \\ C_6H_6 & + 7.5O_2 & \rightarrow 6CO_2 & + 3H_2O \\ 78.12 & & & \end{array}$$

Here 78.12 and 11.7 are the molecular weight and heat of formation of liquid benzene, from Table 8.3, and −57.8 and −68.3 are the heat of formation of gaseous water (steam) and liquid water, respectively, and −94.1 is the heat of formation of carbon dioxide from Table 8.5.

Assuming all water that is produced is maintained as steam, and bringing the values of the heat of formation of all compounds multiplied by the moles of each compound present to the right-hand side of the equation, we calculate the sum of the heat of formation ΔH_f:

$$\begin{aligned} \Delta H_f\ H_2O(g)&: \quad -57.8 \times 3 = -173.4 \\ \Delta H_f\ CO_2&: \quad -94.1 \times 6 = -564.6 \\ \Delta H_f\ C_6H_6&: \quad (-)11.7 \times 1 = \underline{-11.7} \\ \Sigma\Delta H_f &= -749.7 \text{ kcal/mol} \end{aligned}$$

Relating $\Sigma\Delta H_f$ to the weight of benzene present, we find

$$\frac{1}{78.12 \text{ g/mol}} \times (-749.7 \text{ kcal/mol}) \times 1802 = -17{,}293 \text{ Btu/lb}$$

If the benzene were converted to liquid water instead of steam (the higher heating value of substance), the heat produced would be calculated as follows (note the use of −68.3 as the heat of formation of liquid water in lieu of −57.8, that of steam):

$$\begin{aligned} \Delta H_f\ H_2O(l)&: \quad -68.3 \times 3 = -204.9 \\ \Delta H_f\ CO_2&: \quad -94.1 \times 6 = -564.6 \\ \Delta H_f\ C_6H_6&: \quad (-)11.7 \times 1 = \underline{-11.7} \\ \Sigma\Delta H_f &= -781.2 \text{ kcal/mol} \end{aligned}$$

Relating $\Sigma\Delta H_f$ to the weight of benzene present gives

$$\frac{1}{78.12 \text{ g/mol}} \times (-781.2 \text{ kcal/mol}) \times 1802 = -18{,}020 \text{ Btu/lb}$$

Note that in the above examples the heating value is negative, indicative of an exothermic or heat-releasing reaction. These values, 18,020 and 17,293 Btu/lb for the high and low heating values of benzene, correlate with the published values as found in Table 8.1, 18,210 and 17,480 Btu/lb, respectively.

This method of analysis can be used to determine the loss of heat in a furnace resulting from ash formations. For instance, if calcium is present as a liquid and, due to furnace heating, loses moisture as steam and forms lime, CaO, the heat loss is calculated as follows:

$$\begin{array}{ccc} -239.6 & -151.8 & -57.8 \\ Ca(OH)_2 \rightarrow & CaO & + H_2O \\ 74.28 & & \end{array}$$

$$\Delta H_f\ H_2O(g) = -57.8$$

$$\Delta H_f\ CaO = -151.8$$

$$\Delta H_f\ Ca(OH)_2 = (+)\underline{239.7}$$

$$\Sigma \Delta H_f = +30.1 \text{ kcal/mol}$$

$$\frac{30.1 \text{ kcal/mol}}{74.28 \text{ g/mol}} \times 1802 = 730 \text{ Btu/lb}$$

The heats of formation for calcium hydroxide and steam were found in Table 8.5 and for lime in Table 8.4. The resultant "heat of combustion," 730 Btu/lb, is a positive quantity, indicating an endothermic reaction, one in which heat must be added for the reaction to occur. Therefore when calcium hydroxide is present in a combustion reaction, 730 Btu must be subtracted from the furnace fuel heating value for each pound of calcium hydroxide.

HEATING VALUE APPROXIMATION FOR HYDROCARBONS

As an approximation for the heating value of a hydrocarbon when the heat of formation or other thermodynamic data are not available, the figure of 184,000 Btu released for every required pound-mole of stoichiometric oxygen is used. (The figures above the chemical formula are the elemental atomic weights. The lower figure is the molecular weight.)

For benzene,

$$\begin{array}{l} 12.01 \\ C_6H_6 + 7.5O_2 \rightarrow 6CO_2 + 3H_2O \\ 78.12 \end{array}$$

$$\frac{7.5 \text{ lb} \cdot \text{mol } O_2}{\text{mol } C_6H_6} \times \frac{\text{mol } C_6H_6}{78.12 \text{ lb}} \times \frac{184{,}000 \text{ Btu}}{1 \text{ lb} \cdot \text{mol } O_2} = 17{,}665 \text{ Btu/lb}$$

For another substance, isobutane,

$$\begin{array}{l} 12.01 \\ C_4H_{10} + 6.5O_2 \rightarrow 4CO_2 + 5H_2O \\ 58.14 \end{array}$$

$$\frac{6.5 \text{ lb} \cdot \text{mol } O_2}{\text{mol } C_4H_{10}} \times \frac{\text{mol } C_4H_{10}}{58.14 \text{ lb}} \times \frac{184{,}000 \text{ Btu}}{1 \text{ lb} \cdot \text{mol } O_2} = 20{,}571 \text{ Btu/lb}$$

The actual heating value of these substances, from the chart, Table 8.1, is 17,480 Btu/lb for benzene and 19,629 Btu/lb for isobutane.

This method of heating value approximation is valid only for hydrocarbons. The presence of oxygen, nitrogen, halogens, or other compounds may lead to erroneous values by use of this calculation.

DU LONG'S APPROXIMATION

An empirical method of determining the heating value of coal, the Du Long formula, can be used as a rough approximation for the heating value of other carbonaceous materials. This is only an approximation, and its validity is questionable:

$$Q = 14{,}544\text{C} + 62{,}028(\text{H}_2 - 0.125\text{O}_2) + 4050\text{S}$$

where Q is in Btu per pound and C, H, O, and S are the weight fractions, respectively, of carbon, hydrogen, oxygen, and sulfur present. The weight fractions should add to 100 percent unless an inert material is present, such as nitrogen. With the presence of nitrogen, for instance, the weight fraction (percent) of C, H_2, O_2, and S will total 100 percent less the fraction of nitrogen present.

Table 8.6 lists combustion parameters including heating value and products of combustion for an organic waste, based on Du Long's approximation. These values are shown in graphic form in Fig. 8.1.

TABLE 8.6 Typical Combustion Parameters for Stoichiometric Burning

Btu	lb air/10 kBtu	lb dry gas/10 kBtu	lb H_2O/10 kBtu	C	H_2	S	O_2	N_2
3,998	6.880	8.711	0.671	0.15	0.030	0.01	0.011	0.799
5,004	6.955	8.328	0.625	0.20	0.035	0.01	0.015	0.740
6,003	7.006	8.076	0.596	0.25	0.040	0.01	0.020	0.680
7,001	7.043	7.897	0.574	0.30	0.045	0.01	0.025	0.620
8,000	7.070	7.762	0.559	0.35	0.050	0.01	0.030	0.560
8,998	7.092	7.657	0.546	0.40	0.055	0.01	0.035	0.500
10,005	7.108	7.571	0.536	0.45	0.060	0.01	0.039	0.441
11,003	7.122	7.503	0.528	0.50	0.065	0.01	0.044	0.381
12,002	7.134	7.445	0.521	0.55	0.070	0.01	0.049	0.321
13,000	7.143	7.397	0.516	0.60	0.075	0.01	0.054	0.261
13,999	7.152	7.356	0.511	0.65	0.080	0.01	0.059	0.201
14,997	7.159	7.320	0.506	0.70	0.085	0.01	0.064	0.141
16,004	7.165	7.287	0.503	0.75	0.090	0.01	0.068	0.082
17,002	7.171	7.260	0.499	0.80	0.095	0.01	0.073	0.022

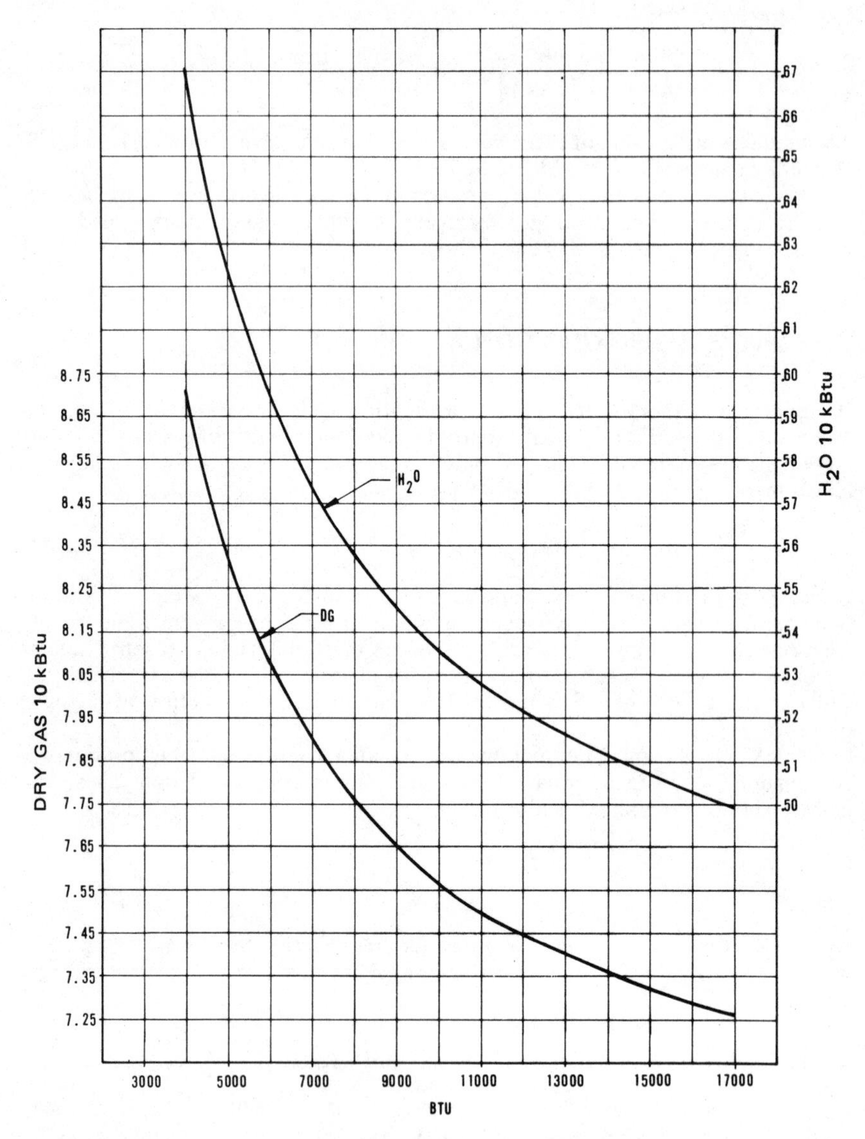

FIGURE 8.1 Combustion parameters.

REFERENCES

8-1. *Incinerator Standards*, Incinerator Institute of America, 1968.

8-2. *Thermodynamic Data for Industrial Incinerators*, National Bureau of Standards, Government Printing Office, Washington, 1972.

8-3. Stull, D., and Prophet, H.: *Thermochemical Tables*, 2d ed., Government Printing Office, Washington, June 1971.

8-4. Perry, R., and Chilton, C.: *Chemical Engineering Handbook*, 5th ed., McGraw-Hill, New York, 1973.

CHAPTER 9
INCINERATOR ANALYSIS

All incinerators utilize air and heat to produce the conditions which will destroy the combustible portion of the waste. Incinerator calculations, therefore, are basically the same for any incineration system or condition.

COMBUSTION PROPERTIES

The efficiency of combustion is related to the combination of three factors, the "three T's": temperature in the furnace, time of residence of the combustion products at the furnace temperature, and turbulence within the furnace.

In general a solid or liquid must be converted to a gaseous phase before burning will occur. (Examine a lit match or a burning log. The flame does not rise directly from the solid. There is a zone immediately above the match, or log, where the gaseous fuel phase has been generated and is mixing with combustion air prior to burning.) The three T's are factors which control the rapidity of conversion of solid and liquid fuel to the gaseous phase.

Flame properties include the following:

Flame

Flame is the envelope or zone where combustion reactions occur and visible radiation is produced.

Flame Front

This is the plane along which combustion starts. It is the dividing line between the fuel-air mixture and the products of combustion.

Flame Speed

This is also called flame velocity, ignition velocity, or rate of propagation.

1. If the flame is stable, the flame front appears stationary. The flame moves toward the burner at the same speed that the fuel-air mixture is injected into the burner.

2. If the fuel-air mixture is too fast, the flame may blow out.
3. If the fuel-air mixture is too slow, a flashback to the burner will occur (consideration of safety factors is necessary).

Flame Stability

This must be maintained after ignition, or combustion may be extinguished. Severe furnace pulsations or explosions could result from instability of the location of the flame front or variations in flame speed.

Flame Temperature

The adiabatic or theoretical temperature assumes no heat losses. Therefore, the enthalpy of products of combustion equals the enthalpy of the reactants. The actual flame temperature (nonadiabatic) differs from the adiabatic temperature. This difference is caused by the following:

1. Radiation losses—radiation from a bunsen burner flame may be 12 to 18 percent.
2. Convection losses—heat to the products of combustion and incoming fuel mixture.
3. Excess air carries heat away.
4. Conduction losses—direct contact with surroundings such as impingement on a furnace wall.
5. All energy in the fuel is not instantaneously released. Therefore the full heating value is not necessarily generated within the flame front.
6. Dissociation of gases will absorb energy (heat) generated during combustion.

The furnace flame temperature can be increased in practice by liberating the heat of combustion as fast as possible by using minimum excess air, intimate mixing, preheating air, or using oxygen or oxygen-enriched air.

Note that organics will not burn in a stable manner with a flame temperature below about 500°F (cigarettes burn steadily at approximately 800°F). Solid waste incinerators operate at internal temperatures of 1200 to 2000°F. At lower temperatures incomplete combustion may occur.

Carbon- and hydrogen-containing organic compounds typically have flame temperatures ranging from 3000 to 4000°F at stoichiometric conditions. (With halogenated hydrocarbons the range would be 3000°F down to nonflammable.)

Temperatures in excess of 4000°F are achieved by blends of hydrocarbon gas and by using more pure oxygen. For instance, the flame temperature of 9 percent acetylene in air is 4200°F, and the flame temperature of 18 percent acetylene in oxygen is 5200°F (stoichiometric conditions).

Solid propellants used in the space program have flame temperatures in the range of 5000 to 6000°F. Flame temperatures of materials normally disposed of by incineration are usually less than 3000°F.

Furnace Temperature

Furnace temperature is a function of fuel heating value, furnace design, air admission, and combustion control. The minimum temperature must be higher than the ignition temperature of the waste. The upper temperature limit is normally a function of the enclosure materials. Over 2400°F operation requires use of special refractory materials. The rates of combustion reactions increase rapidly with increased temperatures. Of the three T's (temperature, time, and turbulence), only temperature can be significantly controlled after a furnace is constructed. Time and turbulence are fixed by furnace design and airflow rate and can normally be controlled only over a limited range.

Furnace Temperature Control

This can be achieved as follows:

Excess air control, i.e., control of the air-fuel ratio. Temperature produced is a direct function of the fuel properties and excess air introduced. Excess air control requires either automatic control or close manual supervision.

Direct heat transfer by the addition of heat-absorbing material within the furnace, such as water-cooled furnace walls. The addition of water sprays in the combustion zone (1000 Btu/lb water evaporated is equivalent to ½ Btu/°F sensible heat in the flue gas) will reduce the furnace temperature. Use of water sprays must be carefully controlled to avoid thermal shock to the furnace refractory.

Furnace Gas Turbulence

Turbulence is an expression relating the physical relationship of fuel and combustion air in a furnace. A high degree of turbulence, intimate mixing of air and fuel, is desirable. Burning efficiency is enhanced with increased surface area of fuel particles exposed to the air. Fuel atomization maximizes the exposed particle surface. Turbulence helps to increase particle surface area by promoting fuel vaporization. In addition, good turbulence exposes the fuel to air in a rapid manner, helping to promote rapid combustion and maximizing fuel release.

A burner requiring no excess air and producing no smoke is said to have perfect turbulence (a *turbulence factor* of 100 percent). If, for instance, 15 percent excess air is required to achieve a no-smoke condition, the turbulence factor is calculated as follows:

$$\text{Turbulence factor} = \frac{\text{stoichiometric air}}{\text{total air}} \times 100\%$$

$$= \frac{1.00}{1.15} \times 100\% = 87\%$$

Fuel gas burners can be designed to produce a turbulence factor close to 100 percent.

Retention Time

Combustion does not occur instantaneously. Sufficient space must be provided within a furnace chamber to allow fuel and combustible gases the time required to fully burn. This factor, termed *dwell time, residence time*, or *retention time*, is a function of furnace temperature, degree of turbulence, and fuel particle size.

Retention time required may be a fraction of a second, as when gaseous waste is burned, or many minutes, as when solid granular waste such as powdered carbon is burned.

THE MASS BALANCE

The flow into an incinerator must equal the flow of products leaving the incinerator. Input includes waste fuel, air (including humidity entrained within the air), and supplementary fuel. The flow exiting the incinerator includes moisture and dry gas in the exhaust as well as ash, both in the exhaust as fly ash and exiting as bottom ash.

Table 9.1, the mass flow table, provides an orderly method of establishing a mass balance surrounding a combustion system. Of initial interest is waste quality: its moisture content, ash content (noncombustible fraction), and heat content. Of prime importance is the generation of moisture and dry flue gas from the combustion process, as described in Chap. 6.

TABLE 9.1 Mass Flow

	Example
Wet feed, lb/h	8000
Moisture, %	22.5
lb/h	1800
Dry feed, lb/h	6200
Ash, %	6.45
lb/h	400
Volatile	5800
Volatile htg. value, Btu/lb	7241
MBtu/h	42.00
Dry gas, lb/10 kBtu	7.5
lb/h	31,500
Comb. H_2O, lb/10 kBtu	0.51
lb/h	2142
Dry gas + comb. H_2O, lb/h	33,642
100% Air, lb/h	27,842
Total air fraction	2.25
Total air, lb/h	62,645
Excess air, lb/h	34,803
Humid/dry gas (air), lb/lb	0.01
Humidity, lb/h	626
Total H_2O, lb/h	4568
Total dry gas, lb/h	66,303

For the purpose of this example, the following waste will be assumed, at the indicated firing rate and other combustion parameters:

8000 lb/h of paper waste

5250 Btu/lb as fired

22.5 percent moisture as fired

5 percent ash as fired

7.50 lb dry flue gas generated per 10,000 Btu released

0.51 lb moisture generated per 10,000 Btu released

Fired with 125 percent excess air

Following Table 9.1:

Wet feed, as received charging rate, is 8000 lb/h.

Moisture. Moisture, by weight, of the wet feed is 22.5 percent. The moisture rate is 22.5 percent of the wet feed rate, 0.225 × 8000 lb/h = 1800 lb/h.

Dry feed equals total wet feed less moisture, 8000 lb/h − 1800 lb/h = 6200 lb/h.

Ash is the percentage of total dry feed that remains after combustion. From the data provided, 5 percent of the total wet feed, 0.05 × 8000 lb/h = 400 lb/h ash. The ash fraction of the dry feed is 400 lb/h ÷ 6200 lb/h = 0.0645.

Volatile is that portion of dry feed that is combusted. It is found by subtracting ash from dry feed: 6200 lb/h − 400 lb/h = 5800 lb/h.

Volatile heating value is the Btu value of the waste per pound of volatile matter. The total heating value as charged is 8000 lb/h × 5250 Btu/lb = 42,000,000 Btu/h. With M representing 1 million, the total waste heat value is 42.00 MBtu/h. On a unit volatile basis, with 5800 lb/h volatile charged, the heating value per pound volatile is 42 MBtu/h ÷ 5800 lb/h = 7241 Btu/lb.

Dry gas produced from combustion of the waste is given for this example as 7.5 lb/10 kBtu, where k represents the number 1000. The dry gas flow is this figure multiplied by the Btu released, or (7.5 ÷ 10,000) lb/Btu × 42 MBtu/h = 31,500 lb/h dry gas.

Combustion H_2O is the moisture generated from burning the waste, in this example 0.51 lb/10 kBtu. Therefore, the combustion moisture flow rate is (0.51 ÷ 10,000) (lb/Btu) × 42 MBtu/h = 2142 lb/h moisture.

Dry gas + combustion H_2O is the sum of the dry gas and the moisture products of combustion, 31,500 lb/h + 2142 lb/h = 33,642 lb/h. This figure is convenient in order to obtain the amount of air required for combustion.

100 percent air is the dry gas and moisture weights that are produced by combustion of the volatile component, which equals the weight of the volatile component plus the weight of the air provided. Likewise, the air requirement is equal to the sum of the dry gas and moisture of combustion less the volatile component. This air requirement is the stoichiometric air requirement, that amount of air necessary for complete combustion of the volatile component of the waste. Using the above figures, the value for 100 percent air is as follows:

$$33{,}642 \text{ lb/h (dry gas + } H_2O) - 5800 \text{ lb/h (volatile)} = 27{,}842 \text{ lb/h}$$

Total air fraction is the air required for effective combustion. This is basically a function of the physical state of the fuel (gas, liquid, or solid) and the nature of the burning equipment. In this example, 125 percent air is required, providing 1.25 + 1.00 = 2.25 total air fraction.

Total air is the stoichiometric requirement multiplied by the total air fraction, 27,842 lb/h × 2.25 = 62,645 lb/h.

Excess air provided to the system is the total air less the stoichiometric air requirement, 62,645 lb/h − 27,842 lb/h = 34,803 lb/h.

Humidity/dry gas (air). The humidity of the air entering the incinerator will have a significant effect on the heat balance to be calculated later in this chapter.

In this case, assume a humidity of 0.01 lb of moisture per pound of dry air. A chart of humidity values is presented in Chap. 4.

Humidity is the flow of moisture into the system with the air supply. Given the airflow and the fractional humidity, the humidity can be calculated as 62,645 lb/h air × 0.01 lb H_2O/lb air = 626 lb/h moisture.

Total H_2O is that moisture exiting the system. It is equal to the sum of three moisture components: moisture in the feed plus moisture of combustion plus humidity, or 1800 lb/h + 2142 lb/h + 626 lb/h = 4568 lb/h total H_2O.

Total dry gas exiting the system is equal to the sum of the dry gas generated by the combustion of the volatiles, with stoichiometric air, plus the flow of excess air into the system. Thus, 31,500 lb/h dry gas + 34,803 lb/h excess air = 66,303 lb/h total dry gas.

HEAT BALANCE

Heat, as mass, is conserved within a system. The heat exiting a system is equal to the amount of heat entering that system. Table 9.2 presents a quantitative means of establishing a heat balance for an incinerator. The heat balance must be preceded by a mass flow balance, as discussed previously. The result of a heat balance is a determination of the incinerator outlet temperature, outlet gas flow, supplemental fuel requirement, and total air requirement. The total heat into the incinerator was calculated previously in the mass flow computations. By determining how much of this total heat produced is present in the exhaust gas, the exhaust gas temperature can be calculated. If the calculated exhaust gas temperature is equal to the desired exhaust gas outlet temperature, the process is autogenous and supplemental fuel is not required. If the desired temperature is lower than the actual outlet temperature, additional air must be added (or additional water, if this is possible) to lower the outlet temperature to that desired. This condition is also autogenous since supplemental fuel is not added to the system.

For the case where the actual outlet temperature is less than the desired outlet temperature, supplemental fuel must be added. The products of combustion must include the products of combustion of the supplemental fuel.

Dry flue gas properties are assumed to be identical to those of dry air.

As with Table 9.1, Table 9.2 will be described on a step-by-step basis.

Cooling air wasted. Assume the incinerator shell is cooled by a flow of 2000 st ft^3/min of air. A standard cubic foot of air weighs 0.075 lb/ft^3, therefore the mass airflow is 2000 ft^3/min × 60 min/h × 0.075 lb/ft^3 = 9000 lb/h. It is further assumed that this flow is wasted, i.e., discharged to the atmosphere.

The temperature of the air at the discharge point is assumed to be 450°F.

From Table 4.1 (Chap. 4) the enthalpy of air at 450°F, the cooling air discharge temperature, is 94 Btu/lb.

The total heat loss due to the wasted cooling air is the quantity of cooling air discharged multiplied by its enthalpy, 9000 lb/h × 94 Btu/lb = 0.85 MBtu/h.

Ash generated is 400 lb/h, from the mass flow table.

The heating value of ash can be calculated from the methods developed in Chap. 8. In lieu of an analysis of the heating value of the ash component, a figure of 130 Btu/lb is reasonable. The heat loss through ash discharge is therefore 400 lb/h ash × 130 Btu/lb ash = 0.05 MBtu/h.

TABLE 9.2 Heat Balance

	Example w/fuel oil	Example w/gas
Cooling air wasted, lb/h	9000	
°F	450	
Btu/lb	94	
MBtu/h	0.85	
Ash, lb/h	400	
Btu/lb	130	
MBtu/h	0.05	
Radiation, %	1.5	
MBtu/h	0.63	
Humidity, lb/h	626	
Correction (@970 Btu/lb), MBtu/h	−0.61	
Losses, total, MBtu/h	0.92	
Input, MBtu/h	42.00	
Outlet, MBtu/h	41.08	
Dry gas, lb/h	66,303	
H_2O, lb/h	4568	
Temperature, °F	1899	
Desired temp., °F	2000	
MBtu/h	43.26	
Net, MBtu/h	2.16	
Fuel oil, air fraction	1.20	1.10
Net Btu/gal	57,578	406
gal/h	37.51	5320
Air, lb/gal	125.06	0.791
lb/h	4691	4208
Dry gas, lb/gal	125.54	0.748
lb/h	4709	3979
H_2O, lb/gal	8.75	0.103
lb/h	328	548
Dry gas w/fuel oil, lb/h	71,012	70,282
H_2O w/fuel oil, lb/h	4896	5116
Air w/fuel oil, lb/h	67,336	66,853
Outlet, MBtu/h	46.32	46.40
Reference t, °F	60	60

Radiation. The heat lost by radiation from the incinerator shell can be approximated as a percentage of the total heat of combustion. Table 9.3 lists typical values of radiation loss. For this case, with a heat release of 42 MBtu/h a radiation loss of 1.5 percent is used.

The total loss by radiation is equal to 42 MBtu/h released × 1.5 percent, or 0.63 MBtu/h.

Humidity is water vapor within the air. The humidity component of the air, 626 lb/h, is taken from the mass flow sheet.

When one is considering the heat absorbed by the moisture or water within the incinerator, exhaust steam humidity has released its heat of vaporization because it is in the vapor phase. The other moisture components, moisture in the feed and moisture of combustion, enter the reaction as a liquid, and the heat of vaporization is released by the reaction. To simplify these moisture calculations,

TABLE 9.3 Furnace Radiation Loss Estimates

Furnace rate, MBtu/h	Radiation loss, % of furnace rate
< 10	3
15	2.75
20	2.50
25	2
30	1.75
> 35	1.50

the heat of vaporization of humidity moisture at 60°F, 960 Btu/lb, is added to the total heat capacity of the flue gas. Therefore the correction factor is 626 lb/h humidity × 970 Btu/lb = 0.61 MBtu/h.

Total losses of an incinerator are the sum of the heat discharged as cooling air and the heat lost in the ash discharge and the radiation loss. To this is added the correction for humidity. In this case the total loss is 0.85 MBtu/h + 0.05 MBtu/h + 0.63 MBtu/h − 0.61 MBtu/h = 0.92 MBtu/h.

Input to the system is the heat generated from the combustible, or volatile, portion of the feed. From the mass flow table this figure is 42.00 MBtu/h.

Outlet heat content is that amount of heat exiting in the flue gas. The heat left in the flue gas is the heat generated by the feed (input) less the total heat loss, 42.00 MBtu/h − 0.92 MBtu/h = 41.08 MBtu/h.

Dry gas is 66,303 lb/h, from the mass flow sheet.

H_2O is 4568 lb/h, from the mass flow sheet.

Temperature is the temperature of the exhaust gas where the heat content of the dry gas flow plus the heat content of the moisture flow exiting the incinerator equals the outlet, MBtu/h. From the methods of Chap. 4:

With 66,303 lb/h dry gas and 4568 lb/h moisture at 1800°F and at 1900°F, the dry gas enthalpy is 453.24 and 481.57 Btu/lb, respectively, and the moisture enthalpy is 1948.02 and 2007.17 Btu/lb, respectively. Therefore the total calculated enthalpy is

1800°F	38.95 MBtu/h
X	41.08 MBtu/h (outlet)
1900°F	41.10M Btu/h

By interpolation,

$$X = 1800 + (1900 - 1800)\frac{41.08 - 38.95}{41.10 - 38.95} = 1899°\text{F}$$

Therefore the exhaust gas temperature is 1899°F.

Desired temperature is the temperature to which it is desired to bring the products of combustion. For this example a temperature of 2000°F will be used.

At this temperature the heat content of the wet gas stream is that of the dry gas and that of moisture at 2000°F or 66,303 lb/h dry gas × 510.07 Btu/lb + 4568 lb/h moisture × 2067.42 Btu/lb = 43.26 MBtu/h.

Net MBtu/h is the amount of heat that must be added to the flue gas to raise its heat content to the desired level. (In this case the desired heat level is evaluated at 2000°F.) The net MBtu/h, therefore, is the desired less the outlet MBtu/h, 43.26 MBtu/h − 41.06 MBtu/h = 2.20 MBtu/h.

Note that this analysis will continue based on the use of No. 2 fuel oil as supplemental fuel. Table 9.4 lists fuel oil combustion parameters. Table 9.5 likewise is a listing of combustion parameters for natural gas. The incinerator analysis of Tables 9.2 and 9.6 includes listings for the use of natural gas as supplemental fuel. The values for natural gas are calculated in similar manner to the calculations for fuel oil.

TABLE 9.4 No. 2 Fuel Oil, 139,703 Btu/gal, 7.6 lb/gal

Total Air:	1.1	1.2	1.3
lb air/gal	114.640	125.062	135.483
lb dry gas/gal	115.115	125.537	135.958
lb H_2O/gal	8.615	8.751	8.886
Temp., °F		Heat available, Btu/gal	
200	126,210	125,707	125,206
300	123,016	122,255	121,495
400	119,802	118,780	117,760
500	116,562	115,277	113,995
600	113,283	111,732	110,184
700	109,965	108,146	106,328
800	106,029	103,885	101,742
900	103,197	100,829	98,463
1,000	99,747	97,099	99,747
1,100	96,255	93,325	90,397
1,200	92,721	89,505	86,291
1,300	89,147	85,643	82,140
1,400	85,535	81,738	77,943
1,500	81,887	77,796	73,707
1,600	78,205	73,817	69,431
1,700	74,487	69,799	65,115
1,800	70,746	65,757	60,771
1,832	69,538	64,452	59,369
1,900	66,971	61,679	56,389
2,000	63,175	57,578	51,984
2,100	59,341	53,445	59,349
2,192	55,813	49,628	44,385
2,200	55,507	49,294	43,084
2,300	41,637	45,114	38,594
2,400	47,750	40,916	34,085
2,500	43,852	36,706	29,562
2,600	39,914	32,453	24,995
2,700	35,938	28,162	20,388

TABLE 9.5 Natural Gas, 1000 Btu/st ft³, 0.050 lb/st ft³

Total Air:	1.05	1.10	1.15
lb air/st ft³	0.755	0.791	0.827
lb dry gas/st ft³	0.712	0.748	0.784
lb H_2O/st ft³	0.103	0.103	0.104
Temp., °F	Heat available, Btu/st ft³		
200	861	860	857
300	839	837	834
400	817	814	810
500	794	790	785
600	772	767	761
700	749	743	736
800	722	715	707
900	702	694	685
1000	678	670	660
1100	654	644	633
1200	629	619	607
1300	605	593	580
1400	579	567	553
1500	554	541	526
1600	529	514	498
1700	503	487	470
1800	477	460	442
1832	468	451	433
1900	450	433	414
2000	424	406	385
2100	397	378	356
2192	372	352	329
2200	370	350	327
2300	343	322	298
2400	316	294	269
2500	289	265	239
2600	261	237	210
2700	233	208	179

Fuel oil, air fraction is the total air fraction required for combustion of supplementary fuel. The total air normally required for efficient combustion of gas fuel is from 1.05 to 1.15 and for light fuel oil is 1.10 to 1.30. In this case assume No. 2 fuel oil is used for supplemental heat, with a total air supply of 1.20.

Net Btu/gal. When fuel is combusted, the products of combustion must be heated to the desired flue gas temperature. The amount of heat required to heat these combustion products must be subtracted from the total heat of combustion to obtain the effective heating value of the fuel. As the temperature to which the fuel products must be raised increases, the net heat available from the fuel decreases. From Table 9.4, bringing the products of combustion of fuel oil, with 1.2 total air, to 2000°F, a net heating value of 57,578 Btu/gal is available.

The gallons of fuel oil required to provide the heat required to bring the exhaust gas temperature from its actual temperature, 1899°F, to the desired temper-

TABLE 9.6 Flue Gas Discharge

	Example w/fuel oil	Example w/gas
Inlet, °F	2000	2000
Dry gas, lb/h	71,012	70,282
Heat, MBtu/h	46.32	46.60
Btu/lb dry gas	652	660
Adiabatic t, °F	176	176
H_2O saturation, lb/lb dry gas	0.5511	0.5511
lb/h	39,135	38,732
H_2O inlet, lb/h	4896	5116
Quench H_2O, lb/h	34,239	33,616
gal/min	68	67
Outlet temp., °F	120	120
Raw H_2O temp., °F	60	60
Sump temp., °F	147	147
Temp. diff., °F	87	87
Outlet, Btu/lb dry gas	111.65	111.65
MBtu/h	7.93	7.85
Req'd. cooling, MBtu/h	38.39	38.55
H_2O, lb/h	441,264	443,103
gal/min	883	886
Outlet, ft^3/lb dry gas	16.515	16.515
ft^3/min	19,546	19,345
Fan press., in WC	30	30
Outlet, actual ft^3/min	21,101	20,884
Outlet, H_2O/lb dry gas	0.08128	0.08128
H_2O, lb/h	5772	5713
Recirc. (ideal), gal/min	68	67
Recirc. (actual), gal/min	548	538
Cooling H_2O, gal/min	883	886

ature, 2000°F, are equal to the net MBtu/h required divided by the net Btu/gal available: 2.16 MBtu/h net ÷ 57,578 Btu/gal = 37.51 gal/h.

Air required for combustion of fuel oil, with 1.2 total air, from Table 9.4, is 125.06 lb/gal of fuel oil.

The fuel combustion airflow is the unit flow multiplied by the fuel quantity, 125.06 lb air/gal × 37.51 gal fuel oil = 4691 lb/h air.

Dry gas. From Table 9.4 the dry gas produced from combustion of fuel oil with 1.2 total air is 125.54 lb/gal of fuel oil.

The dry gas flow rate is 125.54 lb dry gas/gal × 37.51 gal/h fuel oil = 4709 lb/h.

H_2O produced from combustion of fuel oil with 1.2 total air and 0.013 humidity is 8.75 lb/gal fuel oil from Table 9.4.

The moisture flow rate from combustion of fuel oil is 8.75 lb H_2O/gal fuel oil × 37.51 gal fuel oil/h = 328 lb/h.

Dry gas with fuel oil is the total quantity of dry gas exiting the system. It is equal to the dry gas produced from combustion of the waste plus the dry gas produced from fuel combustion: 66,303 lb/h + 4709 lb/h = 71,012 lb/h dry gas.

H_2O with fuel oil is the total quantity of moisture exiting the system, that cal-

culated in the mass flow sheet plus the contribution from combustion of supplementary fuel, 4568 lb/h + 328 lb/h = 4896 lb/h.

Air with fuel oil is the total amount of air entering the incinerator, calculated from the mass flow sheet, plus that needed for supplemental fuel combustion, 62,645 lb/h + 4691 lb/h = 67,336 lb/h.

Outlet MBtu/h, the total heat value of the flue gas exiting the incinerator, is the sum of the heat content of the gas prior to adding supplemental fuel and the heat addition of the supplemental fuel. The supplemental fuel adds 37.51 gal/h × 139,703 Btu/gal = 5.24 MBtu/h to the flue gas. Therefore, the flue gas outlet contains 41.08 MBtu/h + 5.24 MBtu/h = 46.32 MBtu/h.

As a check on this figure, the outlet temperature will be calculated by using the flue gas flow (71,012 lb/h dry gas and 4896 lb/h moisture):

1900°F	44.02 MBtu/h
X	46.32 MBtu/h
2000°F	46.34 MBtu/h

$$X = 1900 + (2000 - 1900)\frac{46.32 - 44.02}{46.34 - 44.02} = 1999°\text{F}$$

This calculation of flue gas outlet temperature, 1999°F, is a good approximation of the desired gas outlet temperature, 2000°F.

Reference t is the datum temperature for enthalpy. It is the temperature at which feed, supplemental fuel, and air enter the system, 60°F for this example.

FLUE GAS DISCHARGE

To meet the rigorous air pollution codes in effect today, wet gas scrubbing equipment is often necessary. Table 9.6, the flue gas discharge table, provides a method of calculating gas flow volumes exiting a wet scrubbing system as well as scrubber flow quantities. This table can also be used when calculating volumetric flow from a dry flue gas system.

Table 9.6 entries are as follows:

Inlet. Insert the incinerator outlet temperature, 2000°F, from the heat balance table. This temperature is the inlet temperature of the flue gas processing system.

Dry gas is the total flow of dry gas exiting the incinerator. From the heat balance table, this figure is 71,012 lb/h.

Heat is the total heat exiting the incinerator in the flue gas, MBtu/h. From the heat balance table this figure is 46.32 MBtu/h.

The heat is calculated in terms of the dry gas component of the flue gas, as Btu/lb dry gas. From the entries for total heat and dry gas flow, the heat is 46.32 MBtu/h ÷ 71,012 lb/h = 652 Btu/lb dry gas.

Adiabatic t. When 1 lb of water evaporates, it absorbs approximately 1000 Btu, without a change in temperature. This heat absorption is called the *heat of vaporization*, or *latent heat*. Latent heat is opposed to sensible heat, which is the heat required for a change in temperature without a change in phase. For evaporated water (steam) the sensible heat is approximately 0.5 Btu/lb steam for every rise of 1°F.

The adiabatic temperature of the flue gas is the quench temperature. Quenching of a gas is defined as use of the latent heat of water (or other liquid) to decrease the gas temperature. This process does not involve the addition or removal of heat, only the use of the heat of vaporization of the quench liquid, i.e., water. The term *adiabatic* defines a process where heat is neither added nor removed from a system.

Considering the properties of dry flue gas equal to the properties of dry air, listed in Table 4.6, note that a maximum amount of moisture can be held in dry air at a particular temperature. This table lists saturation moisture quantities, volume, and enthalpy at saturation as a function of temperature. The temperature at which the enthalpy of the saturated dry flue gas (dry air) is equal to the enthalpy calculated above, Btu/lb dry gas, is the adiabatic temperature of the system. There has been no transfer of heat from the system, only the conversion of latent heat in the quench water to sensible heat in the dry flue gas.

In this example the adiabatic temperature is found in Table 4.5 as that temperature where the dry flue gas (saturated mixture) will have an enthalpy of approximately 652 Btu/lb, 176°F.

H_2O saturation. The quenched flue gas, at the adiabatic temperature (176°F in this example) will contain an amount of moisture equal to the maximum amount of moisture that it can hold, saturation. From Table 4.5 the saturation moisture, in lb H_2O/lb dry gas (air), is read opposite the adiabatic temperature. In this case, for 176°F adiabatic temperature the saturation moisture is 0.5511 lb H_2O/lb dry gas.

The moisture flow is the saturation moisture multiplied by the dry gas flow, 0.5511 lb H_2O/lb dry gas × 71,012 lb dry gas/h = 39,135 lb H_2O/h.

H_2O inlet. The moisture component of the flue gas exiting the incinerator is inserted here from the heat balance table: 4896 lb/h.

Quench H_2O is the moisture required for quenching the incoming flue gas to its adiabatic temperature. This is equal to the saturated moisture content of the flue gas less the moisture initially carried into the system with the flue gas. This figure is equal to H_2O saturation (39,135 lb/h) less H_2O inlet (4896 lb/h), which in this example is equal to 34,239 lb/h.

The conversion factor from lb/h of water to gal/min (8.34 lb/gal × 60 min/h) is 500. The quench water required is equal to 34,239 lb/h ÷ 500 = 68 gal/min.

Outlet temperature is that temperature entering the low (negative) pressure side of the induced-draft (ID) fan, or, with no ID fan, the temperature within the stack. This temperature is normally selected in the range of 120 to 160°F. The lower this temperature, the smaller the size of the outlet plume and the lower the volumetric flow of flue gas. For this example 120°F was chosen as the outlet temperature. As can be seen below, a lower outlet temperature would require additional amounts of cooling water.

Raw H_2O temperature. This entry is the temperature of the water available for cooling the flue gas from the adiabatic temperature to the outlet temperature. In this example a raw water temperature of 60°F was chosen.

Sump temperature. Normally a quantity of water in excess of that calculated for quenching and for cooling the flue gas is provided for particulate removal. The excess water is generally collected in a sump where a quiescent period is allowed to permit larger particles within the spent water to settle to the sump floor, eventually to be drained. The temperature of the water in the sump must be ascertained to determine the effective cooling rate of the water flow. The sump temperature is a practical impossibility to forecast accurately, but an empirical relationship has been established. The sump temperature is assumed equal to the adiabatic temperature divided by 1.2. In this example the sump temperature is 176 ÷ 1.2 = 147°F.

Temperature differential. The temperature differential of note is the difference in temperature between the raw water entering the cooling tower (or scrubber) and the temperature exiting the tower, the sump temperature. Sump temperature less raw H_2O temperature is in this example 147 − 60 = 87°F, the temperature differential.

Outlet. The gas exiting the scrubber system is designed to be at the outlet temperature, 120°F in this example, saturated with moisture. The outlet enthalpy is inserted from Table 4.5 for the outlet temperature chosen: 111.65 Btu/lb dry gas.

The total heat in the outlet flue gas is its enthalpy multiplied by its flow, that is, 111.65 Btu/lb dry gas × 71,012 lb dry gas/h = 7.93 MBtu/h.

Required cooling. As noted previously, the flue gas is initially quenched, without heat addition or removal, to its adiabatic temperature. To reduce the adiabatic temperature of the flue gas to the desired outlet temperature, a supply of cooling water is required. This cooling water must remove the heat content at adiabatic conditions relative to the heat content at outlet conditions. The required cooling is therefore the heat inlet less the outlet MBtu/h. For this example 46.32 MBtu/h heat inlet less 7.93 MBtu/h outlet equals 38.39 MBtu/h required cooling; i.e., with removal of 38.39 MBtu/h from the saturated flue gas stream the flue gas temperature will fall from 176 to 120°F.

H_2O. The moisture flow referred to is that required to achieve the desired cooling effect. With $Q = WC\,\Delta t$ or $W = Q/C\,\Delta t$ and W the cooling water in lb/h, Q the cooling load in Btu/h, C the specific heat of water [1 Btu/(lb°F)], and Δt the temperature difference of the cooling water across the flue gas stream (sump water temperature less raw water temperature), the required cooling water flow can be calculated. For this example:

$$W = 38.39 \text{ MBtu/h} \div 1 \text{ Btu/(lb}\cdot\text{°F)} \div 87\text{°F}$$

$$= 441{,}264 \text{ lb/h}$$

The flow in gallons per minute is that in lb/h divided by 500, 441, 264 ÷ 500 = 883 gal/min.

Outlet. The outlet volumetric flow is obtained with use of Table 4.5. The specific volume, ft^3 mixture/lb dry gas (air), is found in this table for the outlet temperature. For this example, with an outlet of 120°F the specific volume is 16.515 ft^3/lb dry gas.

The volumetric flow is equal to the specific volume multiplied by the dry gas flow divided by 60 min/h; that is, 16.515 ft^3/lb dry gas × 71,012 lb/h ÷ 60 min/h = 19,546 ft^3/min.

Fan pressure. Induced-draft fans used to clean the incinerator of gases may require a relatively high pressure. The actual differential pressure across the far., inserted here, will be used to modify the volumetric flow value. For this example the fan pressure is 30 in WC.

Outlet actual ft^3/min. The volumetric flow immediately prior to entering the induced-draft (ID) fan will experience an expansion because of the fan suction. The volumetric flow correction is as follows: Multiply the value in ft^3/min determined above by the ratio 407 ÷ (407 − p), where p is the pressure across the fan. The figure 407 is atmospheric pressure (14.7 lb/in^2 absolute) expressed in inches of water column (14.7 lb/in^2 absolute ÷ 62.4 lb H_2O/ft^3 × 1728 in^3/ft = 407 in WC). In this example the corrected, or actual, volumetric flow entering the ID fan is [407 ÷ (407 − 30)] × 19,547 ft^3/min = 21,101 ft^3/min actual flow.

Outlet. From Table 4.5, insert the saturation humidity, lb H_2O/lb dry gas (dry air), corresponding to the outlet temperature. For this example, with an outlet temperature of 120°F, the humidity is 0.08128 lb H_2O/lb dry gas.

The total moisture exiting the stack is the saturation humidity multiplied by the dry gas flow. For this case, 0.08128 lb H_2O/lb dry gas × 71,012 lb dry gas/h = 5772 lb H_2O/h.

Recirculation (ideal). The recirculation flow is that amount of flow required for quenching, as calculated above (68.5 gal/min for this example). The temperature of this water flow is not critical. This flow is used adiabatically, where the latent heat (not a temperature change) in the water flow reduces the temperature of the flue gas. It is termed a recirculation flow because spent scrubber water from the scrubber sump can be recirculated to the Venturi for use at 140°F or greater, instead of a cooler flow of water.

Recirculation (actual). In practice the ideal flow is inadequate to fully clean the gas stream of particulate. Ideal quenching requires intimate contact of each molecule of water with gas, instantaneous evaporation, and instantaneous heat transfer between the moisture and the gas, none of which occurs. To compensate for actual versus ideal conditions, an empirical factor is used. In this case this factor is 8. Therefore, for actual recirculation flow use the ideal flow multiplied by 8, 68.5 gal/min × 8 = 548 gal/min.

Cooling H_2O. Insert the flow of cooling water calculated above, in this case, 883 gal/min.

COMPUTER PROGRAM

The mass flow, heat balance, and flue gas discharge analyses presented in this chapter have been developed into a computer program written in the BASIC language and menu-driven. The program accepts up to four different waste streams as feed concurrently. It considers individual gas components and has waste heat boiler and wet scrubber systems options. The program displays and prints out more comprehensive information than is immediately available from the analysis sheets in this chapter. This program and related programs are available from Incinerator Consultants Incorporated, 11204 Longwood Grove Drive, Reston, Virginia 22094 (703/437-1790).

BIBLIOGRAPHY

Achinger, W., and Giar, J.: *Testing Manual for Solid Waste Incinerators*, USEPA, 1973.

Corey, R.: *Principles and Practice of Incineration*, John Wiley & Sons, New York, 1969.

Disposal of Industrial Wastes by Combustion, American Society of Mechanical Engineers, January 1971.

Rimer, A.: *Solid Waste Engineering Laboratory Manual*, Duke University, 1978.

Steam, 38th ed., Babcock & Wilcox, 1972.

CHAPTER 10
ON-SITE SOLID WASTE DISPOSAL

The disposal of solid waste is the most visible of problems associated with waste in this country. The per capita generation of solid waste is approximately 5 lb/day in the United States, over 0.5 Mton/day from domestic sources alone. The many varieties of incinerators in general use for the on-site disposal of solid waste are presented in this chapter. Solid waste incinerators designed for central-station disposal are discussed in a later chapter.

SOLID WASTE INCINERATION

Solid waste incinerators are usually categorized according to the nature of the material which they are designed to burn, i.e., refuse or industrial waste. However, more than one waste type can often be burned in a given unit.

Incinerators for destruction of solid waste are the most difficult class of incinerators to design and operate, primarily because of the nature of the waste material. Solid waste can vary widely in composition and physical characteristics, making the effects of feed rates and parameters of combustion very difficult to predict. Solid waste incinerators most often burn wastes over a range of low and high heat values, i.e., from wet garbage with an as-received heat value as low as 2500 Btu/lb, to plastic wastes, over 19,000 Btu/lb. Materials handling, firing, and residue removal equipment are more critical, cumbersome, expensive, and difficult to control with these than with other types of incinerators.

TYPES OF SOLID WASTE INCINERATORS

There are eight main types of solid waste incinerators:

1. Open-burning
2. Single-chamber incinerators
3. Teepee burners
4. Open-pit incinerators

5. Multiple-chamber incinerators
6. Controlled air incinerators
7. Central-station disposal
8. Rotary kiln incinerators

Controlled air incineration, central-station disposal, and rotary kilns will be discussed in subsequent chapters.

OPEN BURNING

Open burning is the oldest technique for incineration of wastes. Basically it consists of placing or piling waste materials on the ground and burning them without the aid of specialty combustion equipment.

This type of system is found in most parts of the United States. It results in excessive smoking and high particulate emission, and it presents a fire hazard.

Open burning has been utilized to dispose of high-energy explosives such as dynamite or TNT. For proper incineration, the waste is placed on an asbestos pad which is in turn placed over gravel, in a cleared location, remote from populated areas.

SINGLE-CHAMBER INCINERATORS

Single-chamber incinerators will, in general, not meet the air pollution emission standards that have been developed over the past 10 to 15 years. A typical single-chamber incinerator is shown in Fig. 10.1. Solid waste is placed on the grate and fired. These incinerators have also been manufactured in top-loading (flue loading) configuration for apartment house waste disposal, as shown in Fig. 10.2, firing waste in 55-gal drums or wire baskets or in a concrete or refractory-lined structure with a cast-iron grate, etc. This equipment may or may not have a firing system to ignite the waste. As with open burning, smoking and excessive air pollution emissions can occur.

Attempts have been made to control emissions to reasonable levels by the addition of an afterburner, as shown in Fig. 10.2. Note that both an afterburner and a draft control damper have been added to control the combustion process. Most of the air emissions in single-chamber incinerators result from incomplete burning of the waste. Normally a temperature of 1400°F is required, at a retention time of 0.5 s, and the afterburner is used to obtain these combustion parameters in the exiting off-gas. Tables 10.1 and 10.2 show stack emissions and other parameters of operation and design from tests performed on flue-fed incinerators similar to that pictured in Fig. 10.2. Table 10.1 represents conditions without an afterburner in operation while Table 10.2 reflects conditions with an afterburner firing. Note the dramatic change in particulate emissions with and without provision of an afterburner.

A *jug incinerator* is another type of single-chamber unit. A typical jug incinerator is shown in Fig. 10.3. This is a specialty incinerator used for the destruction of cotton waste and other waste agricultural products. It is a brick-lined vertical cylindrical or conical structure. Waste is fed through the top section of the

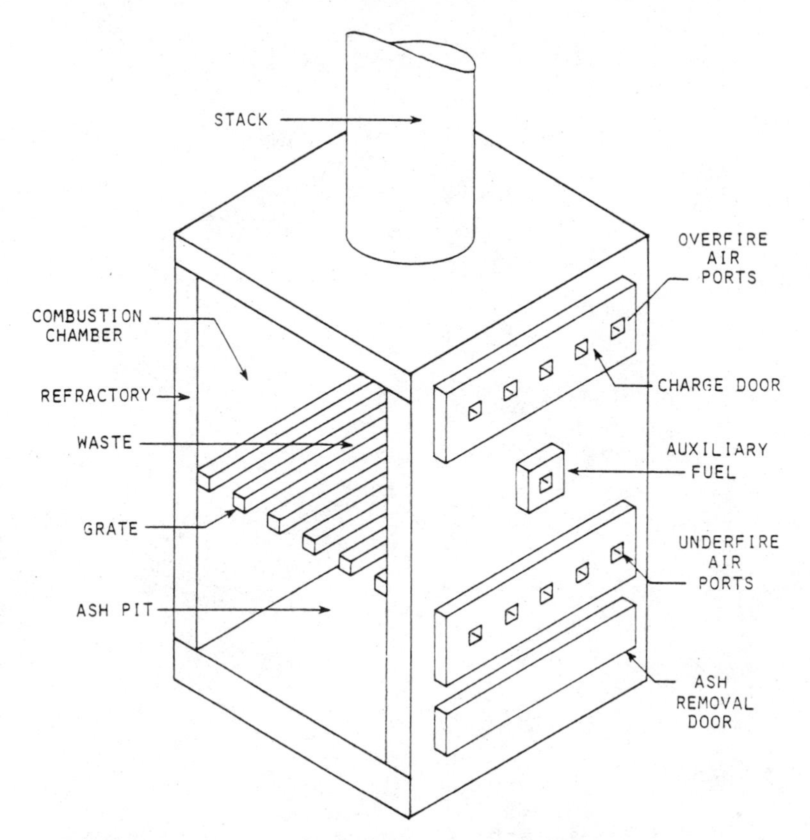

FIGURE 10.1 Single-chamber incinerator. *Source:* Ref. 10-1.

incinerator and falls to its floor, which may or may not be provided with grates. Waste is pneumatically conveyed to the incinerator charging system, and the transfer air is the only combustion air supplied to the incinerator. Afterburners are provided in the stack to control air emissions, although many such incinerators discharge from their conical top, without provision of a stack or afterburning equipment.

OPEN-PIT INCINERATORS

Open-pit incinerators have been developed for controlled incineration of explosive wastes, wastes which would create an explosion hazard or high heat release in a conventional, enclosed incinerator. They are constructed as shown in Fig. 10.4 with an open top and a number of closely spaced nozzles blowing air from the open top down into the incinerator chamber. Air is blown at high velocity, creating a rolling action, i.e., a high degree of turbulence. Burning rates within the incinerator provide temperature in excess of 2000°F with low smoke and relatively low particulate emissions discharges.

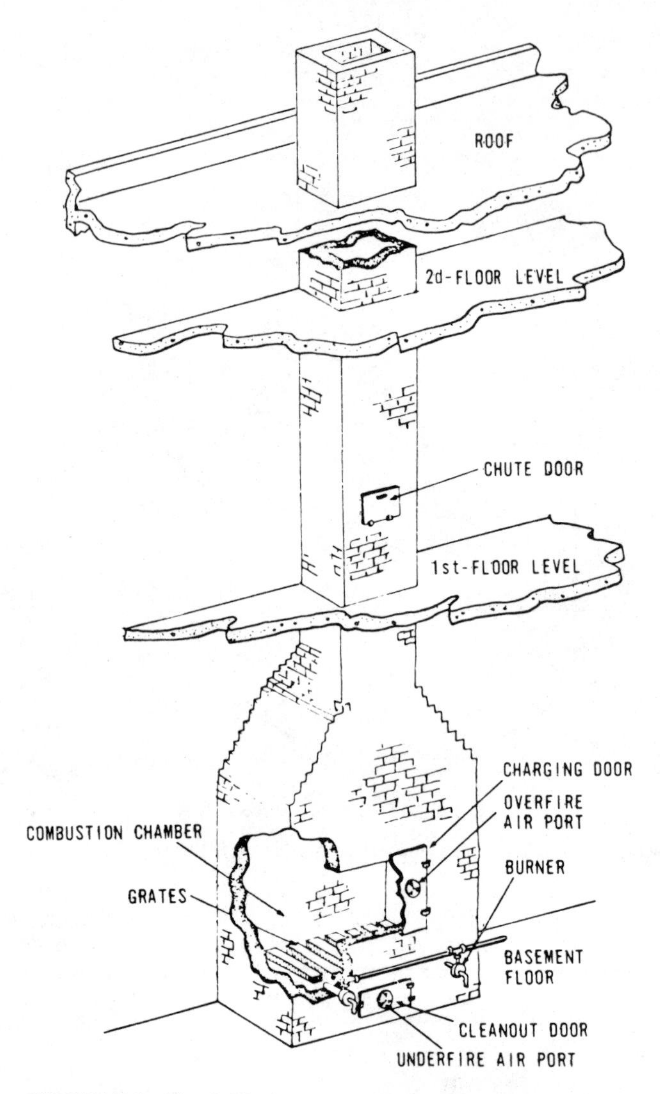

FIGURE 10.2 Flue-fed incinerator modified by a roof afterburner and a draft control damper. *Source:* Ref. 10-2.

Incinerators of this type may be built either above- or below-ground. They are constructed with refractory walls and floor or as earthen trenches. The width of an open-pit incinerator is normally on the order of 8 ft, with a depth of approximately 10 ft. The length varies from 8 to 16 ft.

Overfire air nozzles are 2 to 3 in in diameter, located above one edge of the pit. They fire down at an angle of 25° to 35° from the horizontal. The incinerator is normally charged from a top-loading ramp on the edge opposite the air nozzles.

TABLE 10.1 Particulate Emissions from Typical Flue-Fed Incinerators

Test designation	Particulate matter lb/ton	gr/st ft^3 at 12% CO_2	gr/st ft^3	Average stack volume, st ft^3/min	Grate area, ft^2	Stack height, ft
1	76	2.27	0.61	458	1.5	25
2	52	1.40	0.13	1,190	16	35
3	48	1.60	0.21	326	9	68
4	37	1.40	—	213	8	32
5	37	1.18	0.20	820	12	80
6	34	1.06	0.21	930	12	80
7	25	0.94	0.18	500	6	54
8	23	0.99	0.10	1,120	12	56
9	23	0.75	0.2	860	12	80
10	19	0.75	0.26	530	4	25
11	17	0.60	0.09	441	20	56
12	7	0.27	0.08	817	8	46

Source: Ref. 10-2.

TABLE 10.2 Particulate Emissions from a Typical Flue-Fed Incinerator Modified with a Draft Control Damper and a Roof Afterburner

Test designation	Burning rate, lb/h	Particulate matter lb/ton	gr/st ft^3 at 12%CO_2	gr/st ft^3	After-burner efficiency, %	Average oxygen content, %	Average stack volume, st ft^3/min	Average outlet temperature, °F
A	100	5.9	0.20	0.004	80	12.1	760	1,130
B	80	5.2	0.18	0.035	82	11.6	690	1,240
C	68	5.6	0.20	0.034	80	12.7	710	1,130
D	49	1.2	0.15	0.027	85	9.5	590	1,560

Source: Ref. 10-2.

Some units have a mesh placed on their top to contain larger particles of fly ash. Residue cleanout doors are often provided on aboveground incinerators.

For a waste with a heating value of 5000 Btu/lb note the following typical parameters of design for open-pit incinerators:

- Heat release of 3.4 MBtu/h per foot of length.
- Provision of 100 to 300 percent excess air.
- Overfire air of 850 st ft^3/min per foot of pit length at 11 in WC.

Particulate emissions are normally below 0.25 gr/dry st ft^3, corrected to 12 percent CO_2, which is unacceptable with regard to current air pollution control standards. (Most current statutes limit air pollution emissions from burning refuse to 0.08 gr/dry st ft^3 corrected to 12 percent CO_2.) Other than combustion control by control of overfire air, there is no mechanism practicable

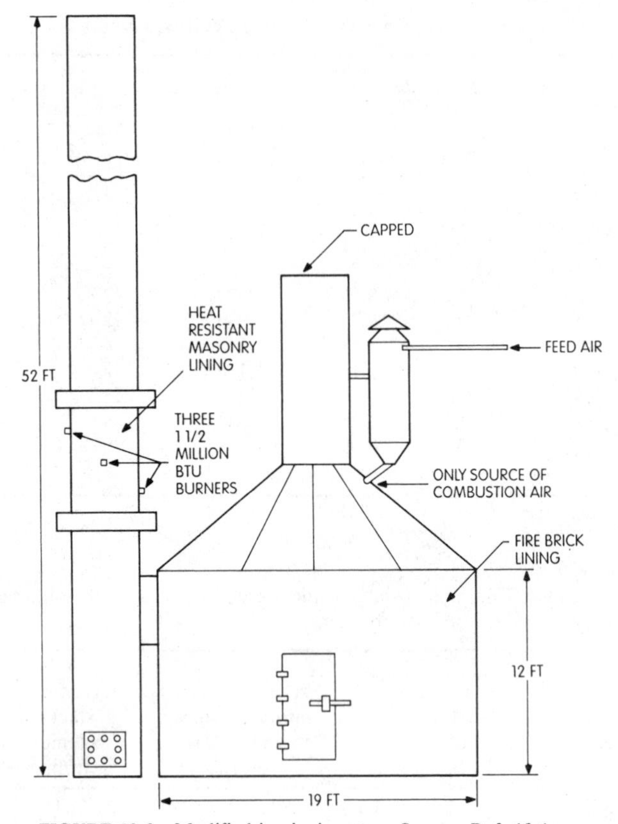

FIGURE 10.3 Modified jug incinerator. *Source:* Ref. 10-1.

for control of exhaust emissions. This incinerator, while effective in destruction of some waste, cannot normally be used without relaxation of local air pollution emissions requirements.

MULTIPLE-CHAMBER INCINERATORS

In an attempt to provide complete burnout of combustion products and decrease the airborne particulate loading in the exiting flue gas, multiple-chamber incinerators have been developed. A first, or primary, chamber is used for combustion of solid waste. The secondary chamber provides the residence time, and supplementary fuel, for combustion of the unburned gaseous products and airborne combustible solids (soot) discharged from the primary chamber.

There are two basic types of multiple-chamber incinerators: the retort and the in-line systems.

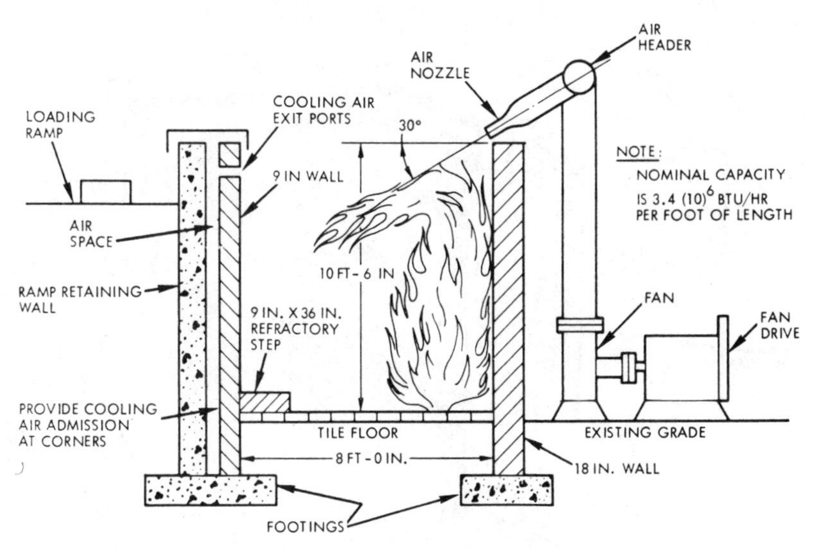

FIGURE 10.4 Open-pit incinerator.

Retort Incinerator

This unit is a compact cubic-type incinerator with multiple internal baffles. The baffles are positioned to guide the combustion gases through 90° turns in both lateral (horizontal) and vertical directions. At each turn ash drops out of the flue gas flow. The primary chamber has elevated grates for discharge of waste and an ash pit for collection of ash residual. A cutaway view of a typical retort-type incinerator is shown in Fig. 10.5. Figure 10.6 gives dimensional data for typical retort units.

Overfire air and underfire air are provided above and below the primary chamber grate. This air is normally supplied by forced-air fans at a controlled rate.

Flue gas exits the primary chamber through an opening, termed a *flame port*, which discharges to the secondary chamber or to a smaller *mixing chamber* immediately before the secondary chamber. The flame port is actually an opening atop the bridge wall, separating the primary from the secondary chamber.

Air ports are provided in the secondary combustion chamber and, when present, in the mixing chamber. Supplemental fuel is provided in the secondary and primary chambers. Depending on the nature of waste charged, the fuel supply in the primary chamber may be unnecessary after start-up, i.e., after bringing the chamber temperature to a level high enough for the waste to ignite and sustain its own combustion. The secondary chamber normally requires a continuous supplemental fuel supply.

As the flue gas enters and exits the secondary combustion chamber, larger airborne particles settle out of the gas stream. Temperatures in the secondary chamber are high enough (in the range of 1400°F for refuse and other carbonaceous waste) to destroy unburned airborne particles. This equipment therefore has relatively low particulate emissions and in many cases can meet an emission standard of 0.08 gr/dry st ft^3 corrected to 12 percent CO_2, without additional air pollution control equipment.

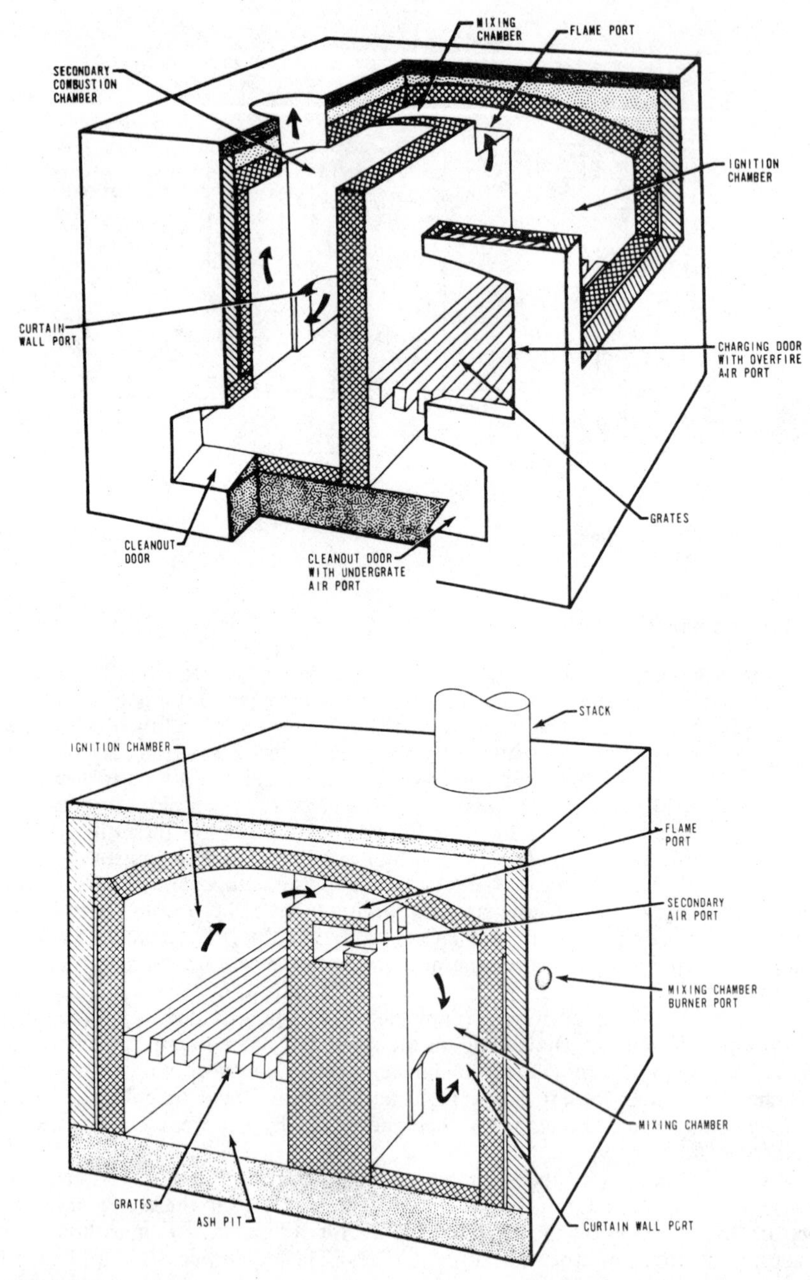

FIGURE 10.5 Cutaway of a retort multiple-chamber incinerator. *Source:* Ref. 10-3.

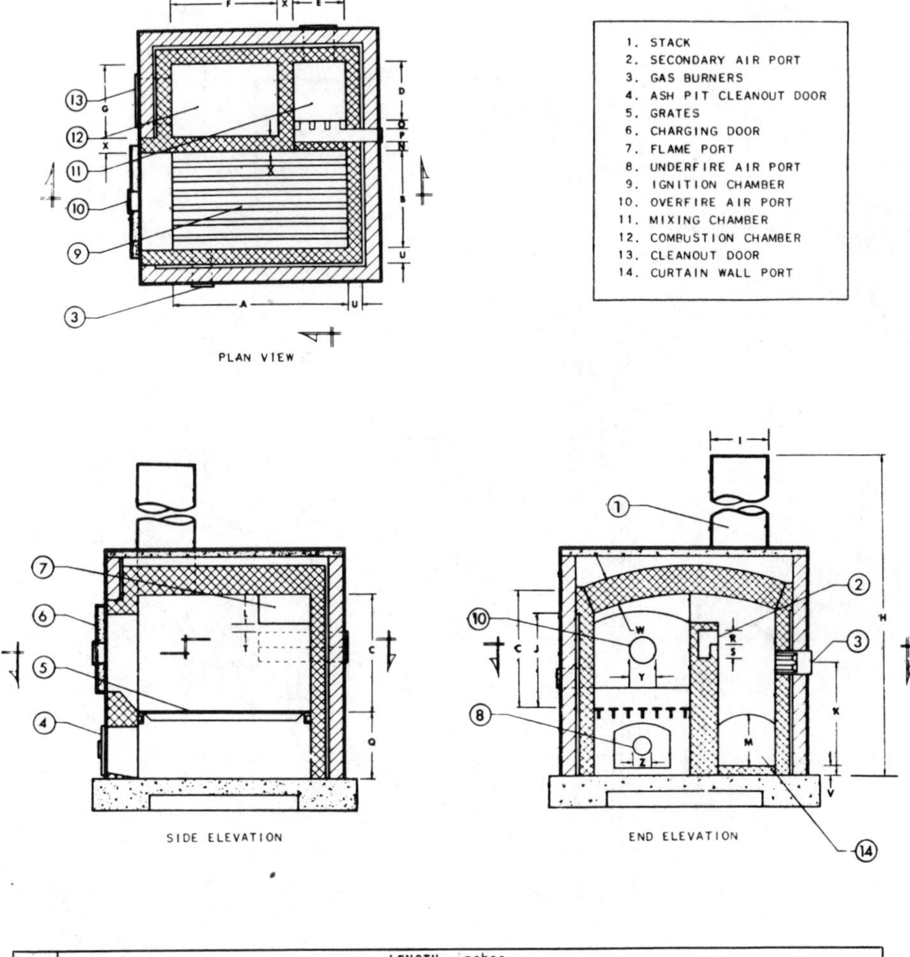

SIZE OF INCINERATOR, lb hr	LENGTH, inches																									
	A	B	C	D	E	F	G	H*	I	J	K	L	M	N	O	P	Q	R	S	T	U	V	W	X	Y	Z
50	31½	13½	22½	9	6¾	20¼	13½	18	8	18½	20	3¾	10	4½	2¼	2¼	9	2½	2½	2½	4½	2½	4½	4½	6	4
100	40½	18	28½	13½	9	27	18	19	12	23	28	5	15	2½	2½	4	14½	5	0	2½	4½	2½	4½	4½	8	5
150	45	22½	33½	15½	11½	29	22½	20	14	27	35½	5	16½	4½	2½	4½	18	5	2½	2½	4½	2½	4½	4½	9	6
250	54	27	37½	18	13½	36	27	22	18	30	40	7½	18	4½	4½	4½	20	5	2½	2½	4½	2½	4½	4½	12	6
500	76½	36	47½	27	18	49½	36	28	24	36½	48½	12½	23	9	4½	4½	26	5	5	2½	9	4½	9	9	16	8
750	85½	49½	54	36	22½	54	45	32	30	40	51½	15	28	9	4½	4½	25	5	10	2½	9	4½	9	9	18	8
1000	94½	54	59½	36	27	58½	45	35	34	45	54½	17½	30	9	4½	4½	27½	7½	12½	2½	9	4½	9	9	22	10

*Dimension "H" given in feet.

FIGURE 10.6 Design standards for multiple-chamber retort incinerators. *Source:* Ref. 10-3.

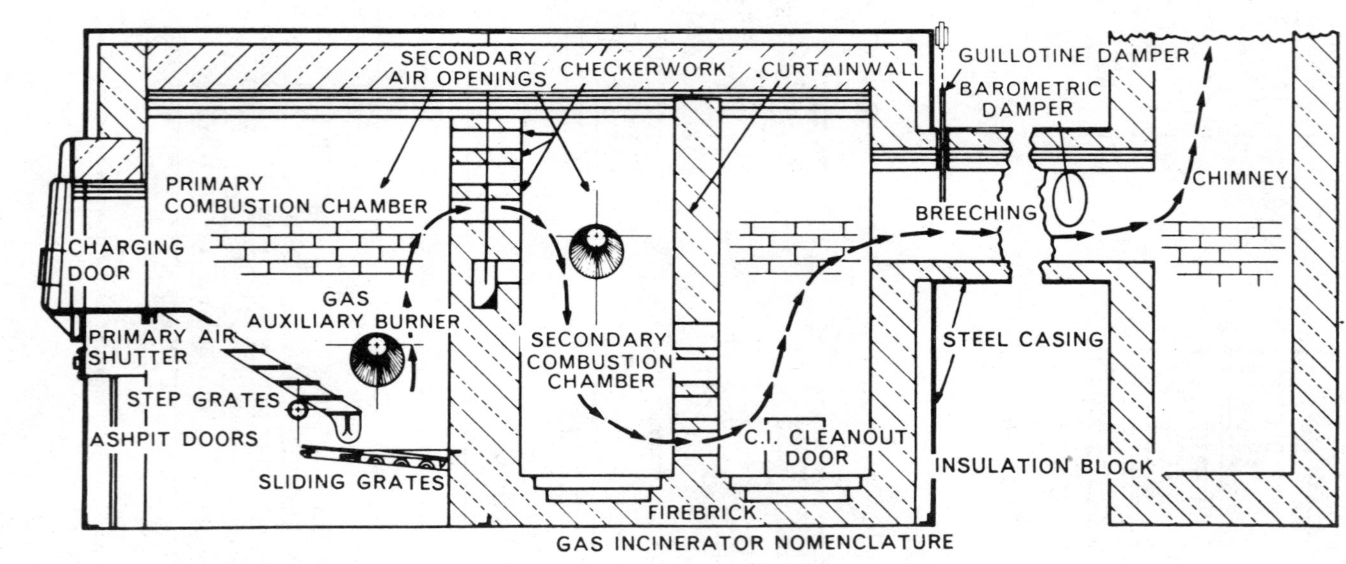

FIGURE 10.7 Gas incinerator. *Source:* Incinerator Committee, Industrial & Commercial Gas Section, American Gas Association, New York, N.Y.

In-Line Incinerator

This is a larger unit than the retort incinerator. Flow of combustion gases is straight through the incinerator, axially, with abrupt changes in the direction of flow only in the vertical direction, as shown in Fig. 10.7, an in-line incinerator using natural gas as supplemental fuel. Waste is charged on the grate which can be stationary or moving. A moving grate lends itself to continuous burning whereas stationary grates, as with the retort incinerator, are used for batch or semicontinuous operation. In-line incinerators are often provided with automatic ash removal equipment or ash discharge conveyers which also contribute to continuous operation of the incinerator.

As with the retort type, changes in the flow path and flow restrictions in an in-line incinerator provide settling out of larger airborne particles and increase turbulence for more effective burning.

Supplemental fuel burners in the primary chamber ignite the waste whereas secondary-chamber supplementary fuel burners provide heat to maintain complete combustion of the burnable components of the exhaust gas.

The retort incinerator is used in the range of 20 lb/h to approximately 750 lb/h. In-line incinerators are normally provided in the range of 500 to 2000 lb/h and greater with automatic charging and/or ash removal equipment not normally provided for units smaller than 1000 lb/h in capacity. Figure 10.7 shows an in-line incinerator with moving grates for charging and ash disposal. Figure 10.8 is another type of in-line incinerator utilizing manual charging, i.e., fixed grates. Typical in-line unit dimensions are shown in Fig. 10.9.

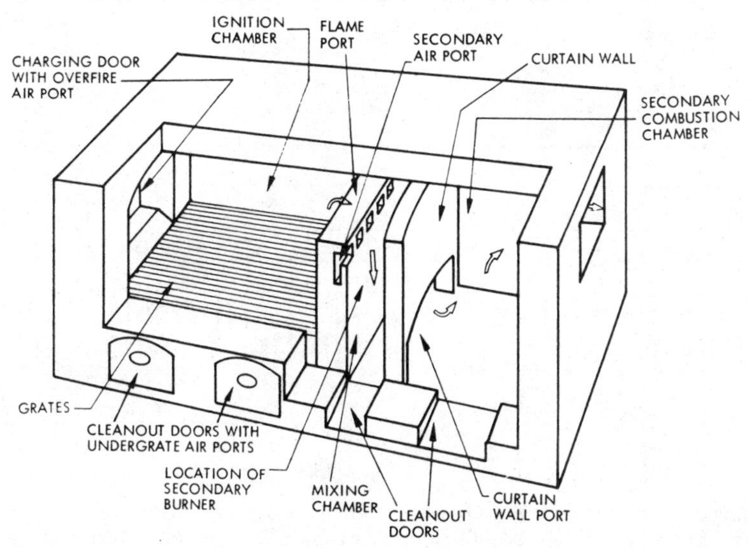

FIGURE 10.8 Cutaway of an in-line multiple-chamber incinerator. *Source:* Ref. 10-4.

Combustion air requirements are the same for either of these incinerators: approximately 300 percent excess air. Approximately half the required air enters as leakage through the charging port and other areas of the incinerator. Of the re-

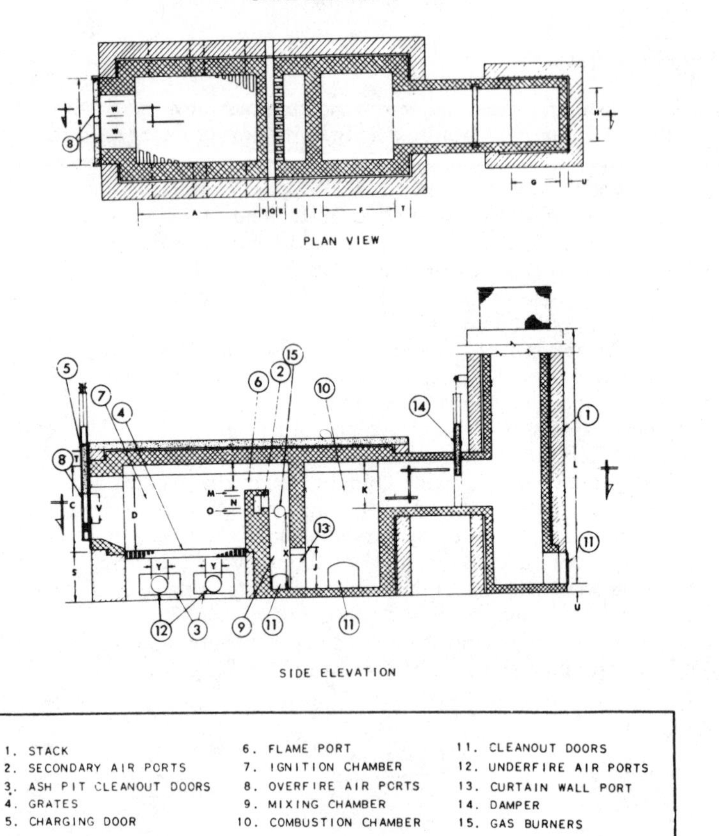

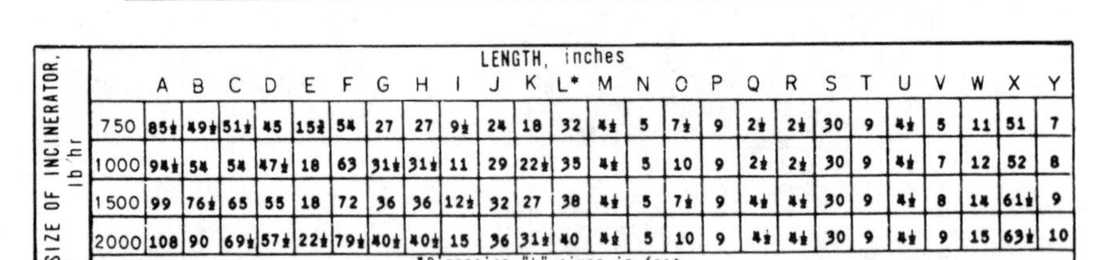

SIZE OF INCINERATOR, lb/hr	LENGTH, inches																								
	A	B	C	D	E	F	G	H	I	J	K	L*	M	N	O	P	Q	R	S	T	U	V	W	X	Y
750	85½	49½	51½	45	15¾	54	27	27	9½	24	18	32	4½	5	7½	9	2½	2½	30	9	4½	5	11	51	7
1000	94½	54	54	47½	18	63	31½	31½	11	29	22½	35	4½	5	10	9	2½	2½	30	9	4½	7	12	52	8
1500	99	76½	65	55	18	72	36	36	12½	32	27	38	4½	5	7½	9	4½	4½	30	9	4½	8	14	61½	9
2000	108	90	69½	57½	22½	79½	40½	40½	15	36	31½	40	4½	5	10	9	4½	4½	30	9	4½	9	15	63½	10

*Dimension "L" given in feet.

FIGURE 10.9 Design standards for multiple-chamber, in-line incinerators. *Source:* Ref. 10-3.

maining air requirement, 70 percent should be provided in the primary combustion chamber as overfire air, 10 percent as underfire air, and 20 percent in the mixing chamber or in the secondary combustion chamber.

Multiple-chamber incinerators will produce significantly lower emissions than single-chamber incinerators, as illustrated in Table 10.3. Water curtains across the path of the flue gases exiting the secondary combustion chamber will decrease emissions even further.

Multiple-chamber incinerators have been designed for specialty wastes. Typ-

TABLE 10.3 Comparison between Amounts of Emissions from Single- and Multiple-Chamber Incinerators

Item	Multiple chamber	Single chamber
Particulate matter, gr/st ft^3 at 12% CO_2	0.11	0.9
Volatile matter, gr/st ft^3 at 12% CO_2	0.07	0.5
Total, gr/st ft^3 at 12% CO_2	0.18	1.4
Total, lb/ton refuse burned	3.50	23.8
Carbon monoxide, lb/ton of refuse burned	2.90	197–991
Ammonia, lb/ton of refuse burned	0	0.9–4
Organic acid (acetic), lb/ton of refuse burned	0.22	<3
Aldehydes (formaldehyde), lb/ton of refuse burned	0.22	5–64
Nitrogen oxides, lb/ton of refuse burned	2.50	1
Hydrocarbons (hexane), lb/ton of refuse burned	<1	—

ical are pathological waste incinerators such as that shown in Fig. 10.10. Table 10.4 lists chemical composition and combustion data for pathological waste. Design factors and gas velocities for pathological waste incinerators are listed in Tables 10.5 and 10.6, respectively. Air emissions from pathological incinerators based on two test runs with and two runs without afterburner firing are listed in Tables 10.7 and 10.8, respectively. Note the significant decrease in emissions with the afterburner in operation.

A crematory retort is shown in Fig. 10.11. Its operating parameters for typical crematory waste are listed in Table 10.9.

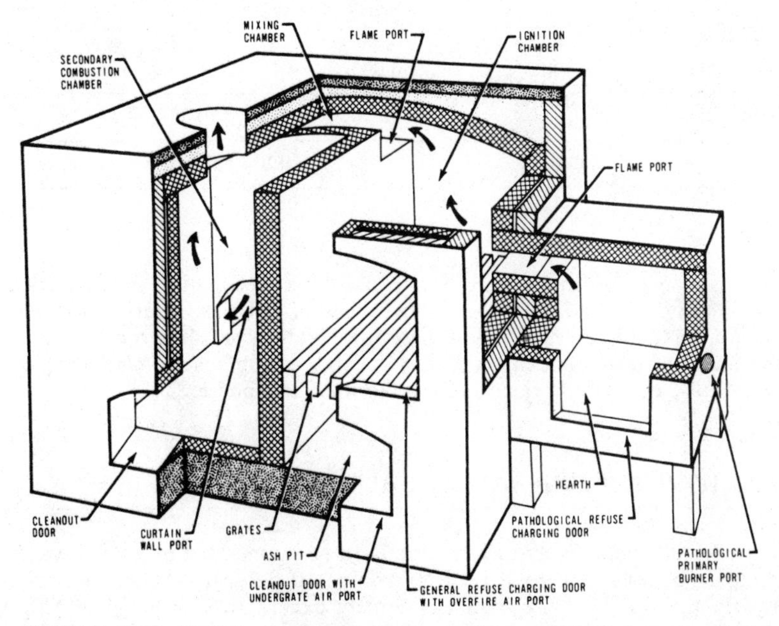

FIGURE 10.10 Multiple-chamber incinerator with a pathological waste retort. *Source:* Ref. 10-3.

TABLE 10.4 Chemical Composition of Pathological Waste and Combustion Data

Ultimate analysis
(whole dead animal)

Constituent	As charged % by weight	Ash-free combustible % by weight
Carbon	14.7	50.80
Hydrogen	2.7	9.35
Oxygen	11.5	39.85
Water	62.1	–
Nitrogen	Trace	–
Mineral (ash)	9	–

Dry combustible empirical formula - $C_5H_{10}O_3$

Combustion data
(based on 1 lb of dry ash-free combustible)

Constituent		Quantity lb	Volume scf
Theoretical air		7.028	92.40
40% sat at 60°F		7.069	93
Flue gas with theoretical air 40% saturated	CO_2	1.858	16.06
	N_2	5.402	73.24
	H_2O formed	0.763	15.99
	H_2O air	0.031	0.63
Products of combustion total		8.054	105.92
Gross heat of combustion		8,820 Btu per lb	

Source: Ref. 10-7, p. 420.

Another specialty incinerator is one used for reclamation of metal drums, shown in Fig. 10.12. Drums contaminated with volatile materials are passed through a high-temperature zone where the volatiles are driven off. These air-borne materials are induced through a secondary combustion chamber, where they are fully destroyed prior to release to the atmosphere.

SOLID WASTE INCINERATION CALCULATIONS

A sugar waste contaminated by caustic is to be incinerated in a solid waste incinerator. For 1000 lb/h total feed containing 10 percent caustic, fuel requirements for heating to 1800°F with 175 percent excess air will be calculated. The cooling water required to exhaust to 90°F will also be determined. This example will be calculated by the methods of Chap. 9. Tables 10.10, 10.11, and 10.12 are the mass balance, heat balance, and flue gas discharge tables, respectively.

TABLE 10.5 Design Factors for Pathological Ignition Chamber (Incinerator Cavity, 25 to 250 lb/h)

Item	Recommended value	Allowable deviation, %
Hearth loading	10 lb/(h · ft^2)	±10
Hearth length-to-width ratio	2	±20
Primary burner design	$\frac{10 \text{ ft}^3 \text{ natural gas}}{\text{lb waste burned}}$	±10

Source: Ref. 10-7, p. 462.

TABLE 10.6 Gas Velocities and Draft (Pathological Incinerators with Hot-Gas Passage below a Solid Hearth)

Item	Recommended values	Allowable deviation, %
Gas velocities		
Flame port at 1600°F, ft/s	20	±20
Mixing chamber at 1600°F, ft/s	20	±20
Port at bottom of mixing chamber at 1550°F, ft/s	20	±20
Chamber below hearth at 1500°F, ft/s	10	±100
Port at bottom of combustion chamber at 1500°F, ft/s	20	±20
Combustion chamber at 1400°F, ft/s	5	±100
Stack at 1400°F, ft/s	20	±25
Draft		
Combustion chamber, in WC	0.25[a]	−0, +25
Ignition chamber, in WC	0.05–0.10	± 0

[a]Draft can be 0.20 in WC for incinerators with a cold hearth.
Source: Ref. 10-7, p. 462.

Heat of Combustion of Glucose

From Chap. 8, assuming the sugar is in the form of glucose, $C_6H_{12}O_6$,

$$\underset{180.18}{\overset{-203.8}{C_6H_{12}O_6}} + 6O_2 \rightarrow \overset{-94.1}{6CO_2} + \overset{-57.8}{6H_2O}$$

$$\Delta H_f \; H_2O(g): \quad -57.8 \times 6 = -346.8$$

$$\Delta H_f \; CO_2: \quad -94.1 \times 6 = -564.6$$

$$\Delta H_f \; C_6H_{12}O_6: \quad -(-203.8) \times 1 = +203.8$$

$$\Sigma \Delta H_f = -707.6 \text{ kcal/mol}$$

TABLE 10.7 Emissions from Two Pathological-Waste Incinerators with Secondary Burners

Test no.	A	B
	Mixing chamber burner operating	Mixing chamber burner operating
Rate of destruction to powdery ash, lb/h	19.2	99
Type of waste	Placental tissue in newspaper at 40°F	Dogs freshly killed
Combustion contaminants		
gr/st ft^3 [a] at 12% CO_2	0.200	0.300
gr/st ft^3	0.014	0.936
lb/h	0.030	0.360
lb/ton charged	3.120	7.260
Organic acids		
gr/st ft^3	0.006	0.013
lb/h	0.010	0.050
lb/ton charged	1.040	1.010
Aldehydes		
gr/st ft^3	N.A.[b]	0.006
lb/h	N.A.[b]	0.020
lb/ton charged	N.A.[b]	0.400
Nitrogen oxides		
ppm	42.70	131
lb/h	0.08	0.099
lb/ton charged	8.84	2
Hydrocarbons	Nil	Nil

[a] CO_2 from burning of waste used only to convert to basis of 12% CO_2.
[b] Not available.
Source: Ref. 10-7, p. 462.

Relating $\Sigma\Delta H_f$ to the weight of glucose present, we get

$$\frac{1}{180.18 \text{ g/mol}} \times (-707.6 \text{ kcal/mol}) \times 1802 = -7077 \text{ Btu/lb}$$

The negative heating value indicates that this is an exothermic reaction; i.e., heat is released.

Caustic Thermodynamics

Caustic is $Ca(OH)_2$, calcium hydroxide. From Chap. 8,

$$\begin{array}{cccc} -239.7 & -151.8 & & -57.8 \text{ (steam)} \\ Ca(OH)_2 \rightarrow & CaO & + & H_2O \\ 74.28 & 56.08 & & 18.02 \end{array}$$

TABLE 10.8 Emissions from Two Pathological-Waste Incinerators without Secondary Burners (Source Tests of Two Pathological-Waste Incinerators)

Test no.	A	B
	Mixing chamber burner not operating	Mixing chamber burner not operating
Rate of destruction to powdery ash, lb/h	26.4	107
Type of waste	Placental tissue in newspaper at 40°F	Dogs freshly killed
Combustion contaminants		
gr/st ft^3 [a] at 12% CO_2	0.500	0.300
gr/st ft^3	0.017	0.128
lb/h	0.030	0.430
lb/ton charged	2.270	8.040
Organic acids		
gr/st ft^3	0.010	0.034
lb/h	0.020	0.110
lb/ton charged	1.514	2.050
Aldehydes		
gr/st ft^3	0.007	0.010
lb/h	0.013	0.033
lb/ton charged	0.985	0.617
Nitrogen oxides		
ppm	14.700	95
lb/h	0.016	0.082
lb/ton charged	1.210	1.550
Hydrocarbons	Nil	Nil

[a] CO_2 from burning of waste only used to convert to basis of 12% CO_2.
Source: Ref. 10-7, p. 461.

$$\Delta H_f\ H_2O(g): \quad -57.8 \times 1 = -57.8$$

$$\Delta H_f\ CaO: \quad -151.8 \times 1 = -151.8$$

$$\Delta H_f\ Ca(OH)_2: \quad -(-239.7 \times 1) = +239.7$$

$$\Sigma \Delta H_f = +30.1 \text{ kcal/mol}$$

Relating $\Sigma \Delta H_f$ to the weight of caustic present gives

$$\frac{1}{74.28 \text{ g/mol}} \times (30.1 \text{ kcal/mol}) \times 1802 = 730 \text{ Btu/lb}$$

This value is positive, indicating that 730 Btu must be added to complete the destruction of each pound of caustic. The contribution of latent heat of moisture must be subtracted from the total heating value, 730 Btu/lb, because the heat required to raise the moisture present to the furnace temperature is included with the heat balance calculations for the incinerator.

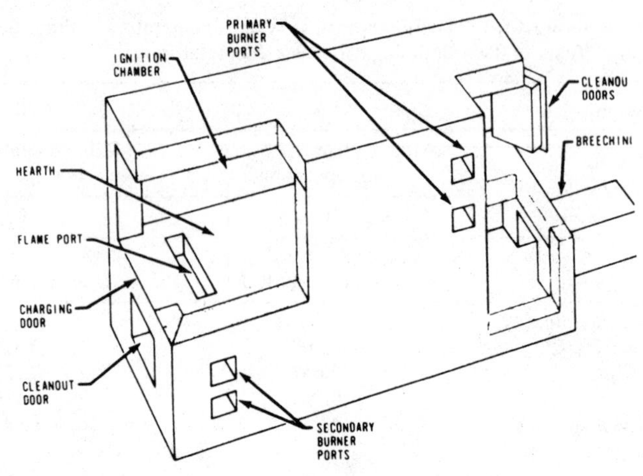

FIGURE 10.11 Crematory retort. *Source:* Ref. 10-3.

Therefore the net heating value of caustic is calculated as follows:

Latent heat of moisture: 970 Btu/lb

$$970\ (\text{Btu/lb } H_2O) \times \frac{18.02\ \text{lb } H_2O}{74.28\ \text{lb caustic}} = 235\ \text{Btu/lb}$$

Net heat required to destruct caustic:

$$730\ \text{Btu/lb gross} - 235\ \text{Btu/lb } H_2O = 495\ \text{Btu/lb}$$

For incinerator calculations a moisture and an ash component will be determined, as follows:

$$\text{Total caustic:}\quad 1000\ \text{lb/h} \times 10\% = 100\ \text{lb/h caustic}$$

The ash generated is CaO, with a molecular weight of 56.08, compared to a molecular weight of 74.28 for caustic. Therefore the quantity of ash produced is

$$\frac{56.08\ \text{lb ash}}{74.28\ \text{lb caustic}} \times 100\ \text{lb/h caustic} = 75.50\ \text{lb/h ash}$$

The ash absorbs heat as follows:

$$495\ \text{Btu/lb caustic} \times \frac{74.28\ \text{lb caustic}}{56.08\ \text{lb ash}} = 656\ \text{Btu/lb ash}$$

The quantity of moisture within the caustic is proportional to atomic weight:

$$\frac{18.02\ \text{lb } H_2O}{74.28\ \text{lb caustic}} \times 100\ \text{lb caustic} = 24\ \text{lb } H_2O$$

TABLE 10.9 Operating Procedures for Crematory

Phase	Duration, 1½ h operation, min	Burner settings	Casket	Body		
				Moisture	Tissue	Bone calcin
Charging[a]	—	Secondary zone on				
Ignition	15	All on	20% burns	—	—	—
Full combustion	30	All on	80% burns	20% evap.	10% burns	—
Final combustion	45	All on	—	80% evap.	90% burns	50%
Calcining	1 to 12 h	All off (or small primary on)	—	—	—	50%
	Duration, 2½ h operation, min					
Ignition	15	All on	20% burns	—	—	—
Full combustion	30	Primary off	60% burns	20% evap.	—	—
Final combustion	15	All on	20% burns	20% evap.	20% burns	50%
	90	All on	—	60% evap.	80% burns	—
Calcining	1 to 12 h	All off (small primary may be on)				

[a]Charge: Casket 75 lb wood
Body 180 lb
Moisture: 108 lb
Tissue: 50 lb
Bone: 22 lb

Source: Ref. 10-7, p. 464.

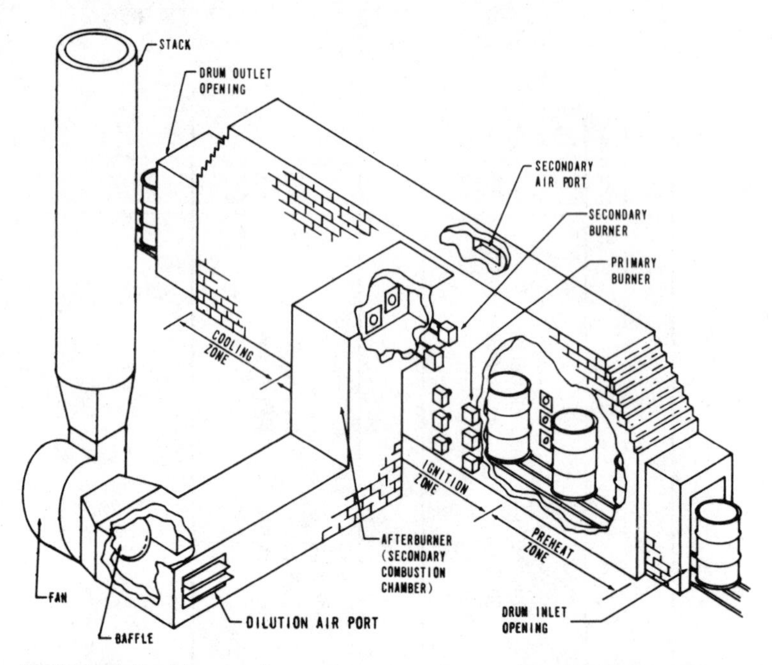

FIGURE 10.12 Diagram of a continuous-type drum reclamation furnace with an afterburner. *Source:* Ref. 10-3.

TABLE 10.10 Mass Flow

	Example (solid)
Wet feed, lb/h	1000
Moisture, %	2.4
lb/h	24
Dry feed, lb/h	976
Ash, %	7.8
lb/h	76
Volatile, lb/h	900
Volatile htg. value, Btu/lb	7077
MBtu/h	6.37
Dry gas, lb/10 kBtu	7.89
lb/h	5026
Comb. H_2O, lb/10 kBtu	0.57
lb/h	363
Dry gas + comb. H_2O, lb/h	5389
100% Air, lb/h	4489
Total air fraction	2.75
Total air, lb/h	12,345
Excess air, lb/h	7856
Humid/dry gas (air), lb/lb	0.01
Humidity, lb/h	123
Total H_2O, lb/h	510
Total dry gas, lb/h	12,882

TABLE 10.11 Heat Balance

	Example (solid)
Cooling air wasted, lb/h	—
°F	—
Btu/lb	—
MBtu/h	—
Ash, lb/h	76
Btu/lb	656
MBtu/h	0.05
Radiation, %	3
MBtu/h	0.19
Humidity, lb/h	123
Correction (@970 Btu/lb), MBtu/h	−0.12
Losses, total, MBtu/h	.12
Input, MBtu/h	6.37
Outlet, MBtu/h	6.25
Dry gas, lb/h	12,882
H_2O, lb/h	510
Temperature, °F	1651
Desired temp., °F	1800
MBtu/h	6.83
Net, MBtu/h	0.58
Fuel oil, air fraction	1.1
Net Btu/gal	70,746
gal/h	8.20
Air, lb/gal	114.64
lb/h	940
Dry gas, lb/gal	115.12
lb/h	944
H_2O, lb/gal	8.62
lb/h	71
Dry gas w/fuel oil, lb/h	13,826
H_2O w/fuel oil, lb/h	581
Air w/fuel oil, lb/h	13,285
Outlet, MBtu/h	7.40
Reference t, °F	60

The values developed for wet feed, moisture, ash content, and heating value are inserted in Table 10.10. Total moisture and dry gas flow, plus air requirements for waste combustion, are calculated by the methods of Chap. 9.

The ash heating value, 656 Btu/lb, is inserted in the heat balance sheet, Table 10.11. This table and the flue gas analysis, Table 10.12, have been completed per Chap. 9 calculations.

TABLE 10.12 Flue Gas Discharge

	Example (solid)
Inlet, °F	1800
Dry gas, lb/h	13,826
Heat, MBtu/h	7.40
Btu/lb dry gas	535
Adiabatic t, °F	171
H_2O saturation, lb/lb dry gas	0.4493
lb/h	6212
H_2O inlet, lb/h	581
Quench H_2O, lb/h	5631
gal/min	11.3
Outlet temp., °F	90
Raw H_2O temp., °F	70
Sump temp., °F	143
Temp. diff., °F	73
Outlet, Btu/lb dry gas	48.212
MBtu/h	0.67
Req'd. cooling, MBtu/h	6.73
H_2O, lb/h	92,192
gal/min	184
Outlet, ft^3/lb dry gas	14.547
ft^3/min	3352
Fan press, in WC	20
Outlet, actual ft^3/min	3525
Outlet, H_2O/lb dry gas	0.03115
H_2O, lb/h	431
Recirc. (ideal), gal/min	11.3
Recirc. (actual), gal/min	91
Cooling H_2O, gal/min	184

REFERENCES

10-1. *Source Category Survey: Industrial Incinerators*, USEPA 450 3-80-013, May 1980.

10-2. MacKnight, R.: "Controlling the Flue-Fed Incinerator," *Journal of the Air Pollution Control Association*, 10:103–9, April 1960.

10-3. Danielson, J.: *Air Pollution Engineering Manual*, County of Los Angeles, Air Pollution Control District, AP-40, May 1973.

10-4. *Recommended Methods of Reduction, Neutralization, Recovery or Disposal of Hazardous Waste*, vol. 3, Disposal Process Descriptions, Ultimate Disposal, Incineration and Pyrolysis Processes, USEPA 670/2-73-053C, August 1973.

BIBLIOGRAPHY

Archinger, W., and Giar, J.: *Testing Manual for Solid Waste Incinerators*, USEPA, 1973.

Hitchock, D.: "Solid Waste Disposal: Incineration," *Chemical Engineers*, May 21, 1979.

Rice, R.: "Solving the Small Company Solid Waste Problem," *Public Works Magazine*, September 1976.

Rubel, F.: *Incineration of Solid Wastes*, Noyes, Park Ridge, N.J., 1974.

CHAPTER 11
PYROLYSIS AND CONTROLLED AIR INCINERATION

Pyrolysis is a thermal process related to oxidation, or burning. In this chapter the mechanisms of pyrolysis are discussed, and equipment utilizing pyrolysis and starved air reactions are presented.

GENERAL DESCRIPTION

Pyrolysis is the destructive distillation of a solid, carbonaceous, material in the presence of heat and in the absence of stoichiometric oxygen. It is an endothermic reaction; i.e., heat must be applied for the reaction to occur.

Ideally a pyrolytic reaction will occur as follows, using cellulose:

$$C_6H_{10}O_5 \xrightarrow{\text{heat}} CH_4 + 2CO + 3H_2O + 3C$$

A gas is produced containing methane, CH_4, carbon monoxide, CO, and moisture. The carbon monoxide and methane components are combustible, providing heating value to the off-gas. The carbon residual, a char, also has heating value. This is an idealized reaction. No oxygen is added, and the original material is pure cellulose, $C_6H_{10}O_5$. In general the initial material is not pure and contains additional components, both organic and inorganic. The off-gas is a mixture of many simple and complex organic compounds. The char is often a liquid which contains minerals, ash, and other inorganics as well as residual carbon or tars.

Pyrolysis as an industrial process has been in use for years and only since the late 1960s has had significant use in destruction of waste materials. The pyrolysis process produces charcoal from wood chips, coke and coke gas from coal, fuel gas and pitch from heavy-hydrocarbon still bottoms, etc.

PYROLYSIS PRODUCTS

As noted previously, an ideal, simple, pyrolytic reaction produces an off-gas with combustible content and a char. In an actual case, however, the feed materials are complex and the products generated from the pyrolytic process are likewise complex and varied. Depending on the temperature of the reaction, the pressure in the reaction chamber, and the amount of air (or oxygen) allowed to enter the chamber, the reaction products will vary in composition.

The product yields from the pyrolysis of typical solid waste charges are listed in Table 11.1. For this particular unit the temperature in the pyrolysis chamber varied from 950 to 1650°F, with a waste retention time of 12 to 15 min. The processed municipal waste had the majority of its glass and metal content removed. Note that the off-gas produced is greatest at the higher temperature whereas tars and other liquors generated are greatest at the lower temperature.

The gas produced in these reactions will have a heating value of approximately 300 to 400 Btu/st ft^3. Ammonia is produced from the nitrogen component of the waste feed, and much of this ammonia combines with sulfur in the waste to produce ammonium sulfate.

Table 11.2 lists the composition of the solid residues, and the residue heating value, from the processes of Table 11.1.

Typical off-gas compositions for various reaction temperatures are given in Table 11.3. This gas has sufficient heating value to maintain combustion, and it can be used as a low-grade fuel. It does contain particulate matter and trace contaminants, such as organic acids, and these components must be removed before the pyrolysis gas can be used remote from the reactor system. In general, pyrolysis gas is utilized solely in the afterburner of the pyrolysis system.

THE PYROLYSIS SYSTEM

An idealized pyrolysis system for disposal of mixed waste is shown in Fig. 11.1. The waste received is sorted for removal of glass, metal, and cardboard, all of which has possible resale value. The waste stream enters a shredder (grinder), and the shredded material passes through a magnetic separator where residual ferrous metal is removed, for resale.

The balance of the waste stream will be fed into the reactor from a feed hopper. The hopper discharge and the feeder must be provided with air locks to minimize the infiltration of air (oxygen) which will degrade the pyrolysis reaction.

Shredding is a necessary step, not only to allow metals removal but also to provide a uniform-size feed of relatively small particles to the reactor.

The converter is heated externally, as shown. Other types of pyrolytic reactors are designed to allow sufficient air infiltration to provide some burning within the reactor, generating enough heat internally to sustain the process.

Gas, exiting the reactor, is collected in a storage tank where organic acids and other organic compounds condense and are eventually discharged. Between 30 and 40 percent of the gas is required to heat the pyrolytic reactor; the balance of the gas stream can be used for other processes. In this generalized scheme a significant portion of the heating value of the off-gas is contained within the condensables. If the gas is heated and the discharged char residue is cooled, the condensables will remain in the gaseous state. As the gas cools and the condensables leave the gas stream, the gas heating value will decrease. Therefore, for maximum energy reclamation from the gas, it is important that the gas be kept in a heated state as long as possible—at least long enough for the gas to reach the farthest gas burner. Storage should be mini-

TABLE 11.1 Yields of Products from Pyrolysis of Municipal and Industrial Waste

Refuse	Pyrolysis temp., °F	Yields, weight percent of waste							Yields per ton of waste				
		Residue	Gas	Tar	Light oil in gas	Free ammonia	Liquor	Total	Gas, ft^3	Tar, gal	Light oil in gas, gal	Liquor, gal	Ammonium sulfate, lb
Raw municipal waste	930	9.3	26.7	2.2	0.5	0.05	55.8	94.6	11,509	4.8	1.5	133.4	17.9
	1380	11.5	23.7	1.2	0.9	0.03	55.0	92.3	9,628	2.6	2.5	131.6	23.7
	1650	7.7	39.5	0.2	0.0	0.03	47.8	95.2	17,741	0.5	0.0	113.9	25.1
Processed municipal waste	930	21.2	27.7	2.3	1.3	0.05	40.6	93.2	11,545	5.6	3.7	96.7	16.2
	1380	19.5	18.3	1.0	0.9	0.02	51.5	91.2	7,380	2.2	2.6	122.6	28.4
	1650	19.1	40.1	0.6	0.2	0.04	35.3	95.3	18,058	1.4	0.6	97.4	31.5
Industrial-sample A	930	36.1	23.7	1.9	0.5	0.05	31.6	93.9	9,563	4.1	1.4	75.2	12.5
	1380	37.5	22.8	0.7	0.9	0.03	30.6	92.5	9,760	1.5	2.6	73.0	19.5
	1650	38.8	29.4	0.2	0.6	0.04	21.8	90.8	12,318	0.5	1.6	51.1	21.7
Industrial-sample B	930	41.9	21.8	0.8	0.6	0.03	29.5	94.6	9,270	1.7	1.6	70.2	20.4
	1380	31.4	25.5	0.8	0.8	0.03	31.5	90.0	10,952	1.8	2.2	74.9	21.2
	1650	30.9	31.5	0.1	0.5	0.03	29.0	92.0	14,065	0.02	1.4	68.5	22.9

Source: Ref. 11-1.

TABLE 11.2 Chemical Analyses[a] of Solid Residues from Pyrolysis of Municipal and Industrial Waste

Refuse	Pyrolysis temp., °F	Proximate, percent				Ultimate, percent						
		Moisture	Volatile matter	Fixed carbon	Ash	Hydrogen	Carbon	Nitrogen	Oxygen	Sulfur	Heating value, Btu/lb	MBtu/ton
Raw Municipal Waste	930	2.6	4.4	29.6	66.0	0.4	32.4	0.5	0.5	0.2	5020	10.04
	1380	2.2	7.4	51.4	41.2	0.8	54.9	1.1	1.8	0.2	8020	16.04
	1650	1.0	4.7	31.7	63.6	0.3	36.1	0.5	0.0	0.2	5260	10.52
Processed Municipal Waste	930	1.7	4.8	56.7	38.5	0.6	57.7	0.8	2.1	0.3	8800	17.60
	1380	1.3	13.4	34.6	52.0	0.8	41.9	0.8	4.4	0.1	6080	12.16
	1650	1.2	3.3	53.5	43.2	0.5	53.4	0.7	1.8	0.4	8090	16.18
Industrial-Sample A	930	0.9	2.6	15.2	82.2	0.3	17.0	0.1	0.2	0.2	2520	5.04
	1380	1.2	5.1	17.9	77.0	0.5	19.4	0.2	1.8	0.2	2900	5.08
	1650	0.1	2.5	12.9	84.6	0.3	14.8	0.2	0.0	0.2	2180	4.36
Industrial-Sample B	930	0.3	3.0	9.7	87.3	0.2	11.8	0.1	0.4	0.2	1660	3.32
	1380	1.0	3.6	16.6	79.8	0.3	19.5	0.2	0.0	0.2	2680	5.36
	1650	0.2	6.4	16.2	77.4	0.4	19.3	0.3	2.4	0.2	2810	5.62

[a]Moisture on as-received basis, all other data on dry basis.
Source: Ref. 11-1.

TABLE 11.3 Pyrolysis Gas Composition

	Pyrolytic temperature, °F			
	900	1200	1500	1700
Gas composition, volume percent				
Carbon monoxide	33.6	30.5	34.1	35.3
Carbon dioxide	44.8	31.8	20.6	18.3
Hydrogen	5.6	16.5	28.6	32.4
Methane	12.5	15.9	13.7	10.5
Ethane	3.0	3.1	0.8	1.1
Ethylene	0.5	2.2	2.2	2.4
Heating value (HHV), Btu/st ft^3	312	403	392	385

Source: Ref. 11-2.

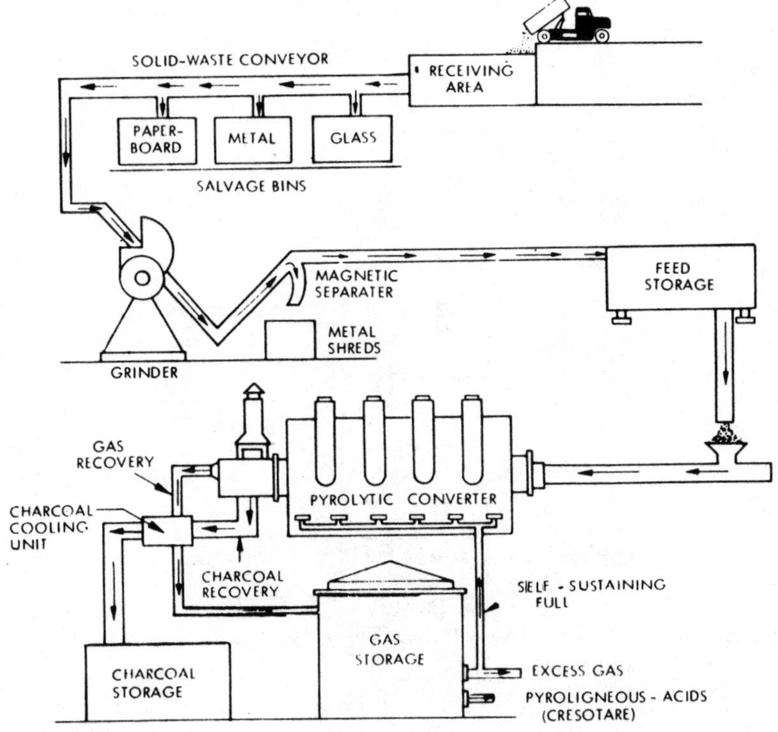

FIGURE 11.1 Pyrolytic waste conversion. *Source:* Ref. 11-3.

mized because the condensables will leave the gas stream relatively readily in any quiescent area.

The residual solid material is termed *charcoal*. This is, ideally, a desired by-product of this reaction.

A stack is shown immediately downstream of the converter. Upon start-up of the process, when an outside source of heat is required to initiate the reaction (not shown), the initial off-gas is basically composed of steam, carbon dioxide, entrapped air, and trace amounts of carbon monoxide. These components can be vented through the stack until the process stabilizes and pyrolysis gas is produced.

PUROX® AND TORRAX®

The Union Carbide Corporation developed a pyrolysis reactor for destruction of refuse or other mixed solid waste, as shown in Fig. 11.2. The reactor is a vertical refractory-lined chamber with waste feed charged from its top.

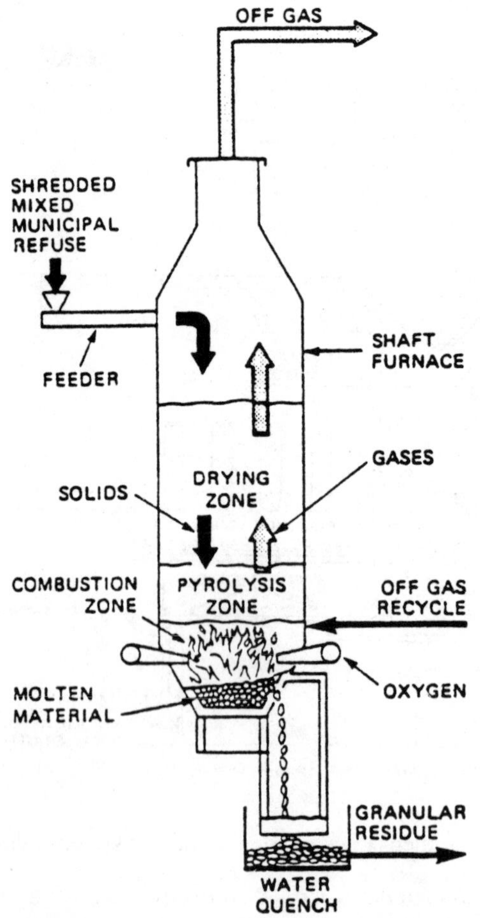

FIGURE 11.2 Purox reactor. *Source:* Union Carbide Corp., Charleston, W.Va.

Oxygen is injected in the bottom of the reactor, hence the name Purox, for pure oxygen. Oxygen reacts with a portion of the waste to produce the heat required to sustain this process. The hot gases generated by burning rise up through the waste and pyrolyze the waste as it cools. In the upper portion of the reactor, the gas is cooled further as it dries the incoming material.

The off-gas exiting the furnace is relatively clean, normally at a temperature of approximately 200°F, and with a heating value of 300 Btu/st ft^3. The use of oxygen instead of air results in a higher heating value (because of the absence of nitrogen which would act as a dilutant) and the absence of nitrogen oxides in the gas stream.

The energy in the off-gas is normally equivalent to approximately 80 percent of the heat energy within the waste. The residue from this process is quenched in a water bath and is characteristically granular. It is sterile and free of any biologically active material because it has gone through a hot molten state. Also, because of the burning reaction included in this process, it contains a relatively small carbon residual, as shown in Table 11.4. The volume of solid residue is 2 to 3 percent of the incoming refuse volume. This reactor has been built in 200 ton/day units.

TABLE 11.4 Purox Granular Residue

Component	Weight Percent
FeO	9.0
Fe_2O_3	1.7
MnO	0.7
SiO_2	63.1
$CaCo_3$	1.6
CaO	13.7
Al_2O_3	9.2
TiO_2	0.1

Source: Union Carbide Corp., South Charleston, Va.

The Torrax (total reduction) system for pyrolysis of solid waste has originally been developed by the Carborundum Corporation and is marketed as the Andco and Andco-Torrax system. The reactor is shown in Fig. 11.3.

Waste is not shredded or otherwise classified. It is charged into the top of the reactor, and by virtue of its weight and compactability it forms a plug, preventing flow of gases through it.

Air is injected at the bottom of the reactor to promote sufficient burning of the waste to generate heat to sustain the process. Gases rise through the reactor, heating and pyrolyzing the charge, drying the fresh material, and exiting the reactor beneath the refuse plug.

Gas exits the reactor at temperatures in the range of 800 to 1000°F and has a heating value of approximately 120 to 150 Btu/st ft^3. Normally a gas cannot sustain combustion with a heating value less than 150 Btu/st ft^3. The low heating value of this gas, therefore, makes it uneconomical for export. Because of its high temperature and its low heating value, it becomes economical to fire this off-gas on-site, adjacent to the reactor. The gas can be used for generation of steam or hot water, although supplemental fuel may be required to sustain combustion. The off-gas can be used to provide heat for the reactor and generate steam, increasing system efficiency.

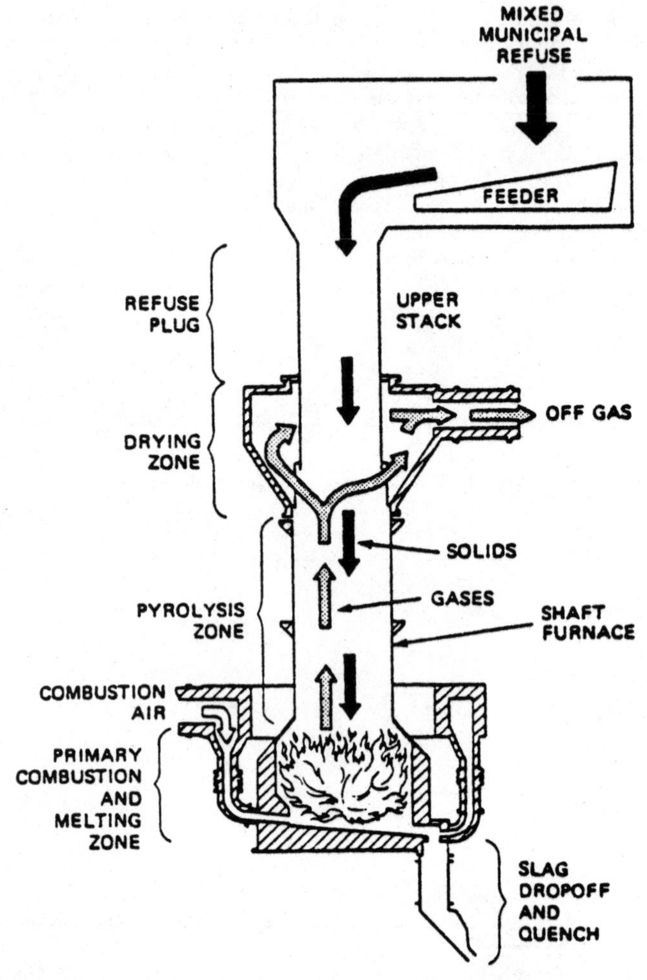

FIGURE 11.3 Torrax reactor. *Source:* Andco-Torrax Corp.

The solid residual normally amounts to only 3 to 5 percent by volume, 15 to 20 percent by weight of the incoming feed. The residual is continuously removed (or tapped) from the reactor and is immediately quenched. It is granular and typically has a composition as listed in Table 11.5.

OTHER PYROLYSIS SYSTEMS

There are many studies in progress not far distant from those of the traditional alchemists. Instead of yellow gold the goal is black gold—petroleum and petroleum-derived products. Currently processes and systems are exemplary if

TABLE 11.5 Torrax Solid Residue

Component	Typical % by Weight	Range, %
SiO_2	45.0	32.0–58.0
Al_2O_3	10.0	5.5–11.0
TiO_2	0.8	0.5–1.3
Fe_2O_3	10.0	0.5–22.0
FeO	15.0	11.0–21.0
MgO	2.0	1.8–3.3
$CaCO_3$	1.1	0–1.5
CaO	8.0	4.8–12.1
MnO	0.6	0.2–1.0
Na_2O	6.0	4.0–8.6
K_2O	0.7	0.4–1.1
Cr_2O_3	0.5	0.1–1.7
CuO	0.2	0.1–0.3
ZnO	0.1	0–0.3
Particle density	174.7 lb/ft^3	
Residue density	87.4 lb/ft^3	
Screen size	4% > 3½ mesh	
	2% < 30 mesh	

Source: Andco-Torrax Corp.

they just dispose of waste, generating innocuous residual materials without creating nuisance odor or budget overruns. But new processes will undoubtedly follow the perfectly sound theory of oil from waste (one hydrocarbon from another), and perhaps a workable system will be found to generate the potential of 1 barrel of oil per ton of waste by pyrolysis. A related process, starved air or controlled air combustion has been successfully developed.

STARVED AIR INCINERATION

In the early 1960s a new type of incinerator started gaining in popularity. The *modular combustion unit* has become an economical and efficient system for on-site and central destruction of waste. These incinerators are also known as *controlled air* units. They can be operated as excess air units (EAU) or starved air units (SAU).

THEORY OF OPERATION

The SAU consists of two major furnace components, as shown in Fig. 11.4, a primary chamber and a secondary chamber. Waste is charged into the primary chamber, and a carefully controlled flow of air is introduced. Only enough air is

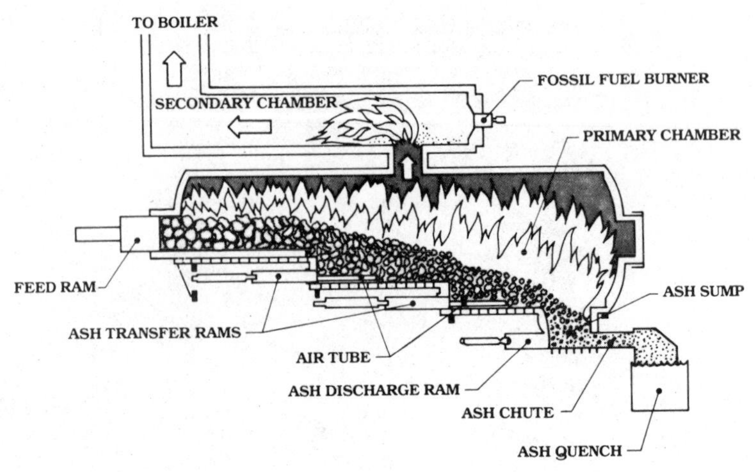

FIGURE 11.4 Starved air incinerator. *Source:* Ref. 11-4.

provided to allow sufficient burning for heating to occur. Typically 70 to 80 percent of the stoichiometric air requirement is introduced into the primary chamber.

The off-gas generated by this starved air reaction will contain combustibles, and this gas is burned in the secondary chamber, which is sized for sufficient residence time to totally destroy organics in the off-gas. As in the primary chamber, a carefully controlled quantity of air is introduced into the secondary chamber, but in this case excess air, 140 to 200 percent of the off-gas stoichiometric requirement, is maintained to effect complete combustion. Gas-cleaning devices such as wet scrubbers or electrostatic precipitators may not be required. The burnout of the off-gas in the secondary chamber is usually sufficient to clean the gas to 0.08 grams/st ft^3.

Figure 11.5 illustrates the variety of configurations currently marketed for controlled air incineration. They all have a starved air primary section and a secondary or afterburner chamber.

CONTROL

As can be seen in Fig. 11.6, the temperature is directly related to the excess air provided. Temperature therefore is normally utilized to control airflow in both primary and secondary chambers.

Below stoichiometric the temperature of the reaction increases with an increase in airflow. As more air is provided, more combustion will occur, so more heat will be released. This heat release will result in higher temperatures produced.

Control of the primary-chamber operation, therefore, where less than complete oxidation is required, is as follows:

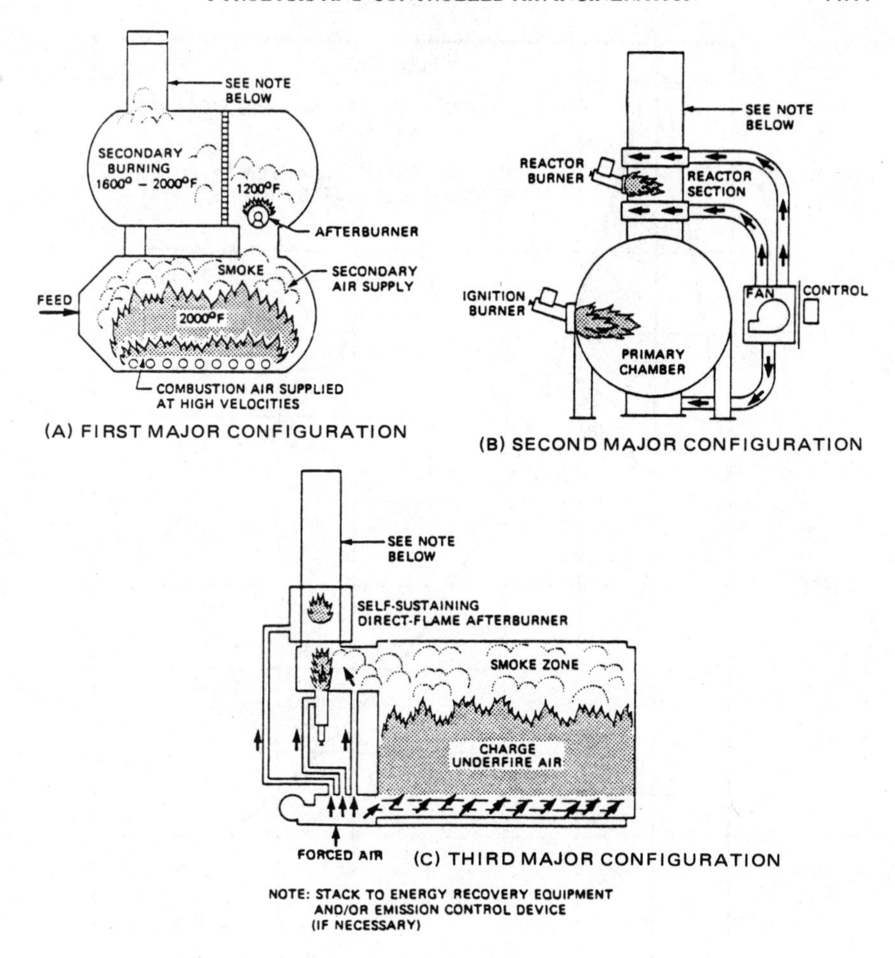

FIGURE 11.5 Starved air incinerator configurations.

- With higher temperatures, decrease airflow.
- With lower temperatures, increase airflow.

The secondary chamber is designed for complete combustion, greater than stoichiometric air supplied. At stoichiometric conditions all the combustible material present will combust completely. Additional air will act to quench the off-gas, i.e., will lower the resulting exhaust gas temperature. Therefore, control of the secondary-chamber operation is as follows:

- With higher temperatures, increase airflow.
- With lower temperatures, decrease airflow.

SAUs are normally provided with temperature detectors which automatically control fan damper positioning to provide the required chamber airflow.

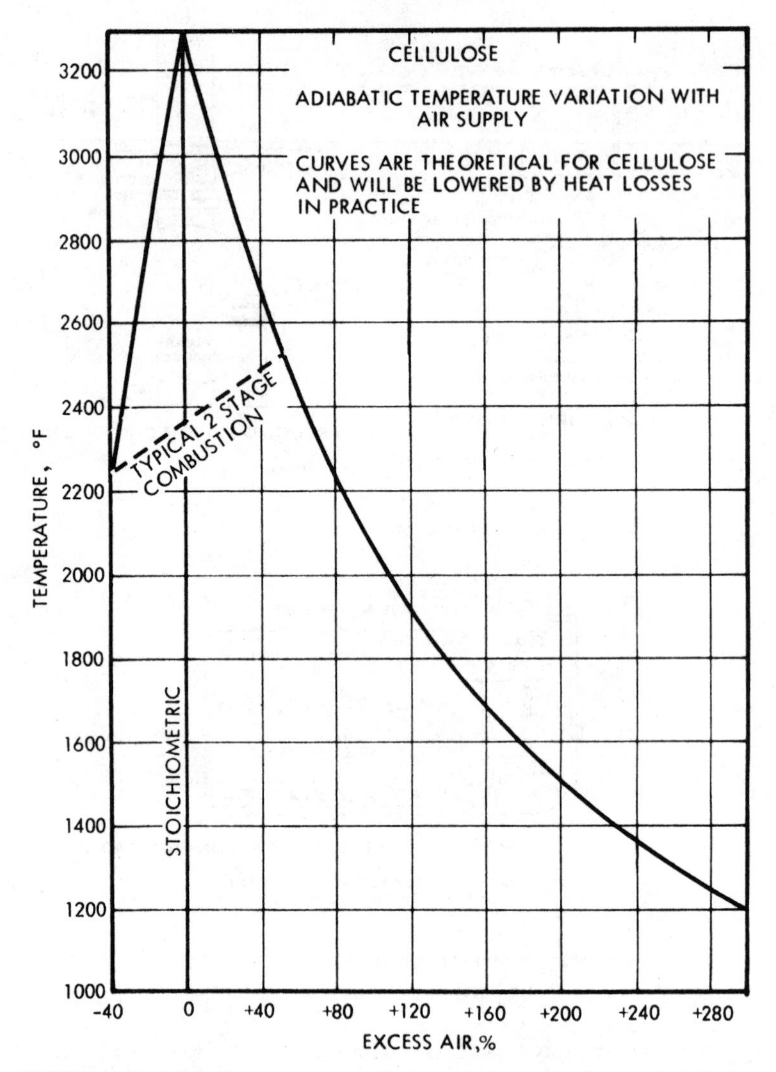

FIGURE 11.6 Adiabatic temperature variation with air supply. *Source:* Ref. 11-3.

INCINERABLE WASTES

The SAUs were originally developed for the destruction of trash. They are applicable for other solid waste destruction, and their secondary chamber can be used for destruction of gaseous or liquid waste in suspension. It is not applicable for incineration of endothermic materials.

The nature of the SAU process is such that turbulence of the waste feed is minimal. Materials requiring turbulence for effective combustion such as powdered carbon or pulp wastes are not appropriate candidates for starved air incineration.

AIR EMISSIONS

Compared to other incineration methods, the airflow in the primary chamber, firing the waste, is low in quantity and is low in velocity. The low velocity and near absence of turbulence of the waste result in minimal amounts of particulate carried along in the gas stream. Complete burning is accomplished in the secondary chamber, and the resulting exhaust gas is clean and practically free of particulate matter, i.e., smoke and soot. The SAU can usually comply with exhaust emissions standards to 0.08 grams/st ft^3 without the use of supplemental gas-cleaning equipment such as scrubbers or baghouses.

WASTE CHARGING

Smaller units, under 750 lb/h, are normally batch-fed. Waste is charged over a period of hours, and after a full load has been placed in the chamber, the chamber is sealed and the waste fired.

Figure 11.7 illustrates a typical hopper-ram assembly designed to minimize the quantity of air infiltration into the primary chamber when charging. Figure 11.8 illustrates a double-ram charging system which allows a more continuous feed than the single-ram. Note that the furnace charging door is not opened until the hopper is sealed by the upper ram, preventing air infiltration from the hopper.

Larger units are usually provided with a continuous waste charging system, a screw feeder, or a series of moving grates.

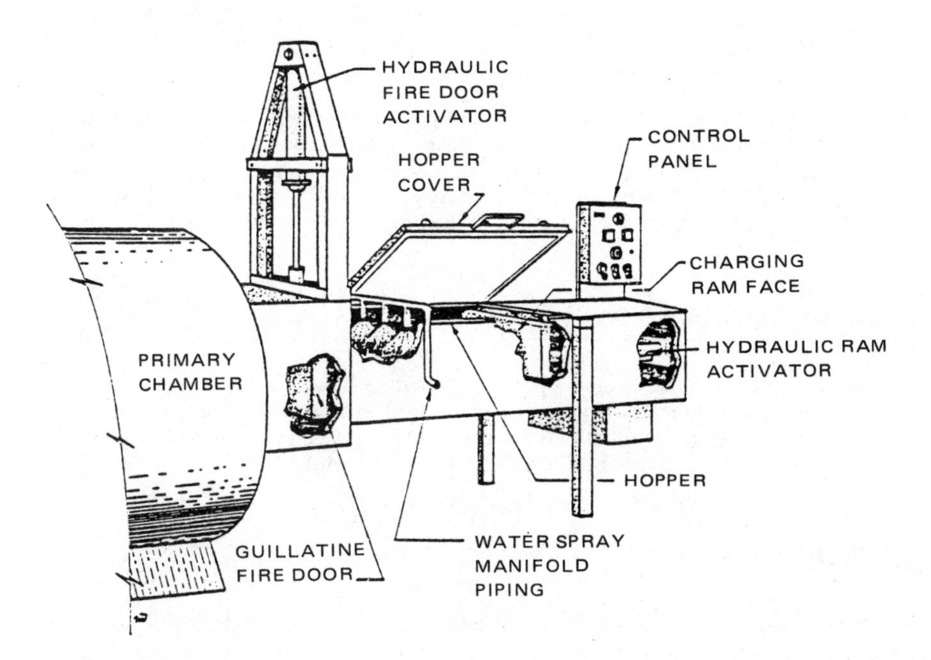

FIGURE 11.7 Typical standard hopper-ram assembly.

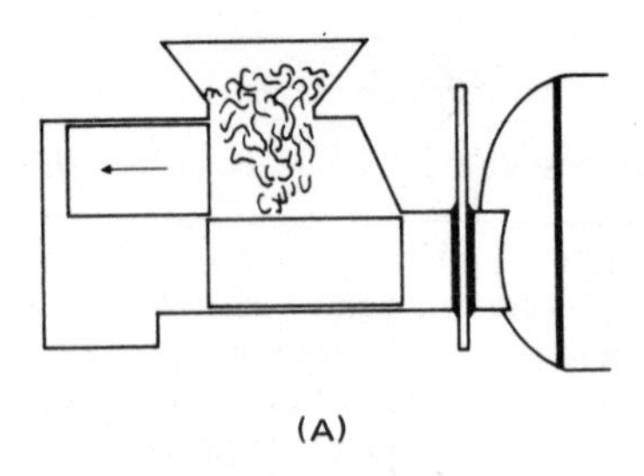

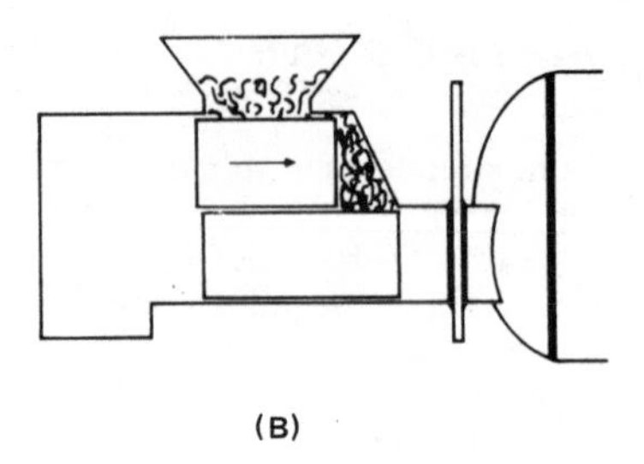

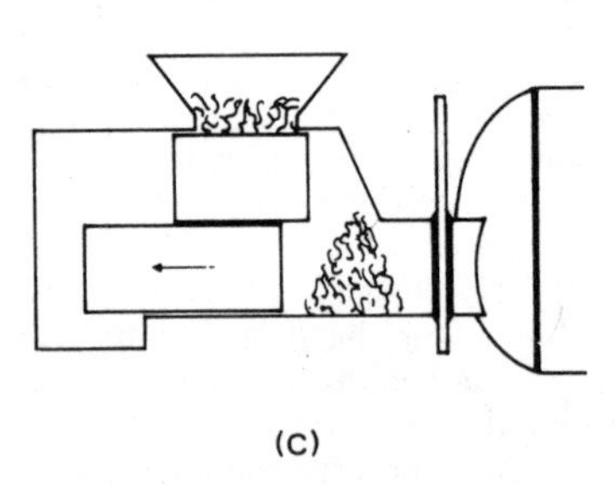

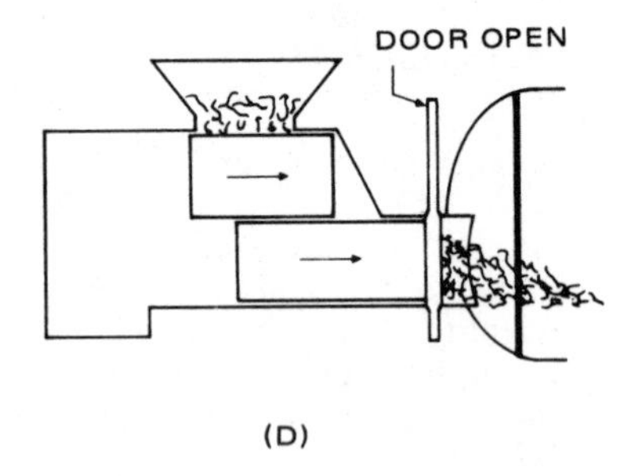

FIGURE 11.8 Double-ram type of charging system.

ASH DISPOSAL

As with waste charging, SAUs are provided with both manual and automatic discharge systems. With smaller units, after burnout the chamber is opened and ash residue is manually raked out. With continuous operating units such as that shown in Fig. 11.4, ash is continually discharged, normally into a wet well, where it is transferred to a container or truck by means of a drag conveyer.

ENERGY RECLAMATION

Waste heat utilization is a viable option provided with SAUs. The hot gas exiting the secondary chamber is relatively clean. Boiler or heat-exchanger surfaces placed within this gas stream will therefore be subject to minimal particulate matter carryover and attendant problems of erosion and plugging.

TYPICAL SYSTEMS

Dimensional data of a typical SAU (Morse Boulger, Inc.) system are shown in Fig. 11.9. Its internal configuration is similar to that of the unit in the upper left of Fig. 11.5. Its charging system is similar to that of Fig. 11.7.

CAPACITY lbs./hr.	A	B	C	CC*	D	E	F	FF*	G	H	J	JJ*	CHUTE CAP. cu. yd.
100-160	8′-6″	4′-6″	5′-0″	9′-0″	6′-4″	3′-6″	9′0″	13′-0″	1′-4″	11′-0″	2′-0″	6′-0″	0.5
220-350	9′-6″	5′-6″	6′-0″	10′0″	7′-0″	4′-6″	11′-0″	15′-0″	1′-7½″	11′-0″	2′-0″	6′-0″	0.5
320-525	9′-6″	6′-0″	6′-6″	10′-6″	7′-0″	5′-6″	12′-6″	16′-6″	1′-9″	11′-0″	2′-6″	6′-6″	1.0
430-700	10′-0″	6′-6″	7′-0″	11′-0″	8′-0″	5′-6″	13′-0″	17′-0″	2′-2½″	11′-0″	2′-6″	6′-6″	1.0
640-1050	11′-6″	7′-6″	8′-0″	12′-0″	8′-0″	6′-0″	15′-0″	19′-0″	2′-6″	13′-0″	2′-6″	6′-6″	2.0
870-1400	12′-0″	8′-0″	8′-6″	12′-6″	9′-0″	6′-6″	16′-0″	20′-0″	2′-9½″	13′-0″	3′-6″	8′-0″	2.0
1300-2100	12′-6″	9′-6″	10′-0″	14′-0″	9′-6″	8′-6″	19′6″	23′-6″	3′-5″	13′-0″	3′-6″	8′-6″	3.0
1950-3200	14′-6″	10′-6″	11′-0″	15′-0″	10′-0″	9′-6″	21′-6″	25′-6″	3′-11″	13′-0″	4′-0″	9′-0″	3.0
2400-3900	16′-0″	11′-0″	11′-6″	15′-6″	12′-0″	9′-6″	22′-0″	26′-0″	4′-5″	14′-0″	5′-6″	9′-6″	4.0
2900-4700	18′-0″	11′-6″	12′-0″	16′-0″	14′-0″	9′-6″	22′-6″	26′-6″	4′-9″	14′-0″	5′-6″	9′-6″	4.0

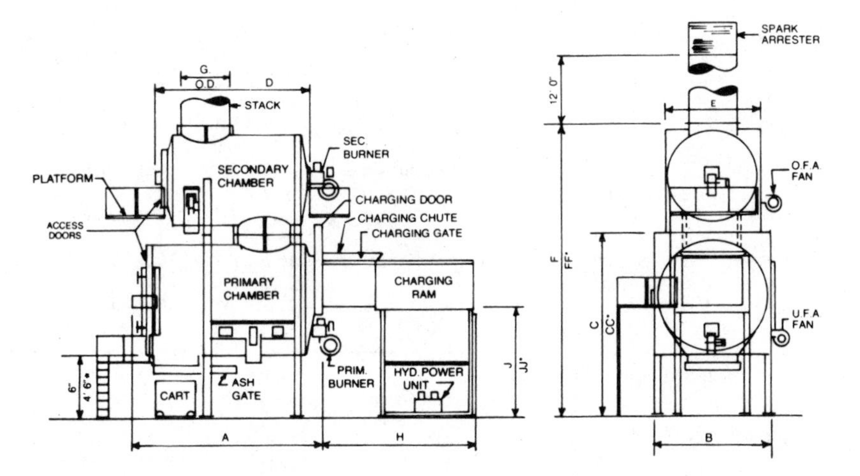

FIGURE 11.9 Typical SAU system. *Source:* Morse Boulger, Inc., Queens, N.Y.

Table 11.6 lists typical SAU systems which are provided with energy generation systems. These are basically "standard" models which are normally modified to the customers' specific waste, heat recovery mode, or other needs.

TABLE 11.6 SAU Energy Generation

Burning rate, lb/h	Waste feed, Btu/lb	Energy generation mode	Energy generation rate	Manufacturer
700	6,000	Steam, 100 lb/in^2 gauge	2,550 lb/h	Morse Boulger
1,000	6,000	Steam, 100 lb/in^2 gauge	3,850 lb/h	Morse Boulger
1,400	6,000	Steam, 100 lb/in^2 gauge	5,100 lb/h	Morse Boulger
3,200	6,000	Steam, 100 lb/in^2 gauge	11,600 lb/h	Morse Boulger
4,700	6,000	Steam, 100 lb/in^2 gauge	17,000 lb/h	Morse Boulger
1,280	6,285	Steam, 160 lb/in^2 gauge	2,025 lb/h	George L. Simonds
1,650	8,500	Steam, 150 lb/in^2 gauge	3,500 lb/h	George L. Simonds
1,050	6,240	Steam, 125 lb/in^2 gauge	1,900 lb/h	George L. Simonds
650	6,500	Steam, 150 lb/in^2 gauge	2,500 lb/h	George L. Simonds
1,800/15 gal/h	8,500 trash/ waste oil	Steam, 150 lb/in^2 gauge	8,000 lb/h	George L. Simonds
1,000	6,500	Steam, 100 lb/in^2 gauge	3,458 lb/h	Smokatrol
		Hot water, 105° Δt	66 gal/min	
1,500	6,500	Steam, 100 lb/in^2 gauge	5,187 lb/h	Smokatrol
		Hot water, 105° Δt	86 gal/min	
2,000	6,500	Steam, 100 lb/in^2 gauge	6,916 lb/h	Smokatrol
		Hot water, 105° Δt	132 gal/min	
2,500	6,500	Steam, 100 lb/in^2 gauge	8,645 lb/h	Smokatrol
		Hot water, 105° Δt	165 gal/min	
1,250	4,500	Steam, 150 lb/in^2 gauge	3,200 lb/h	Consumat
	7,000		4,950 lb/h	
2,100	4,500	Steam, 150 lb/in^2 gauge	5,400 lb/h	Consumat
	7,000		8,400 lb/h	
6,250	4,500	Steam, 150 lb/in^2 gauge	16,100 lb/h	Consumat
	7,000		25,000 lb/h	
8,400	4,500	Steam, 150 lb/in^2 gauge	21,600 lb/h	Consumat
	7,000		33,600 lb/h	

Source: Selected manufacturers' data.

REFERENCES

11-1. Sanner, W., and Ortuglio, C.: *Conversion of Municipal and Industrial Refuse into Useful Materials by Pyrolysis*, U.S. Department of the Interior, Bureau of Mines, Report of Investigations 7428, Government Printing Office, Washington, August 1970.

11-2. Drobuy, N., Hull, H., and Testin, R.: *Recovery and Utilization of Municipal Solid Wastes*, U.S. Public Health Service Publication 1908, 1971.

11-3. *Recommended Methods of Reduction, Neutralization, Recovery or Disposal of Hazardous Waste*, vol. 3, Disposal Process Descriptions, Ultimate Disposal, Incineration and Pyrolysis Processes, USEPA 670/2-73-053C, August 1973.

11-4. *Source Category Survey: Industrial Incinerators*, USEPA 450/3-80-013, May 1980.

BIBLIOGRAPHY

Galandak, F., and Racstain, M.: "Design Considerations for Pyrolysis of Sewage Sludge," *Journal of the Water Pollution Control Federation*, February 1979.

King, G.: *Producing Clean Water and Energy from Pharmaceutical Wastewater*, Midland Ross Report, 1980.

Mallan, G., and Finney, C.: "New Techniques in the Pyrolysis of Solid Waste," *Journal of the Water Pollution Control Federation*, February 1979.

"Modular Incinerators Made for the Eighties," *Waste Age*, May 1980.

Rice, R.: "Solving the Small Company Solid Waste Problem," *Public Works Magazine*, July 1976.

Ross, R.: *Industrial Waste Disposal*, Van Nostrand Reinhold, New York, 1968.

Sieger, R., and Maroney, P.: *Incineration, Pyrolysis of WWTP Sludges*, USEPA, 1977.

CHAPTER 12
ROTARY KILN TECHNOLOGY

The rotary kiln incinerator is the most universal of thermal waste disposal systems. It can be used for the disposal of a wide variety of solid and sludge wastes and for the incineration of liquid and gaseous waste. The rotary kiln system has found application in industrial waste incineration and municipal waste incineration, and more recently it has been applied to the cleanup of sites with organic contaminated soils.

KILN SYSTEM

A rotary kiln system used for waste incineration is shown in Fig. 12.1. It includes provisions for feeding, air injection, the kiln itself, an afterburner, and an ash collection system. The gas discharge from the afterburner is directed to an air emissions control system. An induced-draft (ID) or exhaust fan is provided within the emissions control system to draw gases from the kiln through the equipment line and discharges through a stack to the atmosphere.

Other designs, such as the one shown in Fig. 12.2, may include a waste heat boiler between the afterburner and the scrubber for energy recovery. The waste heat boiler reduces the temperature of the gas stream sufficiently to allow the use of a fabric filter, or baghouse, for particulate control. The scrubber in this illustration utilizes water and alkali injection. It is used for acid gas control. Dry scrubbing may be used in lieu of the wet scrubbing system shown.

The kiln system in Fig. 12.3 illustrates the variety of wastes that can be processed. Note that the kiln is used for the destruction of solid waste (charged by the open bucket shown), waste packs or drums (dropped into the kiln vertically above the kiln entrance), and tar waste (a sludge pumped to the kiln entrance). As pictured, sprays inject water into the exiting flue gas in an effort to reduce the particulate emissions level.

There are a number of areas within the kiln system where leakage can occur, as can be seen in each of the figures cited above. The feeding ports cannot be completely sealed, and the kiln seals are areas of potential leakage. The ash system is normally provided with a water seal, but for dry ash collection there will always be some leakage. To ensure that the leakage is into the system, that no hot, dirty gases leak out of the kiln to the surrounding areas, the kiln is maintained with a negative draft. The ID fan is sized to maintain a negative pressure throughout the system so that leakage is always into, not out of, the kiln system.

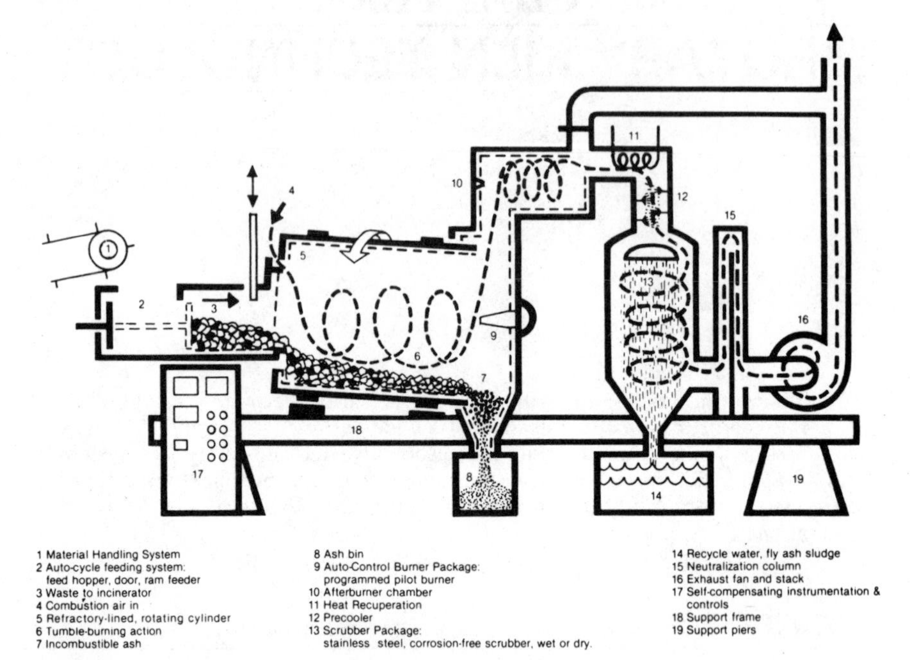

FIGURE 12.1 Rotary kiln systems. *Source:* C-E Raymond, Combustion Engineering, Inc., Chicago, Ill.

KILN APPLICATION

The rotary kiln can incinerate a wide variety of wastes; however, its application has limitations. Advantages and disadvantages in the use of a rotary kiln as an incinerator can be summarized as follows:

Advantages:

- Able to incinerate a wide variety of waste streams
- Minimal waste preprocessing required
- Existing techniques (slagging) for the direct disposal of wastes in metal drums
- Able to incinerate varied types of wastes (solids, liquids, sludges, etc.) at the same time
- Availability of many types of feed mechanisms (ram feeder, screw, gravity feed, direct injection of liquids and sludges, etc.)
- Easy control of residence time of waste in the kiln
- Provision of high turbulence and effective contact with air within the kiln

Disadvantages:

- There is relatively high particulate carryover to the gas stream because of turbulence of the waste stream.
- Normally a separate afterburner is required for the destruction of volatiles.

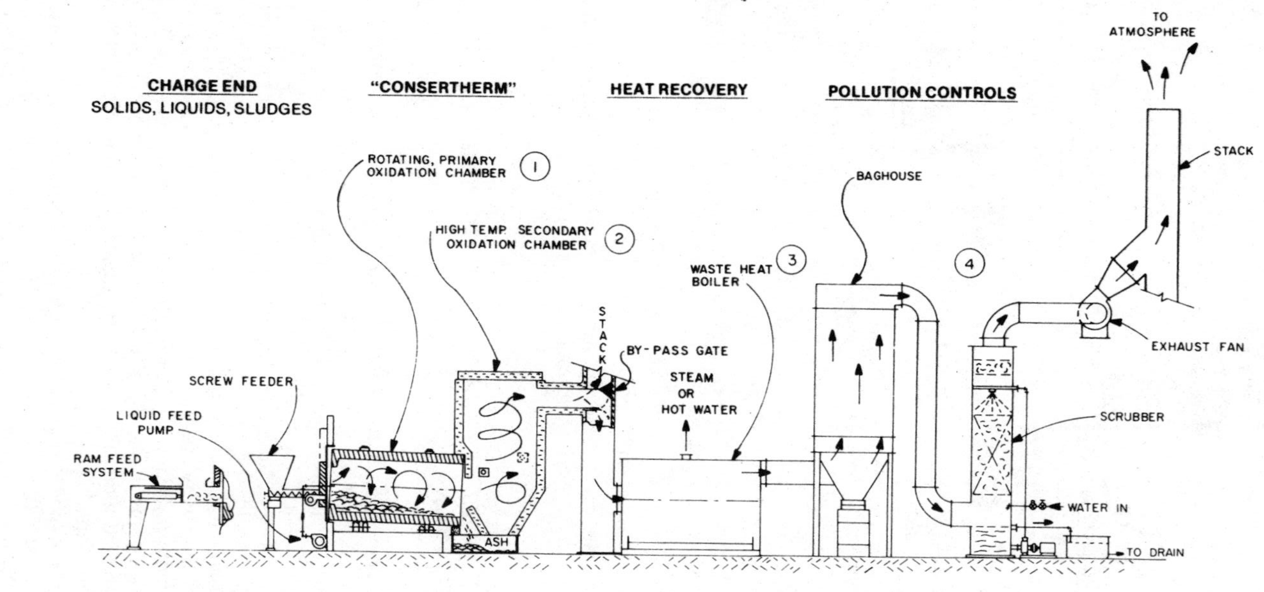

FIGURE 12.2 Rotary kiln with waste heat boiler. *Source:* R. Rayve, Consertherm, East Hartford, Conn.

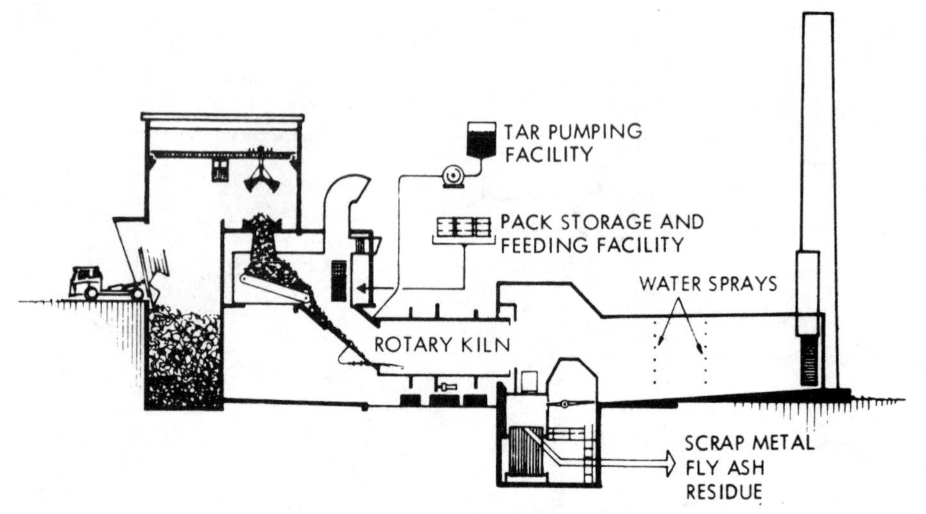

FIGURE 12.3 Typical major industrial rotary kiln facility. *Source:* Ref. 12-1.

- There is little or no ability to control conditions along the kiln length in a conventional kiln.
- It requires a relatively high amount of excess air, nominally 100 to 150 percent of stoichiometric.
- An effective kiln seal is difficult to maintain.
- A significant amount of heat is lost in the ash discharge.
- Operation in a slagging mode to process inorganic wastes or metal drums increases the kiln maintenance requirements.

THE PRIMARY COMBUSTION CHAMBER (ROTARY KILN)

The conventional rotary kiln is a horizontal cylinder, lined with refractory, which turns about its horizontal axis. Waste is deposited in the kiln at one end, and the waste burns out to an ash by the time it reaches the other end. Kiln rotational speed is variable, in the range of ¾ to 2½ r/min. The ratio of length to diameter of a kiln used for waste disposal is normally in the range of 2:1 to 5:1.

Most kiln designs utilize smooth refractory on the kiln interior. Some designs, particularly those for the processing of granular material (dirt or powders), may have internal vanes or paddles to encourage motion along the kiln length and to promote turbulence of the feed. Care must be taken in the provision of internal baffles of any kind. With certain material consistencies, such as soil of from 10 to 20 percent moisture content, baffles may tend to retard the movement of material through the kiln.

The kiln is supported by at least two trunnions. One or more sets of trunnion rollers are idlers. Kiln rotation can be achieved by a set of powered trunnion rollers, by a gear drive around the kiln periphery, or through a chain driving a large sprocket around the body of the kiln.

The kiln trunnion supports are adjustable in the vertical direction. The kiln is normally supported at an angle to the horizontal, or rake. The rake will normally vary from 2 to 4 percent (¼ to ½ in/ft of length), with the higher end at the feed end of the kiln. Other kiln designs have a zero or slightly negative rake, with lips at the input and discharge ends. These kilns are operated in the slagging mode, as discussed subsequently, with the internal kiln geometry designed to maintain a pool of molten slag between the kiln lips.

A source of heat is required to bring the kiln up to operating temperature and to maintain its temperature during incineration of the waste feed. Supplemental fuel is normally injected into the kiln through a conventional burner or a ring burner when gas fuel is used.

There are a number of variations in kiln design, including the following:

- Parallel flow or counterflow
- Slagging or nonslagging mode
- Refractory or bare wall

The more commonly used kiln design, referred to as the *conventional kiln*, is a parallel-flow system, nonslagging, lined with refractory.

KILN EXHAUST GAS FLOW

When gas flow through the kiln is in the same direction as the waste flow, the kiln is said to have *parallel* or *cocurrent flow*, as indicated in the kiln in Fig. 12.1. With countercurrent flow, the gas flows opposite to the flow of waste. The burner(s) is(are) placed at the front of the kiln, the face of the kiln from which air or gas originates.

Generally, a countercurrent kiln is used when an aqueous waste, one with at least 30 percent water content, is to be incinerated. Waste is introduced at the end of the kiln far from the burner. The gases exiting the kiln will dry the aqueous waste, and its temperature will drop. If aqueous waste were dropped into a kiln with cocurrent flow, water would be evaporated at the feed end of the kiln. The feed end would be the end of the kiln at the lowest temperature, and a much longer kiln would be required for burnout of the waste.

Wastes with a light volatile fraction (containing greases, for instance) should utilize a kiln with cocurrent flow. These volatiles will likely be released from the feed immediately upon entering the kiln. Use of a cocurrent kiln provides a higher residence time than use of a countercurrent kiln for the effective burnout of these volatiles.

SLAGGING MODE

At temperatures in the range of 2000 to 2200°F, ash will start to deform for many waste streams; and as the temperature increases, the ash will fluidize. The actual temperatures of initial deformation and subsequent physical changes to the ash are a function of the chemical constituents present in the waste residual. They are also a function of the presence of oxygen in the furnace. The ash deformation temperatures will vary with reducing vs. oxidation atmospheres, as noted in Table 13.3, Ash Fusion Temperatures, which lists deformation temperatures for coal and a typical refuse mix. Eutectic properties can be controlled by the use of additives to the molten material.

A kiln can be designed to generate and maintain molten ash during operation. Operation in a slagging mode provides a number of advantages over nonslagging operation. When a kiln is operating in a nonslagging mode, however, and slagging occurs, slagging is undesirable and must be eliminated.

Differences in slagging vs. nonslagging kilns are outlined in Table 12.1. As noted, the construction of a slagging kiln is more complex than that of a nonslagging kiln, requiring provision of a lip at the kiln exit to contain the molten material. A nonslagging kiln will normally have a smooth transition with no impediments to the smooth discharge of ash.

Slagging kilns have been designed and operated with a negative rake; e.g., the outer surface of the kiln at the feed end is lower than the kiln surface at the discharge end. This will permit the accumulation of more slag in the kiln than with zero or positive rake. The kiln internal surface must be designed for this operating mode. For instance, as noted previously, an internal refractory lip is required on the kiln feed end.

The slagging kiln can accept metal drums. The ash eutectic properties at the molten slag temperatures will tend to dissolve a ferrous metal drum placed in the kiln. The placing of drums containing waste in a kiln may be undesirable from a safety and maintenance standpoint (even with the tops of the drums removed, localized heating of the drum surface may occur, causing an explosion, and the impact of a dropping drum will eventually damage kiln refractory). However, if drums are to be placed in a kiln (they should be quartered), slagging kilns are able to absorb the drum into a homogeneous residue discharge. The nonslagging kiln can only move the drum through the unit and must include specialty equipment for handling the drum body as it exits the kiln.

Salt-laden wastes will tend to melt in the range of 1300 to 1600°F and can produce severe caking, or deposits, in a nonslagging kiln. Often salt-bearing wastes are prohibited from kilns because they will produce an unacceptable buildup on the kiln surface which can eventually choke off the kiln. In a slagging kiln, how-

TABLE 12.1 Slagging vs. Nonslagging Kiln

Factor	Effect
Construction	More complex with slagging kiln
Duty	Slagging kiln can accept drums, salt-laden wastes; nonslagging kiln is limited
Temperature	Higher with slagging kiln
Retention time	Greater residence required in nonslagging kiln
Process control	Thermal inertia or forgiveness in slagging kiln
Emissions	Less particulate, greater NO_x in slagging kiln
Slag	Slagging kiln may require CaO, Al_2O_3, SiO_2 additives; dissolves drums, salts
Ash	Wet, less leachable with slagging; wet or dry with nonslagging kiln
Maintenance	Higher with slagging kiln
Refractory	More critical with slagging kiln

ever, the temperature is kept high enough to keep the salts in a molten state. The salts combine with the molten ash in the pool at the bottom of the kiln and are maintained in their molten state until quenched.

The temperature in a slagging kiln must be sufficiently high to maintain the ash as a molten slag. Temperatures as high as 2600 to 2800°F are not uncommon. A nonslagging kiln will normally operate at temperatures below 2000°F.

The destruction of organic compounds is achieved by a combination of high temperature and residence time. Generally, the higher the temperature, the shorter the residence time required for destruction. Conversely, the higher the residence time, the lower the required temperature. The use of higher temperatures in the slagging kiln reduces the residence time requirements for the off-gas. The afterburner associated with a slagging kiln can often be smaller than that required for a nonslagging kiln.

The molten slag can weigh hundreds or even thousands of pounds. As a concentrated material, a liquid, it represents a significant thermal inertia within the kiln. The molten slag tends to act as a heat sink which provides thermal stability to the system. The slagging kiln is much less subject to temperature extremes than the nonslagging kiln because of the presence of this massive melt. It will maintain a relatively constant-temperature profile under rapid changes in kiln loading. This stability leads to more predictable system behavior. Safety factors employed in the design and operation of downstream equipment (such as an exhaust gas scrubber or the induced-draft fan) can be reduced when a slagging kiln is used.

The tumbling action of a rotating kiln encourages the release of particulate to the gas stream. From 5 to 25 percent of the nonvolatile solids in a feed stream may become airborne with the use of a conventional nonslagging kiln. The presence of the molten slag in a slagging kiln acts much like the fluid ash in a pulverized-coal (PC) burner. The slag will absorb particulate matter from the gas stream and can reduce particulate emissions from the kiln to 25 to 75 percent of the emissions from a nonslagging kiln. However, emissions of NO_x are greater with a slagging than with a nonslagging kiln. The generation of NO_x is generally not significant until the temperature of the process increases above 2000°F. Above this temperature the formation of NO_x will increase substantially. At 2600°F the generation of NO_x is almost 10 times as great as at 1800°F.

A danger in slagging kiln operation is that the melt will solidify. When this happens the kiln will be off-balance. With an eccentric-turning kiln, if rotation of the kiln is not stopped, damage to kiln supports and to the kiln drive may occur. In addition, the incineration process will degrade under a melt freeze. Operating stability will be lost, and demands on downstream equipment (the gas scrubbing system, for instance) may be too severe. One reason for the loss of a molten slag, besides a drop in temperature, is a change in the feed quality. To ensure the maintenance of an adequate melt, additives may have to be employed. These additives may include CaO, Al_2O_3, SiO_2, or another compound or set of compounds, depending on the nature of the waste. Additives will help maintain the eutectic, to ensure that the melt will remain in a molten state.

The molten slag from a slagging kiln is dropped into a wet sump. (The hot slag can "pop" or explode as it contacts the cooling water in the sump.) The slag immediately hardens into a granular material (termed *frit*) with the appearance of gravel or dark glass. The ash from a nonslagging kiln can be collected wet or dry.

Refractory for a slagging kiln will experience more severe duty than that for nonslagging kiln service. The higher operating temperatures will directly affect refractory life, as will the corrosive effect of the melt. In addition, if steel drums are dropped into the kiln, the physical impact of the drum on the kiln surface will be damaging. The molten slag will absorb the steel and ferrous metals, as well as other metals, which are highly corrosive to the refractory. The refractory must resist this corrosive attack, high temperatures, and impact loading. The resulting refractory system will be expensive and will require frequent maintenance.

OPERATION

The waste retention time in a kiln can be varied. It is a function of kiln geometry and kiln speed, as shown in the following equation:

$$t = \frac{2.28\ L/D}{SN}$$

where: t = mean residence time, min
L/D = internal length-to-diameter ratio
S = kiln rake slope, in/ft of length
N = rotational speed, r/min

For a given L/D ratio and rake, the solids residence time within the unit is inversely proportional to the kiln speed. By doubling the speed, the residence time will halve. An example of this calculation is as follows:

> Calculating the residence time for a kiln rotating at N = 0.75 r/min with a 1 percent slope (S = 0.12 in/ft of length), with a 4-ft inside diameter and 12-ft length L, we get
>
> $$t = \frac{2.28(12/4)}{0.12 \times 0.75} = 76 \text{ min}$$
>
> By inspection note that a doubling of rotation N would halve the retention time, and halving the rake S would double the retention time.

The above calculation was for the residence time of solids or other materials within the kiln, not the kiln exhaust gas. The off-gas residence time can be determined by the application of the heat balance and flue gas analyses developed previously in this text.

KILN SEAL

Sealing a kiln is a difficult task. Efficient kiln operation requires that kiln seals be provided and maintained to control the infiltration of unwanted airflow into the system. With too much air, fuel usage increases and process control deteriorates.

The kiln turns between two stationary yokes. The kiln diameter, which can vary from 4 to 20 ft, will have a periphery of from 12 to 60 ft. At 1 ft/min velocity, the kiln surface is moving at a rate of up to 60 ft/min. A seal must close this gap between the yoke and the kiln surface while the kiln is moving at this surface velocity. The kiln surface is not a machined surface and will have variations in texture and dimension, making the task of sealing very difficult. A further problem is that the kiln interior is normally at relatively high temperatures, which tend to encourage wear of the kiln surface.

Two types of seals are illustrated in Fig. 12.4. The rotating portion of this seal, a T-ring in this illustration, is mounted on the kiln surface. There are as many variations in kiln seal designs as there are kiln and kiln seal manufacturers.

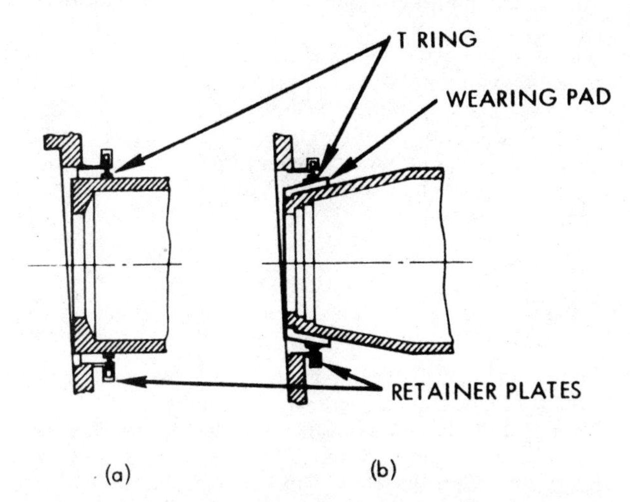

FIGURE 12.4 Kiln seal arrangements. (*a*) Single floating-type feed-end air seal. (*b*) Single floating-type air seal on air-cooled tapered feed end. *Source:* Ref. 12-2.

LIQUID WASTE INJECTION

Liquid waste will likely be either aqueous (at least 30 percent water content) or organic. This waste stream can be injected into or external to the flame envelope, depending on its heating value.

Supplemental fuel that will be heating or maintaining heat within the kiln should be allowed to burn completely. An aqueous waste stream should be injected downstream of the supplemental-fuel flame envelope. If it were injected within this envelope, it would quench the flame and the flame would not develop its full temperature and heat release.

Liquid waste with a significant heat content, at least 6000 Btu/lb, will supplement the heat content of the fuel and can be injected into the flame envelope.

SYSTEM SELECTION

The kiln will burn out solids and will volatilize organics. Generally not all the organics will be incinerated in the kiln, and so a high temperature must be maintained at a specific residence time for their destruction. This is the purpose of a secondary combustion chamber, or an afterburner. An afterburner is normally placed immediately downstream of the kiln. It is stationary, designed to maintain the temperature of the gas stream exiting the kiln at a preselected temperature level for organics destruction for a specific time. The released volatiles will exit the kiln in the flue gas and will enter an afterburner in the kiln systems illustrated in Figs. 12.1 and 12.2. The afterburner will normally have at least a single burner to provide the supplemental fuel required for burnout of the organics in the gas stream.

There are kiln system designs in which the volatiles released from a kiln have a high enough heating value that they require no external source of supplemental fuel for complete combustion. In these instances the afterburner does not require supplemental fuel, and it acts as an extension of the kiln, an unfired secondary combustion chamber providing the residence time for burnout of the organics. This secondary combustion chamber would be initially heated by hot gases from the kiln.

DESIGN VARIATIONS

In an effort to control the air distribution and temperature profile along the length of the kiln, a rotary kiln was developed with air injection ports. This kiln, shown schematically in Fig. 12.5, has a combustion air plenum inserted high throughout its length. Combustion air will cool the plenum as well as provide air for feed volatiles. The airflow within the plenum can be directed to any of a number of zones within the kiln. The control of air has been found to allow low- or substoichiometric operation in portions of the kiln or throughout the entire length of the kiln.

The rotary kiln system illustrated in Fig. 12.6 has been developed for municipal solid waste application. The kiln, or rotary combustor, has no refractory. Without refractory, it is believed that the system maintenance cost is reduced. It is constructed of water tubes which absorb from 25 to 35 percent of the heat generated by the burning waste.

Burning begins in the kiln, with air injected through openings in the tube wall construction. Burning of the off-gas is completed within the boiler, which is also constructed of water tubes, with no refractory.

A rotary joint in the kiln hot-water circulation system maintains a water seal under the physical motion of the tubes and the relative high-pressure demands of the hot fluid.

KILN SIZING

A rotary kiln will release 15,000 to 30,000 Btu/(ft^3 · h). To illustrate sizing of a kiln which is to burn 2000 lb/h of sludge cake with an as-received heating value of 2200 Btu/lb as fired, note the following calculations:

Heat release:

$$2000 \text{ lb/h} \times 2200 \text{ Btu/lb} = 4.4 \text{ MBtu/h from waste}$$

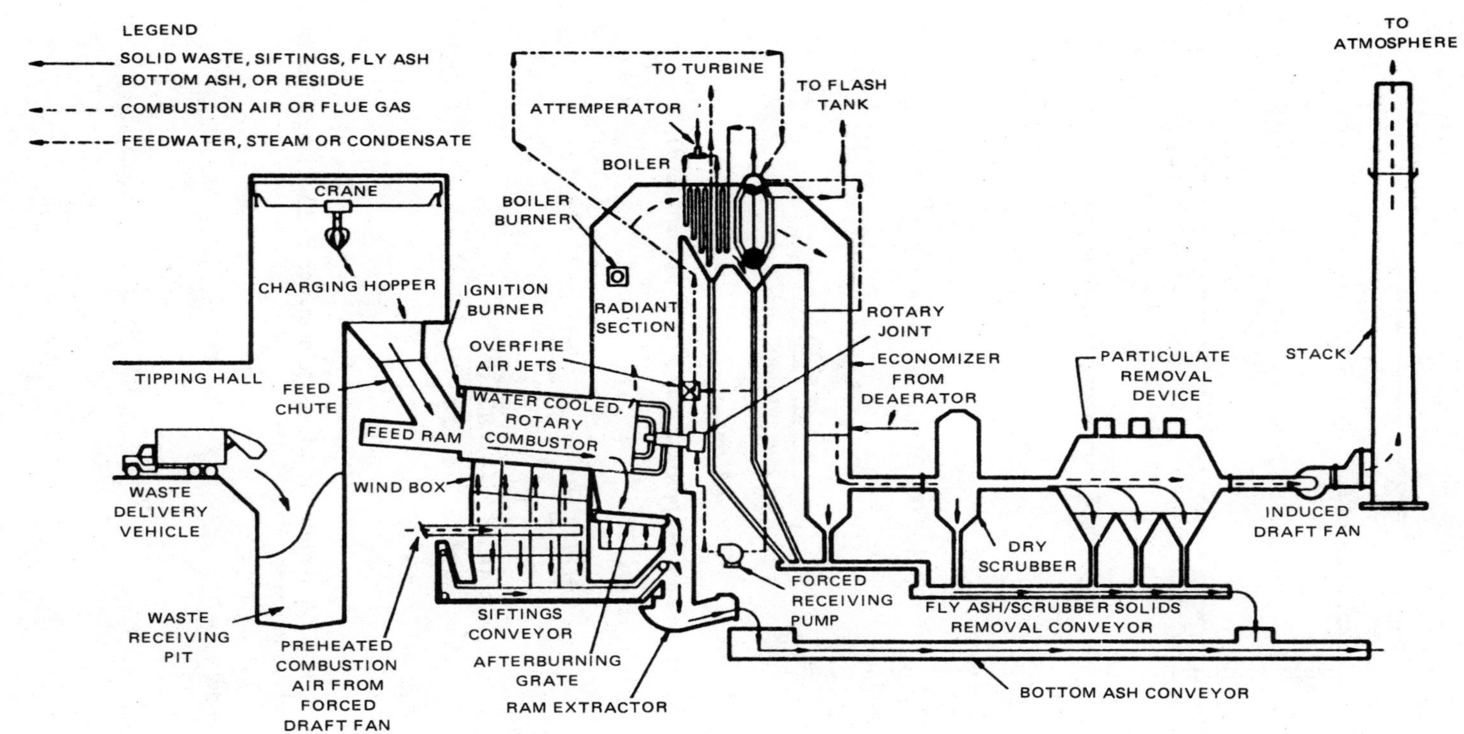

FIGURE 12.5 Air injection rotary kiln. *Source:* J. F. Angelo, Universal Energy International, Inc., Little Rock, Ark.

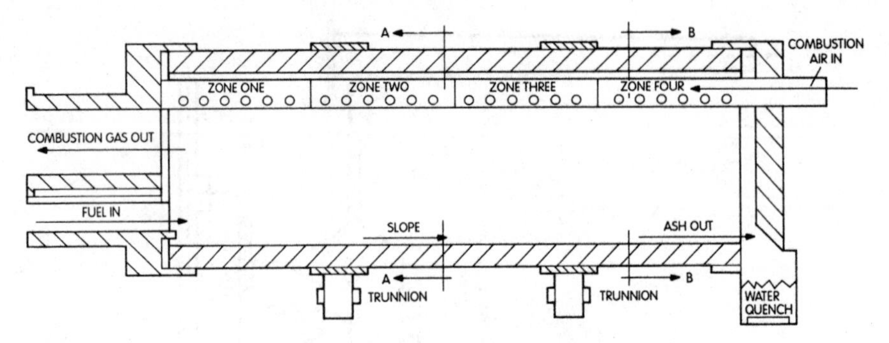

FIGURE 12.6 Rotary combustor. *Source:* Westinghouse/O'Connor Combustor Corp., Pittsburgh, Pa.

Heat release (assumed):

$$25{,}000 \text{ Btu/(h} \cdot \text{ft}^3\text{) in kiln}$$

Required volume:

$$V = \frac{4.4 \times 10^6 \text{ Btu/h}}{25{,}000 \text{ Btu/(h} \cdot \text{ft}^3\text{)}}$$

$$= 176 \text{ ft}^3$$

For a kiln whose length L is approximately 3 times its diameter D,

$$L = 3D$$

$$V = \frac{\pi D^2 L}{4} = \frac{\pi D^2}{4} \times 3D = \frac{3\pi D^3}{4}$$

$$D = \left(\frac{4V}{3\pi}\right)^{1/3} = \left(\frac{4 \times 176 \text{ ft}^3}{3\pi}\right)^{1/3} = 4.21 \text{ ft}$$

Take $D = 4$ ft:

$$L = 3D = 3 \times 4 \text{ ft} = 12 \text{ ft}$$

Therefore the kiln size required is 4-ft inside diameter by 12 ft long. The actual heat release is as follows:

Actual volume:

$$\frac{\pi(4)^2 \times 12}{4} = 150.796 \text{ ft}^3$$

Release:

$$\frac{4.4 \times 10^6 \text{ Btu/h}}{150.796 \text{ ft}^3} = 29{,}178 \text{ Btu/(h} \cdot \text{ft}^3\text{)}$$

REFERENCES

12-1. *Recommended Methods of Reduction, Neutralization, Recovery or Disposal of Hazardous Waste*, vol. 3, Disposal Process Descriptions, Ultimate Disposal, Incineration and Pyrolysis Processes. USEPA 670/2-73-053C, August 1973.

12-2. Lyon, S.: "Incineration of Raw Sludges and Greases," *Journal of the Water Pollution Control Federation*, April 1973.

BIBLIOGRAPHY

Allis Chalmers Corporation: *Kilns Flow of Material,* Allis Chalmers Publication Number 22B1212-2, Milwaukee, Wis.

Brunner, C. R.: *Handbook of Hazardous Waste Incineration,* Tab Reference Books, New York, 1989.

Dillon, A. P.: *Hazardous Waste Incineration Engineering,* Noyes, Park Ridge, N.J., 1981.

Freeman, H. F.: *Standard Handbook of Hazardous Waste Treatment and Disposal,* McGraw-Hill, New York, 1988.

CHAPTER 13
CENTRAL DISPOSAL SYSTEMS

In this chapter burning systems that have been developed for centralized disposal of waste are presented. These facilities include mass burning systems for municipal solid waste disposal and large-scale industrial waste disposal systems. Controlled air incineration systems, which are used also in central disposal facilities, are covered in another chapter.

THE NEED FOR CENTRAL DISPOSAL

Central disposal of waste is a consideration in the disposal of municipal solid waste. In recent years industrial firms with many plants have looked toward incineration of their wastes in a central plant location as an efficient and economical means of disposal.

In Europe, after World War II, the disposal of municipal solid waste at a central location, by incineration, was given impetus by the following factors:

1. Population concentrations and increases required the use of more land for housing and for farming. The use of land for burying refuse was becoming impractical.
2. Technology developed to the point where it became economical to generate energy, i.e., steam and or electric power, from incineration. Economics of scale dictated that the larger the facility, the more efficient would be its energy generation potential.
3. In general all utilities, including refuse disposal and electric or steam power generating industries, were state-owned. The interests of the electric utility and the refuse disposal authority were therefore common. This conflicts with conditions in the United States, where refuse collection is a public or government function and electric power generation is generally a private-sector function. Cooperation between these agencies in Europe has promoted the development of energy-producing incinerators. The power utility readily purchases energy from an incineration facility, providing revenues for the incinerator operation.
4. The higher cost of fossil fuel, particularly fuel oil, has helped promote energy generation, hence, central disposal facilities.

Only in the past decade, when the United States came to the realization that the cost and availability of energy were unreliable and out of its control, has a serious attempt been made to generate energy from waste in central collection and incineration facilities.

Table 13.1 lists information on selected mass burning central facilities in the

TABLE 13.1 Typical Central Mass Burning Facilities, 1988

Location	Start-up	Design rate, tons/day	Initial cost, $	Process equipment	Comments
Baltimore, Md.	1985	2250	170,000,000	Signal	This plant has been in continuous operation since late 1985. It generates electricity and includes ferrous recovery from ash. This facility has had vigorous public support and is looked upon as a model for resource recovery for large cities.
Chicago, Ill.	1971	1600	25,000,000	Martin	This plant has been in continuous operation burning an average of 1200 tons/day since 1971. It has been designed to generate steam. However, a reliable steam customer has only been found in the past few years. A steam pipeline had to be built for this customer, and steam is scheduled to be delivered the end of 1981. Early problems included freezing of the air-cooled steam condensers during winter months. Ferrous metals are recovered from the incinerator ash. There is no tipping fee; disposal costs are paid by the city of Chicago.
Marion County, Oreg.	1986	550	47,500,000	Ogden/Martin	This is the first waste-to-energy plant in the United States utilizing a dry scrubbing system with a baghouse. It generates 11 MW of electric energy, which it sells to Portland General Electric Co.
North Andover, Mass.	1985	1500	185,000,000	Wheelabrator	In operation since late 1985, this facility generates power for sale to the New England Electric Co.
Panama City, Fla.	1987	510	38,000,000	Westinghouse	Two rotary combustors are utilized for mass burning. Steam is available for sale to a future industrial park, but currently electric power is generated for sale to the local utility.
Peekskill, N.Y.	1984	2250	239,000,000	Wheelabrator	Located in Westchester County, this facility reclaims ferrous metals and sells electricity to Con Edison.
Pinellas County, Fla.	1986	3150	152,000,000	Wheelabrator	The facility started initial operations in 1983 and was upgraded from 2000 tons/day to current capacity in 1986. Ferrous and nonferrous metals are reclaimed, and electricity is sold to Florida Power Corp.
Quebec City, PQ	1974	720	25,000,000	Von Roll	This plant has been processing an average of 600 tons/day since start-up. Until recently the plant experienced severe air emissions problems. However, new equipment has been installed, and all the air pollution emissions codes are satisfied. Steam is sold to a local paper company at $3.00 per 1000 lb. Ferrous metals are recovered from incinerator ash. The current tipping fee is $20.00 per ton.
Tampa, Fla.	1985	1000	70,000,000	Waste Mgmt.	This facility, located at McCay Bay, includes four rotary kiln incinerators. It reclaims scrap metals and generates electricity for sale to Tampa Electric Co.

United States. Cost data are included for reference only. The tipping fee (the cost of dumping refuse at the facility) may reflect conditions other than cost. For instance, a tipping fee may have been established to attract customers from an existing landfill. Likewise, the steam charge may have been established at a lower rate than economics would dictate to encourage sales. A low steam cost would promote steam sales, and such promotion would be necessary, because in the United States incineration is considered an unconventional, untried source of steam.

MUNICIPAL SOLID WASTE

Central-station incineration is usually applied to municipal solid waste destruction. The average characteristics of refuse and other wastes are listed in Chap. 8. The actual variation in average waste composition from one country to another is listed in Table 13.2. The Ash column represents the residual from coal or wood burning for domestic heat in the winter months. For instance, 43 percent of the composition of refuse in the United States was ash due to household coal burning in 1939, whereas 30 years later this component, i.e., ash from coal burning, was absent from the refuse.

When burning refuse, the generation of dry gas and moisture from combustion can be estimated as follows:

Dry gas 7.5 lb/10,000 Btu fired
Moisture 0.51 lb/10,000 Btu fired

TABLE 13.2 A Summary of International Refuse Composition (in percentages)

	Ash	Paper	Organic matter	Metals	Glass	Miscellaneous
United States (1939)	43.0	21.9	17.0	6.8	5.5	5.8
United States (1970)	0	44.0	26.5	8.6	8.8	12.1
Canada	5	70	10	5	5	5
United Kingdom	40–40	25–30	10–15	5–8	5–8	5–10
France	24.3	29.6	24	4.2	3.9	14
West Germany	30	18.7	21.2	5.1	9.8	15.2
Sweden	0	55	12	6	15	12
Spain	22	21	45	3	4	5
Switzerland	20	40–50	15–25	5	5	—
Netherlands	9.1	45.2	14	4.8	4.9	22
Norway (summer)	0	56.6	34.7	3.2	2.1	8.4
Norway (winter)	12.4	24.2	55.7	2.6	5.1	0
Israel	1.9	23.9	71.3	1.1	0.9	1.9
Belgium	48	20.5	23	2.5	3	3
Czechoslovakia (summer)	6	14	39	2	11	28
Czechoslovakia (winter)	65	7	22	1	3	2
Finland	—	65	10	5	5	15
Poland	10–21	2.7–6.2	35.3–43.8	0.8–0.9	0.8–2.4	—
Japan (1963)	19.3	24.8	36.9	2.8	3.3	12.9

Source: Ref. 13-1.

GRATE SYSTEM

The grate system is one of the most crucial systems within the mass burning incinerator. The grate must transport refuse through the furnace and, at the same time, promote combustion by adequate agitation and good mixing with combustion air. Abrupt tumbling caused by the dropping of burning solid waste from one tier to another will promote combustion. This action, however, may contribute to excessive carryover of particulate matter in the exiting flue gas. A gentle agitation will decrease particulate emissions.

Combustion is largely achieved by injection of combustion air below the grates, i.e., underfire air. Underfire air is also necessary to cool the grates. It is normally provided at a rate of approximately 40 to 60 percent of the total air entering the furnace. Too low a flow of underfire air will inhibit the burning process and will result in high grate temperatures.

Note the ash fusion temperatures listed in Table 13.3. These temperatures limit the operating temperatures of the grate areas. With insufficient air a reducing atmosphere will result, and the ash deformation temperature can be as low as 1800°F. If the refuse reaches this temperature, slagging will begin, further reducing the air supply by clogging the grates and forming large, unwieldy clinkers. The ash properties of coal are listed for comparison.

TABLE 13.3 Ash Fusion Temperatures

	Reducing Atmosphere, °F	Oxidizing Atmosphere, °F
Refuse		
Initial deformation	1880–2060	2030–2100
Softening	2190–2370	2260–2410
Fluid	2400–2560	2480–2700
Coal		
Initial deformation	1940–2010	2020–2270
Softening	1980–2200	2120–2450
Fluid	2250–2600	2390–2610

Source: Ref. 13-2.

Overfire air is injected above the grates. Its main purpose is to provide sufficient air to completely combust the flue gas and flue gas particulate rising from the grates. Numerous injection points are located on the furnace walls above the grates to provide a turbulent overfire air supply along the furnace length.

Ash and other particles dropping through the grates are termed *siftings*, and they must be effectively removed from the system. Siftings can readily clog grate mechanisms, generate fires, and create housekeeping problems if not attended to. Siftings, due to their small particle size, have been found to be more dense than incinerator ash, approximately 1780 versus 1040 lb/yd^3 for typical incinerator ash.

GRATE DESIGN

A number of different types of grate designs are used in central waste burning facilities. Each grate system manufacturer provides a unique grate feature, attempting to obtain a competitive edge in the marketplace.

The grate system manufacturer should be contacted for design and sizing information for a particular grate design. The following listing describes typical grate systems, both generic grate types and grates specific to certain manufacturers:

Circular Grate

This type is no longer in common usage. As shown in Fig. 13.1, a rotary cone slowly moves a series of rabble arms across a circular grate. The circular grate furnace is charged at short intervals with the fresh charge falling directly on top of a pile of burning refuse. The refuse forms a cone-shaped pile that gradually burns down with combustion largely on the surface of the pile. Combustion air is provided through the grate and through the center cone by forced-draft fans. Refuse is moved slowly down to the peripheral dumping grates by the rotating rabble arms. The refuse will burn out to ash on these dumping grates, which must then be cleaned. They can be cleaned manually, through clean-out doors on the periphery of the furnace, or they can be provided with power cylinders. These power cylinders will lower and raise the individual dumping grate sections to ex-

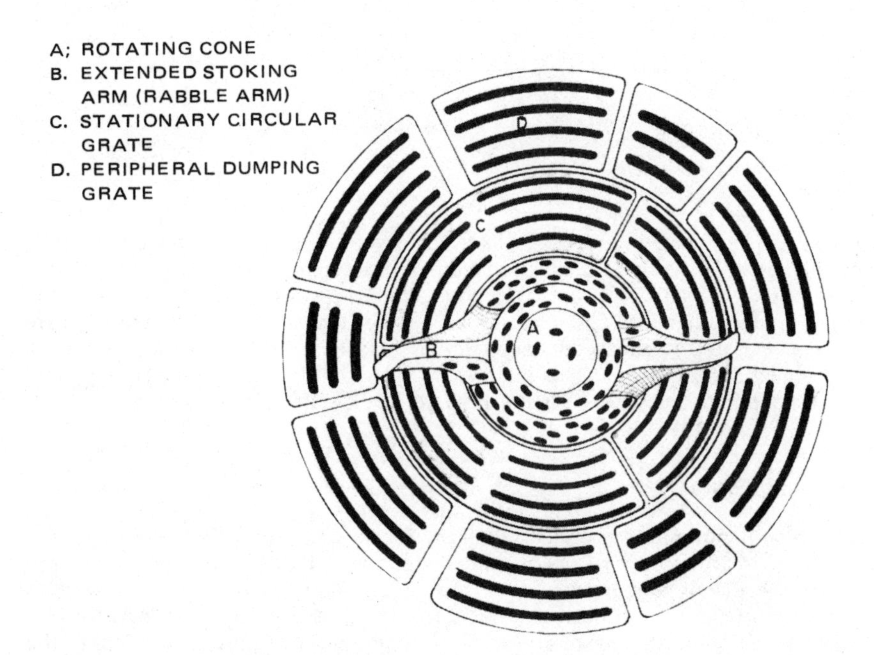

FIGURE 13.1 Circular grates.

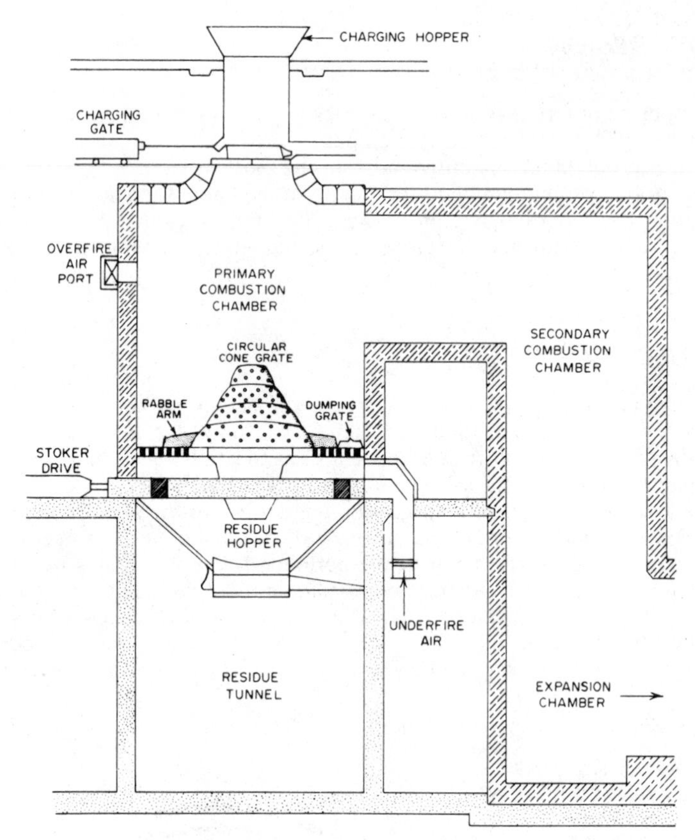

FIGURE 13.2 Vertical circular furnace. *Source:* Ref. 13-3.

pedite ash removal. Figure 13.2 illustrates a vertical circular furnace utilizing a circular grate. This furnace is automated with automatic ash (residue) collection in a residue hopper, for eventual removal from the system. Waste is combusted in the primary chamber, above the circular grate. The products of combustion are fully incinerated in the secondary combustion chamber, which may contain a supplemental fuel burner. This incinerator system discharges relatively high levels of pollutants and as such is not favored for new installations.

Traveling Grate

This type is no longer in common usage. As shown in Fig. 13.3, it is normally not a single grate but a series of grates which are placed in a manner that separates the drying and burning functions of the incinerator. The angled grate receives refuse on a continuous basis from a charging hopper. The refuse is normally dried on this grate and does not begin to burn until it drops onto the horizontal, or burning, grate. The speed of the burning grate is chosen so that

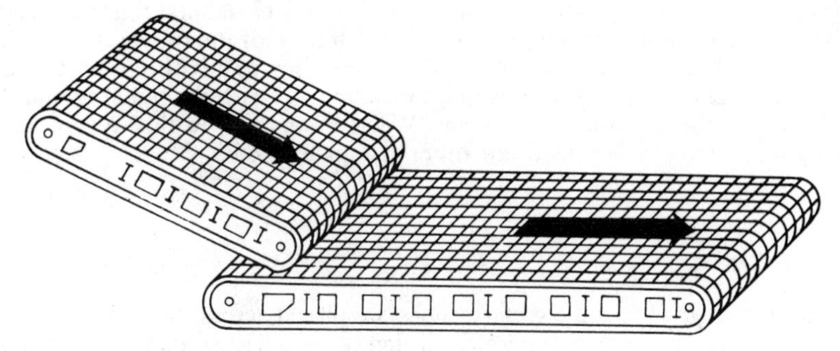

FIGURE 13.3 Traveling grates.

by the time the burning refuse reaches the end of the grate it is combusted to ash. The ash falls from this grate to an ash hopper. The grates themselves are continuous metal-belt conveyers or interlocking linkages. Refuse is not agitated on the grate, but it does experience some turbulence when falling from one grate to another. Figure 13.4 shows a typical system utilizing a traveling grate, a feeder stoker unit. Refuse is dropped by a grapple bucket into a hopper feeding a charging chute. An inclined stoker feeder moves the waste to a second, horizontal stoker, while the waste is being dried by the surrounding hot flue gases. Waste burns to ash on the horizontal stoker, and the ash res-

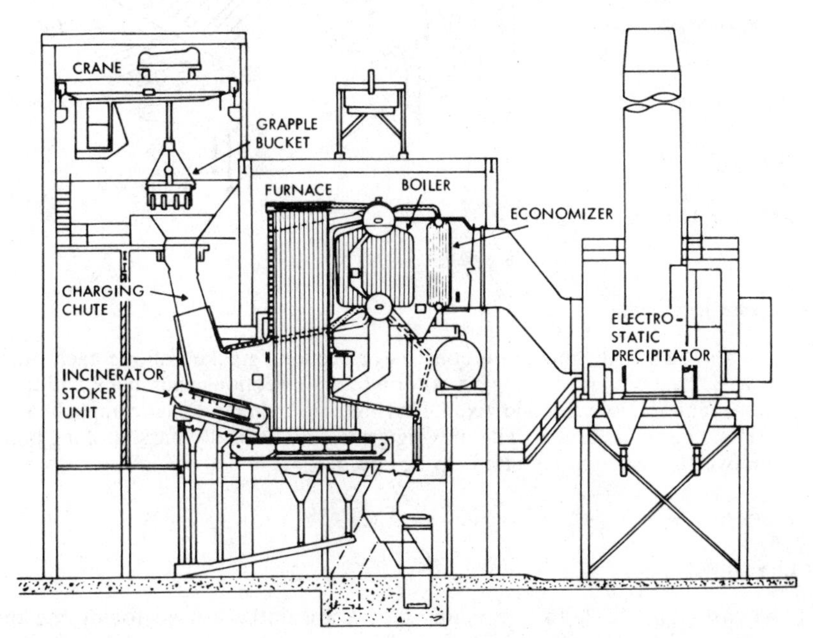

FIGURE 13.4 Traveling grate system. *Source:* Ref. 13-4.

idue falls into either of two pits, each serviced by a horizontal flight conveyer. Note that sifting hoppers are provided beneath each of the two grates. Siftings are conveyed to the main ash conveyers where they exit with the bottom ash. The furnace is lined with water walls, tubes integral with the boiler which capture heat for the generation of steam. Waste is charged to this system as received. It is neither shredded nor otherwise processed.

Rocking Grate

As shown in Fig. 13.5, these grate sections are placed across the width of the furnace. Alternate rows are mechanically pivoted or rocked to produce an upward and forward motion, advancing and agitating the waste. The stroke of the grate sections is 5 to 6 in. This grate will handle refuse on a continuous basis.

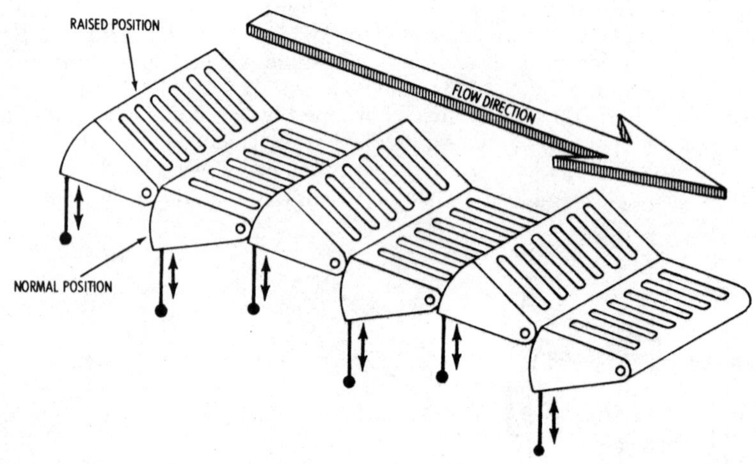

FIGURE 13.5 Rocking grates.

Reciprocating Grate

As shown in Fig. 13.6, this grate consists of sections stacked above each other similar to overlapping roof shingles. Alternate grate sections slide back and forth while adjacent sections remain fixed. Drying and burning are accomplished on single, short but wide grates. The moving grates are basically bars, stoking bars, which move the waste along and help agitate it.

Rotary Kiln

As shown in Fig. 13.7, two traveling grates are initially used for drying the incoming refuse and for initial ignition. The kiln is at the heart of this system.

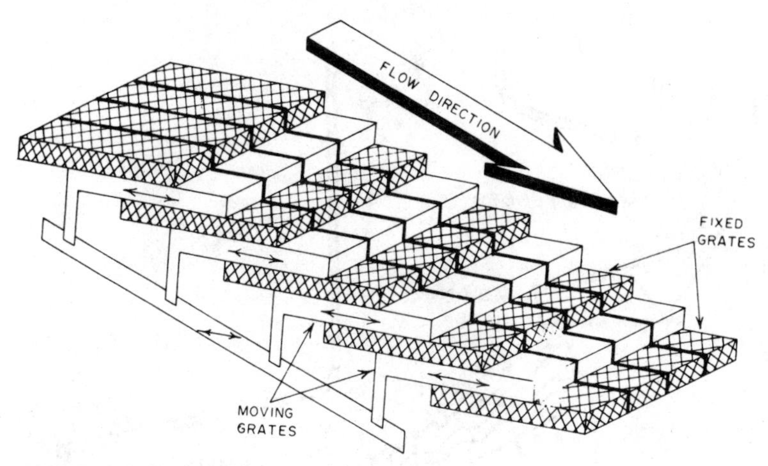

FIGURE 13.6 Reciprocating grates.

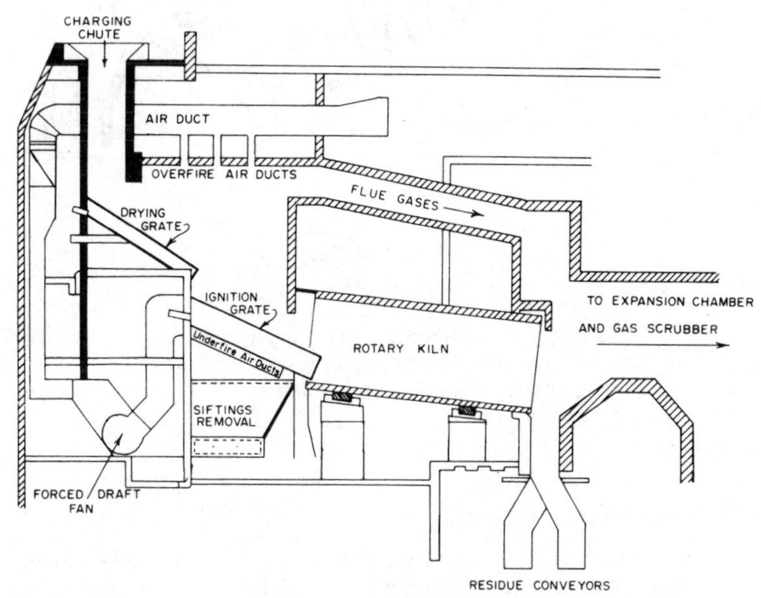

FIGURE 13.7 Municipal rotary kiln incineration facility. *Source:* Ref. 13-5.

By varying the kiln rotational speed, burnout of the refuse is accurately controlled. The refuse burns out in the kiln, and ash is discharged from the end of the kiln to residue conveyers. Some of the flue gases are diverted for drying the incoming refuse. Flue gas can be passed through a waste heat boiler for energy recovery.

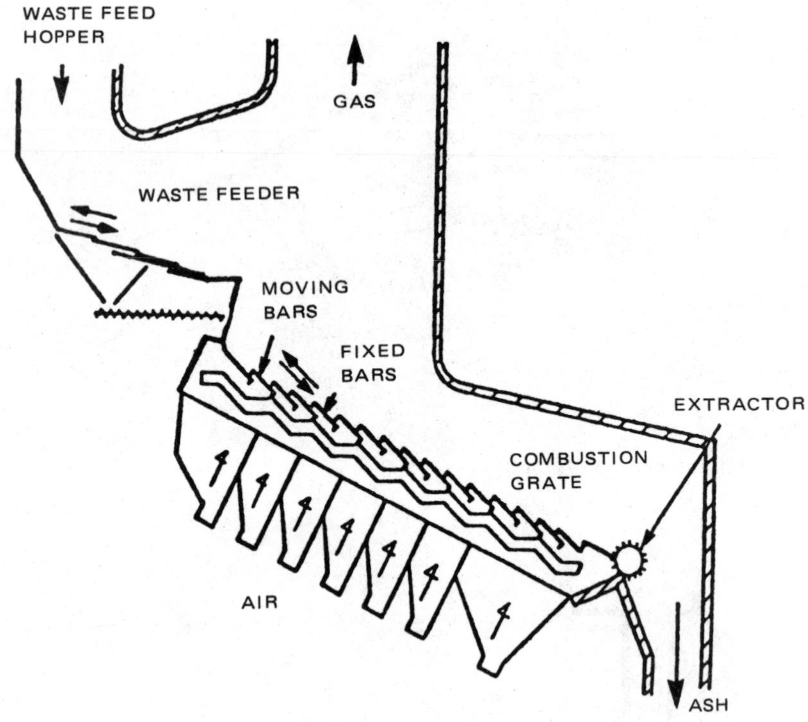

FIGURE 13.8 Martin system. *Source:* Ref. 13-6.

Martin System[1]

As shown in Fig. 13.8, this system utilizes reverse reciprocating grates. As the grates move forward and then reverse, there is continuous agitation of the waste. The bars making up the grates are hollow, allowing air to circulate within them and keep them relatively cool.

Von Roll System[2]

As shown in Fig. 13.9, a series of reciprocating grates are used to move refuse through the furnace. The first grate section dries the refuse, the second is a burning grate, and burnout to ash takes place on the third grate.

[1]Proprietary system, marketed in the United States by Ogden Corporation.
[2]Proprietary system, marketed in the United States by Wheelabrator.

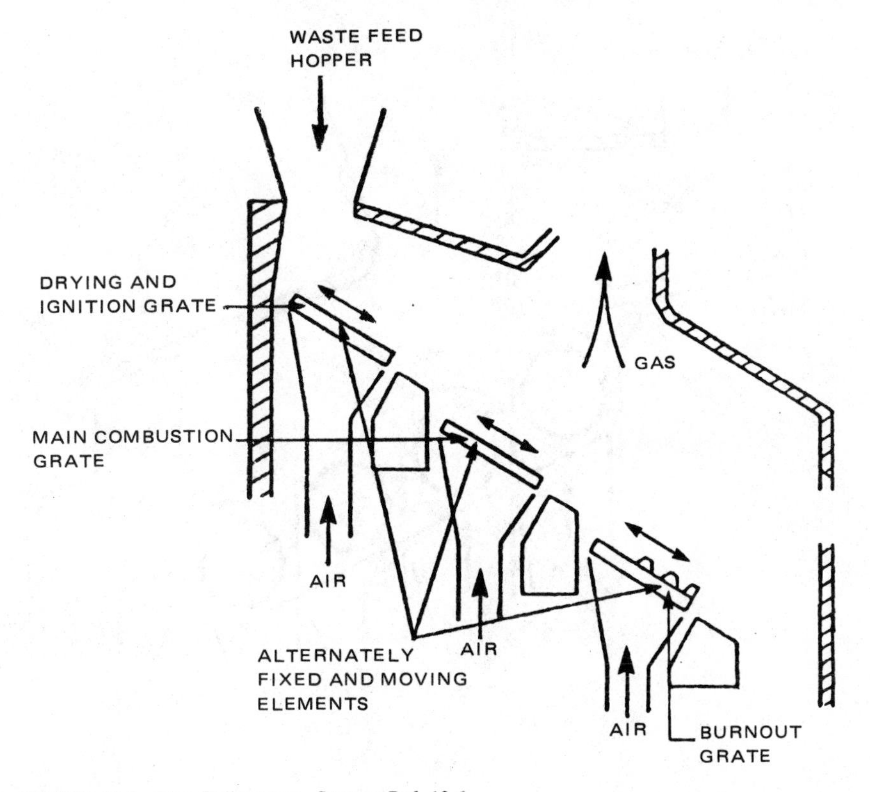

FIGURE 13.9 Von Roll system. *Source:* Ref. 13-6.

VKW System[3]

Shown in Fig. 13.10 is a variation of the traveling grate concept. A series of drums are utilized as grates. The drums rotate slowly, agitating the waste and moving it along to subsequent drums. Air passes through openings in these drums, or roller grates, as underfire air. Both speed of rotation on the roller grates and quantity of underfire air per roller grate are variable.

Alberti System[4]

As illustrated in Fig. 13.11, this grate system has a single-section grate constructed of fixed and moving elements arranged, as shown, in a series of steps. Feed is rammed from the feed hopper to the grates. The fixed grate contains the refuse while the moving elements agitate the waste, driving it down to the next grate.

---

[3]Proprietary system of VKW, Düsseldorf, West Germany, marketed in the United States by Browning-Ferris Industries, Houston, Texas.

[4]Proprietary system.

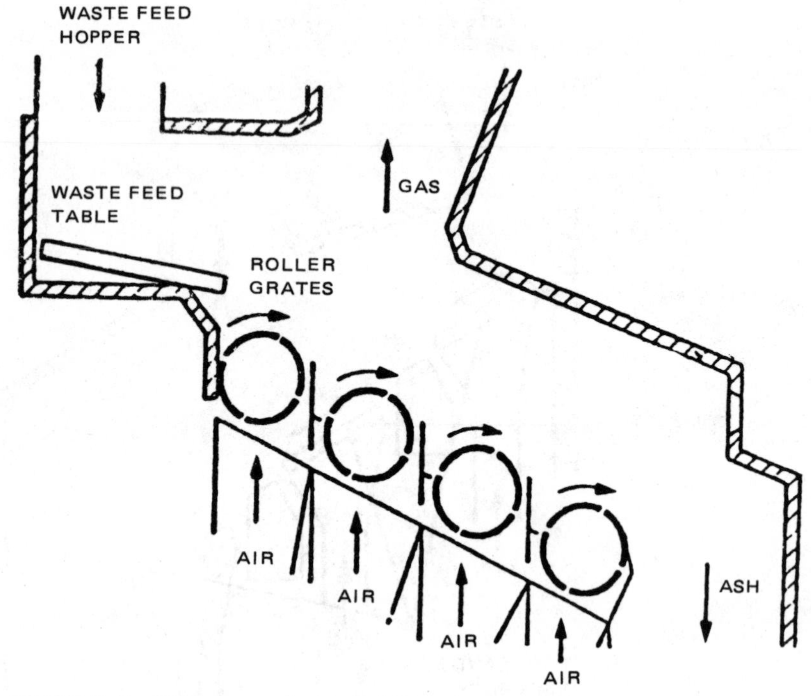

FIGURE 13.10 VKW system. *Source:* Ref. 13-6.

Esslingen System[5]

As shown in Fig. 13.12, a traveling grate is used to feed a rocking grate system. It is normally provided with a single-grate section composed of semicircular rocking elements. Each movement of these elements promotes transport and agitation of the waste. Underfire air passes through the rocking elements, keeping them cool and providing for combustion of the waste.

The Heenan Nichol System[6]

As shown in Fig. 13.13, this system utilizes grates composed of three or more sections which are arranged in steps. Each pair of elements moves in a rocking manner so that at any moment half of the elements are moving. All odd-numbered elements are linked to each other, as are all the even-numbered elements. The rocking action moves and agitates the waste.

---

[5]Proprietary system.
[6]Proprietary system.

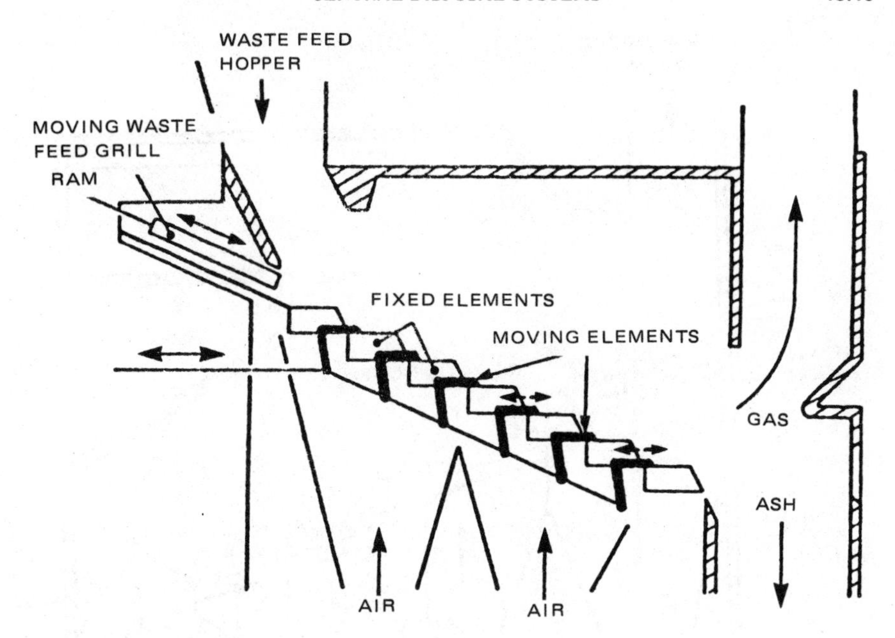

FIGURE 13.11 Alberti system. *Source:* Ref. 13-6.

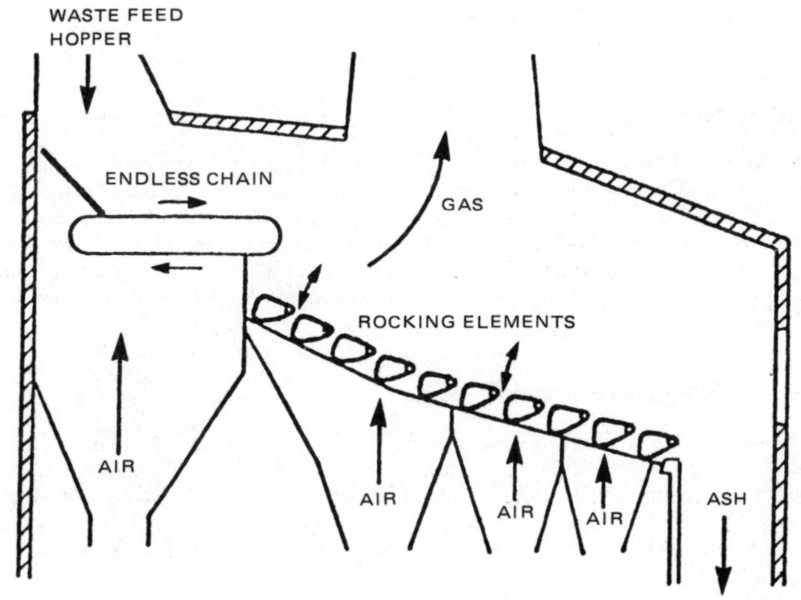

FIGURE 13.12 Esslingen system. *Source:* Ref. 13-6.

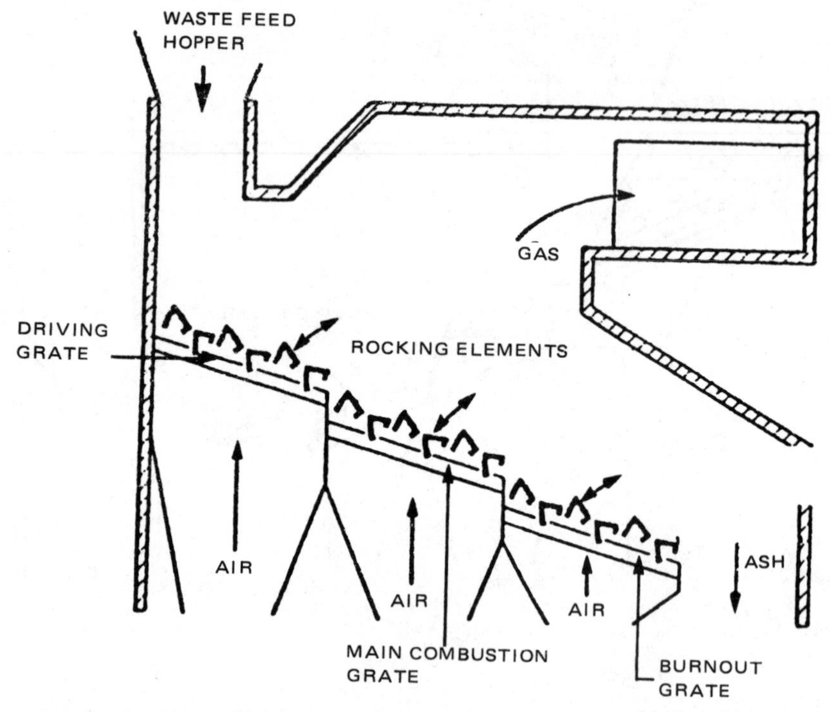

FIGURE 13.13 Heenan Nichol system. *Source:* Ref. 13-6.

CEC System[7]

This system, illustrated in Fig. 13.14, utilizes a single-section grate. Two or more grate sections can be arranged in parallel. The grate is constructed of successive sliding, rocking, and fixed elements. The sliding and rocking elements are synchronized so that the sliding elements move over the rocking elements when the rocking elements are retracted. The sliding elements, therefore, promote transport of the waste while the rocking elements provide the required agitation.

Bruun and Sorensen System[8]

As shown in Fig. 13.15, this system utilizes a series of rollers, up to six in each of its three sections. Odd-numbered rollers turn clockwise while even-numbered rollers turn counterclockwise. Underfire air passes through the rotary grates, or drums. The action of the drums provides good agitation, and the slope of the grate to the horizontal promotes the transport of waste along the grate.

---

[7]Proprietary system of Carbonisation Enterprise et Céramique.
[8]Proprietary system.

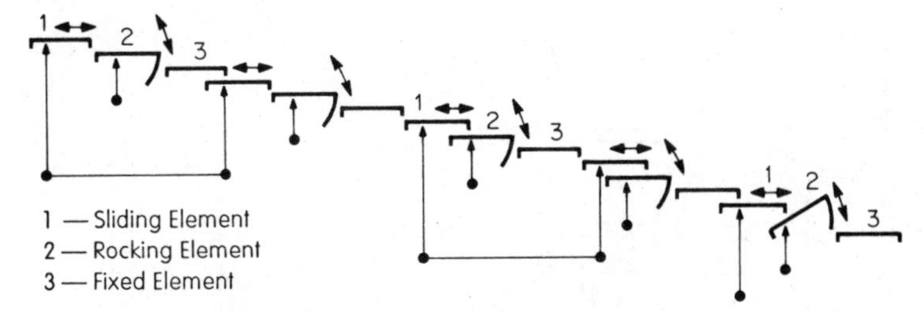

FIGURE 13.14 CEC system. *Source:* Ref. 13-6.

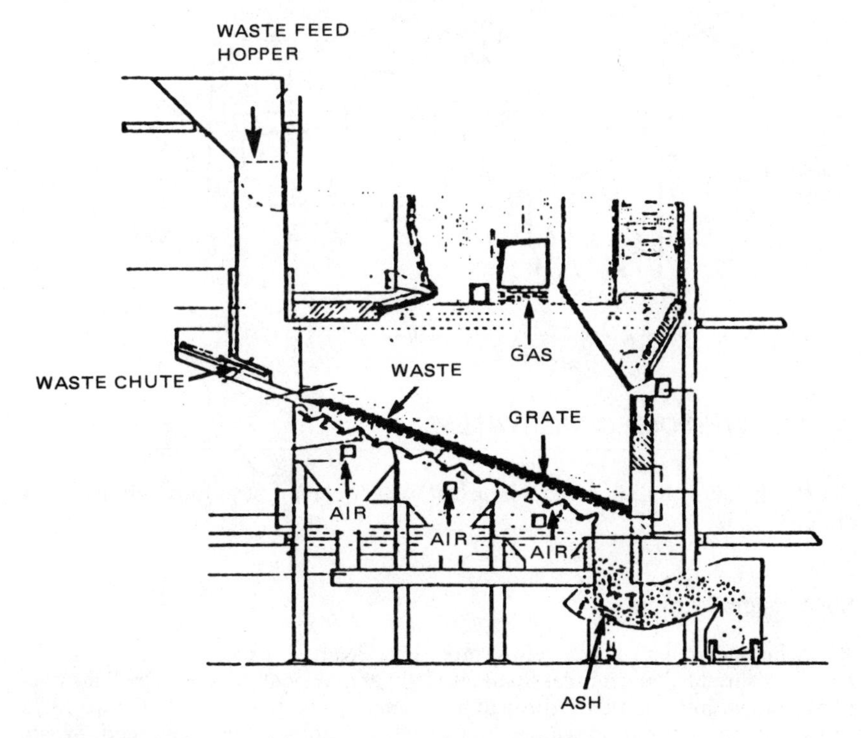

FIGURE 13.15 Bruun and Sorensen system. *Source:* Ref. 13-6.

Volund System[9]

This system is shown in Fig. 13.16. It utilizes a rotary kiln for controlled burning of waste. Reciprocating grates are used for waste drying and initial combustion. Burnout takes place within the kiln.

[9]Proprietary system, marketed in the United States by Waste Management, Inc., Chicago. A similar system is manufactured by International Incinerators, Columbus, Ga.

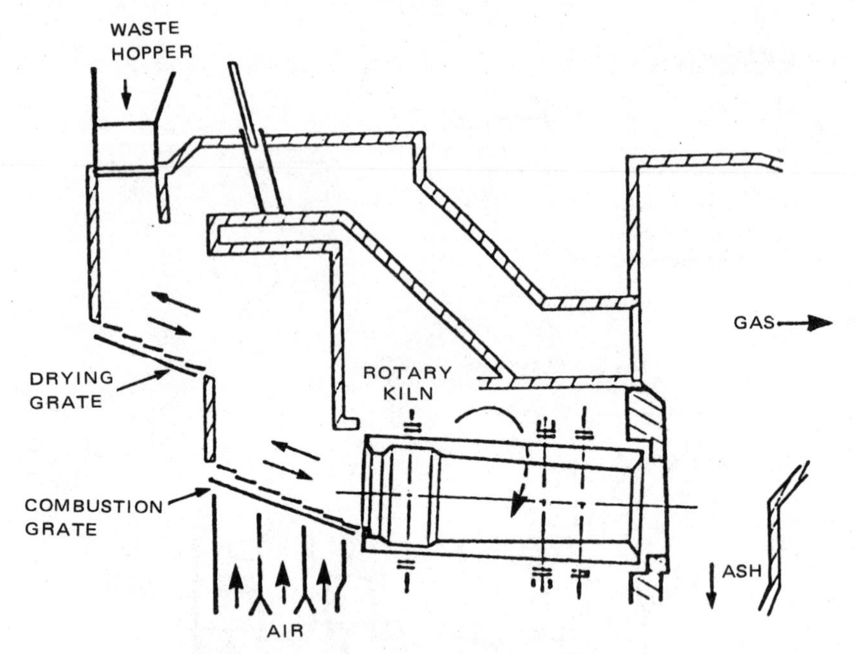

FIGURE 13.16 Volund system. *Source:* Ref. 13-6.

ASSOCIATED DISPOSAL SYSTEMS

There are refuse burning systems in use which cannot strictly be classified as grate systems.

Suspension Burning

An incinerator coupled to a refuse processing system is pictured in Fig. 13.17. Refuse is shredded and air classified into light and heavy fractions. The light fraction is blown into the boiler through a pneumatic charging system. Figure 13.18 shows the air distribution within the furnace and around the waste feed. Waste that is not burned in suspension will drop onto the shredder stoker, a variation of the traveling grate, and will burn out. Ash not airborne, produced by suspension burning, will also drop onto the spreader stoker. The stoker moves slowly, discharging its ash load to an ash hopper for ultimate disposal. Heavier components of the refuse that are not incinerated are composed mainly of metals and glass. These materials can, in certain instances, be marketed.

FLUID BED INCINERATION

There have been some attempts to adapt limited European experience with fluid bed incineration of municipal solid waste to the United States. This technology,

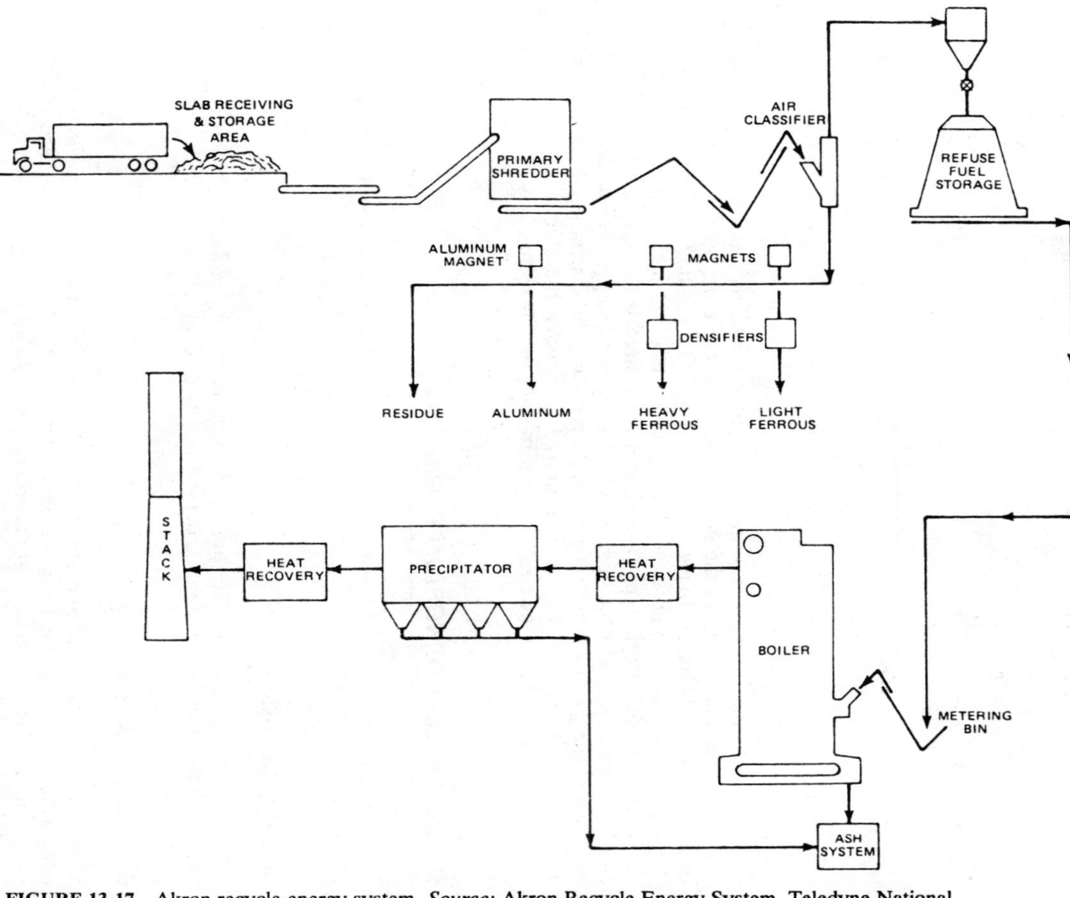

FIGURE 13.17 Akron recycle energy system. *Source:* Akron Recycle Energy System, Teledyne National, Akron, Ohio.

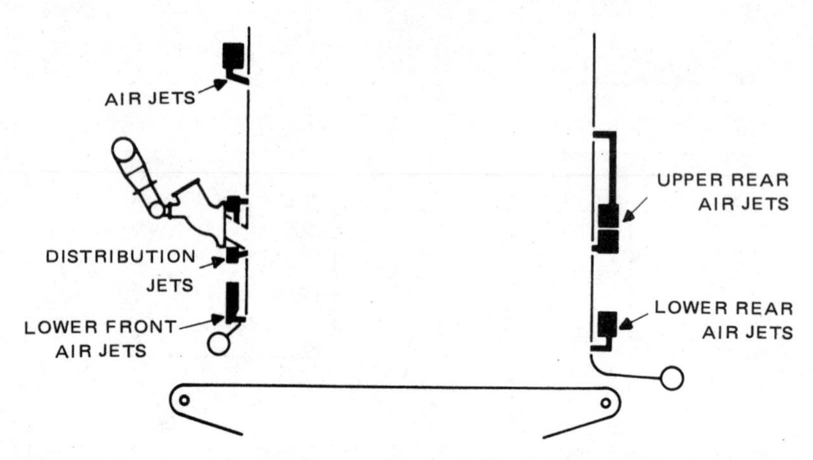

FIGURE 13.18 Air and feed distribution. *Source:* Babcock & Wilcox, North Canton, Ohio.

described in Chap. 15, requires that glass and low-melting-point metals (such as aluminum) be removed from the waste stream. These components, in even relatively small quantities, will slag the furnace bed. In addition, the feed must be reduced to uniform size, no larger than 1- to 1½-in mean particle size.

The advantage of this type of incineration system is the ability to add limestone (or other alkali) to the bed, which will capture halogens (chlorides and fluorides) and other compounds, significantly reducing the discharge of acid gases. The effort and resultant high cost required to remove the aluminum and glass from the waste stream, however, restrict the use of this technology.

INCINERATOR CORROSION PROBLEMS

Severe corrosion has been found in three major areas of the mass burning incinerator system.

Scrubber Corrosion

The acidic components of the flue gas present corrosion problems in wet scrubbing equipment, which will be discussed in a later chapter.

Corrosion of Grates

This problem was alluded to earlier in this chapter. With insufficient airflow, high temperatures and a reducing atmosphere can occur and ash can soften or fluidize. Fluid ash can be exceedingly corrosive, readily attacking cast iron or steel.

Table 13.4 illustrates the corrosion rate as a function of temperature for two steel alloys in widespread use in grate construction. Temperature alone greatly increases the corrosive rate, the amount of material lost per month of service. At 1200°F over ⅜ in of material is "wasted" or lost from the steel structures, for instance.

TABLE 13.4 Gas-Phase Corrosion at Elevated Temperatures

Alloy	Temperature, °F	Corrosion rate, mils/month
A106	800	0.9
A106	1000	8
A106	1200	36
T11	800	0.8
T11	1000	6
T11	1200	29

Source: Ref. 13-7.

Fireside Corrosion

There are two modes of corrosion that affect boiler tubes. Low-temperature or dewpoint corrosion is metal wastage caused by sulfuric or hydrochloric acid condensation. Chlorides and sulfides within the refuse (chlorides are present in plastics) will partially convert to free chlorine, hydrogen chloride, sulfur dioxide, and sulfur trioxide in the exhaust gas stream. These gases will condense at temperatures below 300°F, and their condensate or liquid phase will be hydrochloric and sulfuric acid, both of which will attack steel. It is important, therefore, that the temperature of the boiler tubes, constructed of steel, be kept above 300°F. This is a function of the temperature of the steam or hot water generated. The boiler tubes will be at a temperature close to that of the circulating fluid, and this 300°F rule therefore limits the minimum temperature of the steam or hot water generated.

High-temperature corrosion is a more complex problem. Table 13.5 lists the

TABLE 13.5 Nominal Operating Conditions of Water-Wall Incinerators

Location	Steam pressure, lb/in^2 gauge	Steam temp., °F	Metal temp., °F (approx.)
Milan, Italy	500	840	890
Mannheim, Germany	1800	980	1030
Frankfurt, Germany	960	930	980
Munster, Germany	1100	980	1030
Moulineaux, France	930	770	820
Essen Karnap, Germany	--	930	980
Stuttgart, Germany	1100	980	1030
Munich, Germany	2650	1000	1050
Rotterdam, Netherlands	400	680	730
Edmonton, England	625	850	900
Coventry, England	275	415	465
Amsterdam, Netherlands	600	770	820
Montreal, Canada	225	395	445
Chicago (N. W.), Illinois	265	410	460
Oceanside, New York	460	465	515
Norfolk, Virginia	175	375	425
Braintree, Massachusetts	265	410	460
Harrisburg, Pennsylvania	275	460	510
Hamilton, Ontario	250	400	450

Source: Ref. 13-7.

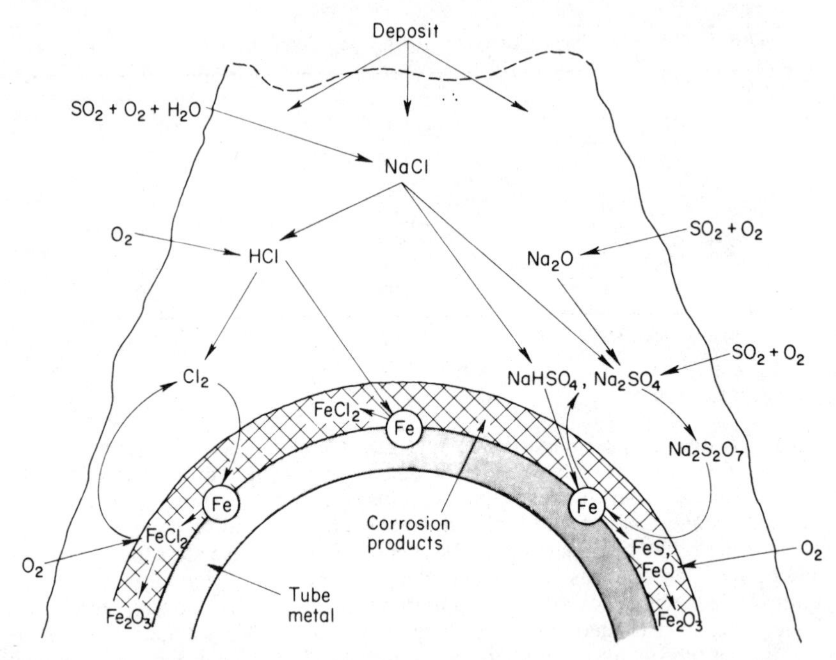

FIGURE 13.19 Sequence of chemical reactions explaining corrosion on incinerator boiler tube. *Source:* Ref. 13-7.

steam pressure, temperature, and external tube temperatures for an assortment of incinerator systems burning municipal solid waste. At temperatures exceeding 700°F a complex reaction takes place between the sulfide and chlorine/chloride-bearing flue gas and the steel boiler tube, as illustrated in Fig. 13.19. Chlorine reacts with the iron in the tube wall to produce ferrous chloride which, upon contact with oxygen in the flue gas, converts to iron oxide. The iron oxide (rust) will leave the surface of the steel, causing wastage of the steel surface. Other components of the refuse which become airborne, such as alkali salts, will promote this corrosion. For incinerators operating above 700°F metal temperatures (most of the incinerator tubes noted in Table 13.5 operate at temperatures in excess of 700°F), special refractory-lined water walls must be utilized to protect the tubes from this metal wastage.

Table 13.6 indicates the relative performance of various alloys in incinerator fireside areas. Although stainless steels appear to have favorable corrosion resistance, the danger of stainless-steel stress corrosion cracking prohibits its use in pressure vessels such as boilers and high-temperature hot-water heaters.

Figure 13.20 further illustrates the rate of corrosion of carbon steel by chloride attack as a function of metal temperature.

INCINERATOR REFRACTORY SELECTION

Table 13.7 illustrates the types of problems that can be expected within the various areas of a large incinerator system associated with refractory selection. The nature of the hot gas stream within the incinerator can create significant detri-

TABLE 13.6 Performance of Alloys in Fireside Areas[a]

Alloy	Resistance to wastage		
	300–600°F	600–1200°F	Moist deposit
Incoloy 825	Good	Fair	Good
Type 446	Good	Fair	Pits
Type 310	Good	Fair	SCC[b]
Type 316L	Good	Fair	SCC
Type 304	Good	Fair	SCC
Type 321	Good	Fair	SCC
Inconel 600	Good	Poor	Pits
Inconel 601	Good	Poor	Pits
Type 416	Fair	Fair	Pits
A106-Grade B (carbon steel)	Fair	Poor	Fair
A213-Grade T11 (carbon steel)	Fair	Poor	Fair

[a]Arranged in approximate decreasing order.
[b]Stress-corrosion cracking.
Source: Ref. 13-7.

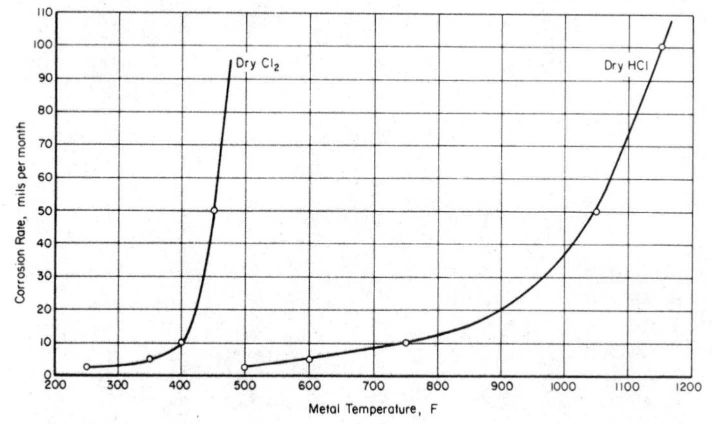

FIGURE 13.20 Corrosion of carbon steel in chlorine and hydrogen chloride. *Source:* Ref. 13-7.

mental effects on grates, walls, ceilings, and other areas within the furnace enclosure. Some of the more common concerns in refractory selection are discussed here.

Abrasion

This is the effect of impact of moving solids within the gas stream or of heavy pieces of materials charged into the furnace upon refractory surfaces. Fly ash also causes abrasive effects. Abrasion is the wearing away of refractory, or any surface, under direct contact with another material with relative motion to its surface.

TABLE 13.7 Suggested Refractory Selection for Incinerators

Incinerator part	Temperature (°F) range	Abrasion	Slagging	Mechanical shock	Spalling	Fly ash adherence	Recommended refractory
Charging gate	70–2600	Severe, very important	Slight	Severe	Severe	None	Superduty
Furnace walls, grate to 48 in. above	70–2600	Severe	Severe, very important	Severe	Severe	None	Silicon carbide or superduty
Furnace walls, upper portion	70–2600	Slight	Severe	Moderate	Severe	None	Superduty
Stoking doors	70–2600	Severe, very important	Severe	Severe	Severe	None	Superduty
Furnace ceiling	700–2600	Slight	Moderate	Slight	Severe	Moderate	Superduty
Flue to combustion chamber	1200–2600	Slight	Severe, very important	None	Moderate	Moderate	Silicon carbide or superduty
Combustion chamber walls	1200–2600	Slight	Moderate	None	Moderate	Moderate	Superduty
Combustion chamber ceiling	1200–2600	Slight	Moderate	None	Moderate	Moderate	Superduty
Breeching walls	1200–3000	Slight	Slight	None	Moderate	Moderate	Superduty
Breeching ceiling	1200–3000	Slight	Slight	None	Moderate	Moderate	Superduty
Subsidence chamber walls	1200–1600	Slight	Slight	None	Slight	Moderate	Medium duty
Subsidence chamber ceiling	1200–1600	Slight	Slight	None	Slight	Moderate	Medium duty
Stack	500–1000	Slight	None	None	Slight	Slight	Medium duty

Source: Ref. 13-8.

Slagging

When a portion of charged material, usually ash, metals, or glass within an incinerator, reaches a high enough temperature, deformation of that material will occur. The material will physically change to a more amorphous state and may begin to flow, as a heavy liquid. When the temperature of the material is then reduced below that required for deformation, the material will solidify into a hard slag. This process can take place when high temperatures are experienced on a grate. When the grate section moves through a lower-temperature zone, a slag may form. Molten ash may become airborne, then attach to a refractory or a metal surface within the furnace which is cooler than the air stream. Slag will then form on this surface. This slag can be acidic (as a result of silicon, aluminum, or titanium oxides released from the burning waste), or it can be basic (due to the generation of oxides of iron, calcium, magnesium, potassium, sodium, or chromium), and the selection of refractory must be compatible with these materials to help ensure long refractory life. (For acidic slag, fireclay or high-alumina and/or silica firebrick would be used. Chrome, magnesite, or forsterite brick would be used for basic slags.)

Mechanical Shock

The impact of falling refuse can cause mechanical shock, as can constant vibration caused by grate bushings or supports and vibration set up by turbulent flow adjacent to air inlet ports.

Spalling

The flaking away of the refractory surface, or spalling, is most commonly caused by thermal stresses or mechanical action. Uneven temperature gradients can cause local thermal stresses in brick which will degrade the refractory surface, causing a spalling condition. The more common type of mechanical spalling is caused be rapid drying of wet brickwork. The steamed water does not have an opportunity to escape the brick surface through the natural porosity of the refractory, but expands rapidly, causing cracking and spalling of the brick.

Fly Ash Adherence

As noted above, fly ash can have fluid properties within the hot gas stream and adhere to refractory or other cooler surfaces within the furnace chamber. Fly ash accumulation can result in corrosive attack on these surfaces. Heavy accumulations will interfere with the normal surface cooling effects within the furnace, resulting in a decrease of furnace refractory life.

As noted above, Table 13.7 lists various areas within an incinerator and describes the severity of the above problems at each of these locations. Normal temperature ranges are noted as well as recommended refractory types.

TABLE 13.8 Typical Steam Generation

Solid waste type	MSW	MSW	MSW	RDF
Steam temperature, °F	620	500	465	400
Steam pressure, lb/in^2 gauge	400	225	260	250
Steam production, tons/ton refuse	3.6	1.4–3.0	1.5–4.3	4.2

Note:
MSW: Municipal solid waste.
RDF: Refuse-derived fuel, i.e., shredded MSW less metals.
Source: Ref. 13-9.

HEAT RECOVERY OR WASTING OF HEAT

Steam can be generated by utilization of incinerator waste heat, as shown in Table 13.8, which lists steam production rates. Variations in waste heating value will produce variations in steam generation, as shown in Table 13.9. But these quantities must be weighed against the overall implications of a boiler installation for waste burning. Note the following comparison between an energy recovery system and an incinerator without provision for heat recovery:

With heat recovery	Without heat recovery
Reduced gas temperatures and volumes due to absorption of heat by heat recovery system	Hotter gas temperatures
Moderate excess air	High excess air required to control furnace temperatures
Moderate-size combustion chamber	Large refractory-lined combustion chamber to handle high gas flow
Smaller air and induced-draft fans required for smaller gas volume	Higher gas volumes due to higher temperatures, requiring larger air and gas flow equipment
Steam facilities including integral water wall, boiler drums, and boiler auxiliary equipment required	No steam facilities required
Operations involve boiler system monitoring, adjustments for steam demand, etc.	Relatively simple operating procedures
Steam tube corrosion is possible as well as corrosion within exhaust gas train	Corrosion possible in the exhaust gas train
Licensed boiler operators are required to operate incinerator	Conventional operators satisfactory
Considerable steam credits possible, including in-plant energy savings in addition to salvage	Only credits are possible salvage of the equipment after its useful life

TABLE 13.9 Steam Production Related to MSW Quality

	As-received heating value, Btu/lb				
	6500	6000	5000	4000	3000
Refuse:					
% moisture	15	18	25	32	39
% noncombustible	14	16	20	24	28
% combustible	71	66	55	44	33
Steam generated, tons/ton refuse	4.3	3.9	3.2	2.3	1.5

Source: Ref. 13-10.

RESOURCE RECOVERY PLANT EMISSIONS

A number of states have established regulations specifically governing emissions of central disposal incinerators firing municipal solid waste (resource recovery facilities). Table 13.10 lists these regulations for six states and the EPA. Most of these criteria are noted as guidelines. A guideline is not necessarily established by statute, but it is used by the regulator as a criterion for the permitting of these facilities.

In the absence of other data, Table 7.8 provides typical emissions factors for resource recovery facilities.

TABLE 13.10 Emissions Limitations for Municipal Waste Incinerators

Pollutant	California guidelines 7% O_2	Illinois guidelines 12% CO_2	New Jersey guidelines 7% O_2	New York guidelines 12% CO_2	Pennsylvania BAT criteria 7% O_2	USEPA guidelines 12% CO	Wisconsin guidelines 12% CO_2
Particulate	0.01	0.01	0.015	0.01	0.015	—	0.015
(below 2 μm), gr/dry st ft^3	0.008	—	—	—	—	—	—
HCl	30 ppm	30 ppm/1 h or 90% reduction	50 ppm/1 h or 90% reduction	50 ppm/8 h or 90% reduction	30 ppm/1 h or 90% reduction	—	50 ppm
SO_2	30 ppm	50 ppm/1 h or 70% reduction	50 ppm/1 h or 80% reduction	0.2–2.5 lb/(MBtu/h)	50 ppm/1 h or 70% reduction	—	—
NO_x	140–200 ppm	100 ppm/1 h	350 ppm/1 h	BACT	—	—	—
Hydrocarbons	70 ppm	—	70 ppm/1 h	—	—	—	—
CO	400 ppm	100 ppm/1 h	400 ppm/1 h 100 ppm/4 days	—	400 ppm/8 h 100 ppm/4 days	50 ppm/4 h	—
Dioxin	—	—	—	2 ng/Nm^3	—	—	3 ng/Nm^3
Furnace temp., design, °F	—	1800	1800	1800	—	—	—
Furnace temp., min., °F	1800 ±200	1500	1500	1500 for 15 min	1800	1800	1500
Residence time, min., s	1	1.2	1	1	1	1	1
Lime injection, min., lb/h	—	100	—	—	—	—	—
Baghouse temp., max., °F	—	—	—	300	—	—	250
Combustion efficiency, %	—	99.9/2 h	—	99.9/8 h 99.95/7 days	99.9/4 days	—	—
Minimum O_2, %	—	—	6	—	—	6–12	—
Opacity, max., %	—	10	20	10/6 min	30/3 min/h	—	20

REFERENCES

13-1. "World Survey Finds Less Organic Matter," *Refuse Removal Journal*, 10:26, 1967.

13-2. Wisely, F., and Hinchman, H.: "Refuse as a Supplementary Fuel," *Proceedings of the Third Annual Environmental Engineering Science Conference*, Louisville, Ky., March 1973.

13-3. *Municipal Incinerator Enforcement Manual*, USEPA 340/1-76-013, January 1977.

13-4. Golembiewski, M., and Baladi, E.: *Environmental Assessment of a Waste-to-Energy Process: Braintree Municipal Incinerator*, USEPA, 1979.

13-5. *Hazardous Material Design Criteria*, USEPA 600/2-79-198, October 1979.

13-6. *European Waste-to-Energy Systems: An Overview*, Energy Research & Development Administration, CONS-2103-6, U.S. Department of Commerce, Springfield, Va., June 1977.

13-7. Miller, P.: *Corrosion Studies in Municipal Incinerators*, USEPA SW-72-3-3, 1972.

13-8. *Municipal Refuse Disposal*, Institute for Solid Wastes, American Public Works Association, 1970.

13-9. Weinstein, N., and Toro, R.: *Thermal Processing of Municipal Solid Waste for Resource and Energy Recovery*, Ann Arbor Science, Ann Arbor, Mich., 1976.

13-10. Popperman, C.: "The Harrisburg Incinerator," *Proceedings of the National Incinerator Conference*, American Society of Mechanical Engineers, Miami, 1974, pp. 247–254.

BIBLIOGRAPHY

Astrom, L., and Harris, D.: "Comparative Study of European and North American Steam Producing Incinerators," *Proceedings of the National Incinerator Conference*, American Society of Mechanical Engineers, Miami, 1974, p. 264.

Battelle Memorial Institute, *Corrosion Studies in Municipal Incinerators*, USEPA, 1972.

The Conversion of Existing Municipal Incinerators to Codisposal, USEPA SW-743, 1979.

DeMarco, J.: *Municipal Scale Incinerator Design and Operation*, USEPA, 1973.

Jahnke, J.: "A Research Study of Gaseous Emissions from a Municipal Incinerator," *Journal of the Air Pollution Control Association*, August 1977.

Kennedy, J.: "The Disposal of Solid Wastes," *Journal of Environmental Health*, September 1968.

Niessen, W.: *Combustion and Incineration Processes*, 1st ed., Marcel Dekker, Inc., New York, 1981.

A Review of Techniques for Incineration of Sewage Sludge with Solid Wastes, USEPA 600/2-76-288, December 1976.

Rubel, *Incineration of Solid Wastes*, Noyes, Park Ridge, N.J., 1974.

CHAPTER 14
BIOMEDICAL WASTE INCINERATION

Biomedical wastes are generated by hospitals, laboratories, animal research facilities, and other institutional sources. The disposal of these wastes is coming under severe public scrutiny, and regulations are being promulgated to control their disposal. Incineration is a favored method of treating these wastes because it is the only commercially available method of treatment which destroys the organisms associated with this waste completely and effectively.

THE WASTE STREAM

Biomedical waste is a term coming into common usage to replace what had been referred to as *pathological* or *infectious* wastes and to include additional related waste streams. Where the term *pathological waste* is used here, it refers to anatomical wastes, carcasses, and similar wastes. Table 14.1 is a listing of wastes classified as biomedical and includes a description of each waste as well as typical characteristics. The bag designations (red, orange, yellow, blue) are used in Canada. In the United States, generally most of these wastes are classified as "red bag."

The hospital waste stream has changed significantly in the last few years. Disposable plastics have been replacing glass and clothing in what appears to be, at first look, a means of cutting costs. They represent a greater cost in their disposal, however, since many of these plastics contain chlorine and with the increase in the use of plastics, the increase in chlorine creates the need for additional equipment in the incineration process. The plastics content of the hospital waste stream has grown from 10 percent to over 30 percent in the past 10 years.

It is rare to find an incinerator designated for biomedical waste destruction to be fired solely on this type of waste. Generally, particularly in hospitals, installation of an incinerator encourages the disposal of other wastes in the unit. Besides the cost savings this represents in not having to cart away this trash, there is the potential for heat recovery. For example, hospitals generally require steam throughout the year for their laundry, sterilizers, autoclaves, and kitchens. As more waste is fired, more heat is produced and more steam is generated.

Another set of wastes includes those generated in hospital laboratories that are hazardous wastes under the Resource Conservation and Recovery Act

TABLE 14.1 Characterization of Biomedical Waste

Waste class	Component description	Typical component weight, % (as fired)	HHV dry basis, Btu/lb	Bulk density as fired, lb/ft	Moisture content of component, wt %	Weighted heat value range of waste component, Btu/lb	Typical component heat value of waste as fired, Btu/lb
A1	Human anatomical	95–100	8,000–12,000	50–75	70–90	760–2,600	1,200
(red bag)	Plastics	0–5	14,000–20,000	5–144	0–1	0–1,000	180
	Swabs, absorbents	0–5	8,000–12,000	5–62	0–30	0–600	80
	Alcohol, disinfectants	0–0.2	11,000–14,000	48–62	0–0.2	0–28	20
	Total bag						1,480
A2	Animal infected anatomical	80–100	9,000–16,000	30–80	60–90	720–6,400	1,500
(orange bag)	Plastics	0–15	14,000–20,000	5–144	0–1	0–3,000	420
	Glass	0–5	0	175–225	0	0	0
	Beddings, shavings, paper, fecal matter	0–10	8,000–9,000	20–46	10–50	0–810	600
	Total bag						2,520
A3a	Gauze, pads, swabs, garments, paper, cellulose	60–90	8,000–12,000	5–62	0–30	3,360–10,800	6,400
(yellow bag)	Plastics, PVC, syringes	15–30	9,700–20,000	5–144	0–1	1,440–6,000	3,250
	Sharps, needles	4–8	60	450–500	0–1	3–5	5
	Fluids, residuals	2–5	0–10,000	62–63	80–100	0–11	30
	Alcohols, disinfectants	0–0.2	7,000–14,000	48–62	0–50	0–28	15
	Total bag						9,700
A3b	Plastics	50–60	14,000–20,000	5–144	0–1	6,930–12,000	9,000
(yellow bag)	Sharps	0–5	60	450–500	0–1	0–3	0
Lab waste	Cellulosic materials	5–10	8,000–12,000	5–62	0–15	340–1,200	650
	Fluids, residuals	1–20	0–10,000	62–63	95–100	0–100	30
	Alcohols, disinfectants	0–0.2	11,000–14,000	48–62	0–50	0–28	20
	Glass	15–25	0	175–225	0	0	0
	Total bag						9,700
A3c	Gauze, pads, swabs	5–20	8,000–12,000	5–62	0–30	280–3,600	1,000
(yellow bag)	Plastics, petri dishes	50–60	14,000–20,000	5–144	0–1	6,930–12,000	9,000
R&D	Sharps, glass	0–10	60	450–500	0–1	0–6	0
	Fluids	1–10	0–10,000	62–63	80–100	0–200	100
	Total bag						10,100
B1	Noninfected						
(blue bag)	Animal anatomical	90–100	9,000–16,000	30–80	60–90	810–6,400	1,400
	Plastics	0–10	14,000–20,000	5–144	0–1	0–20,000	1,000
	Glass	0–3	0	175–225	0	0	0
	Beddings, shavings, fecal matter	0–10	8,000–9,000	20–46	10–50	0–810	600
	Total bag						3,000

Source: Ref. 14-1.

(RCRA) (refer to Chap. 3 for a discussion of these requirements). Table 14.2 lists some of these wastes. If more than 220 lb/month of these wastes is generated, the incinerator must be permitted under the provisions of RCRA, which are additional to state requirements.

TABLE 14.2 Hazardous Wastes under RCRA Typically Generated by In-Hospital Laboratories

Acetone	Methyl alcohol
Antineoplastics	Methyl cellosolve
Butyl alcohol	Pentane
Cyclohexane	Petroleum ether
Diethyl ether	Tetrahydrofuran
Ethyl alcohol	Xylene

Source: Ref. 14-2.

Waste generation rates will vary from one hospital to another, as a function of the number of hospital beds, the number of intensive-care beds, and the presence of other specialty facilities. In the absence of specific generation data, the figures in Table 14.3 can be used as an estimate of waste generation rates.

TABLE 14.3 Estimated Waste Generation Rates

Hospital	13 lb/(occupied bed · day)
Rest home	3 lb/(person · day)
Laboratory	0.5 lb/(patient · day)
Cafeteria	2 lb/(meal · day)

Source: Ref. 14-3.

REGULATORY ISSUES

Biomedical waste incinerators are generally small, much smaller than the central disposal incinerators that have been in the public eye in many of the densely populated areas of the country. Regulations have addressed the larger municipal solid waste incinerators. Smaller units, such as 2 to 10 tons/day biomedical waste incinerators, have generally not been subject to rigorous regulatory attention in the past. The only restriction on their operation in many parts of the country is that they not create a public nuisance. That has meant that no odors are to be generated and that the opacity is to be low, i.e., no greater than Ringleman no. 1 for more than, for instance, 5 min/h. Incinerators have been designed to this standard, which is virtually no standard at all. As public attention is starting to focus on hazardous, dangerous, and toxic wastes, the regulatory attitude toward biomedical waste incinerators is starting to change. These incinerators are not addressed by the federal government yet, but many states are moving in the direction of regulation. In some states these wastes are classified as hazardous; in others they are regulated as a unique waste stream with its own set of regulations; and in still others there is still no regulation of biomedical wastes per se.

Table 14.4 lists the current status of state regulations of biomedical waste in-

TABLE 14.4 State Requirements for Incineration of Hospital Wastes

State	Summary of requirements
Alabama	There are no regulations specifically for hospital incinerators. All incinerators must be permitted by the Department of Environmental Management, Air Division, prior to installation. Incinerators with a charging rate less than 50 tons/day may not discharge particulate matter at a rate greater than 0.20 lb/100 lb refuse charged. Incinerators with a charging rate greater than or equal to 50 tons/day may not discharge more than 0.1 lb/100 lb refuse charged. The discharge may not exceed 20 percent opacity except during one 6-min period per hour, when it may not exceed 40 percent. More stringent standards may be specified in the permit depending on the source location and other factors. For information call the Department of Environmental Management, Air Division, in Montgomery at (205) 271-7700.
Alaska	There are no regulations specifically for hospital incinerators. No permit is needed for incinerators with capacities smaller than 1000 lb/h. The only regulation governing these small incinerators is a visual emissions standard of 20 percent opacity. New regulations for incinerators have been proposed. For more information call the Department of Environmental Conservation, Air and Solid Waste Management Section, in Juneau at (907) 465-2666.
Arizona	There are no regulations specifically for hospital incinerators. All incinerators must have two permits: an air quality permit and a solid waste permit. All incinerators must meet a particulate emissions standard of 0.1 gr/dry st ft^3 (corrected to 12 percent CO_2) and a visual emissions standard of 20 percent opacity. For more information call the Department of Environmental Quality, Air Quality Section in Phoenix at (602) 257-2277 and Solid Waste Section at (602) 257-6989.
Arkansas	There are no regulations specifically for hospital incinerators. All incinerators must be permitted. Incinerators with capacities less than 200 lb/h may not emit more than 0.3 gr/dry st ft^3 (corrected to 12 percent CO_2), and larger incinerators may not emit more than 0.2 gr/dry st ft^3 (corrected to 12 percent CO_2). The visual emissions may not exceed 20 percent opacity. For more information call the Department of Health, Division of Health Facility Services, in Little Rock at (501) 661-2201.
California	There are no regulations specifically for hospital incinerators. Air quality is regulated by Air Quality Management Districts, and regulations vary from district to district. All districts require incinerators to be permitted. Each district has particulate, visual, and air toxics emissions standards. The air toxics emissions requirements are highly dependent on the location of the facility and vary significantly from district to district. There are many other requirements. If the incinerator treats wastes generated off-site, it must be permitted by the State Health Services Department. For more information call the California Department of Health Services, Hazardous Materials Management Section, in Sacramento at (916) 324-9611.

TABLE 14.4 State Requirements for Incineration of Hospital Wastes (Continued)

State	Summary of requirements
Colorado	There are no regulations specifically for hospital incinerators. All incinerators must be permitted. Particulate emissions may not exceed 0.1 gr/dry st ft^3 (corrected to 12 percent CO_2). The visual emissions standard is 20 percent opacity. Guidelines for hospital wastes require a secondary-chamber temperature of 1800°F with a retention time of 2 s. If off-site wastes are treated, a certificate of designation is required by the Solid Waste Management Department. For more information call the Department of Health, Air Pollution Control Division, in Denver at (303) 331-8591.
Connecticut	There are guidelines for hospital incinerators. All incinerators must be permitted. Incinerators are required to meet BACT (best-available control technology) standards, which include the following for hospital incinerators: particulate emissions may not exceed 0.15 gr/dry st ft^3 (corrected to 12 percent CO_2); there must be a 90 percent reduction of hydrogen chloride or less than 4 lb/h hydrogen chloride emitted, whichever is less; carbon dioxide emissions may not exceed 100 ppmdv (parts/million dry volume) (corrected to 7 percent O_2); and the combustion efficiency must be at least 99.8 percent. In addition, the primary-chamber temperature must be at least 1800°F, and the secondary-chamber temperature must be at least 2000°F with a retention time of at least 2 s. The visual emissions may not exceed 10 percent opacity. There are other requirements including monitoring and recording procedures. For more information call the Department of Environmental Protection, Solid Waste Management Unit, in Hartford at (203) 566-5847.
Delaware	There are no regulations specifically for hospital incinerators. All incinerators must be permitted by the Air Quality Section of the Department of Natural Resources and Environmental Control. Incinerators must be double-chambered with a secondary-chamber temperature of at least 1400°F. The visual emissions standard is 20 percent opacity. Other requirements, including a higher secondary-chamber temperature, are usually specified in the permit. Other permits may be required depending on the source location. Coastal zone, wetlands, or water pollution regulations may apply. The Solid Waste Section is drafting regulations for infectious waste incinerators. For more information call the Department of Natural Resources and Environmental Control, Air Resources Section, in Dover at (302) 323-4558.
District of Columbia	There are specific air pollution control requirements for pathological waste incinerators. Incinerators must be of multiple-chamber design. Particulate emissions may not exceed 0.03 gr/dry st ft^3 (corrected to 12 percent CO_2). Pathological waste incinerators may only be operated between the hours of 10:00 a.m. and 4:00 p.m. Two interlocking devices are required: one that prevents operation of the primary chamber when the

TABLE 14.4 State Requirements for Incineration of Hospital Wastes (Continued)

State	Summary of requirements
	secondary-chamber outlet is less than 1800°F and another that prevents operation of the primary chamber when the primary-chamber charging door is not closed. There are several other requirements. For more information call the Department of Consumer and Regulatory Affairs, Environmental Control Division, at (202) 767-7370.
Florida	There are no statewide regulations specifically for hospital incinerators; however, some districts do have guidelines for infectious waste incinerators, and these requirements are usually written into air pollution permits. Secondary-chamber temperature and retention time are frequently specified by districts. All incinerators must be permitted. For incinerators with capacities less than 50 tons/day, the visual emissions standard is 5 percent opacity, and there is no particulate emissions standard. The regulations are expected to become more stringent within the next 6 months. For more information call the Department of Environmental Regulation, Solid Waste Planning and Regulation Program, in Tallahassee at (904) 488-0300.
Georgia	There are no regulations specifically for hospital incinerators. Incinerators must be of multiple-chamber design. The primary chamber must reach 800°F before charging, and the secondary-chamber temperature must be at least 1500°F. Incinerators handling less than 500 lb/h may not emit more than 1 lb/h of particulate matter. Incinerators handling 500 lb/h or more may not emit more than 0.2 lb/100 lb charged. The visual emissions standard is 20 percent opacity. For more information call the Department of Natural Resources, Air Quality Section, in Atlanta at (404) 656-4867.
Hawaii	There are no regulations specifically for hospital incinerators. Permits are required for all new or modified incinerators. Requirements, including secondary-chamber temperature and retention time, are determined on a case-by-case basis and specified in the permits. For small incinerators, particulate matter emissions may not exceed 0.2 lb/100 lb refuse charged, and the opacity may not exceed 20 percent. Contact the Department of Health for the definition of small incinerators. All the hospital incinerators in Hawaii are small. For more information call the Department of Health, Air and Solid Waste Permit Section, in Honolulu at (808) 548-6410.
Idaho	There are no regulations specifically for hospital incinerators. Air pollution permits are required for all incinerators. Requirements for hospital incinerators are determined on a case-by-case basis and specified in the permits. The particulate emissions limit is 0.2 lb/100 lb refuse charged, and the visual emissions standard is 20 percent opacity. The Hospital Facilities Section has proposed new regulations for hospital incinerators. For more information on air pollution permits call the Department of Health and Welfare, Air Quality Bureau, in Boise at (208) 334-5898, and for information on the proposed regulations call the Hazardous Materials Bureau at (208) 334-5879.

TABLE 14.4 State Requirements for Incineration of Hospital Wastes (Continued)

State	Summary of requirements
Illinois	There is a rule specifically for hospital incinerators. Requirements include obtaining an air pollution control permit and special ash disposal procedures. The air quality regulations apply to all incinerators. The particulate emissions standard for incinerators with capacities less than 2000 lb/h is 0.1 gr/dry st ft^3 (corrected to 12 percent CO_2). Carbon monoxide emissions may not exceed 500 ppmdv (parts/million dry volume). Other requirements, including secondary-chamber temperature and retention time, are determined on a case-by-case basis and specified in the permits. For more information call the Environmental Protection Agency, Division of Air Pollution Control, in Springfield at (217) 782-2113.
Indiana	There are no regulations specifically for hospital incinerators. All incinerators must have air pollution control permits. Particulate emissions standards depend on the size of the incinerator and location of the facility. The visual emissions standards depend on whether the facility is in an attainment or nonattainment area for particulate matter. All incinerators must be multiple-chamber. The secondary-chamber temperature must be at least 1800°F with a retention time of 1 s. If antineoplastics are treated, the retention time must be at least 1.5 s. There are also monitoring and recording requirements. By the end of 1988 new solid waste rules with additional requirements will be in effect. For more information call the Department of Environmental Management, Office of Solid and Hazardous Waste Management in Indianapolis at (317) 232-8842 and the Office of Air Management at (317) 232-8459.
Iowa	There are no regulations specifically for hospital incinerators. All incinerators must be permitted. Incinerators with a capacity less than 100 lb/h may not emit more particulate matter than 0.35 gr/dry st ft^3 (corrected to 12 percent CO_2), and those with larger capacities may not emit more than 0.2 gr/dry st ft^3 (corrected to 12 percent CO_2). The opacity may not exceed 40 percent. Emissions testing during start-up is required. Operators must evaluate the emissions for many contaminants including particulates, polyaromatic hydrocarbons, chlorinated hydrocarbons, dioxins, furans, and metals. Other requirements are determined on a case-by-case basis and specified in the permits. For more information call the Department of Natural Resources, Air Quality Section, in Des Moines at (515) 281-8935.
Kansas	Two permits are required for hospital incinerators: a solid waste processing permit and an air quality permit. There are many solid waste processing requirements, including waste handling and storage, and ash disposal requirements. Air quality regulations include particulate emissions standards which range from 0.10 to 0.30 gr/dry st ft^3 (corrected to 12 percent CO_2) depending on the capacity of the incinerator. The visual emissions standard is 20 percent opacity. For

TABLE 14.4 State Requirements for Incineration of Hospital Wastes (Continued)

State	Summary of requirements
	more information call the Department of Health and Environment, Solid Waste Management Section in Topeka at (913) 296-1590 and the Air Quality and Radiation Section at (916) 296-1572.
Kentucky	There are no regulations specifically for hospital incinerators. Permits are required for incinerators larger than 500 lb/h or incinerators that emit significant amounts of air toxics. The limits for air toxics emissions are based on site location, stack height, the contaminant, and several other factors. The particulate standard for incinerators with capacities less than 50 tons/day is 0.1 gr/dry st ft^3 (corrected to 12 percent CO_2). The standard for larger incinerators is 0.08 gr/dry st ft^3 (corrected to 12 percent CO_2). Incinerators that do not require permitting are exempt from the particulate standard. The visual emissions standard is 20 percent opacity. For more information call the Cabinet of Natural Resources, Air Quality Permit Branch, in Frankfort at (502) 564-3382.
Louisiana	There are no regulations specifically for hospital incinerators. Permits are required for all incinerators. All requirements are determined on a case-by-case basis and specified in the permits. Typical requirements for hospital incinerators are a minimum secondary-chamber temperature of 1800°F with a retention time of 1 s, a visual emissions standard of 20 percent opacity, and a grain loading limit. For more information call the Department of Environmental Quality, Office of Air Quality and Nuclear Energy, in Baton Rouge at (504) 342-1201.
Maine	There are no regulations specifically for hospital incinerators; however, there are BPT (best practical technology) guidelines for hospital incinerators. The secondary-chamber temperature must be at least 1800°F with a retention time of 1 s. If antineoplastics are treated, the secondary-chamber temperature must be at least 2000°F with a retention time of 2 s. The particulate emissions rate may not exceed 0.1 gr/dry st ft^3 (corrected to 12 percent CO_2). Opacities in excess of 10 percent may trigger a stack test requirement. There are other requirements. For more information call the Department of Environmental Protection, Bureau of Air Quality Control, in Augusta at (207) 289-2437.
Maryland	There are no regulations specifically for hospital incinerators. All incinerators must be permitted. A full air toxics analysis is required during permitting as well as a stack test after construction. The secondary-chamber temperature must be at least 1800°F with a retention time of 2 s. The particulate emission rate may not exceed 0.1 gr/dry st ft^3 (corrected to 12 percent CO_2), and no visual emissions are allowed (0 percent opacity). For more information call the Department of Environment, Air Management Administration, in Baltimore at (301) 225-5260.
Massachusetts	There are no regulations specifically for hospital incinerators. All incinerators must be permitted. BACT analysis is carried out on a case-by-case basis to determine the specifics of each per-

TABLE 14.4 State Requirements for Incineration of Hospital Wastes (Continued)

State	Summary of requirements
	mit. Each incinerator must go through performance testing. The particulate emissions limit is determined as part of BACT, but would never exceed 0.02 gr/dry st ft^3 (corrected to 12 percent CO_2). The visual emissions standard is 20 percent opacity. If the incinerator has a capacity larger than 1 ton/h, then it must be permitted by the local authorities as well. For more information call the Department of Environmental Quality, Division of Air Quality Control, in Boston at (617) 292-5619.
Michigan	All incinerators must be permitted. The term *pathological* refers to carcasses and body parts whereas *infectious* refers to other contaminated wastes including needles and plastic containers. Infectious waste incinerators must have a minimum secondary-chamber temperature of 1800°F with a retention time of 1 s. There is no temperature requirement for pathological incinerators. All incinerators must meet particulate standards which depend on the incinerator capacity and air toxics in the emissions. There are many additional requirements. Off-site facilities must be part of a county waste management plan. For more information call the Department of Natural Resources, Air Quality Division, in Lansing at (517) 373-7023.
Minnesota	There are no regulations specifically for hospital incinerators. Permits are required for incinerators with capacities greater than 1000 lb/h. The particulate emissions standard is 0.2 gr/dry st ft^3 (corrected to 12 percent CO_2) for incinerators with capacities less than 200 lb/h, 0.15 gr/dry st ft^3 (corrected to 12 percent CO_2) for incinerators rated between 200 and 2000 lb/h, 0.10 gr/dry st ft^3 (corrected to 12 percent CO_2) for those rated between 2000 and 4000 lb/h, and 0.08 gr/dry st ft^3 (corrected to 12 percent CO_2) for larger incinerators. The visual emissions limit is 20 percent opacity. The regulations state that the secondary-chamber temperature must be at least 1200°F with a retention time of 0.3 s; however, higher temperatures and longer retention times are specified in permits. These as well as other requirements are determined on a case-by-case basis. Proposed regulations are expected to be final in 1.5 to 2 years. For more information call the Pollution Control Agency, Division of Air Quality, in Roseville at (612) 296-7711.
Mississippi	There are no regulations specifically for hospital incinerators. All incinerators must be permitted. Incinerators must be of multiple-chamber design. The particulate emissions standard is 0.1 gr/dry st ft^3 (corrected to 12 percent CO_2) in residential or developed areas, and it is 0.2 gr/dry st ft^3 (corrected to 12 percent CO_2) in remote areas. Other requirements, including air toxics emissions limits, are determined on a case-by-case basis and specified in permits. For more information call the Bureau of Pollution Control, Air Section, in Jackson at (601) 961-5171.
Missouri	All incinerators must be permitted. Infectious waste incinerators must have a minimum secondary-chamber temperature of 1800°F with a retention time of 0.5 s. This temperature requirement does not apply to pathological incinerators. The par-

TABLE 14.4 State Requirements for Incineration of Hospital Wastes (Continued)

State	Summary of requirements
	ticulate emissions standard is 0.3 gr/dry st ft^3 (corrected to 12 percent CO_2) for incinerators smaller than 200 lb/h and 0.2 gr/dry st ft^3 (corrected to 12 percent CO_2) for larger incinerators. Visual emissions may not exceed 20 percent opacity. There are special requirements for off-site facilities. The regulations will be revised within 1 year. For more information call the Department of Natural Resources, Air Pollution Control Program, in Jefferson City at (314) 751-4817.
Montana	There are no regulations specifically for hospital incinerators. Permits are required for incinerators that process more than 25 tons/yr. Incinerators must be multiple-chamber. Particulate matter emissions may not exceed 0.1 gr/dry st ft^3 (corrected to 12 percent CO_2), and the opacity may not exceed 10 percent.All incinerators must meet BACT. Permit requirements are determined on a case-by-case basis. For more information call the Department of Health and Environmental Sciences, Air Quality Bureau, in Helena at (406) 444-3454.
Nebraska	There are no regulations specifically for hospital incinerators. Incinerators with capacities less than 1 ton/h may not emit particulate matter at a rate greater than 0.2 gr/dry st ft^3 (corrected to 12 percent CO_2). Larger incinerators may not emit more than 0.1 gr/dry st ft^3 (corrected to 12 percent CO_2). Visual emissions may not exceed 20 percent opacity. Incinerators must meet BACT if they emit more than 2.5 tons/yr of any of 309 specified contaminants. Other requirements, including secondary temperature and retention time, are written into permits. These requirements are determined on a case-by-case basis. For more information call the Department of Environmental Control, Air Quality Division, in Lincoln at (402) 471-2189.
Nevada	There are no regulations specifically for hospital incinerators. All incinerators must be permitted. The particulate emissions standard for incinerators with capacities less than 2000 lb/h is 3 lb/ton dry charge. The allowable particulate emissions rate for larger incinerators e, lb/h, is determined by the following formula: $$e = (40.7 \times 10^{-5}) \times c$$ where c is the charge rate in pounds per hour. The visual emissions standard is 20 percent opacity. The secondary-chamber temperature must be at least 1400°F with a residence time of 0.3 s. There are also air toxics emissions limits. A county commission permit is required and is often difficult to obtain. For more information call the Department of Conservation and Natural Resources, Division of Environmental Protection, in Carson City at (702) 885-5065.
New Hampshire	There are no regulations specifically for hospital incinerators. Hospital incinerators with capacities greater than 200 lb/h must be permitted. An air quality analysis is required during the permitting process. Incinerators with capacities less than or

TABLE 14.4 State Requirements for Incineration of Hospital Wastes (Continued)

State	Summary of requirements
	equal to 200 lb/h may not emit particulate matter at a rate greater than 0.3 gr/dry st ft^3 (corrected to 12 percent CO_2), and visual emissions may not exceed 20 percent opacity. There are numerous requirements for larger incinerators including secondary-chamber temperature and retention time and air toxics emissions limits. These requirements are determined on a case-by-case basis and specified in the permits. For more information call the Department of Environmental Services, Division of Air Resources, in Concord at (603) 271-1390.
New Jersey	There are no regulations specifically for hospital incinerators. Each incinerator must be part of the State Solid Waste Management Plan and be permitted as a solid waste facility. The regulations state that the secondary-chamber temperature must be at least 1500°F with a residence time of 1 s. The state requires "state of the art air pollution control equipment" which is defined in Department of Health Guidelines. These guidelines include a minimum secondary-chamber temperature of 1800°F and a particulate emissions limit for incinerators with capacities larger than 800 lb/h of 0.010 gr/dry st ft^3 (corrected to 7 percent O_2). Many other requirements are determined on a case-by-case basis and specified in the permits. For more information call the Department of Environmental Protection, Bureau of Engineering and Regulatory Development, in Trenton at (609) 984-0491.
New Mexico	There are no regulations specifically for hospital incinerators. A permit is required for incinerators that emit any criteria pollutant at a rate greater than 10 lb/h for controlled emissions and greater than 100 lb/h for uncontrolled emissions. Visual emissions may not exceed 20 percent opacity. Registration of air toxics emitted above certain levels is required. Other requirements are determined on a case-by-case basis and specified in the permits. For more information call the Health and Environment Department, Air Quality Bureau, in Santa Fe at (505) 827-0070.
New York	There are no regulations specifically for hospital incinerators. Incinerators must be permitted. Incinerators with capacities of 2000 lb/h or greater must meet particulate emissions standards. The particulate emissions limits are based on refuse charging rate and are determined by a graph established by the state. The particulate standard is 0.5 lb/h for a charging rate of 100 lb/h or less and about 12 lb/h for a charging rate of 4000 lb/h. The visual emissions standard is 20 percent opacity. The regulations are more stringent for facilities in New York City and Nassau and Westchester Counties. For more information call the Department of Environmental Conservation, Division of Air Resources, in Albany at (518) 457-2044.
North Carolina	There are no regulations specifically for hospital incinerators. All incinerators must be permitted. The particulate emissions standard for incinerators with capacities less than 100 lb/h is 0.2lb/h. The particulate standard for incinerators rated between

TABLE 14.4 State Requirements for Incineration of Hospital Wastes (Continued)

State	Summary of requirements
	100 and 2000 lb/h is 0.002 times the amount charged in pounds per hour. Incinerators with larger capacities may not emit more than 4 lb/h. Visual emissions may not exceed 20 percent opacity for more than 6 min/h. For more information call the Department of Natural Resources and Community Development, Air Quality Section, in Raleigh at (919) 733-3340.
North Dakota	There are no regulations specifically for hospital incinerators. Incinerators must be permitted. The allowable particulate emission rate e is determined by the following formula: $e = 0.00515 \times R \times 0.9$ where R is the refuse burning rate in pounds per hour. Visual emissions may not exceed 20 percent opacity. The regulations state that the secondary-chamber temperature must be at least 1500°F with a retention time of 0.3 s; however, more stringent requirements are usually written into permits. A minimum secondary-chamber temperature of 1800°F and a retention time of 1 s are usually specified. There are no restrictions on air toxics emissions. For more information call the Department of Health, Air Quality Management Branch, in Bismarck at (701) 224-2348.
Ohio	There are no regulations specifically for hospital incinerators. All incinerators must have a permit that meets BAT requirements. Incinerators with capacities less than 100 lb/h may not emit more than 0.2 lb particulate matter per 100 lb charged, and larger units may not emit more than 0.1 lb/100 lb charged. The maximum opacity allowed is 20 percent. The secondary chamber must be at least 1600°F. Incinerators may not emit more than 4 lb/h of hydrogen chloride. There are several additional requirements. The BAT policy will be revised in the near future. For more information call the Environmental Protection Agency, Division of Air Pollution Control, in Columbus at (614) 644-2270.
Oklahoma	There are no regulations specifically for hospital incinerators. Permits are required for all incinerators. Incinerators must be of multiple-chamber design with a primary-chamber temperature of at least 800°F. The particulate emissions standards vary depending on the charge rate. Visual emissions may not exceed 20 percent opacity for more than 5 min in an hour and not more than 20 min in 24 h. For information call the Department of Health, Division of Air Quality Service, in Oklahoma City at (405) 271-5220.
Oregon	There are no regulations specifically for hospital incinerators. All incinerators must have an air pollution permit. In addition, if the incinerator accepts off-site wastes, a solid waste permit is required. The secondary-chamber temperature must be at least 1600°F. Visual emissions may not exceed 20 percent. Particulate emissions limits and other requirements are determined on a case-by-case basis and specified in the permits. The maximum particulate emission limit ever allowed is 0.1 gr/dry st ft^3

TABLE 14.4 State Requirements for Incineration of Hospital Wastes (Continued)

State	Summary of requirements
	(corrected to 12 percent CO_2) for small incinerators. New incinerator regulations have been proposed. For more information call the Department of Environmental Quality, Air Quality Division, in Portland at (503) 229-5186.
Pennsylvania	Both air quality and waste management permits are required for hospital incinerators. There are BAT requirements for hospital incinerators. Facilities with capacity less than or equal to 500 lb/h must meet a particulate emissions standard of 0.08 gr/dry st ft^3 (corrected to 7 percent O_2). Facilities rated between 500 and 2000 lb/h must meet a standard of 0.03 gr/dry st ft^3 (cor-rected to 7 percent O_2), and larger facilities must meet a standard of 0.015 gr/dry st ft^3 (corrected to 7 percent O_2). Hydrogen chloride emissions from incinerators with capacities less than or equal to 500 lb/h must not exceed 4 lb/h or shall be reduced by 90 percent. Hydrogen chloride emissions from larger incinerators may not exceed 30 ppmdv (corrected to 7 percent O_2) or shall be reduced by 90 percent. Carbon monoxide emissions may not exceed 100 ppmdv (corrected to 7 percent O_2). Visual emissions may not exceed 10 percent opacity for a period or periods aggregating more than 3 min in any hour and may never exceed 30 percent opacity. There are many other requirements including secondary-chamber temperature, ash residue testing, and monitoring and recording procedures. For more information call the Department of Environmental Resources, Bureau of Air Quality Control at (717) 787-9256 and the Bureau of Waste Management in Harrisburg at (717) 787-1749.
Puerto Rico	There are no regulations specifically for hospital incinerators. All incinerators must be permitted. An environmental impact statement is required for all facilities. The particulate emissions standard for incinerators with capacities less than 50 tons/day is 0.4 lb/100 lb refuse burned. Visual emissions may not exceed 20 percent opacity. There are additional requirements. For more information call the Environmental Quality Board, Air Quality Area, in San Juan at (809) 722-0077.
Rhode Island	There are no regulations specifically for hospital incinerators. All incinerators must be permitted. Single-chamber incinerators may not be used. Hazardous material incinerators must meet a particulate emissions standard of 0.08 gr/dry st ft^3 (corrected to 12 percent CO_2). For more information call the Department of Environmental Management, Division of Air and Hazardous Materials, in Providence at (401) 277-2808.
South Carolina	There are no regulations specifically for hospital incinerators; however, there is a hospital incinerator policy which all new or modified units must comply with. Incinerators must be of dual-chamber design. The secondary-chamber temperature must be at least 1800°F with a retention time of 2 s. The visual emissions standard is 10 percent opacity. New regulations for hospital incinerators are expected to go into effect in 1 year. For more information call the Department of Health and Environmental Control, Bureau of Air Quality Control, in Columbia at (803) 734-4750.

TABLE 14.4 State Requirements for Incineration of Hospital Wastes (Continued)

State	Summary of requirements
South Dakota	There are no regulations specifically for hospital incinerators. Permits are required for incinerators with capacities larger than 100 lb/h. The particulate standard for incinerators with capacities greater than or equal to 50 tons/day is 0.18 mg/dry st m^3 (corrected to 12 percent CO_2). There is no particulate emissions standard for smaller incinerators. Visual emissions may not exceed 20 percent opacity. For more information call the Department of Water and Natural Resources, Office of Air Quality and Solid Waste, in Pierre at (605) 773-3153.
Tennessee	There are no regulations specifically for hospital incinerators. All incinerators must be permitted. The particulate emissions standard is 0.2 percent of the charging rate for incinerators with capacities less than or equal to 2000 lb/h and 0.1 percent of the charging rate for larger incinerators. Visual emissions may not exceed 20 percent opacity. Other requirements, including air toxics emissions limits and secondary-chamber temperature and retention time, are determined on a case-by-case basis and specified in the permits. For more information call the Department of Health and Environment, Division of Air Pollution Control, in Nashville at (615) 741-3931.
Texas	All incinerators burning general hospital waste must be permitted. Incinerators with capacity less than 200 lb/h burning only carcasses and body parts, blood, and nonchlorinated containers do not need permits. Incinerators must be of dual-chamber design with a secondary-chamber temperature of at least 1800°F and a retention time of 1 s. The visual emissions standard is 20 percent opacity. If hydrogen chloride emissions exceed 4 lb/h, the incinerator must be equipped with a scrubber. Other requirements, including air toxics emissions limits, are determined on a case-by-case basis and specified in the permits. For more information call the Air Pollution Control Board, Combustion Section, in Austin at (515) 451-5711.
Utah	There are no regulations specifically for hospital incinerators. All new air pollutions sources must be permitted. All sources must meet BACT. Requirements, including air toxics emissions limits, secondary-chamber temperature and retention time, and particulate standards, are determined on a case-by-case basis and specified in the permits. For more information call the Department of Health, Bureau of Air Quality, in Salt Lake City at (801) 538-6108.
Vermont	There are no regulations specifically for hospital incinerators. The secondary chamber temperature must be at least 1600°F with a retention time of 1 s. The particulate emissions standard for incinerators with capacities smaller than 50 tons/day is 0.1 lb/100 lb refuse burned. The amount of refuse burned is equal to the amount charged minus the amount of ash generated. The standard for larger incinerators is 0.08 gr/dry st ft^3 (corrected to 12 percent CO_2). The visual emissions may not exceed 20 percent except for 6 min/h when they may not exceed 60 percent opacity.

TABLE 14.4 State Requirements for Incineration of Hospital Wastes (Continued)

State	Summary of requirements
	Guidelines for hospital incinerators have been drafted. For more information call the Agency of Natural Resources, Air Pollution Control Division, in Montpelier at (802) 244-3731.
Virginia	There are no regulations specifically for hospital incinerators. Two permits are required for hospital incinerators: a waste management permit and an air quality permit. All incinerators must be of multiple-chamber design. Particulate emissions may not exceed 0.14 gr/dry st ft^3 (corrected to 12 percent CO_2). Visual emissions may not exceed 20 percent opacity except for one 6-min period per hour when they may not exceed 60 percent. There are also monitoring and reporting requirements as well as emissions limits for air toxics. New regulations for infectious waste management have been proposed. For more information call the Department of Air Pollution Control in Richmond at (804) 786-5478.
Washington	There are no regulations specifically for hospital incinerators. All incinerators must meet BACT. The particulate emissions standard for incinerators with capacity less than 250 tons/day is 0.030 gr/dry st ft^3 (corrected to 7 percent O_2), and the particulate standard for larger incinerators is 0.020 gr/dry st ft^3 (corrected to 7 percent O_2). Hydrogen chloride emissions may not exceed 50 ppmdv (corrected to 7 percent O_2), and sulfur dioxide emissions also may not exceed 50 ppmdv (corrected to 7 percent O_2). Visual emissions may not exceed 10 percent opacity. The average secondary-chamber temperature must be at least 1800°F and may never be less than 1600°F. For more information call the Department of Ecology, Air Quality Program, in Olympia at (206) 459-6256.
West Virginia	There are no regulations specifically for hospital incinerators. All incinerators must be permitted. The particulate emissions standard for incinerators with capacities less than or equal to 15,000 lb/h is 5.43 lb/ton burned, and for larger incinerators the standard is 2.72 lb/ton burned. Opacity may not exceed Ringleman no. 1 except during 6 min/h when it may not exceed Ringleman no. 2. Other requirements are determined on a case-by-case basis and specified in the permits. Generally the secondary-chamber temperature must be at least 1600°F with a retention time of 0.5 s. If antineoplastics are generated, a higher temperature is specified, usually 1800°F. For more information call the Air Pollution Control Commission in Charleston at (304) 348-4022.
Wisconsin	All infectious waste incinerators must be permitted. The particulate standard for incinerators with capacities less than 200 lb/h is 0.00 gr/dry st ft^3 (corrected to 7 percent O_2). This standard also applies to infectious waste incinerators with capacities no greater than 400 lb/h which are operated no more than 6 h/day. The particulate standard for incinerators rated between 200 and 1000 lb/h is 0.03 gr/dry st ft^3 (corrected to 7 percent O_2), and the standard for larger incinerators is 0.015 gr/dry st ft^3 (corrected to 7 percent O_2). Incinerators larger than 200 lb/h may not emit more than 50 ppmdv (parts/million dry volume) hydrogen chloride or 4 lb/h,

TABLE 14.4 State Requirements for Incineration of Hospital Wastes (Continued)

State	Summary of requirements
	whichever is less restrictive. All incinerators must limit carbon monoxide emissions to 75 ppmdv. The visual emissions standard is 5 percent opacity. Stack testing is required for all facilities. For more information call the Department of Natural Resources, Air Management Bureau, in Madison at (608) 266-7718.
Wyoming	There are no regulations specifically for hospital incinerators. All incinerators must be permitted. Two-stage combustion is required. Incinerators must meet BACT. The particulate emissions standard is 0.2 lb/100 lb refuse charged. Visual emissions may not exceed 20 percent opacity. For more information call the Department of Environmental Quality, Air Quality Division, in Cheyenne at (307) 777-7391.

cinerators in the United States today. In many cases, regulations are under review and are expected to change. The regulations for some states are much more extensive than can be included in this book. The telephone numbers of the cognizant agency in each state, the District of Columbia, and Puerto Rico are included in this listing.

The state agency should be contacted for more complete and up-to-date information on regulations and the permitting process.

HAZARDOUS WASTE INCINERATION

Where hazardous regulations must be complied with, the incinerator design and operation must be subject to the RCRA regulations for handling and disposal. Incineration regulations under RCRA require an extensive analytical and compliance process, as described in Chap. 3.

In addition to operating requirements, the RCRA incinerator regulations mandate extensive record-keeping and reporting procedures. A detailed, comprehensive operator training program must also be implemented.

COMBINED HAZARDOUS WASTE SYSTEMS

Hazardous waste incineration systems require an RCRA permit and strict operating controls and reporting standards. Another significant issue associated with hazardous waste incinerators is that the ash is always considered hazardous.

Procedures exist for delisting ash (declaring ash nonhazardous), but this requires extensive testing and administrative activity (filings and petitions) which represent at least 18 months of reporting and review.

If, for example, 1000 lb of biomedical waste were incinerated, approximately

200 lb of ash would be generated. In a state not classifying such waste as hazardous, the ash could be deposited directly in a nonhazardous (municipal waste) landfill. If 100 lb of a hazardous waste were fired in the secondary chamber of this same incinerator, all the ash would be considered hazardous and would have to be deposited in a hazardous waste landfill (unless it were delisted). Where 100 lb of hazardous waste was originally present, now at least 200 lb of hazardous waste must be disposed of.

As a general rule, it is impractical and uneconomical to incinerate hazardous and nonhazardous waste in the same incinerator.

WASTE COMBUSTION

Hospital wastes will contain paper and cardboard, plastics, aqueous and nonaqueous fluids, anatomical parts, animal carcasses, and bedding, glass bottles, clothing, and many other materials. Much of this waste is combustible. Lighting a match to a mixed assortment of hospital waste will generally result in a sustained flame, unless it contains a high proportion of liquids, anatomical, or other pathological waste materials.

Thermal treatment technologies include starved air and excess air combustion processes, as described previously in this text.

The main advantage of starved air is a low air requirement in the primary chamber. With little air passing through the waste there is less turbulence within the system and less particulate carryover from the burning chamber. This low airflow also results in very low nitrogen oxide generation, although this is not normally a concern in hospital incinerators. Less supplemental fuel is required than with excess air systems, where the entire airflow must be brought to the operating temperature of the incinerator. With the lower airflow, fans, ducts, flues, and air emissions control equipment can be sized smaller than in excess air systems.

Starved air operation is difficult, if not impossible, to achieve for two reasons. It is difficult to control air leakage into the system, and it is not possible to determine an accurate waste heating value on which to base a definition of the stoichiometric air requirement (see Table 14.5). As listed in Table 14.5, the stoichiometric air requirement can vary by a factor of 4 depending on the types of materials normally found in a biomedical waste stream.

TABLE 14.5 Waste Combustion Characteristics

Waste constituent	Btu/lb	lb air/lb waste[a]
Polyethylene	19,687	16
Polystyrene	16,419	13
Polyurethane	11,203	9
PVC	9,754	8
Paper	5,000	4
Pathological	Will not support combustion	

[a]Stoichiometric requirement.

WASTE DESTRUCTION CRITERIA

Generally, paper waste (cellulosic materials) requires that a temperature of 1400°F be maintained for a minimum of 0.5 s for complete burnout. The temperature–residence time requirement for biomedical waste destruction must be at least equal to the requirements for paper waste; however, the specific relationship between temperature and residence time must be determined for the specific waste. Many states require that a temperature of 1800°F be maintained for a minimum of 1 or 2 s, and some states require a 2000°F off-gas temperature.

On the high-temperature side, it is necessary to consider the general nature of much of the biomedical waste stream. It has a high proportion of organic material, including cellulosic waste. The noncombustible portion of this waste (ash) will begin to melt, or at least desolidify, as the temperature increases. Table 13.3 lists the ash fusion temperature of refuse, which represents the same constituents as much of the biomedical waste stream. Above 1800°F, the ash produced will begin to deform in a reducing atmosphere, i.e., where there is a lack of oxygen in the furnace. When ash starts to deform and then is moved to a cooler portion of the incinerator, or to an area of the furnace where additional oxygen is present, the ash will harden into slag or clinker. This hardened ash can clog air ports, disable burners, corrode refractory, and interfere with the normal flow of material through the furnace. To prevent slagging, the temperature within the incinerator should never be allowed to rise above 1800°F. Higher temperatures also encourage the discharge of heavy metals to the gas stream, which is another reason not to impose an arbitrarily high temperature on the process.

PAST PRACTICES

Modular units have been popular in the past because of their relatively low cost. A major factor contributing to their cost advantage over rotary kilns and other equipment is that they require no external air emissions control equipment to produce a fairly clean stack discharge. When properly designed and operated, they can achieve a particulate emissions rate of 0.08 gr/dry st ft^3 (corrected to 50 percent excess air). As new regulations are promulgated, however, lower emissions limitations will make the use of baghouses or electrostatic precipitators mandatory, and the modular incinerator (with inclusion of control equipment in the system package) will likely lose its price advantage over other systems.

The higher cost of rotary kilns was due to their need for external air emissions control equipment and to inclusion of the drive mechanism necessary for its operation.

STARVED AIR PROCESS LIMITATIONS

The most important issue associated with starved air combustion is the nature of the waste. (Note that there may be references to pyrolysis in some literature, including sales brochures for such equipment, but this term is usually used in error. The process is starved air combustion.)

For any starved air reaction to occur, the waste must be basically organic and able to sustain combustion without the addition of supplemental fuel, a definition of *autogenous* combustion. Without sufficient heat content to sustain combustion, the concept of substoichiometric burning has no meaning.

A waste with a moisture content in excess of 60 percent will not burn autogenously at 1600°F. As noted in Table 14.1, the moisture contents of red bag, orange bag, and blue bag wastes are generally in excess of this figure. Starved air combustion will not work when wastes with this moisture fraction are placed in the furnace.

Usually a starved air incinerator is designed to fire a paper waste, and the burners in both the primary and secondary chambers of the incinerator are sized appropriately: for a relatively small supplemental-fuel requirement. When a bag of pathological waste is placed in the primary combustion chamber, the waste will not burn autogenously and supplemental fuel must be added. In most present starved air incinerator designs, the burners have too low a capacity to provide the heat required when organics released by a starved air process are not present. Without unburned organics in the secondary chamber (when starved air combustion has not occurred in the primary combustion chamber), the sizing of the secondary burner is generally inadequate (the burner is too small) to provide the fuel required for complete burnout of the gas stream.

On the other end of the operating range, when a waste with a very high heating value is introduced (a plastic material such as polyurethane or polystyrene, for instance, as noted in Table 14.5), a good deal of air is required to generate even the substoichiometric requirement necessary to generate heat for the process to advance. This air quantity is often much greater than the airflow present from the fans provided with the unit. If a high-quality paper waste (waste paper, boxes, cartons, cardboard, etc.) is introduced into an incinerator, starved air combustion will generally work, assuming there is good control of infiltration air. The incinerator should be designed, however, not for paper waste, but for the firing of pathological waste, which requires a relatively large heat input and a coordinated increase in airflow (primary and secondary combustion air and burner air fans) in both the primary and the secondary combustion chambers.

Starved air operation of an operating incinerator can be easily checked. Increase the airflow in the primary combustion chamber, and if the temperature increases, the incinerator is operating in the starved air mode. If the temperature decreases, the incinerator is operating as an excess air unit; i.e., starved air operation is not occurring.

One must expect, in the design of biomedical waste incinerators, that although the incinerator charge may be sufficiently large to preclude swings from very high to very low heat value wastes, this is not always the case. It is likely that a single incinerator charge can contain a polystyrene mattress (and very little else) that will start to burn almost at once upon insertion into the incinerator.

Likewise, since much of the waste charged into an incinerator is in opaque bags, and the waste cannot be identified, a charge can consist almost wholly of anatomical waste, or animal carcasses, or liquid (aqueous) materials which have a very low heat content. An incinerator must be designed for this certain variation in waste stream quality. Of the incinerators on the market today, the starved air unit is least able to adapt to changes in waste constituents.

REFERENCES

14-1. Ontario Ministry of the Environment, "Incinerator Design and Operating Criteria," vol. II, *Biomedical Waste Incinerators,* October 1986.

14-2. Doyle, B. W.: "The Smoldering Question of Hospital Wastes," *Pollution Engineering,* July 1985.

14-3. Brunner, C. R.: "Biomedical Waste Incineration," Monograph, presented at the Air Pollution Control Association Annual Conference, New York, June 1987.

BIBLIOGRAPHY

Brunner, C. R.: "Hospital Waste Disposal by Incineration," *Journal of the Air Pollution Control Association,* Vol. 38, No. 10, October 1988, pp. 1297–1309.

Brunner, C. R.: *Hazardous Air Emissions from Incineration,* Chapman & Hall, 2d ed., New York, 1986.

EPA Guide for Infectious Waste Management, USEPA, EPA/530-SW-86-014.

Powell, F. C.: "Air Pollutant Emissions from the Incineration of Hospital Wastes: The Alberta Experience," *Air Pollution Control Association Journal,* Vol. 33, No. 7, July 1987.

CHAPTER 15
SLUDGE INCINERATION

Sludges are nonnewtonian liquids with relatively high solids content. The particle sizes of the solids in a sludge are very small compared to those in what is defined as a slurry, which is basically a liquid containing solids of large particle size. The types of incineration equipment utilized for sludge disposal are presented in this chapter.

SLUDGE INCINERATION EQUIPMENT

Incineration provides ultimate volume reduction when used in combination with other treatment processes, particularly dewatering equipment (vacuum filters, belt filters, centrifuges, or filter presses). Until recently the cost of incineration has been prohibitive at many installations because of other, more economical disposal options: readily available land disposal, sludge lagoons, and ocean disposal. Increasing environmental constraints and decreasing land availability have increased the viability of sludge incineration.

Sludge parameters that have the most influence over incineration are moisture content, percentages of volatiles and inerts (or noncombustibles), and calorific value. Moisture content is important because of its thermal load on the incinerator. Volatiles and inerts, which affect the net heating value of sludge, can be controlled to some extent by treatment processes such as degritting, mechanical dewatering, and sludge digestion. The majority of combustibles in biological or organic sludges are present as volatiles, in the form of grease or light hydrocarbons, with the balance of the combustibles appearing as fixed carbon. Volatile percentage and moisture content can vary a great deal, so that equipment must normally be designed to handle a wide range of values. The types of incinerators commonly in use burning sludge wastes include

- Multiple-hearth
- Fluid bed
- Conveyer furnace
- Cyclone furnace

THE MULTIPLE-HEARTH INCINERATOR

The multiple-hearth incinerator (also known as the *Herreshoff furnace*) is the most prevalent incinerator for the disposal of sewage sludge in this country. It was developed at the turn of the century for ore roasting, and it has been adapted to carbon regeneration and recalcining and has many other industrial applications.

This incinerator will handle sludges in the range of 50 to 85 percent moisture. It is generally not applicable to the incineration of solid materials.

Multiple-hearth furnaces range from 6 to 25 ft in diameter and from 12 to 65 ft in height. The number of hearths is dependent on the waste feed and processing requirements but generally varies between 5 and 12 hearths. Waste retention time is controlled by the rabble tooth pattern and the rotational speed of the center shaft.

Sludge cake is introduced at the top of the furnace (see Fig. 15.1). The furnace interior consists of a series of circular refractory hearths, one above the other. The hearths are arch structures, self-supporting off the refractory-lined cylindrical wall of the furnace. They are, in sequence, no. 1 as the top hearth, no. 2 as the next-to-top, etc. Where the top hearth is used as an afterburner section, with

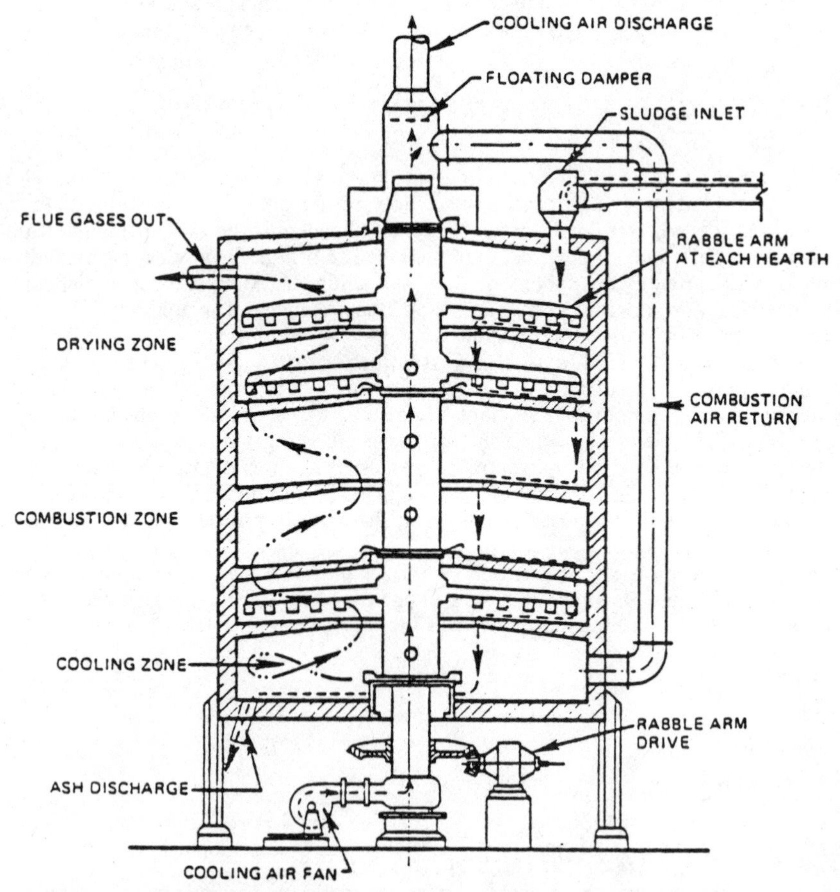

FIGURE 15.1 Cross section of a typical multiple-hearth incinerator. *Source*: Ref. 15-1.

sludge deposited on the hearth beneath the top hearth, the top hearth may be referred to as a *zero hearth*. In this case, the next-to-top hearth would be referred to as hearth no. 1, and so on.

A vertical shaft is positioned in the center of the furnace. Rabble arms are attached to the center shaft above each hearth. The center shaft, along with the rabble arms, rotates relatively slowly, at approximately 1 r/min. Teeth on the rabble arms are positioned to move sludge across its hearth and through the furnace.

Every other hearth has a large annular opening between the hearth and the center shaft. These are called *in hearths*. The teeth on the rabble arms of these hearths will "rabble" or wipe sludge to the center of the hearth, where the sludge will fall off the edge of the inner refractory ring, landing on the hearth below, an *out hearth*.

There are a series of openings on the outside or periphery of the out hearth. The inside of an out hearth is fairly close to the center shaft. A *lute ring cover* is a collar located above the hearth, attached to and moving with the center shaft, preventing sludge from dropping adjacent to the center shaft. Teeth on the rabble arms on the out hearths move sludge to the drop holes on the outside of the hearth, where the sludge drops to the hearth below, an in hearth. This process repeats until sludge, or ash, reaches the bottom hearth, the floor of the furnace, where it discharges from the furnace.

Teeth on each hearth agitate the sludge, exposing new surfaces of the sludge to the gas flow within the furnace. As sludge falls from one hearth to another, it again has new surfaces exposed to the hot gas.

The upper hearths of the furnace comprise a drying zone where the sludge cake gives up moisture (evaporation) while cooling the hot flue gases. Flue gas exits the top hearth of the furnace at 800 to 1200°F. The center hearths are the burning zone where gas temperatures can reach 1700 to 1800°F. Burnout of sludge to ash occurs in the lower hearths of the furnace.

The center shaft and rabble arms are hollow. Air is passed through the center shaft, which is constructed to distribute this air to each of the rabble arms and discharge it through the top of the shaft. After passing through the rabble arms and center shaft, this cooling air reaches temperatures of 200 to 450°F. Often this heated air is recycled into the furnace as preheated sludge combustion air.

Excess air of 75 to 125 percent of the stoichiometric (ideal) air requirement must be provided to ensure adequate burnout of sludge. There are places within the furnace that are not in the direct path of airflow where sludge and gases can "hide out" and not burn effectively. The use of excess air will tend to encourage burning in these furnace areas.

From 10 to 20 percent of the ash content of the sludge is airborne, and exhaust gas cleaning equipment must be provided for its capture. This airborne particulate loading is highly erosive. Materials in contact with this dirty exhaust should be relatively dense with good resistance to erosion and the corrosive effect of the gas.

Occasional odor problems can exist that may require installation of afterburner equipment. This is of particular concern where there is a significant industrial component to the waste stream, in excess of 20 percent of the total influent flow to the treatment plant.

Multiple-hearth incinerators are designed to burn wastes with a low gross (as-received) heating value, such as sewage sludge and other high-moisture-content wastes. Their design includes drying as well as burning sections. Materials with lower moisture content, such as coal or solid waste, will start to burn too high in the furnace. There would be insufficient residence time above these top hearths, and the temperatures in this volume of the furnace would be too low for effective burnout.

This incinerator system is relatively complex with fans, burners, shaft speed, and feed charging all being within the operator's control, and all affecting combustion efficiency and burnout. Particular attention must be paid to operator training in the proper control of these various equipment items and parameters.

Gas is drawn through the multiple-hearth system with an induced-draft fan. This fan is sized to create a negative pressure, or draft, within the furnace during operation. Besides draft, its pressure draw must account for flue losses, losses across flue dampers, the pressure differential across the exhaust gas cleaning system (usually a wet scrubber), and the drop across the waste heat boiler when a boiler is included in the system.

The temperature above at least two hearths should be maintained at approximately 1600°F at all times when sludge cake is burned. The off-gas temperature can range from 800 to 1400°F. In general, the exit temperature of the multiple-hearth furnace is from 400 to 600°F lower than the maximum furnace temperature. This is an important consideration. The maximum allowable temperature anywhere within the furnace is 1800°F. Above this temperature the sludge ash fusion temperature will be approached, and ash clinkering will occur. The formation of clinker and slag can clog drop holes, interfere with rabble arm motion, and create other problems within the furnace that can result in shutdown and damage.

The multiple-hearth furnace is a flexible piece of equipment. There is a limitation in the sludge consistency that it can process (generally from 15 to 50 percent solids content), but the nature of the sludge is not necessarily limiting, such as with the fluid bed furnace. Another feature of the multiple-hearth furnace is that it has a relatively constant fuel-use curve; i.e., the use of fuel is directly proportional to the amount of sludge burned. If the sludge feed is doubled, the fuel required to incinerate the sludge is roughly doubled. This is not necessarily true with other types of incinerator systems.

Ash exits this system dry, from beneath the furnace. It can be collected and disposed of dry, or it can be dropped into a wet hopper, mixed with water, and pumped to a lagoon for dewatering and ultimate disposal. The multiple-hearth furnace is flexible enough to allow wet or dry ash handling.

Feeding a multiple hearth is relatively simple. Feed can be dropped onto the top hearth by gravity. It can also be deposited on the top hearth or onto a lower hearth by means of a screw conveyer. Attempts to feed sludge to a multiple-hearth furnace pneumatically have met with failure. Pneumatic systems necessarily inject air into the furnace in strong discrete bursts which makes maintenance of constant furnace draft difficult, if not impossible.

Generally, grease (scum) should not be added to sludge feed. If grease is to be incinerated in the multiple-hearth furnace, it should be added at a lower hearth (a burning hearth) via separate nozzle(s). If it is within the sludge feed, it will tend to volatilize upon entering the furnace, at too high a hearth. The released grease would not experience a high enough temperature or long enough residence time for destruction of volatiles.

Dry solids or solid fuels should not be fed to a multiple-hearth furnace. Exceptions may be considered if the solids are added beneath the drying zone, in the burning zone of the furnace.

It is virtually impossible to maintain heat on standby (without burning sludge) in existing multiple-hearth furnace designs without firing supplemental fuel. This furnace, as noted above, has many paths of air leakage, and therefore heat cannot be effectively maintained within the units as in, e.g., a fluid bed furnace. Keeping a multiple-hearth furnace as hot as 1000°F above most of its hearths for more than 1 or 2 h requires the burning of supplemental fuel.

Fuel oil or gas can be used as supplemental fuel. Generally solid fuels, such as coal or wood chips, should not be placed on a hearth and used as supplemental fuel. They are relatively dry and will start burning on the top hearth, encouraging premature release of volatiles from the waste stream, and inadequate burnout of the off-gas can result.

Where sludge contains grease or other volatile components an afterburner may be required for effective burnout, i.e., elimination of smoke and odor. The afterburner can be an expanded top hearth, or it can be provided as a separate piece of equipment.

The oxygen content of the flue gas exiting the furnace should be continuously measured at the breeching leaving the top hearth. A sample should be continuously extracted from the gas stream, passed through a water bath to clean the sample and to reduce its moisture content, and then measured for oxygen content. The oxygen content should be in the range of 6 to 10 percent by volume.

The multiple-hearth furnace should always be run at a negative pressure (draft) to prevent external leakage of hot, toxic flue gas. If a separate afterburner is needed and if an emergency discharge stack is provided, the stack should always be located downstream of the afterburner. Under no circumstances should an emergency discharge stack be located directly above the top hearth, in the roof of the furnace. This design, where the stack is supported from the top of the furnace, always provides short-circuiting of the incinerator gases. The emergency stack is normally a source of air leakage in the system, and placing it above the top hearth, rather than downstream of the furnace breeching, will cause incinerator gases to cool prematurely.

FLUID BED INCINERATION

The fluid bed furnace was developed for catalyst recovery by Exxon Corporation during World War II. The first fluid bed used for incineration of sewage sludge was installed in 1962. Its use is gaining in popularity in the United States.

As illustrated in Fig. 15.2, the fluid bed furnace is a cylindrical refractory-lined shell with grid structure (either a refractory arch or an alloy steel plate) on its bottom surface to support a sand bed.

Air is introduced at the fluidizing air inlet at pressures in the range of 3.5 to 5 lb/in^2 gauge. The air passes through openings (tuyeres) in the grid supporting the sand and generates a high degree of turbulence within the sand bed. The sand undulates and has the appearance of a fluid in motion.

Air can be introduced cold or, as is usually the case, preheated by the exiting flue gas (see Fig. 15.3). The sand bed is normally maintained at approximately 1300 to 1400°F. It expands 30 to 60 percent in volume when fluidized compared to the unfluidized condition.

Sludge cake is normally introduced within or just above the fluid bed. Fluidization provides maximum contact of air with the sludge cake surface for optimum burning. The drying process is practically instantaneous. Moisture flashes into steam upon entering the hot bed.

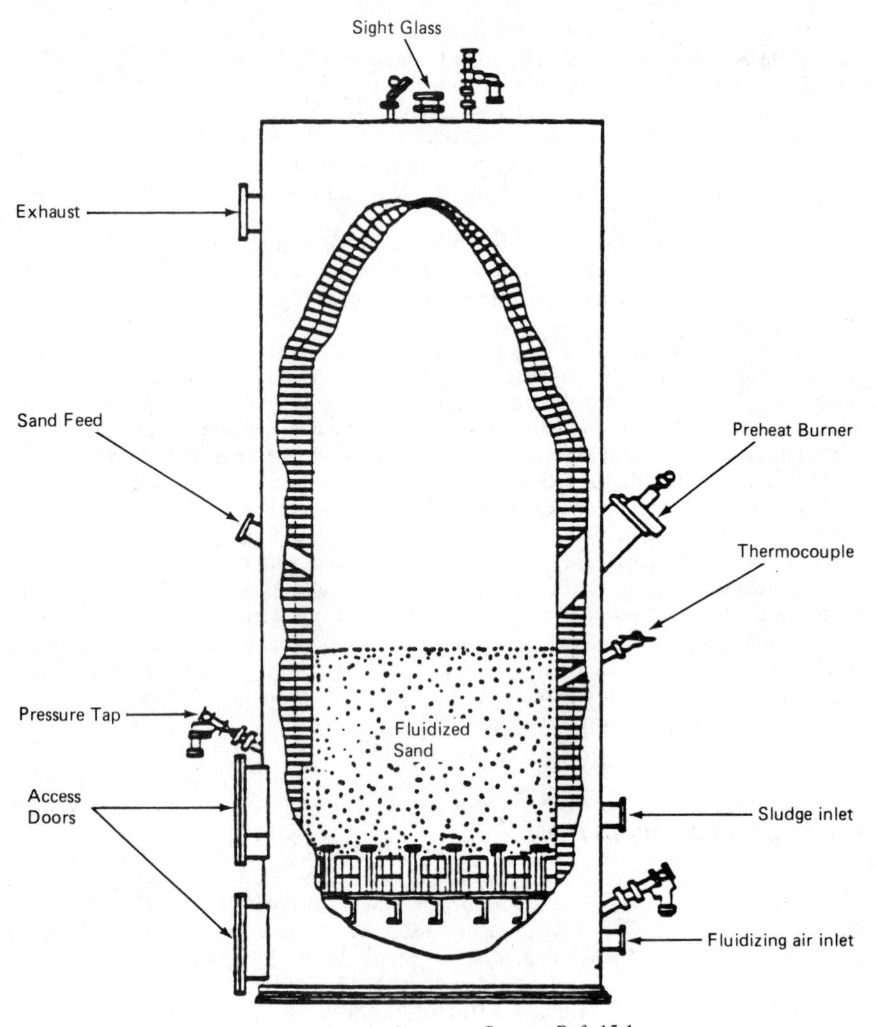

FIGURE 15.2 Cross section of a fluid bed reactor. *Source*: Ref. 15-1.

The furnace itself is an extremely simple piece of equipment with few moving parts. The fluid bed furnace has one major item of air-moving equipment, the forced-draft fan (or fluidizing air blower). The fan is sized to move flue gas through the gas scrubbing systems which necessitates that the reactor be pressurized and that it be airtight, to prevent leakage of flue gas.

The large amount of sand within the furnace is a heat sink which provides a significant thermal inertia within the system. This allows the furnace to be shut down with minimal heat loss. It is a relatively tight system, and the sand will retain heat, e.g., to allow quick start-up after an overnight shutdown. A weekend shutdown will require only 2 or 3 h of heating. The sand bed temperature should be at least 1200°F before sludge is introduced.

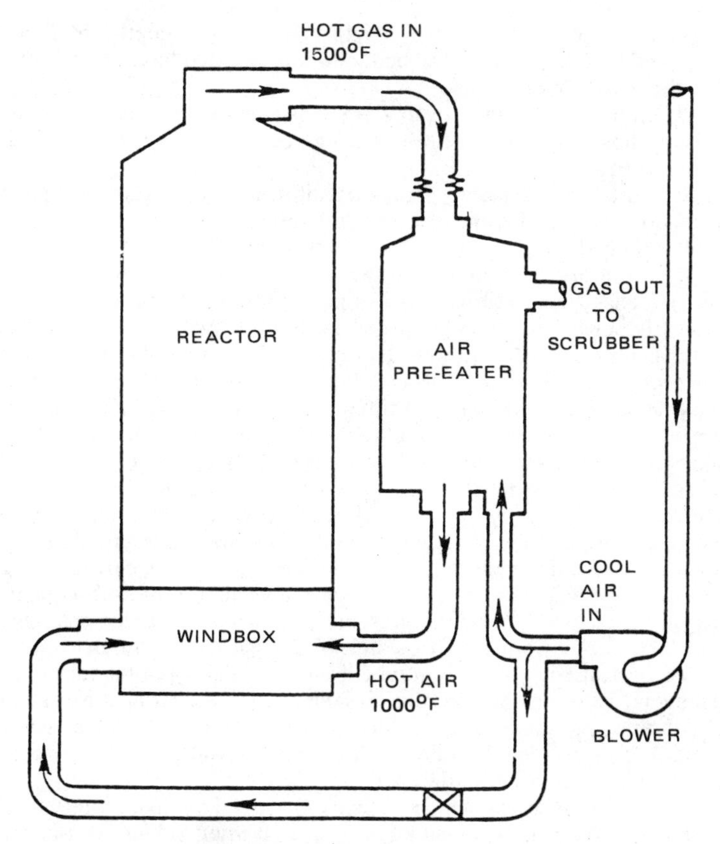

FIGURE 15.3 Fluidized bed system with air preheater. *Source*: Ref. 15-1.

All the sludge ash and some sand become airborne and exit the furnace within the flue gas stream. The gas cleaning system has to be sized for this relatively high particulate loading. Generally, sand has to be made up at the rate of approximately 5 percent of the bed volume every 100 h of operation with sewage sludges.

Because of the intimate mixing of air and sludge in the fluid sand bed, excess air requirements are low, generally 40 percent. The large volume (termed *freeboard*) within the furnace above the sand bed is maintained at 1500 to 1600°F. Residence time of the flue gases at these temperatures is normally sufficient to obtain complete burnout and elimination of odors.

Fuel is used for start-up, reheat, and, depending on the net heat content of the sludge, maintenance of temperature during incineration. It can be injected directly into the bed or sprayed on top of the bed when the bed is fluidized and is at a temperature high enough to ensure ignition.

The bed material is most commonly silica sand, but may also be comprised of limestone, alumina, or ceramic material. The bed expands about 30 to 60 percent in volume from a cold condition, when fluidized with air at a velocity of approximately 2 to 3 ft/s.

Residence time within the furnace is determined by bed depth, the fluidizing air velocity, and freeboard above the bed, and it is selected based on the characteristics of the waste being burned.

The fluid bed furnace is sensitive to waste consistency. Wastes containing soluble alkalis or phosphates, lead, or low-melting-point eutectics may cause seizure or fusion of the bed.

The fluid bed furnace system is compact, of airtight construction, and (with most manufacturers) is maintained as a positive-pressure system. By maintaining the entire system under positive pressure, the furnace is necessarily airtight. This feature is useful in applications where the furnace is required to operate on a noncontinuous basis. If the furnace is to operate 5 days/week, for instance, its airtight construction allows it to be sealed fairly effectively by the sand bed at rest. The refractory and sand within the unit provide thermal inertia. As a result of its construction, the fluid bed furnace after a shutdown will lose as little as 10 to 20°F/h. It can be shut down on a Friday evening and will require only a few hours of heat-up on a Monday morning to be available for waste feeding.

A major area of maintenance in this system is the recuperator, the gas-to-air tube-and-shell heat exchanger that is usually provided with this equipment. The heat-exchanger tubes, usually constructed of 300 series stainless steel, are sensitive to temperature and are generally limited to a sustained operating temperature of 1600°F. Above this temperature severe corrosion will occur, reducing the heat-exchanger tube wall dimension to the point of weakness and subsequent failure. Corrosion is encouraged by the pressure of oxygen in the flue gas. Oxygen will decarburize steel (carbon steel or stainless steel) at temperatures in excess of 1650°F. Decarburization will result in tube wastage, a flaking of the tube surface. When significant chlorides are present in sludge from industrial waste streams, chemical conditioning, or seawater infiltration, exotic materials of constitution may be required to protect the tubes and exposed metal parts from attack by chlorine or hydrogen chloride in the exhaust gases.

Other incinerator heat-exchanger designs, particularly insertion-type units such as the ones of European manufacture, use an intermediate heat-exchange medium such as heated oil, hot water, or steam to provide process heat. These insertion units are less costly than a tube-and-shell exchanger and can be replaced in less time and at less cost than a tube-and-shell unit.

An internal water spray system is often employed in fluid bed systems to help protect the heat exchanger from high-temperature excursions. Water normally is automatically injected into the freeboard when the heat-exchanger inlet temperature is above a preset figure, generally no higher than 1600°F. The location and design of these sprays are critical. Caution is required to avoid their damage from the hot incinerator temperatures.

A fluid bed furnace requires a minimum amount of air to maintain bed fluidization, regardless of the waste feed. The fluidizing air requirement is based on a furnace gas velocity. When the bed is operating at a feed rate below the design capacity, approximately the same air quantity is required as is required at the design load to maintain fluidization. This results in a unit that may have good fuel consumption at its design point, but a relatively high fuel consumption at lesser feeds.

Ash from a fluid bed furnace normally exits with the flue gas. Some materials, such as grit, when burned within a fluid bed furnace will not be airborne but will build up within the bed. This requires periodic bed tapping. Provisions should be included within the fluid bed system to accommodate the possibility of excessive bed buildup.

The air emissions control system must be designed to remove 100 percent of the sludge ash and elutriated sand from the gas stream. This requires high-energy venturi scrubbing systems with higher pressure requirements than many other incineration systems. Typical fluid bed systems need scrubbers rated from 30 to 60 in WC.

Feeding of a fluid bed furnace requires particular attention. Except for furnaces which use an induced-draft fan in which negative pressure is maintained in the freeboard and where waste can be dropped into the furnace by gravity, most fluid bed furnaces are under positive pressure. Waste must be forced into the furnace, and gravity feed cannot be utilized. Positive-displacement pumps or screws are normally used for feeding fluid bed furnaces. Sludge feeding is a potential problem area because of the tendency of the sludge within the feeder to dry during periods when the furnace is maintained hot without sludge feed (hot standby conditions). Hard piping should be used from the pumps to the furnace, and a hot-water purge system should be included to help prevent caking and to provide clearing of the piping when not in use.

A major issue associated with a fluid bed furnace is its ability to handle a wide variety of waste streams. Certain wastes, particularly those containing clays or salts, will tend to slag (seize) the bed, preventing fluidization. Test burns on materials that have not previously been fired in a fluid bed furnace must be made to determine the bed reaction to those materials. If bed seizure is indicated, bed additives may be available to help eliminate this problem by changing the physical properties (eutectic) of the waste feed. If not, a fluid bed unit may not be the one to use. A related issue is agglomeration. Waste materials may build up on individual sand particles within the bed. With changes in operation (such as continuous bed withdrawal), this could be controlled. If not addressed, agglomeration could result in defluidization.

It has been found that natural gas will not necessarily burn well within a fluid bed. It will tend to pass vertically through the bed, without mixing within the sand. It can be fired in a windbox burner beneath the sand bed or injected directly into the bed as low as possible. Sludge, liquids, and prepared solid waste can be fed directly to the bed of the fluid bed reactor. Sludges with a moisture content in excess of 80 percent and aqueous liquids with high water content can be fed into the top of the furnace rather than into the bed. Top injection tends to reduce the bed area otherwise required for sludge moisture evaporation because much of the moisture is evaporated by the time the feed reaches the bed. On the other hand, top injection (feeding at a single point) can result in incomplete combustion of a portion of the feed because of short-circuiting.

The temperature within the bed should be maintained at approximately 1300 to 1400°F for sewage sludges. The introduction of sludges or other wastes containing inorganic materials should be subject to a test burn before injection into a fluid bed furnace.

The oxygen content of the flue gas exiting the furnace should be continuously measured at the breeching leaving the furnace, upstream of the recuperator (if one is provided). A sample should be continuously extracted from the gas stream, passed through a water bath to clean the sample and to reduce its moisture content, and then measured for oxygen content. The oxygen content should be in the range of 4 to 8 percent by volume.

THE CONVEYER FURNACE

There are two types of conveyer furnaces in use for sludge incineration. One is the NASS-type furnace, and the other is an electric or infrared conveyer furnace.

The NASS furnace, shown in Fig. 15.4, utilizes a conveyer belt passing through an enclosed chamber. Sludge, dropping through a feed hopper, is distributed on the conveyer belt by a series of leveling screws. Any organics that may be released as the sludge drops to the belt are caught by a vent hood and are directed to the furnace exhaust, where the temperature is raised high enough to ensure destruction.

Sludge, resting on the conveyer belt, passes under fuel-fired radiant tubes (natural gas, fuel oil, or other source of supplemental heat) where it is heated to and maintained at 1400 to 1800°F. The sludge will burn on the belt, and at the end of its travel the sludge ash falls into a discharge hopper where it is cooled and discharged, dry, from the system.

The conveyer belt itself passes through a water seal after discharging its ash load. The water seal cools the belt, reducing the time of its exposure to the hot furnace temperatures.

The electric or radiant heat (or infrared) furnace is basically a conveyer belt system passing through a long rectangular refractory-lined chamber, as shown in Fig. 15.5.

Combustion air is introduced at the discharge end of the belt, shown as the viewport. It is often heated with an external preheater recuperating exhaust heat. However, the air will pick up heat from the hot burned sludge as sludge and air travel in countercurrent to each other.

Supplemental heat is provided by electric infrared heating elements within the furnace. Cooling air prevents local hot spots in the immediate vicinity of the heaters and is used as secondary combustion air within the furnace.

The conveyer belt is continuous woven wire mesh made of steel alloy chosen to withstand the 1300 to 1500°F encountered within the furnace. The refractory is not brick but ceramic felt, as shown in Fig. 15.6. It does not have a high capacity for holding heat and can therefore be started up from a cold condition relatively quickly.

Sludge fed onto the belt is immediately leveled to a depth of approximately 1 in. There is no other sludge handling mechanism. The belt speed and travel are sized to provide burnout of the sludge without agitation. This feature results in a very low level of particulate emissions.

Usually a low-energy gas scrubber, such as a cyclonic scrubber, is all that is required to clean the flue gas. Excess air requirements are 20 to 30 percent.

Supplemental fuel, in this case electric, is required for start-up and, with wet sludges, to maintain combustion. Unfortunately, the power needed for start-up results in a large connected load. In areas of the country where there are high demand charges for electric power, this system is economically impractical. It is competitive where demand charges are low and where sludge burning is autogenous, requiring no supplemental power.

CYCLONIC FURNACE

The cyclonic furnace is a single-hearth unit where the hearth moves and the rabble teeth are stationary (see Fig. 15.7). Sludge is rabbled toward the center of the hearth where, as ash, it is discharged.

Air is introduced at tangential burner ports on the shell of the furnace. The furnace is a refractory-lined cylindrical shell with a domed top. The air, heated

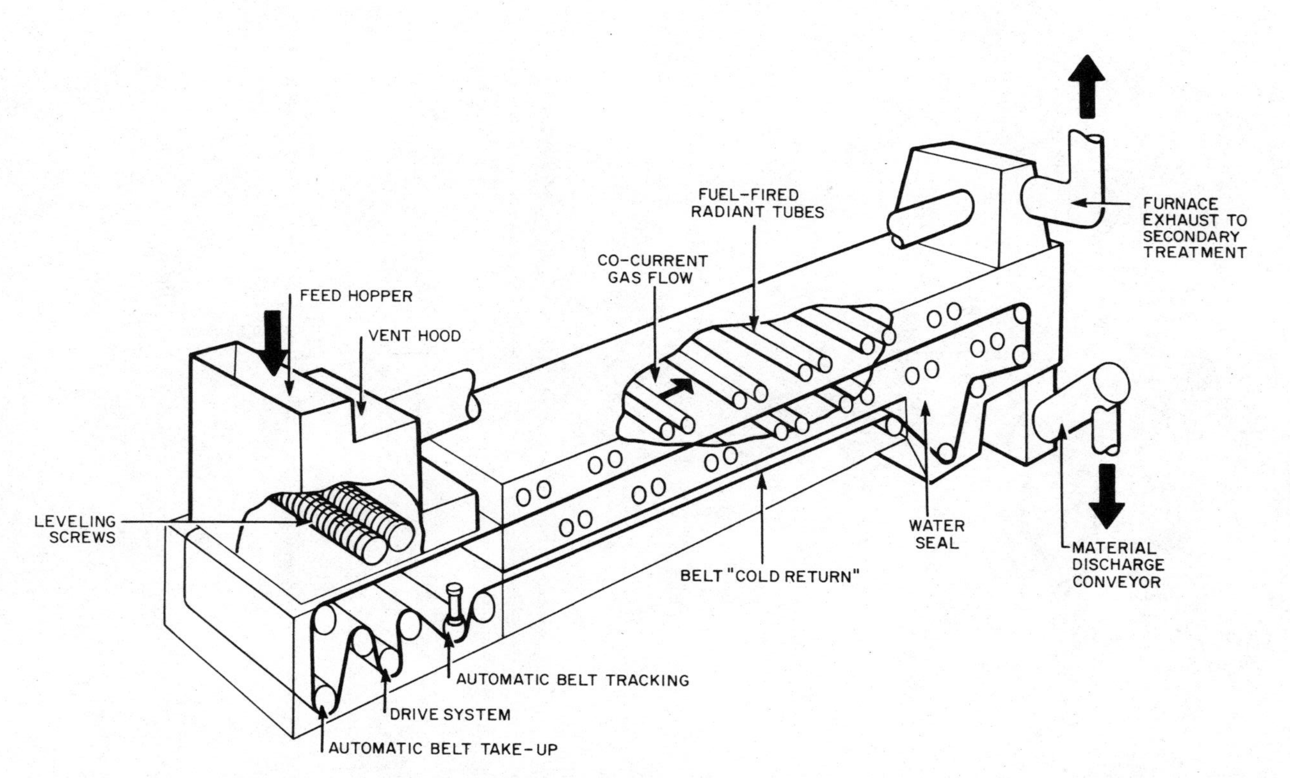

FIGURE 15.4 NASS conveyer furnace. *Source*: F. McManus, NASS Inc.

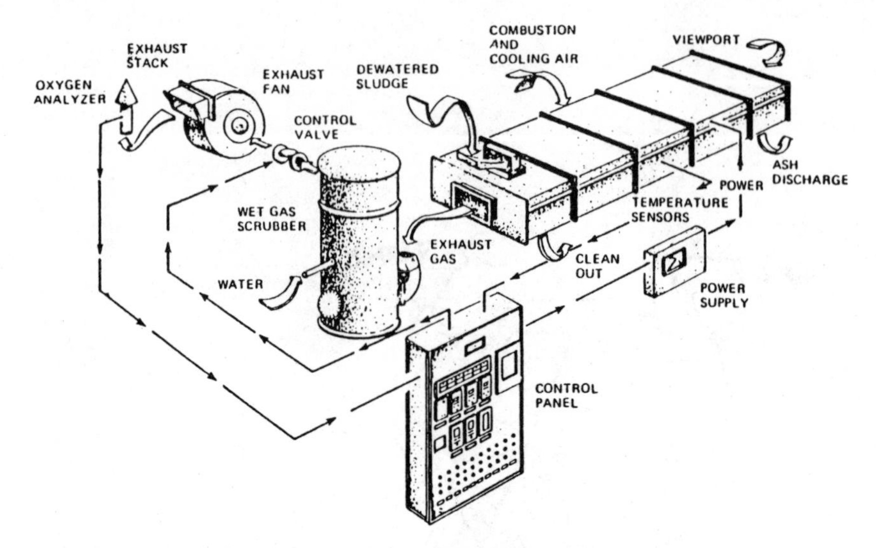

FIGURE 15.5 Radiant heat furnace. *Source*: SHIRCO, Inc., Dallas, Tex.

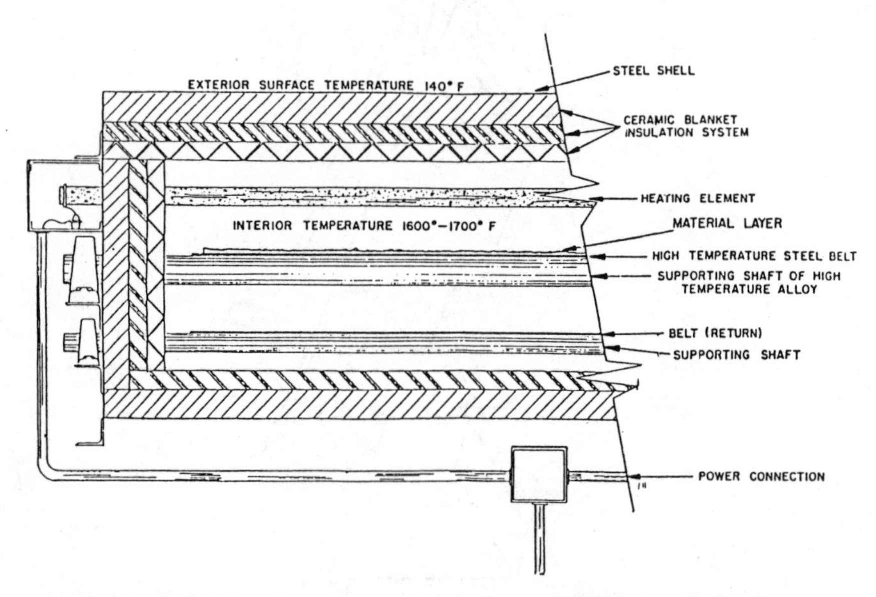

FIGURE 15.6 Radiant heat furnace, cross section. *Source*: SHIRCO, Inc., Dallas, Tex.

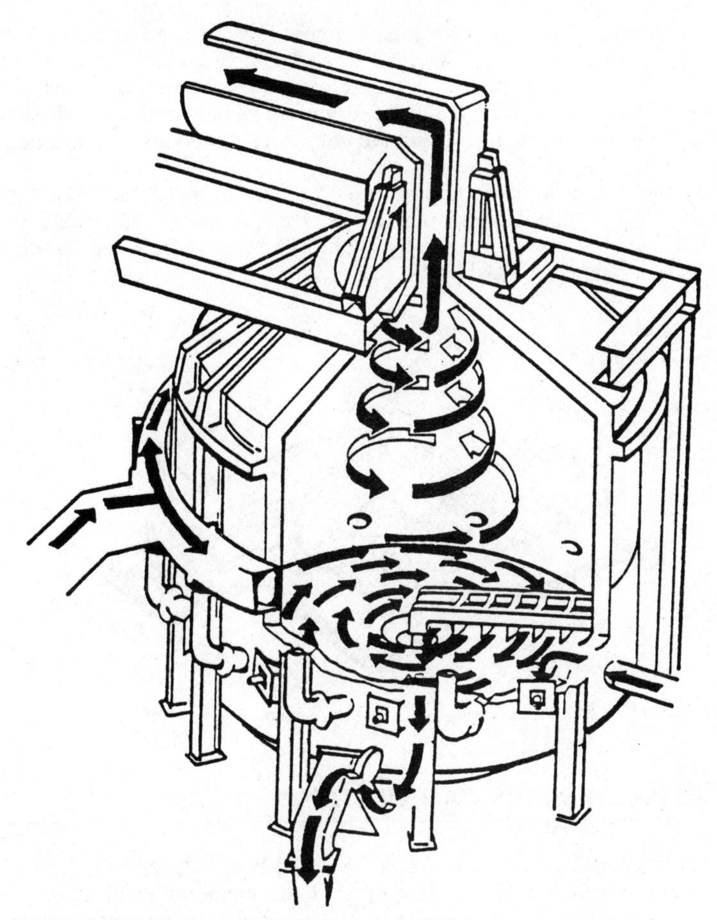

FIGURE 15.7 Cyclonic furnace. *Source*: Ref. 15-1.

with the immediate introduction of supplemental fuel, creates a violent swirling pattern which provides good mixing of air and sludge feed. The air, later flue gas, swirls up vertically in cyclonic flow through the discharge flue in the center of the domed roof.

Sludge is fed into the furnace with a screw feeder and is deposited on the periphery of the rotating hearth. A progressive cavity pump is normally used to feed sludge.

Temperatures within the furnace are 1500 to 1600°F. These furnaces are relatively small and can normally be placed in operation, at operating temperature, within an hour's time.

A variation of the cyclonic furnace is shown in Fig. 15.7. This is a horizontal cyclone furnace. Ash is discharged with the flue gas. Sludge is pumped into the furnace tangentially from the furnace wall. Air, as above, is introduced at tangential burner ports, creating a cyclonic effect.

There is no hearth, only the furnace shell and refractory. The sludge detention time is greater than 10 s in this furnace. The products of combustion exit the furnace in vortex or cyclonic flow at 1500°F, and complete combustion is ensured.

Cyclonic furnaces are suited for sludge generated from relatively small influent flows when used for sewage sludge, 2 million gal/day and less. They are relatively inexpensive, mechanically simple units, well suited for on-site sludge and/or liquid disposal.

Horizontal furnaces, as shown in Fig. 15.8, can be purchased skid-mounted for installation as a complete independent package, requiring only utility and feed connections and a stack. Many packaged units like this are sold for on-site sludge and other waste liquid disposal.

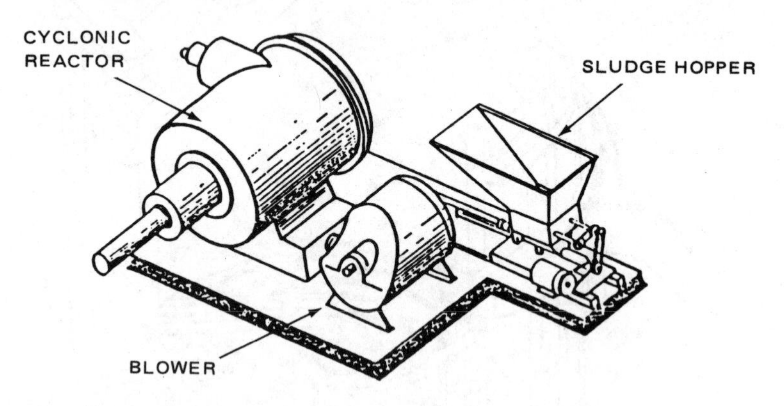

FIGURE 15.8 Skid-mounted cyclonic reactor system. *Source*: Ref. 15-1.

SLUDGE INCINERATOR CALCULATIONS

Table 15.1 (the mass flow sheet), Table 15.2 (the heat balance sheet), and Table 15.3 (the flue gas discharge sheet) represent the incineration of 12,000 lb/h of sludge in an electric furnace. The wet feed has a moisture content of 74 percent, an ash content of 43 percent, and has the following constituents:

Carbon	64.3%
Hydrogen	8.2%
Sulfur	2.2%
Oxygen	21.0%
Nitrogen	4.3%

From Chap. 8, using Du Long's approximation, the heating value is calculated as follows:

$$Q = 14{,}544(C) + 62{,}028(H_2 - 0.125O_2) + 4050(S)$$

$$= 14{,}544(0.643) + 62{,}068(0.082 - 0.125 \times 0.21) + 4050(0.022)$$

$$= 12{,}898 \text{ Btu/lb}$$

Products of combustion are calculated as follows:

TABLE 15.1 Mass Flow

	Example (elect.)
Wet feed, lb/h	12,000
Moisture, %	74
lb/h	8880
Dry feed, lb/h	3120
Ash, %	43
lb/h	1342
Volatile, lb/h	1778
Volatile htg. value, Btu/lb	12,898
MBtu/h	22.93
Dry gas, lb/10 kBtu	7.494
lb/h	17,184
Comb. H_2O, lb/10 kBtu	0.568
lb/h	1302
Dry gas+comb. H_2O, lb/h	18,486
100% air, lb/h	16,708
Total air fraction	1.2
Total air, lb/h	20,050
Excess air, lb/h	3342
Humid/dry gas (air), lb/lb	0.01
Humidity, lb/h	201
Total H_2O, lb/h	10,383
Total dry gas, lb/h	20,526

From the methods of Chap. 4, to calculate combustion parameters:

$$\underset{12.01}{C} + \underset{32.00}{O_2} + 3.7619\overset{28.02}{N_2} \rightarrow \underset{44.01}{CO_2} + \underset{105.408}{3.7619N_2}$$

For 0.643 lb C,

$$\frac{44.01}{12.01} \times 0.643 = 2.356 \text{ lb } CO_2$$

$$\frac{105.408}{12.01} \times 0.643 = 5.643 \text{ lb } N_2$$

$$\underset{32.06}{S} + \underset{32.00}{O_2} + 3.7619\overset{28.02}{N_2} \longrightarrow \underset{64.06}{SO_2} + \underset{105.408}{3.7619N_2}$$

For 0.022 lb S,

$$\frac{64.06}{32.06} \times 0.022 = 0.044 \text{ lb } SO_2$$

TABLE 15.2 Heat Balance

	Example (elect.)
Cooling air wasted, lb/h	—
°F	—
Btu/lb	—
MBtu/h	—
Ash, lb/h	1342
Btu/lb	130
MBtu/h	0.17
Radiation, %	3
MBtu/h	0.69
Humidity, lb/h	201
Correction (@970 Btu/lb), MBtu/h	−0.19
Losses, total, MBtu/h	0.67
Input, MBtu/h	22.93
Outlet, MBtu/h	22.26
Dry gas, lb/h	20,526
H_2O, lb/h	10,383
Temperature, °F	1166
Desired temp., °F	1200
MBtu/h	22.63
Net, MBtu/h	0.37
Fuel oil, air fraction	—
Net Btu/gal	—
gal/h	(108)
Air, lb/gal	—
lb/h	—
Dry gas, lb/gal	—
lb/h	—
H_2O, lb/gal	—
lb/h	—
Dry gas w/fuel oil, lb/h	20,526
H_2O w/fuel oil, lb/h	10,383
Air w/fuel oil, lb/h	20,050
Outlet, MBtu/h	22.63
Reference t, °F	60

$$\frac{105.408}{32.06} \times 0.022 = 0.072 \text{ lb } N_2$$

$$\begin{array}{llll} 2.02 & & & 28.02 \\ 2H_2 = & O_2 + 3.7619N_2 \longrightarrow & 2H_2O + & 3.7619N_2 \\ 4.04 & 32.00 & 36.04 & 105.408 \end{array}$$

For 0.082 lb H_2,

$$\frac{36.04}{4.04} \times 0.082 = 0.732 \text{ lb } H_2O$$

$$\frac{105.408}{4.04} \times 0.082 = 2.139 \text{ lb } N_2$$

TABLE 15.3 Flue Gas Discharge

	Example (elect.)
Inlet, °F	1200
Dry gas, lb/h	20,526
Heat, MBtu/h	22.63
Btu/lb dry gas	1103
Adiabatic t, °F	187
H_2O saturation, lb/lb dry gas	0.9271
lb/h	19,030
H_2O inlet, lb/h	10,383
Quench H_2O, lb/h	8647
gal/min	17.3
Outlet temp., °F	120
Raw H_2O temp., °F	60
Sump temp., °F	156
Temp. diff., °F	96
Outlet, Btu/lb dry gas	111.65
MBtu/h	2.29
Req'd. cooling, MBtu/h	20.34
H_2O, lb/h	211,875
gal/min	424
Outlet, ft^3/lb dry gas	16.515
ft^3/min	5650
Fan press., in WC	28
Outlet, actual ft^3/min	6067
Outlet, H_2O/lb dry gas	0.08128
H_2O, lb/h	1668
Recirc. (ideal), gal/min	17
Recirc. (actual), gal/min	138
Cooling H_2O, gal/min	424

The total dry gas produced per pound of fuel (or sludge volatiles), including the nitrogen in the fuel, is equal to

$$N_2 = 0.043 + 5.643 + 0.072 + 2.139$$

$$= 7.897 \text{ lb}$$

$$CO_2 = 2.356 \text{ lb}$$

$$\underline{SO_2} = \underline{0.044 \text{ lb}}$$

$$\text{Total} = 10.297 \frac{\text{lb dry gas}}{\text{lb fuel}} \times \frac{\text{lb fuel}}{12{,}898 \text{ Btu}}$$

$$= 7.494 \text{ lb dry gas/10 kBtu}$$

The moisture produced is equal to

$$H_2O = 0.732 \frac{\text{lb } H_2O}{\text{lb fuel}} \times \frac{\text{lb fuel}}{12{,}898 \text{ Btu}}$$

$$= 0.568 \text{ lb } H_2O/10 \text{ kBtu}$$

An electric incinerator can require only 20 percent excess air, which is used in this illustration. There is no cooling air with this type of furnace, and the radiation loss was assumed to be equal to 3 percent of the sludge cake heat input.

By completing Table 15.1 per the methods of Chap. 8, insertions can be made in Table 15.2. For a desired outlet temperature of 1200°F the equivalent of 0.37 MBtu/h of supplemental fuel is required. Converting to electric energy

$$0.37 \times 10^6 \text{ (Btu/h)} \times \text{(kWh/3412)} = 108 \text{ kW}$$

Therefore 108 kW of electric energy is required to heat the sludge cake and its products of combustion to 1200°F. Electric energy requires no additional air, unlike fossil fuel; therefore the products of combustion are the same as those prior to the introduction of supplemental heat.

The remainder of Tables 15.2 and 15.3 was completed in accordance with the methods of Chap. 8.

REFERENCE

15-1. Brunner, C.: *Design of Sewage Sludge Incineration Systems*, Noyes, Park Ridge, N.J., 1980.

BIBLIOGRAPHY

Air Pollution Aspects of Sludge Incineration, USEPA 625/4-75-009, June 1975.

Baturay, A.: "Latest Developments in Fluidized Bed Development Technology," *Proceedings of the Puerto Rico Water Pollution Control Federation*, February 1979.

Brunner, C.: "Sewage Sludge Incineration at the Cleveland Southerly Wastewater Treatment Center," American Society of Mechanical Engineers, Intersociety Energy Conference, Atlanta, Ga., August 1981.

The Conversion of Existing Municipal Sludge Incinerators to Codisposal, USEPA SW-743, 1979.

Hazardous Material Incinerator Design Criteria, USEPA 600/2-79-198, October 1979.

Jacknow, J.: "Environmental Aspects of Acceptable Sludge Disposal Techniques," Fifth Conference on Acceptable Sludge Disposal Techniques, January 1978.

Jacknow, J.: "Thermal Sludge Processing Technology," *Sludge Magazine*, July 1979.

Kenson, R.: "Rotary Kiln Incinerators for Sludge Disposal," *Pollution Engineering*, December 1980.

Krindler, E., Youngs, P., and Burkhardt, G.: "Modifying Existing Multiple Hearth Incinerators to Burn Thermal Conditioned Wastewater Sludge," *Journal of the Water Pollution Control Federation*, September 1980.

Lyon, S.: "Incineration of Raw Sludges and Greases," *Journal of the Water Pollution Control Federation*, April 1973.

Recommended Methods of Reduction, Neutralization, Recovery or Disposal of Hazardous

Waste, vol. 3, Disposal Process Descriptions, Ultimate Disposal, Incineration and Pyrolysis Processes, USEPA 670/2-73-053C, August 1973.

A Review of Techniques for Incineration of Sewage Sludge with Solid Wastes, USEPA 600/2-76-288, December 1976.

CHAPTER 16
LIQUID WASTE DESTRUCTION

Increasingly, public attention is focusing on liquid wastes as the most objectionable of waste streams. The fear of buried drums leaking liquid waste or liquid waste seeping into surface waters has resulted in statutes limiting the disposal of many of these waste streams. The existing technology for incineration of liquid waste is presented in this chapter.

LIQUID WASTE PROPERTIES

As discussed in a previous chapter, the line between liquid and nonliquid is not always well defined. A material is considered a liquid, for purposes of incinerator design, if it can be pumped to a burner and atomized, i.e., fired in suspension. In general a material can be pumped if its viscosity is less than 10,000 SSU. Atomization is a function of nozzle type. With the appropriate nozzle design liquids with up to 5000 SSU viscosity can be fired in suspension.

Besides viscosity, other factors are important in liquid incinerator selection and design:

- *Heating value.* Can the liquid sustain combustion, or is auxiliary fuel required?
- *Aqueous content.* A liquid composed of over 60 percent water is considered an aqueous waste.
- *Halogen fraction.* A liquid with a chloride, bromide, or fluoride component requires careful attention to material selection and gas cleaning system design.
- *Metallic salts.* Firing of wastes with metallic salt components often produces salt residue in the furnace and exhaust gas train and may cause severe corrosion of refractory.
- *Sulfonated waste.* Sulfur in a waste will produce acidic corrosion, and material selection and exhaust gas cleaning must be carefully controlled.
- *Organic waste.* This contains carbon and hydrogen and can also contain oxygen, nitrogen, sulfur, and halogens.
- *Cyclic and polycyclic organics.* These are organic compounds characterized by the presence of benzene-type rings. They are difficult to effectively incinerate because of the thermal stability of this ring structure.

FURNACE INJECTION

Liquid waste can be the prime fuel source, injected into a cold furnace through standard nozzles. Its heating value would have to be sufficiently high to maintain combustion at desired furnace temperatures.

With aqueous waste or other low-heating-value waste, injection is normally outside the flame envelope. The prime fuel—that fuel firing to establish and maintain furnace temperatures—is allowed to burn completely and release its complete heating value. In-flame injection of a low-Btu waste would tend to cool the flame and thus interfere with efficient fuel burning.

Liquid waste can be injected within a flame front if its heating value is high enough to add and not remove a net heating value to the flame, normally a minimum of 5000 Btu/lb.

In the injection of fuel or waste within a furnace, care must be taken to avoid flame impingement on furnace walls. Flame impingement indicates that excessively high temperatures are reaching refractory surfaces. This will tend to reduce their life. In addition, unburned residue will collect on impingement areas. For every pound of carbon left as an unburned residual on a furnace wall over 14,000 Btu is lost to furnace heat release. Also the presence of a residual coating will promote refractory corrosion.

LIQUID INJECTION NOZZLES

Liquid fuels must be vaporized before combustion can occur. The degree of atomization and fuel-air mixing is directly related to burning efficiency. A number of different types of nozzles or burners have been developed for efficient burning of the varying types of fuels and liquid waste streams generated today.

Mechanical Atomizing Nozzles

These are the most common types of burner nozzles in current use. Typical mechanical atomizing nozzles are illustrated in Figs. 16.1 and 16.2. Fuel is pumped into the nozzle at pressures of 75 to 150 lb/in^2 through a small fixed-orifice discharge. The fuel is given a strong cyclonic or whirling velocity before it is released through the orifice. Combustion air is provided around the periphery of the conical spray of fuel produced. The combination of combustion air introduced tangentially into the burner and the action of the swirling fuel produce effective atomization. Normal turndown ratios are in the range of 2.5:1 to 3.5:1. By utilizing a return flow line for fuel oil the turndown ratio can be increased to as high as 10:1. Typical burner capacities are in the range of 10 to 100 gal/h. A major disadvantage of this type of burner-nozzle is its susceptibility to erosion and pluggage from solids components of the fluid stream. Flames tend to be short, bushy, or low-velocity, and this results in slower combustion, requiring relatively large combustion chamber volumes. This burner is applicable for fluids with relatively low viscosity, under 100 SSU.

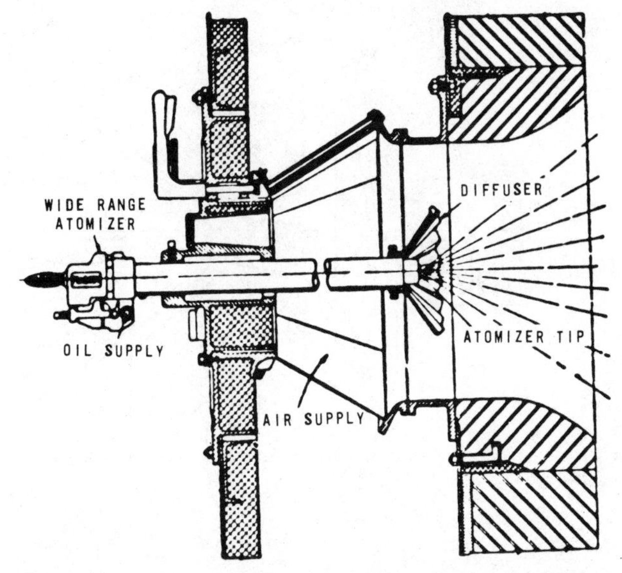

FIGURE 16.1 Mechanical atomizing nozzle. *Source*: Combustion Engineering-Superheater, Inc., New York.

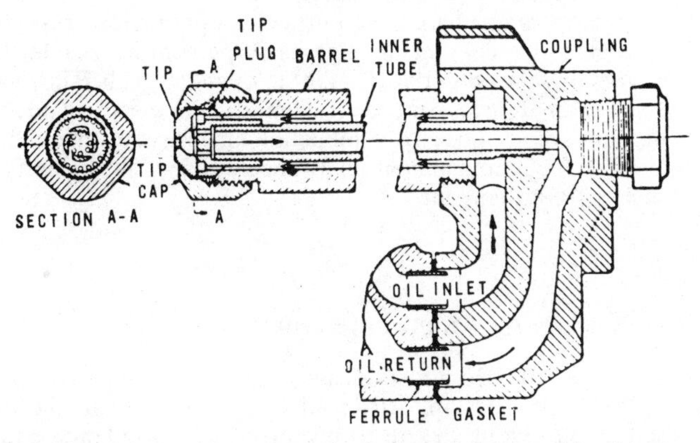

FIGURE 16.2 A wide-range mechanical atomizing assembly with central oil return line. *Source*: Combustion Engineering-Superheater, Inc., New York.

Rotary Cup Burners

As shown in Fig. 16.3, atomization is provided by throwing fuel centrifugally from a rotating cup or plate. Oil is thrown from the lip of the cup in the form of conical sheets which break up into droplets by the effect of surface tension. No air is mixed with fuel prior to atomization. Instead, air is introduced through an annular space around the rotary cup. Normally a common motor drives the oil

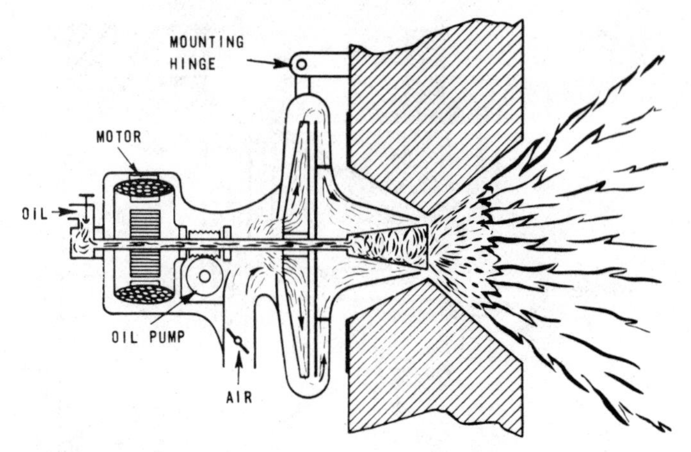

FIGURE 16.3 Rotary cup oil burner. *Source*: Preferred Utilities, Inc., Danbury, Conn.

pump, rotating cup, and combustion air blower. The liquid pressure required for this burner is relatively low, since atomization is a function of cup rotation and combustion air injection, not fuel pressure. This low-pressure requirement and the relatively large openings within the burner fuel path allow passage of fluids with relatively high solids content, as high as 20 percent by weight. Burner capacities range from low flows (under 10 gal/h) to over 250 gal/h. They have a turndown ratio of approximately 5:1 and can fire liquids with viscosity up to 300 SSU. Rotary cup burners are sensitive to combustion airflow adjustment. Insufficient airflow will result in fuel impingement on furnace walls, while excessive combustion airflow will cause a flameout.

External Low-Pressure Air-Atomizing Burners

The major portion of the combustion air requirement is provided at 1 to 5 lb/in^2 gauge near the burner tip. Air is injected externally to the fuel nozzle and is directed to the liquid stream to produce high turbulence and effective atomization. The liquid pressure necessary for operation is only enough for positive delivery, normally less than 1½ lb/in^2 gauge. The quantity of atomizing air required decreases with increased atomization pressure and may range from 400 to 1000 st ft^3/gal of fuel. Secondary combustion air is provided around the periphery of the atomized liquid mixture. The flame is relatively short because of the high amount of air provided at the burner (atomization and secondary combustion air). The short flame allows design of smaller combustion chambers. These burners normally operate with liquids in the range of 200 to 1500 SSU and can handle solids concentrations in the liquid of up to 30 percent. Figure 16.4 illustrates a low-pressure air-atomizing burner. A small quantity of the airflow passes around the fuel discharge to aid in optimization of the fuel flow pattern.

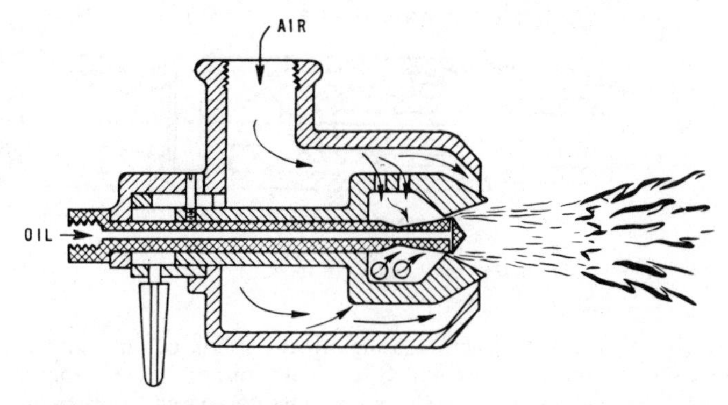

FIGURE 16.4 Low-pressure air-atomizing oil burner. *Source*: Hauck Manufacturing Company, Lebanon, Pa.

External High-Pressure Two-Fluid Burners

The atomizing fluid, air or steam (or nitrogen or other gas), impinges the fuel stream at high velocity to generate small particles that encourage quick vaporization and effective atomization of tars and other heavy liquids. A typical burner is shown in Fig. 16.5. The steam requirement is 2 to 5 lb/gal of fuel, whereas the air requirement is 50 to 200 st ft^3/gal of fuel. The required atomization pressure varies from 30 to 150 lb/in^2 gauge. Turndown is in the range of 3:1 to 4:1. The flame produced is relatively long, requiring appropriately constructed combustion chambers. The fuel viscosity normally handled by these burners ranges from 150 to 5000 SSU for either air or steam atomization. A solids content of up to 70 percent can be accommodated by these burners.

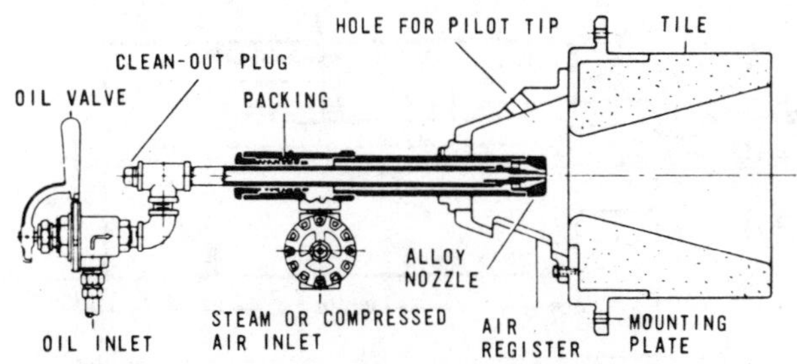

FIGURE 16.5 High-pressure steam- or air-atomizing oil burner. *Source*: North American Manufacturing Company, Cleveland.

Internal Mix Nozzles

Air or steam is introduced within the nozzle, as in Fig. 16.6, to provide impingement of atomization fluid on the fuel stream prior to spraying. Atomization air is provided at pressures less than 30 lb/in^2 gauge, and steam is normally introduced at 90 to 150 lb/in^2 gauge. The turndown ratio for this type burner is from 3:1 to

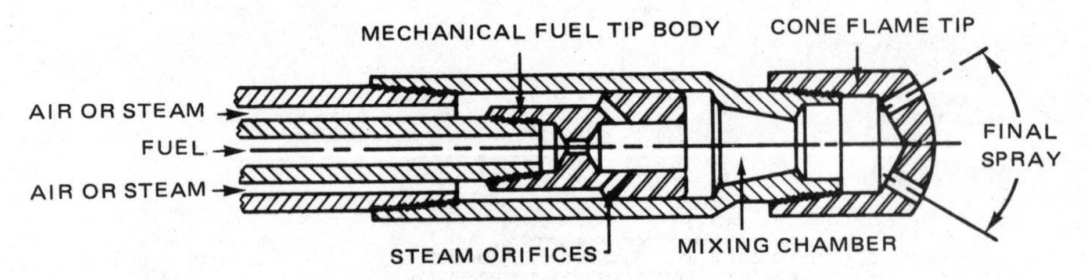

FIGURE 16.6 Internal mix nozzle. *Source*: Ref. 16-1.

4:1. These nozzles cannot tolerate a significant solids content and can handle only low-viscosity fuels, under 100 SSU. This burner is used for clean, low-viscosity liquids. Its advantage is in its low cost compared to other burners.

Sonic Nozzles

These nozzles utilize a compressed gas such as air or steam to create high-frequency sound waves which are directed at the fuel stream. This acoustic energy is transferred to the liquid stream and creates an atomizing force, breaking the stream into minute particles. The fuel nozzle diameter is relatively large, allowing passage of solid particulate streams such as slurries and sludges with high particulate content. Little fuel pressurization is required. The spray pattern is not well defined, with finely atomized, uniformly distributed droplets traveling at low velocities. These nozzles are difficult to adjust, have low turndown, and generate an extremely high noise level during operation. A typical sonic nozzle is shown in Fig. 16.7.

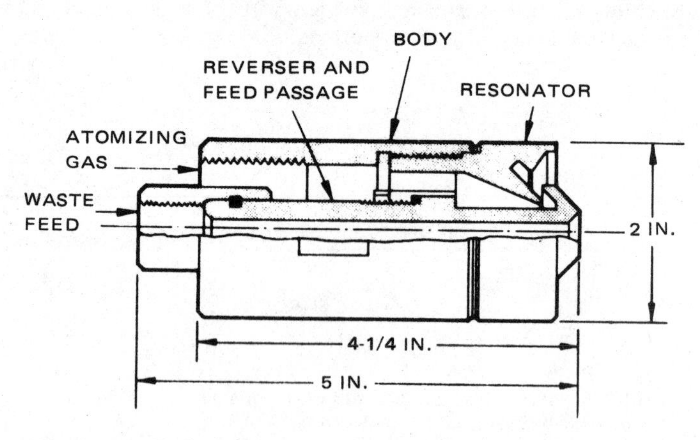

FIGURE 16.7 Typical sonic nozzle. *Source*: Ref. 16-2.

LIQUID DESTRUCTION FURNACES

The type, size, and shape of a furnace are a function of waste characteristics, burner design, air distribution, and furnace wall design.

As discussed previously, flame impingement on a furnace wall is undesirable, creating the potential for refractory corrosion and resulting in lost energy. A furnace must be designed to avoid impingement. Impingement is a function of liquid atomization and vaporization which in turn are dependent on nozzle design, velocity of fluid exiting the burner, air distribution within the furnace, and furnace temperature.

Primary combustion air is that airflow supplied at the fuel burner to combust the prime fuel within the furnace. It is normally distributed through a burner register, an open fan-shaped component normally surrounding the burner nozzle which imparts a circular velocity to the airflow. The register either is fixed or can be adjustable with adjustment arms often located immediately outside the furnace. Secondary air is that airflow necessary for waste combustion and is normally introduced into the furnace downstream of the main flame front. In liquid injection furnaces, the secondary air supply is often used not just as combustion air, but to create turbulence within the furnace and to provide a relatively cool flow on the inside refractory surface, keeping the refractory temperature cooler than that of the center of the furnace. The primary and secondary airflows are also introduced in a manner that aids fuel atomization and helps prevent any unburned materials from impinging on the furnace wall. (Spurious impingement often creates *sparklers*, luminous burnout of volatile particles on the furnace wall.)

Liquid destruction furnaces require 5 to 30 percent excess air to ensure adequate combustion. Another furnace parameter is heat release. Most liquid burners have a heat release rate of 20,000 to 30,000 Btu/(ft^3 · h). Vortex burners, burners where the primary airflow creates a high-velocity cyclonic vortex prior to injection into the furnace chamber, will release 700,000 to 1,000,000 Btu/(ft^3 · h).

Liquid waste furnaces are normally cylindrical, either horizontal or vertical, lined with refractory. Typical furnace types are described as follows:

Nonswirling Type

This is a furnace, such as the one in Fig. 16.8, where the burner(s) is (are) mounted axially or on the side, firing along a radius. These furnaces are relatively inexpensive to build and require minimal combustion air pressure (smaller blowers). Typical heat release rates are 10,000 to 30,000 Btu/(ft^3 · h). A typical furnace size calculation is as follows:

Heat release: $Q = 10{,}000{,}000 \text{ Btu/h}$

Furnace release rate: $F = 20{,}000 \text{ Btu/(h·ft}^3\text{)}$

Furnace volume:
$$V = \frac{Q}{F} = \frac{10{,}000{,}000 \text{ Btu/h}}{20{,}000 \text{ Btu/(h·ft}^3\text{)}}$$
$$= 500 \text{ ft}^3$$

For a chamber 8 ft long (L), the internal diameter D is

$$V = \frac{\pi}{4} D^2 L$$

$$D = \left(\frac{4V}{\pi L}\right)^{1/2} = \left(\frac{4 \times 500 \text{ ft}^3}{\pi \times 8 \text{ ft}}\right)^{1/2}$$

$$D = 8.92 \text{ ft} = 8 \text{ ft } 11 \text{ in}$$

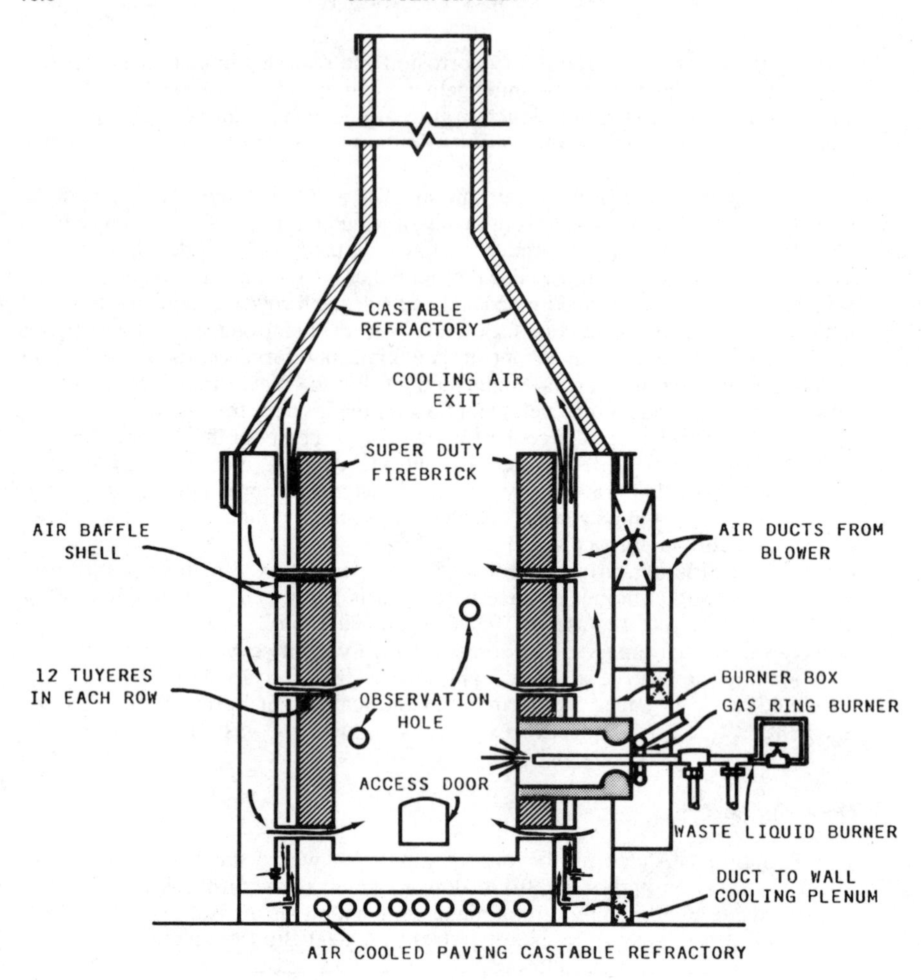

FIGURE 16.8 Liquid waste incinerator.

This incinerator, or furnace, is little more than an empty chamber, refractory-lined, with neither baffles nor other changes in the direction of flow. Air jets are often placed in the chamber side walls; these inject compressed air into the furnace, creating increased turbulence. Increasing turbulence within the combustion chamber will increase the burning efficiency, which will be reflected in an increased burning rate.

Vortex Furnace

Swirl burners or burners firing tangentially into the combustion chamber create a cyclonic or vortex flow within the furnace. Secondary combustion air is also injected tangentially into the furnace to increase the turbulent flow. Figure 16.9 illustrates a furnace with a series of vortex burners. Secondary combustion air is introduced

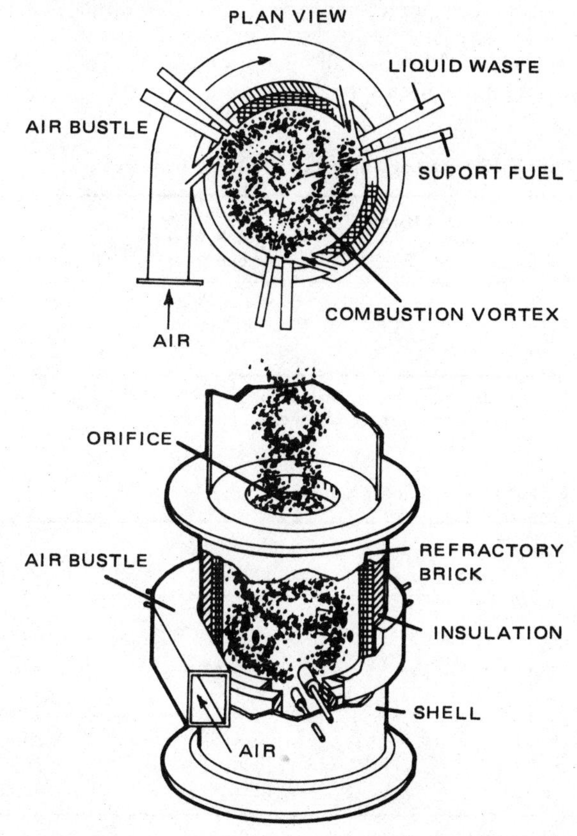

FIGURE 16.9 Vortex combustion incinerator. *Source*: Ref. 16-3.

from an air bustle. The high turbulence within the combustion chamber results in high heat release rates, 40,000 to 100,000 Btu/(h · ft³). Calculating a furnace diameter for a higher rate, say, 60,000 Btu/(h · ft³), the furnace diameter will be as follows:

$$V = \frac{D}{F} = \frac{10{,}000{,}000 \text{ Btu/h}}{60{,}000 \text{ Btu/(h} \cdot \text{ft}^3)}$$

$$= 167 \text{ ft}^3$$

$$V = \left(\frac{4V}{\pi L}\right)^{1/2} = \left(\frac{4 \times 167 \text{ ft}^3}{\pi \times 8 \text{ ft}}\right)^{1/2}$$

$$D = 5.2 \text{ ft} = 5 \text{ ft } 2 \text{ in}$$

Comparing axial furnace to vortex furnace size, 8 ft 11 in versus 5 ft 11 in diameter, respectively, the vortex furnace is over 40 percent smaller. The vortex burner design is more complex than that of the axially fired burner, and high-pressure blowers are required; however, the smaller furnace chamber represents a significantly lower cost.

Tables 16.1 and 16.2 list overall dimensions for typical liquid waste incinerators, axially fired and vortex types.

TABLE 16.1 Axially and Radially Fired Incinerator

Heat release, MBtu/h		Length × width × height,[a] ft	Total weight, lb
Waste	Fuel		
6.3	3.0	9 × 5 × 26	16,000
14.3	7.0	11 × 7 × 26	23,000
24.7	12.0	14 × 8 × 26	32,000

[a]Overall outside dimensions including stack.
Source: T-Thermal, Conshohocken, Pa.

TABLE 16.2 Vertical Vortex Incinerator

Heat release, total MBtu/h	Length × width × height,[a] ft	Total weight, lb
3.0	9 × 6 × 47	7,000
7.0	10 × 7 × 48	9,500
10.0	11 × 7 × 48	13,500
14.0	12 × 8 × 50	17,000
18.0	12 × 8 × 51	19,000
24.0	13 × 9 × 51	23,500
30.0	13 × 10 × 52	29,000
48.0	15 × 12 × 53	41,500

[a]Overall outside dimensions including stack.
Source: T-Thermal, Conshohocken, Pa.

REFERENCES

16-1. *Engineering Handbook for Hazardous Waste Incineration*, USEPA draft, November 1980.

16-2. *Recommended Methods of Reduction, Neutralization, Recovery or Disposal of Hazardous Waste*, vol. 3, Disposal Process Descriptions, Ultimate Disposal, Incineration and Pyrolysis Processes, USEPA 670/2-73-053C, August 1973.

16-3. Witt, P.: "Disposal of Solid Wastes," *Chemical Engineering*, October 4, 1971.

BIBLIOGRAPHY

Brunner, C.: "Industrial Waste Disposal at Cincinnati," in *The Environmental Professional*, Pergamon Press, Elmsford, N.Y., 1981.

Brunner, C.: "Industrial Wastes Incineration at Treatment Plant," *Water Engineering & Management*, June 1981.

Brunner, C., and Trapp, J.: "Progress Report of Industrial Liquid/Fluid Thermal Processing System," *American Society of Mechanical Engineers, Journal of the Solid Waste Processing Division*, Washington, 1980.

Dillon, A.: *Hazardous Waste Incineration Engineering*, Noyes, Park Ridge, N.J., 1981.

Dunn, K.: "Incineration's Role in Ultimate Disposal of Process Wastes," *Chemical Engineering*, October 10, 1975.

Hazardous Material Incinerator Design Criteria, USEPA 600/2-79-198, October 1979.

Perkins, B.: *Incineration Facilities for Hazardous Wastes*, Los Alamos Laboratories, July 1976.

At Sea Incineration, USEPA 600/2-79-137, July 1979.

Shen, T., Chen, M., and Lauber, J.: "Incineration of Toxic Chemical Wastes," *Pollution Engineering*, October 1978.

CHAPTER 17
INCINERATION OF GASEOUS WASTE

Many gaseous wastes are characterized by odor or color. Often these characteristics result from the presence of organic compounds which, when properly incinerated, will destruct. In this chapter techniques for destruction of gaseous waste by incineration are presented.

COMBUSTIBLE GAS

Gases having combustible constituents will have a lower explosive limit (LEL) and an upper explosive limit (UEL). The LEL is the least concentration of the gas in air which will sustain gas combustion. At a concentration below the LEL there is insufficient gas present to generate the heat required to sustain combustion.

The UEL is the highest concentration of gas in air at which combustion can be sustained. At a higher concentration of gas to air there will be insufficient air present to sustain combustion.

Table 17.1 lists lower and upper explosive limits of various liquid and gaseous substances in air. Included in this table is the flash point for some of those materials which are in liquid form. At the flash point, the substances will self-ignite with a spark. The flash point, therefore, is a measure of volatility for liquids. The autoignition temperature (see Table 17.2) is that temperature at which a gas will combust. Bringing a gas to its autoignition temperature in a mixture of air between the LEL and UEL of that gas will result in sustained combustion, perhaps rapid combustion (an explosion).

The method of incineration chosen is affected by the LEL, UEL, and ignition temperature. Flares or direct flame incinerators operate best with gases just above the UEL or just below the LEL. Catalytic incinerators are normally used with concentrations no more than 25 percent of the LEL. The lower the flash point, or autoignition temperature, the less excess air is normally required.

FLARES

Flares are used as a low-cost means of disposal of relatively large amounts of gas containing combustible components. They are suited to processes which are not

TABLE 17.1 Combustibility Characteristics of Pure Gases and Vapors in Air

Gas or Vapor	Lower Limit, % by volume	Upper Limit, % by volume	Closed Cup Flash Point, °F
Acetaldehyde	4.0	57	-17
Acetone	2.5	12.8	0
Acetylene	2.5	80	–
Allyl alcohol	2.5	–	70
Ammonia	15.5	26.6	–
Amyl acetate	1.0	7.5	77
Amylene	1.6	7.7	–
Benzene (benzol)	1.3	6.8	12
Benzlyl chloride	1.1	–	140
Butene	1.8	8.4	–
Butyl acetate	1.4	15.0	84
Butyl alcohol	1.7	–	–
Butyl cellosolve	–	–	141
Carbon disulfide	1.2	50	-22
Carbon monoxide	12.5	74.2	–
Chlorobenzene	1.3	7.1	90
Cottonseed oil	–	–	486
Cresol, *m*- or *p*-	1.1	–	202
Crotonaldehyde	2.1	15.5	55
Cyclohexane	1.3	8.4	1
Cyclohexanone	1.1	–	111
Cyclopropane	2.4	10.5	–
Cymene	0.7	–	117
Dichlorobenzene	2.2	9.2	151
Dichloroethylene (1,2)	9.7	12.8	57
Diethyl selenide	2.5	–	57
Dimethyl formamide	2.2	–	136
Dioxane	2.0	22.2	54
Ethane	3.1	15.5	–
Ether (diethyl)	1.8	36.5	-49
Ethyl acetate	2.2	11.5	28
Ethyl alcohol	3.3	19.0	54
Ethyl bromide	6.7	11.3	–
Ethyl cellosolve	2.6	15.7	104
Ethyl chloride	4.0	14.8	-58
Ethyl ether	1.9	48	-49
Ethyl lactate	1.5	–	115
Ethylene	2.7	28.6	–
Ethylene dichloride	6.2	15.9	56
Ethyl formate	2.7	16.5	- 4
Ethyl nitrite	3.0	50	-31
Ethylene oxide	3.0	80	–
Furfural	2.1	–	140

TABLE 17.1 Combustibility Characteristics of Pure Gases and Vapors in Air (Continued)

Gas or Vapor	Lower Limit, % by volume	Upper Limit, % by volume	Closed Cup Flash Point, °F
Gasoline (variable)	1.4-1.5	7.4-7.6	-50
Heptane	1.0	6.0	25
Hexane	1.2	6.9	-15
Hydrogen cyanide	5.6	40.0	–
Hydrogen	4.0	74.2	–
Hydrogen sulfide	4.3	45.5	–
Illuminating gas (coal gas)	5.3	33.0	–
Isobutyl alcohol	1.7	–	82
Isopentane	1.3	–	–
Isopropyl acetate	1.8	7.8	43
Isopropyl alcohol	2.0	–	53
Kerosene	0.7	5	100
Linseed oil	–	–	432
Methane	5.0	15.0	–
Methyl acetate	3.1	15.5	14
Methyl alcohol	6.7	36.5	52
Methyl bromide	13.5	14.5	–
Methyl butyl ketone	1.2	8.0	–
Methyl chloride	8.2	18.7	–
Methyl cyclohexane	1.1	–	25
Methyl ether	3.4	18	–
Methyl ethyl ether	2.0	10.1	-35
Methyl ethyl ketone	1.8	9.5	30
Methyl formate	5.0	22.7	- 2
Methyl propyl ketone	1.5	8.2	–
Mineral spirits No. 10	0.8	–	104
Naphthalene	0.9	–	176
Nitrobenzene	1.8	–	190
Nitroethane	4.0	–	87
Nitromethane	7.3	–	95
Nonane	0.83	2.9	88
Octane	0.95	3.2	56
Paraldehyde	1.3	–	–
Paraffin oil	–	–	444
Pentane	1.4	7.8	–
Propane	2.1	10.1	–
Propyl acetate	1.8	8.0	58
Propyl alcohol	2.1	13.5	59
Propylene	2.0	11.1	–
Propylene dichloride	3.4	14.5	60
Propylene oxide	2.0	22.0	–
Pyridine	1.8	12.4	74
Rosin oil	–	–	266

TABLE 17.1 Combustibility Characteristics of Pure Gases and Vapors in Air (Continued)

Gas or Vapor	Lower Limit, % by volume	Upper Limit, % by volume	Closed Cup Flash Point, °F
Toluene (toluol)	1.3	7.0	40
Turpentine	0.8	–	95
Vinyl either	1.7	27.0	–
Vinyl chloride	4.0	21.7	–
Water gas (variable)	6.0	70	–
Xylene (xylol)	1.0	6.0	63

Source: Ref. 17-1.

TABLE 17.2 Autoignition Temperature of Some Common Organic Compounds

Compound	Temperature, °F	Compound	Temperature, °F
Acetone	1000	Hydrogen	1076
Ammonia	1200	Hydrogen cyanide	1000
Benzene	1075	Hydrogen sulfide	500
Butadiene	840	Kerosene	490
Butyl alcohol	693	Maleic anhydride	890
Carbon disulfide	257	Methane	999
Carbon monoxide	1205	Methyl alcohol	878
Chlorobenzene	1245	Dichloromethane	1185
Cresol	1038	Methyl ethyl ketone	960
Cyclohexane	514	Mineral spirits	475
Dibutyl phthalate	760	Petroleum naphtha	475
Ethyl ether	366	Nitrobenzene	924
Methyl ether	662	Oleic acid	685
Ethane	950	Phenol	1319
Ethyl acetate	907	Phthalic anhydride	1084
Ethyl alcohol	799	Propane	874
Ethyl benzene	870	Propylene	940
Ethyl chloride	965	Styrene	915
Ethylene dichloride	775	Sulfur	450
Ethylene glycol	775	Toluene	1026
Ethylene oxide	804	Turpentine	488
Furfural	739	Vinyl acetate	800
Furfural alcohol	915	Xylene	924
Glycerin	739		

Source: Ref. 17-2.

continuous; continuous gas generation often lends itself to heat recovery. Incineration by flaring is simply controlled discharge into the atmosphere. Heat recovery, almost by definition, is not possible with a flare.

There are two types of flares, ground-level and elevated, or tower flares, in current use. Ground flares can be used where there is sufficient space around the flare to provide for safety of personnel and equipment. The tower flare is used to keep the flame above the level of surrounding equipment and personnel and to promote dilution of its products of combustion into the air. Radiation from flares should be such that surrounding equipment will not receive more than 3000 Btu/($ft^2 \cdot h$) and that personnel will not receive more than 440 Btu/($ft^2 \cdot h$) continuously, 1500 Btu/($ft^2 \cdot h$) for short-term exposure. Temperatures developed in flare systems normally range from 2000 to 2500°F.

As shown in Fig. 17.1, a flare is basically a stack, or open pipe, discharging a combustible gaseous waste directly to the atmosphere with the end of the stack containing a pilot flame (continuously firing), a source of steam, and an exit nozzle. Combustion air is provided by the surrounding atmosphere.

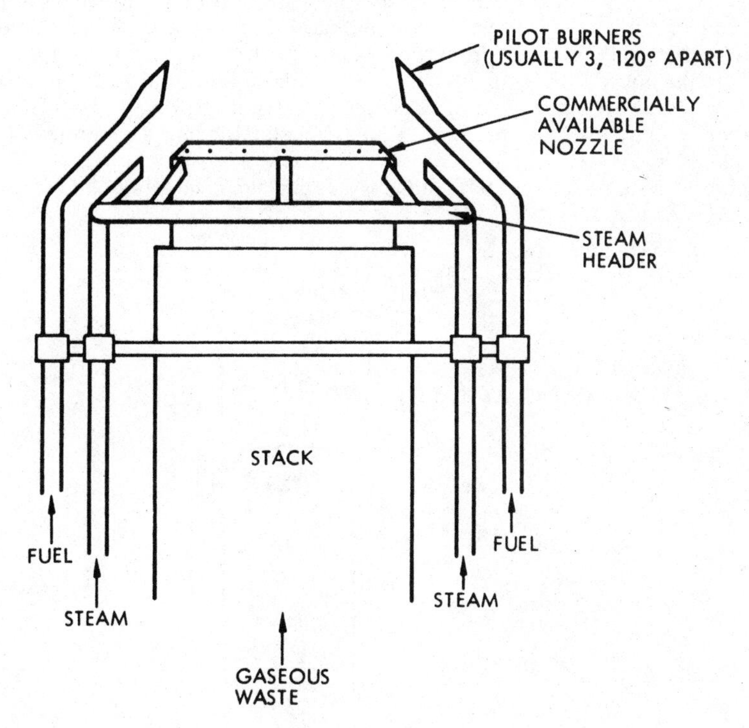

FIGURE 17.1 Stack flare equipped with mixing nozzle. *Source*: Ref. 17-1.

Steam is provided for a number of reasons, as follows:

- To generate turbulence and momentum, promoting good mixing with the surrounding air.

- As a source of heat to help in cracking complex molecules within the gas stream.
- As a reactant in the combustion process to help oxidize carbon to a gaseous state, simplified as

$$C + H_2O(\text{steam}) \rightarrow CO + H_2$$

The CO and H_2 produced will readily combust, burning clean in the presence of air.

Steam flow will generally be in the range of 0.15 to 0.50 pound per pound of hydrocarbon in the gas stream.

Gases with heating values as low as 150 Btu/ft^3 can normally be flared without the addition of supplemental fuel. Below 150 Btu/ft^3 supplementary fuel is normally provided in the form of a gas at the nozzle tip, as shown in Fig. 17.1 as fuel.

Some gases will burn relatively clean and require little or no steam to promote combustion. Such gases include methane, hydrogen, carbon monoxide, and coke oven gas. Most hydrocarbon gases which are normally flared, however, are heavier, i.e., with two or more carbon atoms in their molecular structure. They invariably require a source of steam in order to burn smokeless (compounds such as olefins, aromatics, and the paraffin group above methane).

Figure 17.2 shows a tower flare utilizing internal steam injection, known as the

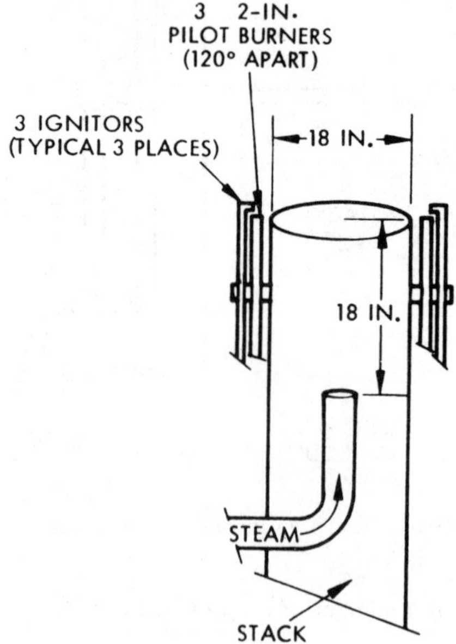

FIGURE 17.2 Esso-type flare. *Source*: Ref. 17-1.

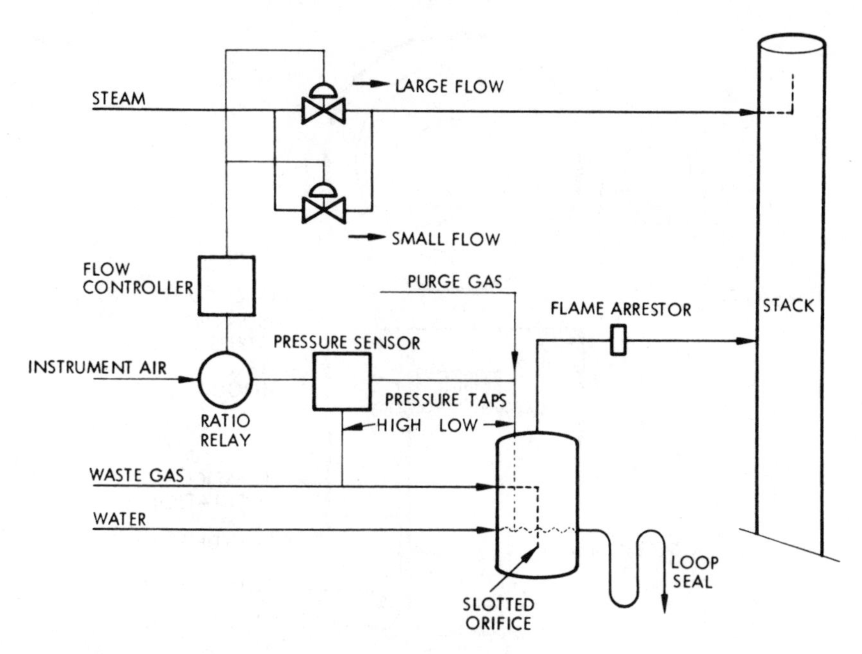

FIGURE 17.3 Waste gas flare system using Esso-type burner. *Source*: Ref. 17-1.

Esso flare. A flare system utilizing the Esso flare is shown in Fig. 17.3. Waste gas is piped to a tank with a water level maintained within it. Liquids and condensable vapors contained within the gas are absorbed in the water which discharges, through a seal, for further processing. An automatic pressure sensor monitors the flow of waste gas to automatically control steam flow to the flare. The waste gas discharges from the tank (commonly referred to as a knockout tank) through a flame arrestor into the stack which feeds the flare. The flame arrestor prevents backflashing of flame from the flare to the upstream equipment. A purge gas, such as nitrogen or argon, is usually provided to help clean the liquid systems during periods when the flare is not in use.

Another type of tower flare is the Sinclair flare shown in Fig. 17.4. Steam is provided through a steam ring with numerous openings, ⅛ in in diameter, for discharging steam into and around the exiting gas stream. These openings are positioned to provide tangential discharge of steam, promoting high turbulence and air inspiration and mixing. A steel shroud, covered in plastic, reduces the noise and radiant heat discharged to the sides and beneath the flare. Another flare, shown in Fig. 17.5, has a steam injector which may or may not be utilized and is provided with insulation for noise reduction.

Figure 17.6 shows a typical ground flare utilizing venturi burner nozzles. Table 17.3 lists capacities of venturi burners as a function of waste gas pressure and burner orifice size. Gas flow properties of natural gas were used to determine the listed figures. Figure 17.7 shows a ground flare which has two sets of burners, one for low and the other for high flow rates of the waste gas. Note the acoustical fence, provided to help attenuate the noise generated by the operating flare.

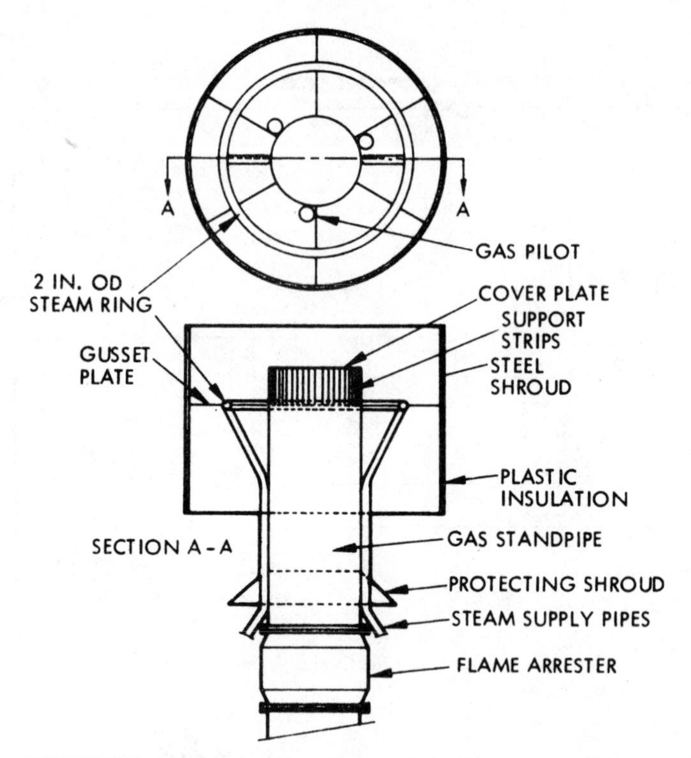

FIGURE 17.4 Sinclair-type flare. *Source*: Ref. 17-1.

Water is sometimes used in lieu of steam, as shown in Fig. 17.8, to reduce smoking. It is less costly than steam supply and injection; however, it is not as effective as steam. It is used with waste gas at low flow rates and where some smoking can be tolerated.

The innermost stack is required for control of mixing of the water, waste gas, and supplemental fuel. The intermediate stack is used to confine the water to aid in mixing with the waste gas. The outer stack confines the flame, directing it upward. Table 17.4 lists water flow requirements for a typical ground flare.

DIRECT FLAME INCINERATION

Direct flame incinerators, also referred to as *fume incinerators* and *gas combustors*, are chambers provided with supplemental fuel burners, which provide heat and retention time to destruct gaseous waste materials. Figure 17.9 is a schematic diagram of a direct flame incinerator. A thermocouple in the combustion chamber measures temperature. Appropriate control circuitry alters the rate of supple-

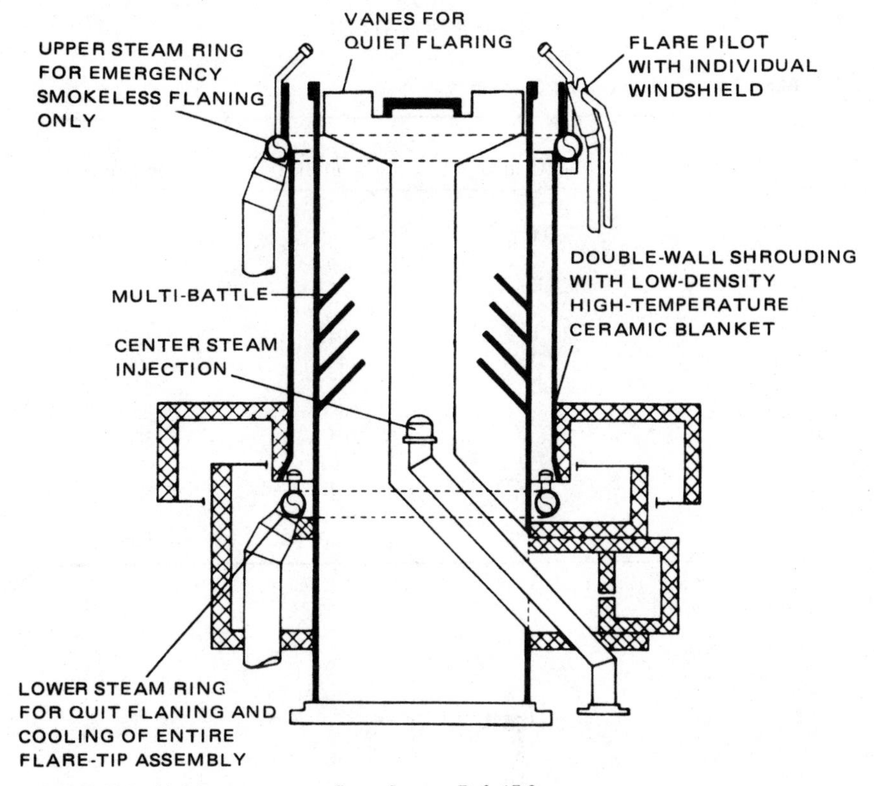

FIGURE 17.5 Multipurpose tower flare. *Source*: Ref. 17-3.

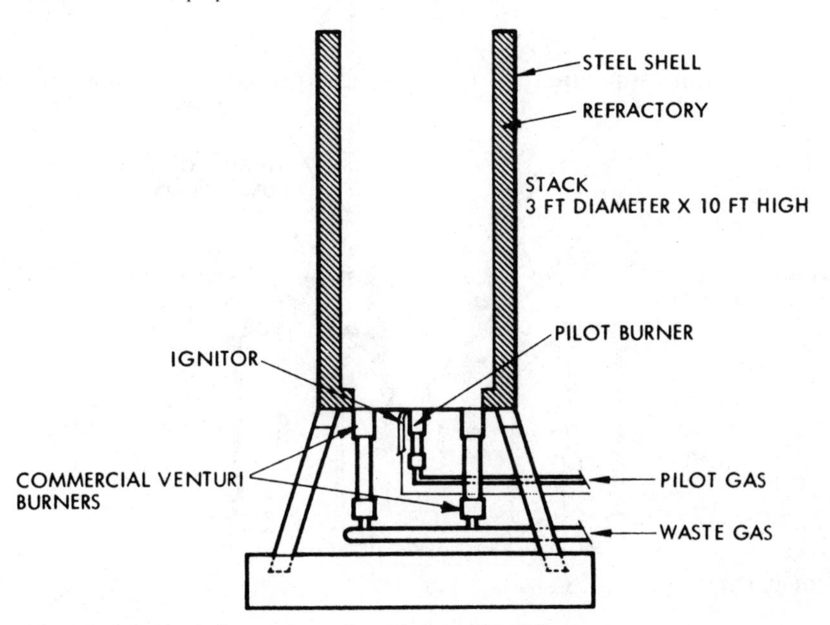

FIGURE 17.6 Vertical venturi-type flare. *Source*: Ref. 17-1.

TABLE 17.3 Venturi Burner Capacities, ft^3/h

Gas pressure, in H_2O	3/16-in Orifice	7/16-in Orifice	1/2-in Orifice
2	70		
4	100		
6	123		
8	142		
10	160		
1/2 lb/in^2 gauge	210	1042	1360
1 lb/in^2 gauge	273	1488	1900
2 lb/in^2 gauge	385	2157	2640
3 lb/in^2 gauge		2654	3200
4 lb/in^2 gauge		3065	3680
5 lb/in^2 gauge		3407	4080
6 lb/in^2 gauge		3742	4480
7 lb/in^2 gauge		4040	4800
8 lb/in^2 gauge		4320	5160

Basis: 1000 Btu/ft^3 natural gas.
Source: Ref. 17-1.

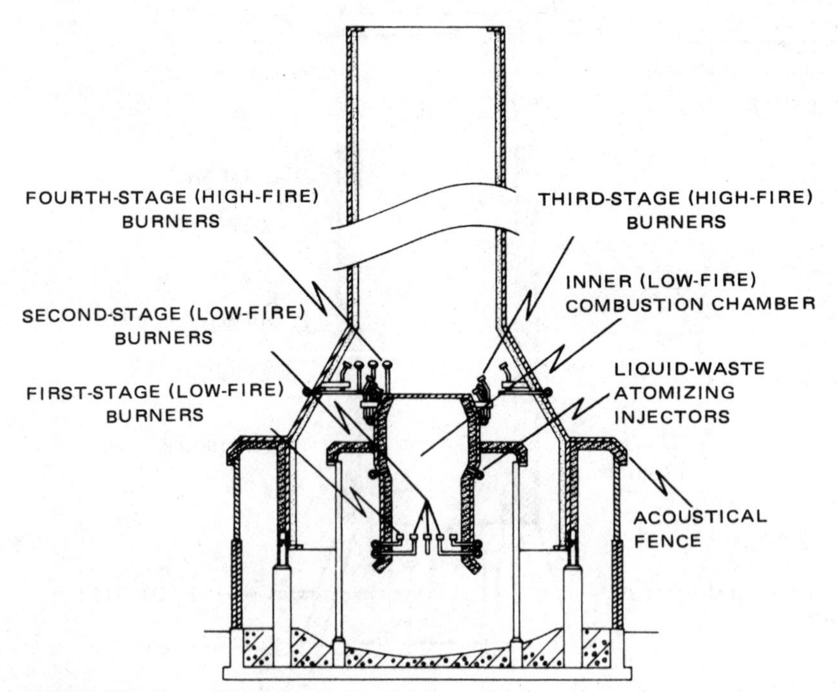

FIGURE 17.7 Ground flare. *Source*: Ref. 17-3.

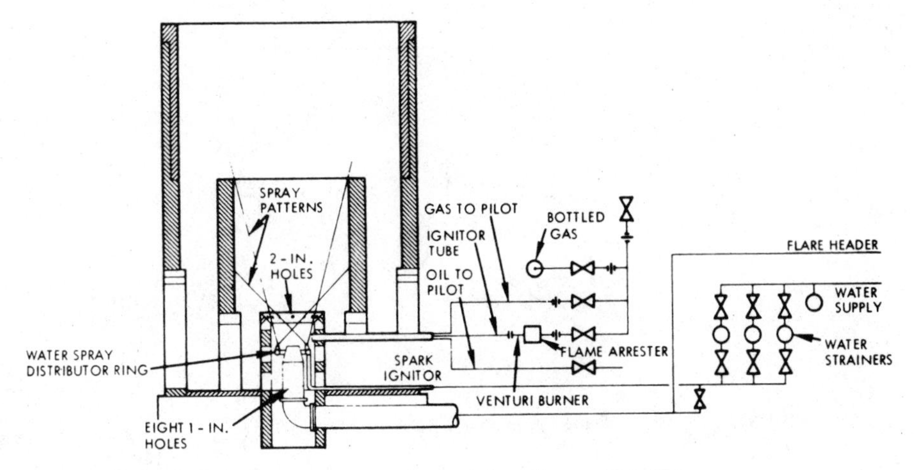

FIGURE 17.8 Typical water spray type of ground flare. *Source*: Ref. 17-1.

mentary fuel entering the furnace to maintain the desired combustion chamber temperature. These incinerators are applicable for most gaseous waste. Their primary use may be for odor control, toxicity elimination, or visible emissions control.

Combustion chamber temperatures are in excess of the autoignition temperature (see Table 17.2) and normally vary, depending on the waste constituents, from 800 to 1500°F. Table 17.5 lists the efficiency of destruction for gaseous waste composed essentially of hydrocarbon compounds.

Retention time is as significant a parameter as temperature. These incinerators are normally designed for combustion chamber sizing to provide 0.25 to 0.50 s retention time, although units have been designed large enough to provide a retention time of 2 to 3 s.

TABLE 17.4 Water Spray Pressures Required for Smokeless Burning[a]

Gas rate, st ft^3/h	Unsaturates, % by volume	Molecular weight	Water pressure, lb/in^2 gauge	Water rate, gal/min
200,000	0–20	28	30–40	31–35
150,000	30	33	80	45
125,000	40	37	120	51

[a]The data in this table were obtained with a 1½-in-diameter spray nozzle in a ground flare with the following dimensions:

	Height, ft	Diameter, ft
Outer stack	30	14
Intermediate stack	12	6
Inner stack	4	12.5

Source: Ref. 17-1.

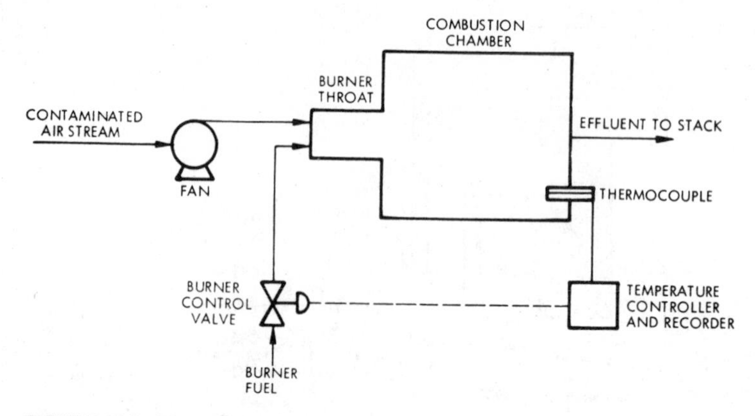

FIGURE 17.9 Direct flame thermal incinerator. *Source*: Ref. 17-1.

The simplicity of automatic direct flame combustion makes it ideal for combustion control. The configuration of this equipment lends itself to heat recovery. Two modes of heat recovery are shown in Fig. 17.10. In one case a heat exchanger utilizes the high temperatures in the combustor exhaust to preheat the incoming combustion air. The second case shows a heat exchanger heating a stream for external use. The stream can be gas, water, or water to steam.

Note that the energy requirement of this or any other heat-generating equipment is a function of the temperature to which the products of combustion must be raised. Burning at 1400°F in the combustion chamber, without heat recovery, the exiting stream will be at 1400°F. All the products of combustion must be brought to this temperature. If a heat exchanger were installed to cool the gas outlet temperature, the temperature within the stack, to 1000°F, the products of combustion would only have to be brought to 1000°F although the combustion

TABLE 17.5 Direct Flame Combustor Efficiency

	Hydrocarbon oxidation	Carbon monoxide oxidation	Odor[a] destruction
Range of temp., °F	1000–1250	1250–1350	1000–1200
Average temp., °F	1100–1200	1300–1350	1100–1150
Efficiency, %	75–85	75–90	50–99
Range of temp., °F	1000–1300	1300–1450	1100–1300
Average temp., °F	1150–1250	1400–1450	1200–1250
Efficiency, %	85–90	90–99	90–99
Range of temp., °F	1100–1500	–	1200–1500
Average temp., °F	1200–1400	–	1350–1400
Efficiency, %	90–100	–	99+

[a]For odor generated from hydrocarbon compounds.

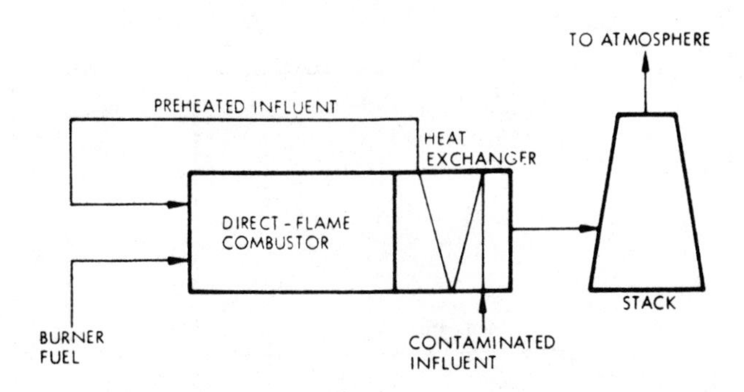

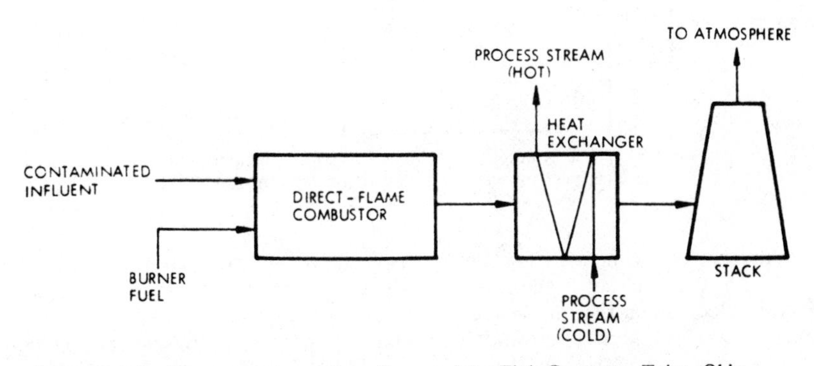

FIGURE 17.10 Heat recovery options. *Source*: John Zink Company, Tulsa, Okla.

chamber would still be maintained at 1400°F. A rough calculation of efficiency, based on absolute temperature, is as follows:

With heat exchanger: 1000°F + 460°F = 1460°R outlet

Without heat exchanger: 1400°F + 460°F = 1860°R outlet

Fuel savings with heat exchanger: $\dfrac{1860 - 1460}{1860} = 22\%$

This figure is a measure of the efficiency of the heat exchanger. It can also be used for cost-effectiveness calculations. For instance, if natural gas at \$6.00 per million Btu were burned without a heat exchanger, and this incinerator were in operation for 2000 h/year burning natural gas at an average rate of 20,000 ft³/h, then 1-year savings with a heat exchanger would be

$$20{,}000 \text{ ft}^3/\text{h} \times 2000 \text{ h/year} \times 1000 \text{ Btu/ft}^3$$
$$\times \$6.00/1{,}000{,}000 \text{ Btu} \times 0.22 \text{ efficiency} = \$52{,}800 \text{ per year}$$

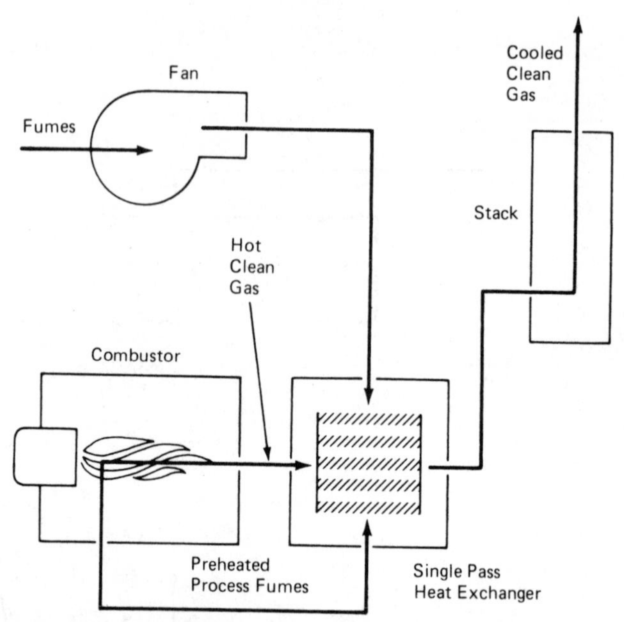

FIGURE 17.11 Forced-draft direct flame fume incineration system, with a single-pass primary heat exchanger. *Source*: Ref. 17-4.

An annual savings of over $50,000 is a significant cost savings. Often more than one heat exchanger section is installed, to further increase the savings realized by decreasing the purchased fuel requirement.

Figure 17.11 shows a fume incinerator with a single-pass heat exchanger which recovers heat by increasing the temperature of the combustion air entering the combustor. A dual heat exchanger system is shown in Fig. 17.12. One heat exchanger is heating the fresh air to provide heat for process equipment while the

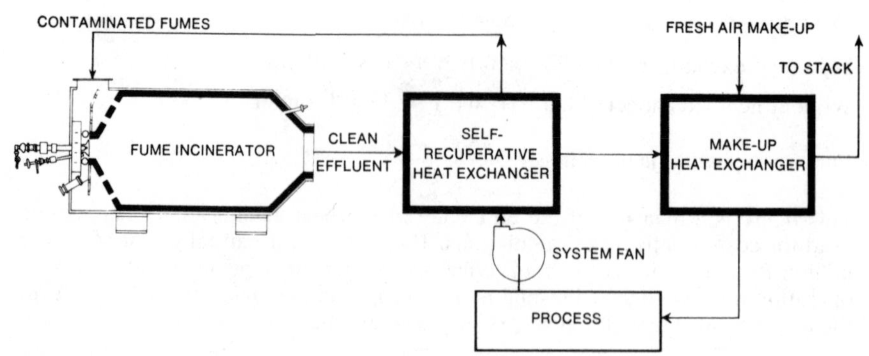

FIGURE 17.12 Dual heat exchanger system. *Source*: Peabody International Corporation, Stamford, Conn.

second heat exchanger is used to heat the combustion air entering the combustion chamber.

The aforementioned fume incineration equipment involved horizontal units where the chamber is horizontal and burners fire along a horizontal axis. Figure 17.13 shows another type of direct flame combustor, a vertical unit, where waste gas, fuel, and combustion air are introduced at the bottom of the incinerator. The chamber immediately above the burner is designed for the required burnout or retention time for the particular waste being incinerated. Vertical units usually are not suited for heat recovery.

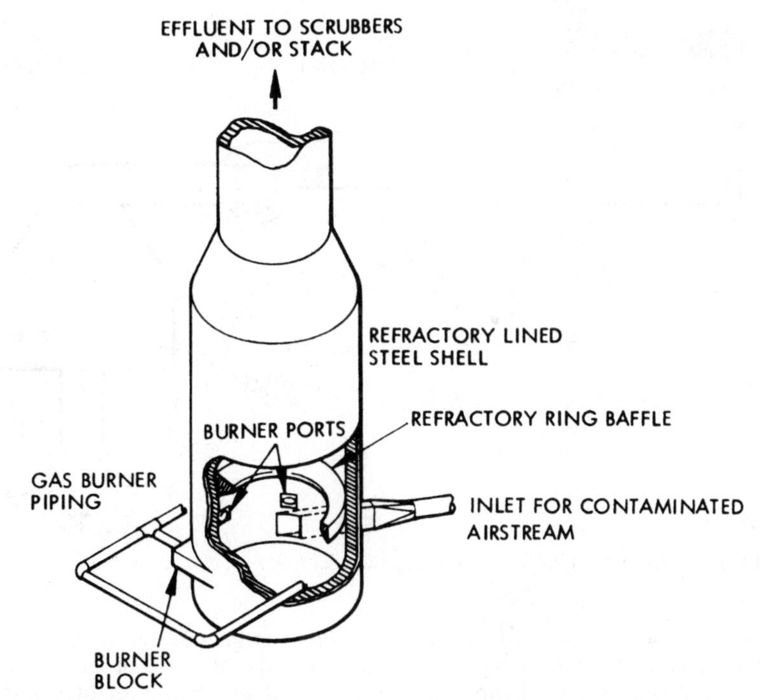

FIGURE 17.13 Vertical incinerator. *Source*: Peabody International Corporation, Stamford, Conn.

Normally the incinerator is designed for complete destruction of organic components by incineration, and particulate matter discharges are almost nonexistent. Where other components are present in the gas, however, such as sulfur or halogens, scrubbers will usually be required.

CATALYTIC INCINERATION

Catalytic incinerators normally destruct gaseous waste at low concentrations, less than 25 percent of the LEL. As shown in Fig. 17.14, heated gas passes

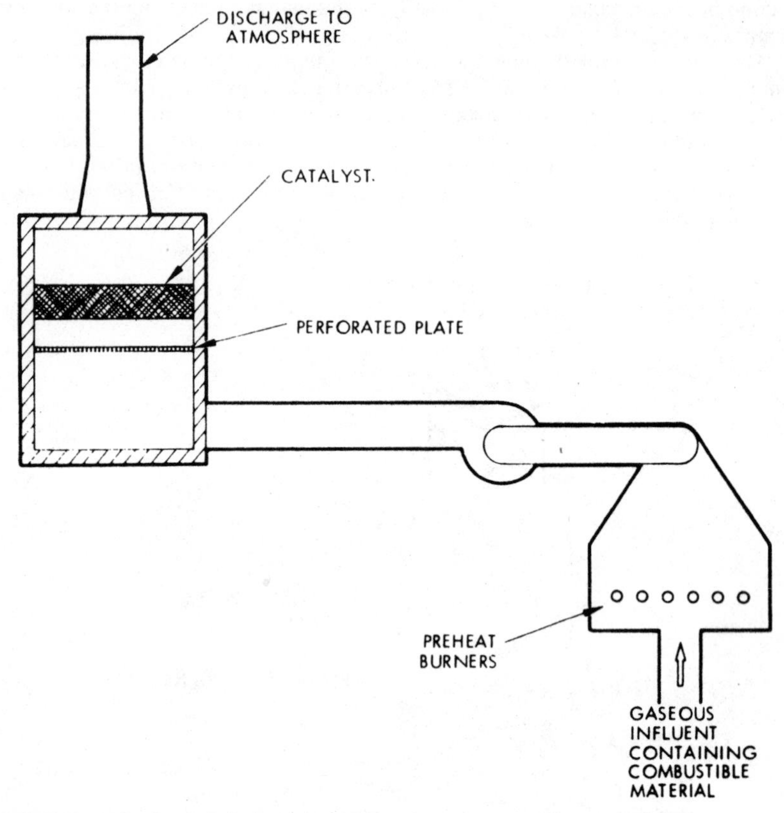

FIGURE 17.14 Catalytic incineration without heat recovery. *Source*: Ref. 17-1.

through a perforated plate to straighten the flow and then passes through a catalytic material prior to discharge. The catalyst has the property of increasing the rate of oxidation at lower temperatures (i.e., use of a catalyst promotes destruction of gaseous waste at lower temperatures).

The actual steps in the catalytic reaction are as follows:

- Diffusion of the reactants within the gas stream, plus the gas stream, through the stagnant fluid surrounding the surface of the catalyst
- Adsorption of the reactants on the catalyst surface
- Reaction of the reactants to form products (usually oxides)
- Desorption of the products from the catalytic surface
- Diffusion of the products from the catalyst pores and surface film to the vapor or gaseous phase outside (downstream) of the catalyst

Catalyst materials normally used are the noble metals, i.e., platinum, palladium, rhodium, etc. Other materials which function as catalyst are copper chromite and the oxides of copper, chromium, manganese, nickel, and cobalt.

Figure 17.15 illustrates types of catalyst sections commonly in use. The mat type is similar in appearance to an air filter. It consists of ribbons of nichrome or stainless-steel wire to which the catalytic material has been applied. Normally an active metal, such as platinum, is applied to this type of carrier.

The porcelain assembly pictured consists of two end plates which are secured by a center post and a number of tear-shaped rods to which the catalyst material is applied. The porcelain is normally coated with aluminum which, in turn, is coated with the catalyst material. Active oxides such as copper oxide or active metals are normally utilized in this configuration.

The third configuration is the honeycomb type, constructed of ceramic or other refractory materials. These can utilize most catalyst materials.

Noting their construction, the catalyst bank offers very little resistance to gas flow, often only a fraction of an inch WC.

Catalyst materials are also available in pellet form, stacked in a chamber within a furnace. When stacked, pellets offer relatively large resistance to gas flow.

Catalytic incineration systems must effect intimate mixing of the combustibles within the system. The gas must be brought to the required ignition temperature (see Table 17.6) for the combustible to be burned, and good temperature control

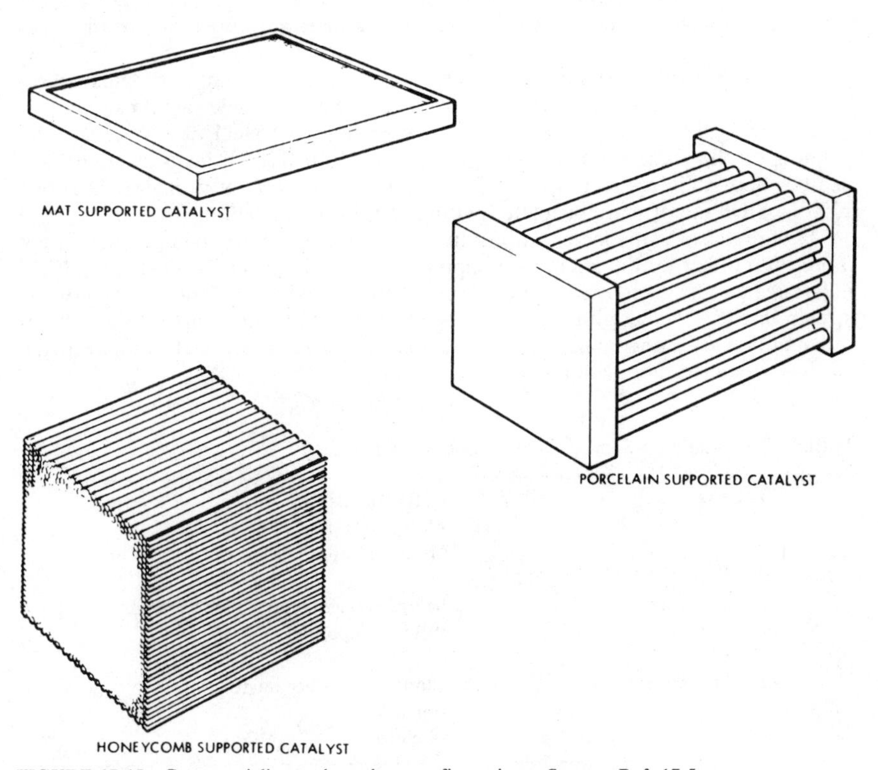

FIGURE 17.15 Commercially used catalyst configurations. *Source*: Ref. 17-5.

TABLE 17.6 Comparison of Temperatures Required for 90 Percent Conversion of Combustibles to CO_2 and H_2O

Combustibles	Ignition Temperature, °F Thermal	Catalytic	Difference, °F	Catalyst/ Thermal, %
Benzene	1,076	575	501	67
Toluene	1,026	575	451	70
Xylene	925	575	163	75
Ethanol	738	575	163	86
Methane	1,170	932	238	85
Carbon monoxide	1,128	500	628	60
Hydrogen	1,065	250	815	47
Propane	898	500	398	71

throughout the catalyst bed is essential. Sufficient oxygen must be present in the gas stream or must be added to it to ensure oxidation of the contaminants.

The gas stream must be free of particulate matter to protect the catalyst from fouling. If particulate matter is present, pretreatment of the gas, such as cyclonic separation or electrostatic precipitators, may be necessary upstream of the catalyst.

Besides particulate fouling, catalysts are sensitive to a number of substances. Table 17.7 lists contaminant components with regard to noble metal catalysts.

The catalytic incinerator chamber is constructed of steel or refractory, depending on the temperatures developed. From 750 to 1100°F heat-resistant steel can be used, stainless steel from 1100 to 1300°F, and refractory materials above 1300°F. Steel is normally protected with 4 to 6 in of insulation.

As with direct flame incineration, the cost of heat exchange equipment is often more than offset by the savings in supplemental fuel consumption. Figure 17.16 illustrates two heat-recovery options, one utilizing a portion of the exiting hot gas stream directly mixed with the incoming stream to raise its temperature and another utilizing a noncontact heat exchanger where heat is transferred from the exiting stream to the incoming air or gas flow.

TABLE 17.7 Platinum-Family Catalyst Contaminants

Poisons	heavy metals phosphates arsenic compounds
Suppressants	halogens (elemental and compounds) sulfur compounds
Fouling Agents	alumina and silica dusts iron oxides silicones

Source: Ref. 17-1.

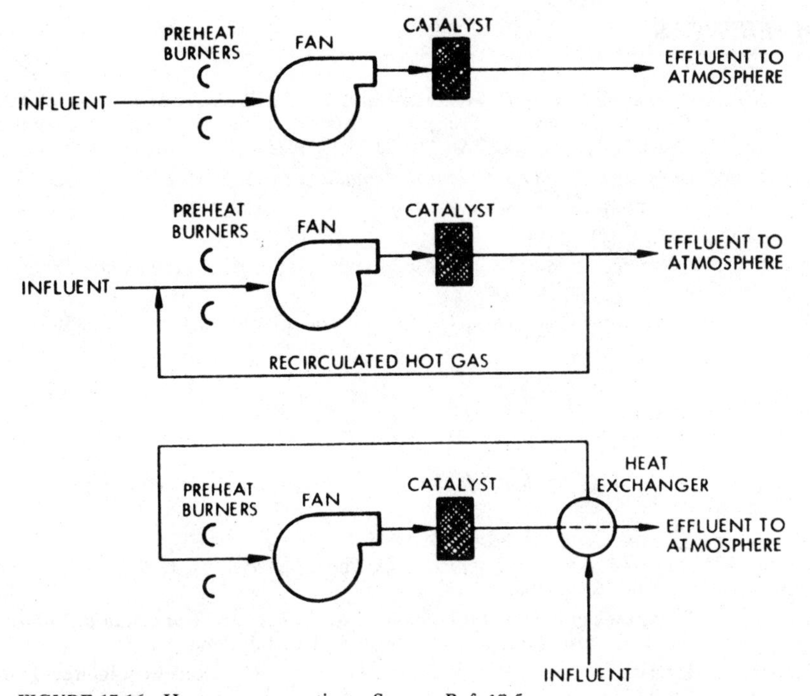

FIGURE 17.16 Heat-recovery options. *Source*: Ref. 17-5.

Figure 17.17 is a schematic of a catalytic incinerator with a self-contained heat exchanger. The incoming flow is blown through the outside of a baffled heat exchanger exiting as heated combustion air or gas. After it is brought to oxidation temperature, the gas passes through the catalyst bank and then through the internal section of the heat exchanger. It will give up heat, or temperature, to the incoming flow in proportion to the size of the heat exchanger.

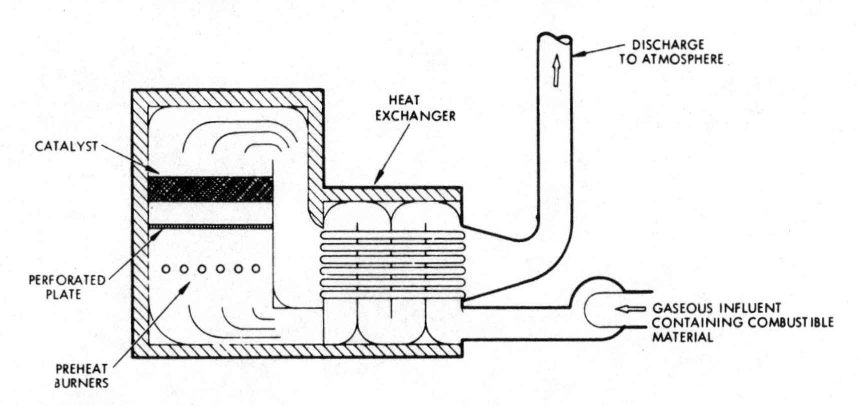

FIGURE 17.17 Catalytic incineration with heat recovery. *Source*: Ref. 17-1.

REFERENCES

17-1. *Recommended Methods of Reduction, Neutralization, Recovery or Disposal of Hazardous Waste*, vol. 3, Disposal Process Descriptions, Ultimate Disposal, Incineration and Pyrolysis Processes, USEPA 670/2-73-053C, Wash., D.C., August 1973.

17-2. Lund, H.: *Industrial Pollution Control Handbook*, McGraw-Hill, New York, 1971.

17-3. Straitz, J.: "Flaring with Maximum Energy Conservation," *Pollution Engineering*, Chicago, Ill., February 1980.

17-4. Ross, R.: "Incineration of Solvent-Air Mixtures," *Chemical Engineering Progress*, New York, N.Y., August 1972.

17-5. Ross, R.: *Industrial Waste Disposal*, Van Nostrand Reinhold, New York, 1968.

BIBLIOGRAPHY

Cheremisinoff, P., and Young, R.: *Pollution Engineering Practice*, 1st ed., Scientific Publishers, Ann Arbor, Mich.

Cross, F.: *Air Pollution Odor Control Primer*, Technomic, Lancaster, Pa., 1973.

Energy Savings of Heat Recovery Equipment for Fume Incinerators, Energy Management Bureau, Chicago, Ill., September 25, 1980.

Meyers, F.: "Fume Incineration with Combustion Air at Elevated Temperature," *Journal of the Air Pollution Control Association*, Pittsburgh, Pa., July 1966.

Pauletta, C., Hazzard, N., and Benfordo, D.: "Economics of Heat Recovery Indirect Fume Incineration," *Air Engineering*, Pittsburgh, Pa., March 1967.

CHAPTER 18
INCINERATION OF RADIOACTIVE WASTE

As the quantity of waste from nuclear power plants, medical applications, civilian and military research, etc., increases and public concern heightens, incineration is becoming a reasonable answer to radioactive waste control. Radioactive waste can be substantially reduced in volume and weight by the incineration techniques presented in this chapter.

RADIOACTIVE WASTE

The vast majority of radioactive waste generated is low-level radioactive waste (LLW), which is defined as follows:

> Low-level waste comprises that radioactive waste which is not spent fuel or high-level waste and which contains less than 10 nanocuries of transuranics per gram of material.

Transuranic (TRU) material refers to uranium or materials, natural or created, with atomic weight equal to or greater than that of uranium.

The International Atomic Energy Agency has defined waste categories, from 1 to 5, based on state (solid, liquid, or gas), the concentration of radioactivity, the shielding required, and the complexity of the treatment method necessary. These categories have not found acceptance in the United States.

Generation rates and current accumulation (mid-1980) of LLW are listed in Table 18.1. Of particular interest is that almost 20 years' worth of material, based on current generation, is accumulated and must eventually be disposed of. Ultimate disposal is currently burial, and increasing public concern over burial of radioactive waste and its transportation to burial sites has severely limited disposal options. In lieu of disposal, therefore, much of this waste has been accumulated and will be accumulated until viable, reliable methods of ultimate disposal can be found, with minimal impact on public health and safety.

With wastes eventually going to landfill, volume reduction becomes of prime importance. Incineration is an effective means of obtaining volume reduction, as noted in Table 18.2.

TABLE 18.1 Generation of Radioactive Waste

	Total Accumulation, 1980, 10^6 ft^3	Annual Generation, 1980, 10^6 ft^3
Government	50.8	1.0
Commercial	15.8	>2.0
Fuel cycle		1.2
reactor operation		1.0
fuel fabrication		0.2
other steps		small
Non-fuel cycle		0.78
industrial		0.67
medical		0.64
academic		0.026
Industrial and other		0.11
Research (including pharmaceutical)		
Total	66.6	3.6

Source: Ref. 18-1.

TABLE 18.2 Development Status of Incinerators Used in the United States to Reduce the Volume of Wastes, 1980

	Development status	
Incinerator type	Nonradioactive wastes[a]	Radioactive wastes
Acid digestion	not applied	pilot plant
Agitated hearth	not applied	under construction[b]
Controlled air	commercial unit	demonstration unit
Cyclone drum	not applied	demonstration unit
Fluidized bed	commercial unit	pilot plant
Microwave/gas plasma	laboratory unit	laboratory unit
Molten glass (Joule heating)	commercial unit	test unit
Molten salt	commercial unit	pilot plant
Moving grate	commercial unit	not applied
Multiple hearth	commercial unit	not applied
Pyrolysis (controlled air)	commercial unit	test unit
Rotary kiln	commercial unit	under construction[b]
Slagging pyrolysis	commercial unit	pilot plant

[a]Includes wastes generated in noncircular applications.
[b]Full-scale units.
Source: Ref. 18-1.

Not all LLW is incinerable. It can be roughly categorized into incinerable and nonincinerable, dry and wet waste. Dry waste includes such items as paper, rags, plastics, rubber, wood, glass, and metals. Dry waste can further be classified as compactible and noncompactible, combustible or incombustible. Most wet wastes are derived from the cleanup of aqueous processes or waste streams prior to recycle or discharge. Such liquid wastes include ion-exchange resin slurries, filter sludges, evaporator concentrates, miscellaneous oils, and solvents. In general, liquid LLW will be concentrated by settling, centrifuging, reverse osmosis, or other means as a first step in volume reduction.

LLW INCINERATION

As would be expected, radioactive waste incineration requires unique considerations not applicable to other waste streams. Safety is of prime concern. Handling, storage, and feeding of material must be carefully controlled. Tight containers and negative ventilation systems exhausting through high-efficiency particulate air (HEPA) filters are necessary prior to feeding and incineration.

Incinerator feeding must be designed to preclude the possibility of the escape of hot gases (which may contain LLW) from the incinerator from backcharging into the feeding room when the incinerator charging door is open. This normally requires that the charging system be provided with a series of air locks and multiple feeding chambers to isolate the incinerator gas under all conditions of operation.

Emergency exhaust systems, such as explosion doors used in conventional incineration systems, cannot normally be used when burning LLW. The opening of a door or stack prior to the exhaust gas cleaning train would discharge LLW to the surrounding air. This requirement may preclude incineration of certain wastes which may cause explosions or the use of supplemental fuel such as natural gas, which is more prone to explosion than other fuels.

Ash and residue from the gas cleaning system will contain LLW more concentrated than when introduced to the system and must be handled with respect to their toxicity. Shielding is necessary when handling these residues, and rapidity in their disposal is also necessary to reduce the possibility of their unwanted release.

Liquid feeding must be at a relatively constant rate. Significant variation in flow (or quality) of liquid LLW can lead to "puffing" or pressure surges which will seek an unwanted exit from the incinerator, bypassing the gas cleaning train.

Controls for an incinerator burning radioactive wastes must necessarily be more encompassing, more detailed, more automated, and more reliable than those for conventional waste burning incinerators.

Another consideration in radioactive waste incinerator design is reclamation. Many radioactive elements and other sources of radioactivity are extremely expensive. Often their reclamation for reuse is economically feasible, even when the waste has a very low level of radioactivity.

DRUM INCINERATOR

The Mound Laboratory, at Miamisburg, Ohio, has developed a number of types of incineration systems for LLW. One of these, a drum incinerator, is shown in Fig. 18.1. A 55-gal drum is used as the burning chamber. A basket, loaded with waste in small cartons, is ignited manually and is lowered into the drum. A source of air is injected into the drum tangentially to promote turbulence and provide a cyclonic effect to the combustion process. Temperatures within the drum reach 2000 to 2200°F. Blowers create sufficient draft to collect all off-gas from the burning material within the drum.

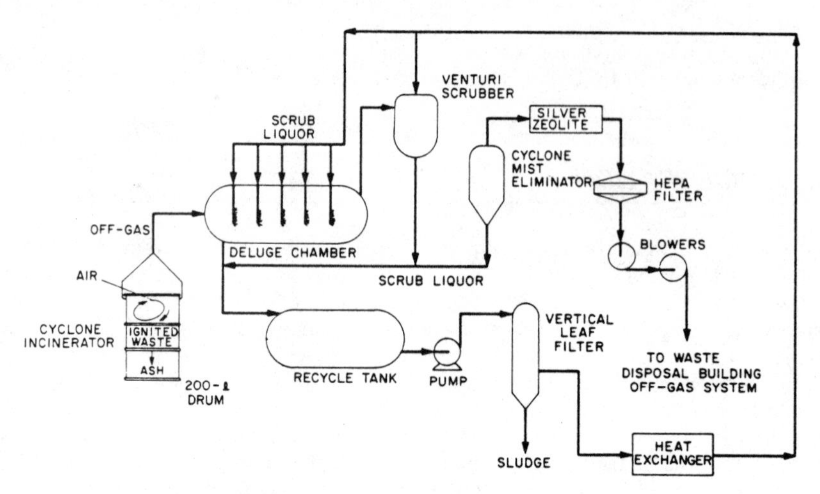

FIGURE 18.1 Mound Facility cyclone incinerator. *Source:* Ref. 18-1.

The off-gas passes through a caustic solution deluge chamber where the gas is cooled and neutralized and most of the particulate matter is removed. The venturi further scrubs the flue gas with caustic, removing the majority of particulate matter remaining in the gas and controlling the tendency for the gas stream to be acidic. An HEPA (high-efficiency particulate air) filter is necessary for essentially complete removal of particulate matter from the gas stream; however, it quickly loses efficiency from moisture within the flowing gas. Therefore it is necessary to dry the flue gas upstream of the HEPA filter. In this case a demister is used to remove entrained moisture particles, and a zeolite contactor removes additional particulate, particularly the caustic residual in the gas.

Liquid (spent caustic water solution) from the deluge tank, the venturi scrubber, and the demister is collected in a recycle tank. The spent liquid, or liquor, passes through a vertical leaf filter which effectively removes the majority of particulate matter in the form of a sludge. A heat exchanger is provided to heat the caustic and maintain it in solution, preventing its collection on pipes and nozzles within the treatment system. The liquor is then ready for recycling.

The initial volume of LLW is significantly reduced in this process. The LLW is concentrated in a sludge and in the HEPA filter, which of course must be disposed of. But the sludge waste and HEPA filter volume is one-tenth the volume of the original waste stream.

STARVED AIR INCINERATOR

A controlled air incinerator for LLW is shown in schematic in Fig. 18.2. Wastes are hand-sorted to combustible and incombustible fractions. The combustible wastes are shredded, placed in plastic bags, and placed in the ram feeding line. Other solid wastes are passed through activity detectors before they are fed to the charging mechanism.

The incinerator operates in the same manner as starved air units described in Chap. 11. An oxygen-deficient atmosphere is provided in the primary chamber, and temperatures are controlled in the range of 1450 to 1750°F. Additional air is

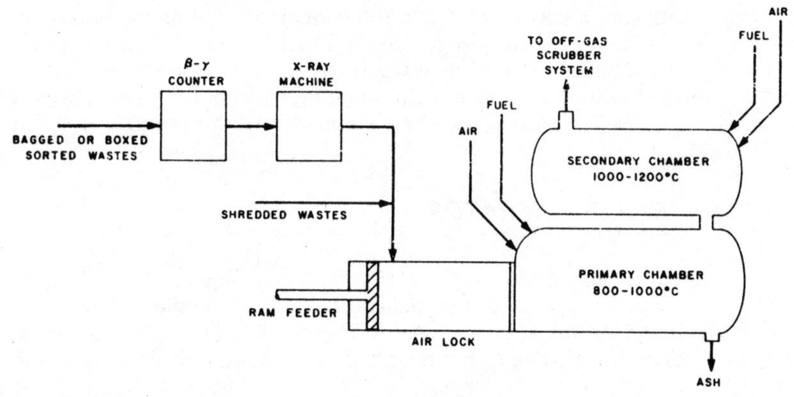

FIGURE 18.2 Los Alamos Scientific Laboratory controlled air incinerator. *Source:* Ref. 18-1.

provided in the secondary chamber to complete combustion, at temperatures from 1700 to 1900°F. Ash burnout in this system is not as complete as with other incineration systems, and the volume reduction therefore is not as high as in other systems. But the air pollution control system is simpler because incinerator emissions are less than in other systems.

AGITATED HEARTH

An agitated-hearth incinerator, similar to the single-hearth incinerator discussed in Chap. 10, has been developed for incineration of LLW. Figure 18.3 illustrates

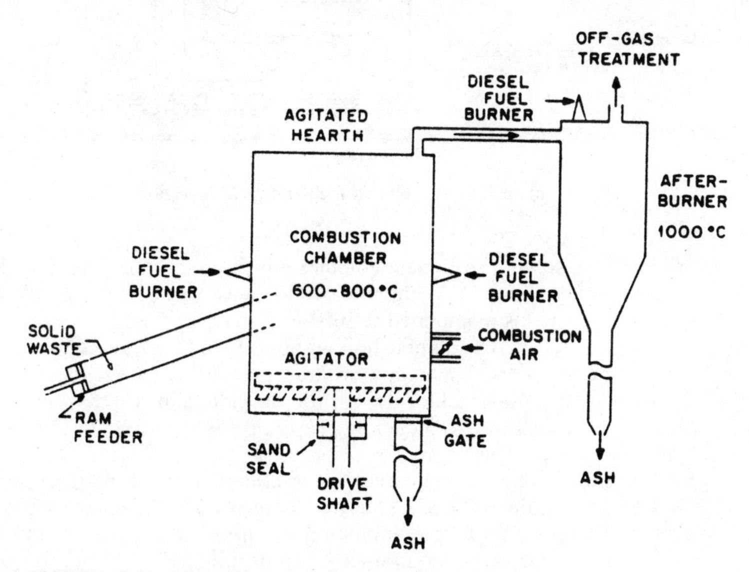

FIGURE 18.3 Agitated-hearth incinerator. *Source:* Ref. 18-1.

this type of unit. Solid waste is fed into the incinerator and drops onto its floor. This incinerator can also burn sludge and liquid LLW.

An agitator rotates slowly, turning the waste material surface to increase its contact with combustion air and to help break up cartons and plastic bags. This incinerator is normally designed as a batch unit with 1 h for burnout plus ½ h for ash removal.

FLUIDIZED BED INCINERATOR

This incinerator can handle solids, sludges, and liquids (see Fig. 18.4). Noncombustibles are separated from the combustible fraction of the waste by an air classifier, after an initial coarse shred. The combustibles (the light fraction) are fine-shredded for feeding to the fluid bed. This fraction is charged to the reactor by means of a screw feeder.

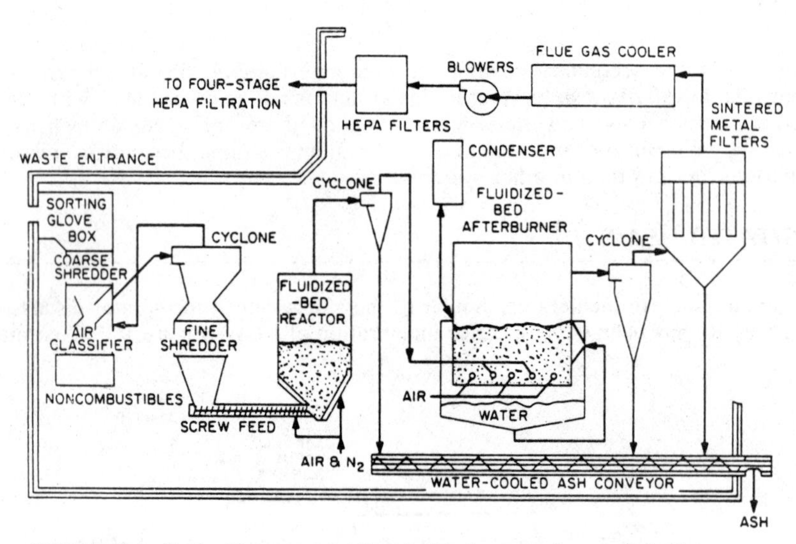

FIGURE 18.4 Rocky Flats plant fluidized bed incinerator. *Source:* Ref. 18-1.

The reactor contains sodium carbonate granules which are fluidized by the injection of compressed air and nitrogen. Within the bed the waste is decomposed by partial combustion, and the bed is maintained at 1050°F. The air-to-nitrogen level of the fluidizing gas is adjusted to permit combustion without open-flame burning.

Chloride components of the waste stream are neutralized within the sodium carbonate bed to produce sodium chloride, carbon dioxide, and water vapor. Off-gas from the reactor passes through a cyclonic separator where most of the larger particulate exits the gas stream.

In the afterburner chamber, combustion air is added to the gas stream as it passes through another fluidized bed. This bed is chromic oxide, an oxidation catalyst which encourages complete combustion of any combustibles within the flue gas. The bed temperature is maintained at approximately 1000°F. A high-pressure water jacket helps control the bed temperature at the desired level.

Flue gas leaving the catalytic afterburner contains fly ash, catalytic dust, and small amounts of sodium carbonate and sodium chloride. This particulate matter is essentially removed by passing the flue gas through another cyclonic separator and a bank of sintered metal filters. The flue gas is then cooled to 120°F in a water-cooled heat exchanger and is then blown through HEPA filters prior to exiting the system. Ash from the cyclones and the sintered metal filters is collected for ultimate disposal.

ELECTROMELTER SYSTEM

Figure 18.5 shows a proposed molten-glass incinerator for a variety of solid and liquid LLW. It uses technology from the glass-making industry, making use of the Joule effect. Use of the electrical conductivity of molten glass at elevated temperatures to maintain the temperature of the melt is known as Joule heating. At start-up a refractory-lined vessel is charged with pieces of glass or glass frit (granular glass). Either a sacrificial wire mounted between two electrodes protruding through the glass layer or a removable electric resistance heater provides the heat necessary for melting the glass. The glass charge heats slowly to the temperature at which the melt becomes electrically conducting. At this point the immersed electrodes provide the electric current necessary for Joule heating, normally at a temperature in the range of 1800 to 2200°F. As material is burned, within the melt, with the injection of combustion air, most of the ash is absorbed into the molten glass, gradually increasing the liquid level. The molten glass and waste mixture is removed through either a gravity drain or an overflow weir into a container where it cools and solidifies.

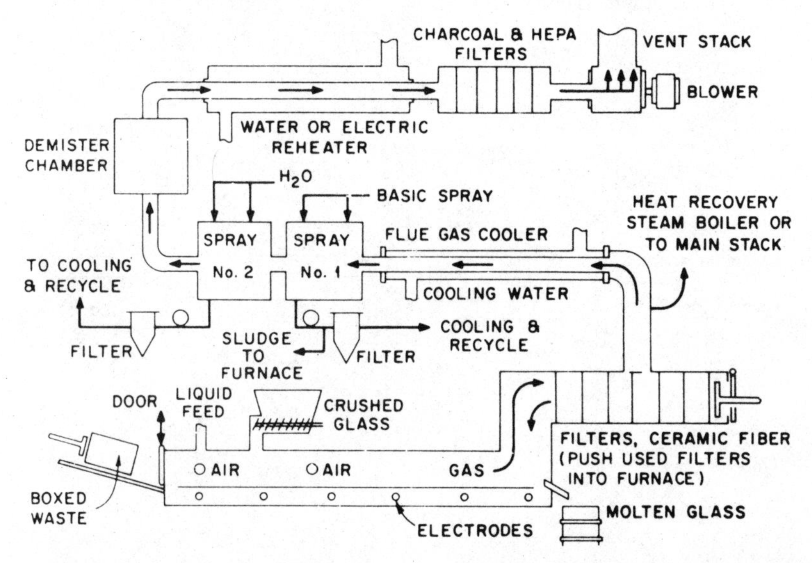

FIGURE 18.5 Sketch of the Penberthy molten glass incinerator (Electromelter) system proposed for treating low-level waste. *Source:* Penberthy Electromelt International, Inc., Seattle, Wash.

Ceramic fiber filters remove the majority of particulate matter from the flue gas stream, and as they become contaminated, they are pushed into the melt where they dissolve into the glass material. The flue gas is cooled in a jacketed-type heat exchanger, and the heated cooling water can be used to generate plant steam or hot water. A series of sprays further cool the flue gas and remove particulate. Alkali can be added to the spray water for acid removal in the flue gas. A demister chamber removes entrained moisture from the flue gas, and the flue gas is heated slightly to reduce its specific humidity in preparation for the HEPA filter sections (they operate best on gas with low humidity). From the HEPA filters the gas stream is discharged to the atmosphere. Spent water from the flue gas coolers and demister is collected, and particulate matter is removed in the form of a sludge. The filter water is reused while the sludge is brought to the furnace, where it is disposed of within the melt.

This incinerator is universal in its application to LLW, producing a sterile product (glass) that is but a fraction of the initial charge. Even the filters used in this process can be disposed of within the melt.

OTHER SYSTEMS

There is no single design or design criterion for LLW. Each incinerator has been designed for a specific waste application and set of circumstances. As noted in Table 18.2, many types of incinerators have applications for LLW. For instance, a rotary kiln LLW incinerator is shown in Fig. 18.6, a molten salt incinerator for LLW is shown in Fig. 18.7, and a pyrolysis unit is shown in Fig. 18.8.

In all these incinerators each discharge is carefully controlled to catch any radioactive component before it can enter the air or water environment.

Table 18.3 lists many LLW facilities in use in the United States, and Table 18.4 lists foreign LLW facilities.

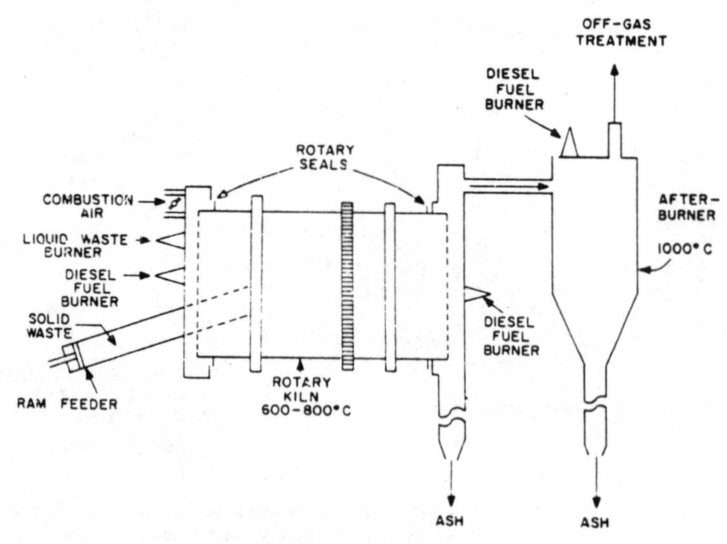

FIGURE 18.6 Rotary kiln incinerator. *Source:* Ref. 18-1.

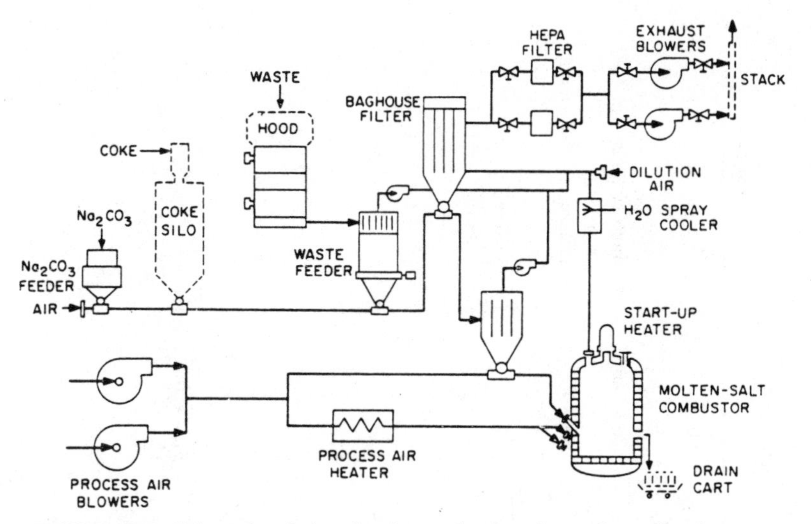

FIGURE 18.7 Schematic of Atomics International molten salt combustion system. *Source:* Rockwell International, Energy Systems Group, Canoga Park, Calif.

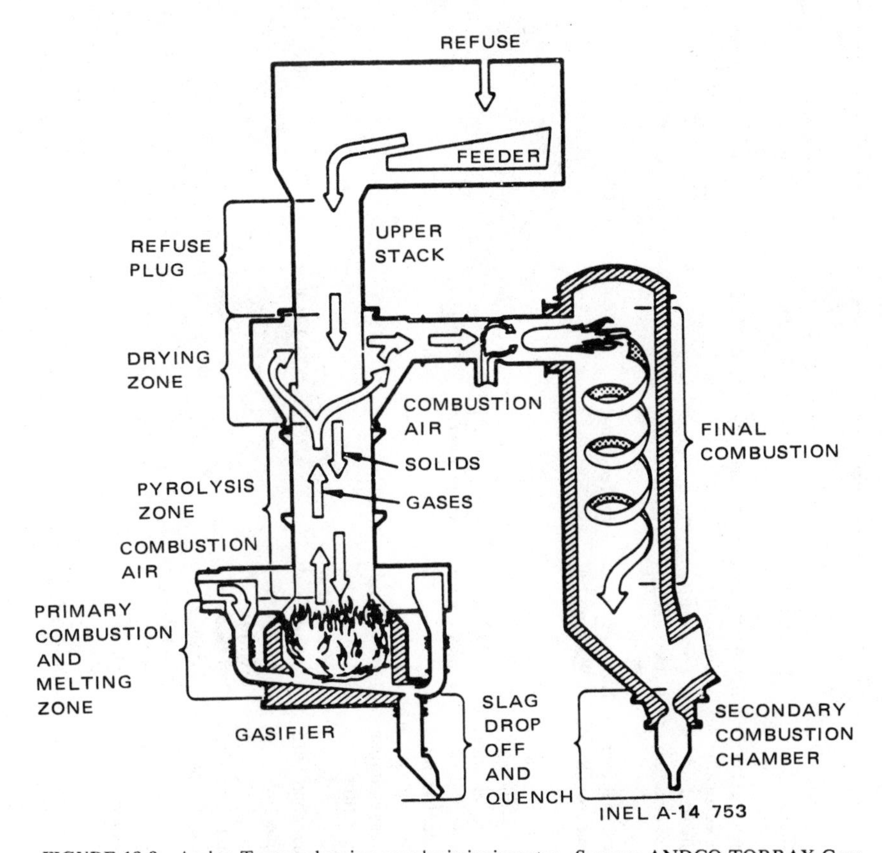

FIGURE 18.8 Andco-Torrax slagging pyrolysis incinerator. *Source:* ANDCO-TORRAX Corp.

TABLE 18.3 U.S. Incinerators Used for Treating Radioactive Wastes

Type of incinerator	Location	Application	Construction date	Pretreatment required	Capacity, kg/h	Operating temperatures, °C	Remarks
Single basket; electrically heated	Los Alamos Scientific Laboratory	Pu recovery	1952		<1	800	
Dual chamber	Rocky Flats Plant	Pu recovery	1959		16	1200–1400	Refractory PuO_2 can form
Dual chamber; electrically heated; horizontal moving belt (woven wire)	Hanford, Washington	Pu recovery	1961	Chopping	2	700–800	Maintenance is different
Single basket; electrically heated	Mound Facility	Pu recovery	~1972		23	800	
Single chamber gas fired	Union Carbide Company (Y-12 Plant)		1955		20	870	
Dual chamber; gas fired	National Lead Company (Ohio)		1954		1000	980	
Dual chamber, gas fired	Gulf General Atomics (California)		1963		20	900–1200	
Dual chamber; gas fired	Kerr-McGee Nuclear (Ohio)		1972		80		

Single chamber	Babcock & Wilcox Company (Virginia)	U recovery	1972		80	1090	
Vortex, gas fired	General Electric Company (North Carolina)	U waste	1972	Shredding to uniform size	450	1000–1100 (815–980)	Feed is blown tangentially into burning chamber
Dual chamber, excess air burner	Westinghouse Nuclear Fuel Division (South Carolina)	Uranium	1974			650–1200	
Dual chamber; gas fired	Nuclear Fuel Services (Tennessee)	Uranium			270		
Dual chamber; gas fired	Goodyear Atomic (Ohio)	U recovery	1971		68	815–1000	
Dual chamber; gas fired	Oak Ridge Gaseous Diffusion Plant	U recovery	1972			930–1100	
Cyclone air feed	Yankee (Rowe) Nuclear Power Plant	Dry low-level waste	1968		18		
Dual chamber; controlled pyrolysis; vertical retort; rotary grate	Battelle Pacific Northwest Laboratory		Designed 1974	Shredding	15	Retort: 100–1200 Afterburner: 800	Not funded in 1976

Source: Ref. 18-1.

TABLE 18.4 Some Incinerators Used outside the United States for Treatment of Radioactive Wastes

Type of incinerator	Location	Application	Construction date	Pretreatment required	Capacity, kg/h	Operating temperature, °C	Remarks
Continuous slagging pyrolysis; movable paddles	Belgium	Solid waste	1975	Sorting and mixing	100	1400–1600	Off-gas passes through sand filters before HEPA
Dual chamber, batch pyrolysis; movable grate	W. Germany (Juelich)	Solid waste	1977	Hand sorting	100	1000	
Vertical retort; afterburning in hot ceramic candle filters; initial gas blast–300 to 850°C	W. Germany (Karleruhe)	Dry low-level waste	1970	Sorting	60	1000–1100	
	Switzerland (EIR-Wuerenlingen	Dry low-level waste		Sorting	25–30	1000	
	Japan (Tauruga)	Power plant waste	1977	Sorting	50	1000	
	(Tokai)		1979	Sorting	100	1000	
	Austria (SEAE-Sieheradorf)	Dry low-level waste	1975 (Controlled)	Sorting	60	1000	Hand loaded; no PVC permitted in feed
Vertical retort; dual chamber; batch operation; oil fired	Sweden (Studuvik)		1976	Sorting	200–400		
Vertical two-zone furnace; continuous operation	France (Cadarache)	Solid and liquid waste	1980	Crushing	30 (solid) 10–15 l/hr (liquid)		

Batch pyrolysis (TRECAN)	Canada (Bruce Nuclear Power Develop)	Low-level solid waste and organic liquids	1971	Sorting	2270 kg/batch (solid) ~45 l/hr (liquid)	870–900 (afterburning)	Bag filter in off-gas stream
	(Chalk River)		Under construction	Sorting	1135 kg/batch		
Dual chamber; controlled air	United Kingdom (Windscale)	Plutonium	1972	Sorting	3–5	900	Wet scrubber system
Continuous dual chamber, electrically heated	France (Marcoule)	Plutonium	1970	Cutting			
Batch dual chamber	France (Marcoule)	Solid waste	Before 1970	Hand sorting	80		Off-gas passes through rotating filter
	(CEM-FAR)	Animal carcasses only			50	900	Bag filter on off-gas stream
Horizontal continuous furnace	France (Gadarache)	Liquids, organochlorides, and organo phosphates	1980		50 l/hr		
Horizontal batch furnace; stationary grate; excess air (Wellman)	United Kingdom (Winkley Point)	Low-level power plant waste	1977	Sorting and bagging	70	900	Off-gas system may need modification
Acid digestion	United Kingdom (Maxwell)	Pu recovery	(1981)	Shredding	10		

Source: Ref. 18-1.

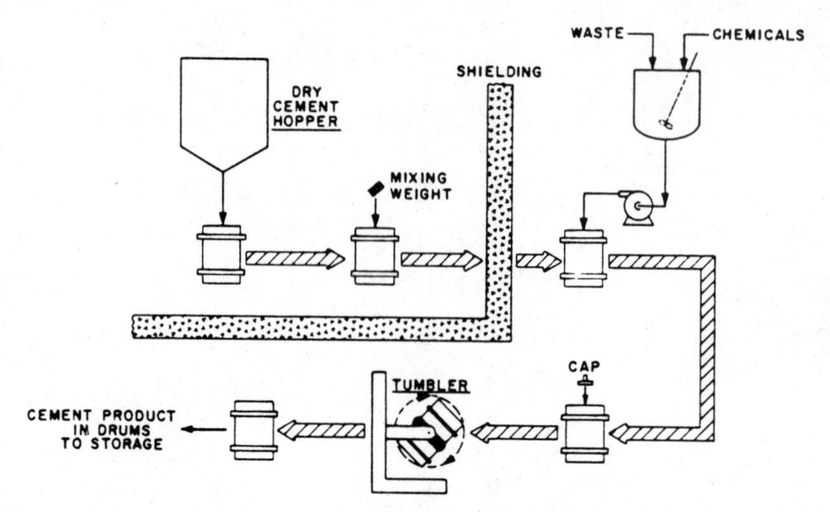

FIGURE 18.9 An in-drum method for mixing radioactive waste with cement. *Source:* Ref. 18-2.

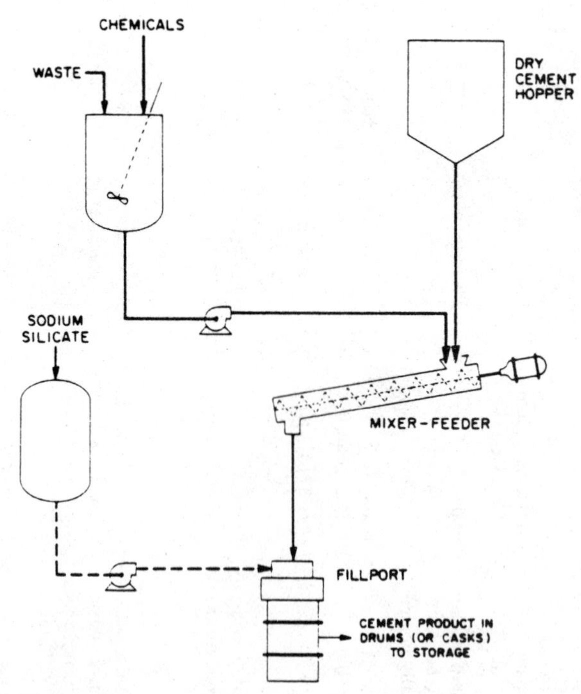

FIGURE 18.10 An in-line method for mixing radioactive cement with waste. *Source:* Ref. 18-3.

Equipment for ultimate disposal, i.e., the mixing of LLW with cement, is shown in Figs. 18.9 and 18.10. Figure 18.8 illustrates an in-drum mixing method where cement and waste are mixed within a container whereas in Fig. 18.10 the waste and cement are mixed external to the drum. Sodium silicate and other chemicals are often used to aid in solidification of liquids and sludges but will normally not be required with ash or other solid wastes.

REFERENCES

18-1. Kibbey, A.: *A State of Art Report on Low Level Radioactive Waste Treatment*, Oak Ridge National Laboratory, September 1980.

18-2. Oyen, L., and Kibbey, A.: "Volume Reduction, Solidification and Packaging of Wastes from Nuclear Power Plants," *Nuclear Power Waste Technology*, chap. 7, American Society of Mechanical Engineers Monograph G00132, New York, 1978.

18-3. Filter, H., and Roberson, K.: "Solidification of Low Level Radioactive Wastes from Nuclear Power Plants," *Management of Low Level Radioactive Waste*, vol. 1, Pergamon Press, Elmsford, N.Y., 1979.

BIBLIOGRAPHY

Aronson, R.: "The Search for Safe Nuclear Waste Disposal," *Machine Design*, March 1980.

Borduin, L., and Taboas, A.: *U.S. Department of Energy Radioactive Waste Incineration Technology*, Los Alamos Laboratories, March 1980.

Hedahl, T., and McCormack, M.: *Research and Development Plan for the Slagging Pyrolysis Incinerator*, U.S. Department of Energy, January 1979.

Hinga, K.: "Disposal of High Level Radioactive Wastes," *Environmental Science and Technology*, January 1982.

Parker, G.: "Incineration of Hazardous and Low Level Radioactive Waste," *Pollution Engineering*, August 1981.

Stretz, L., Allem, C., and Crippen, M.: "Combustible Radioactive Waste Treatment by Incineration and Chemical Digestion," *Journal of the American Institute of Chemical Engineers*, May 28, 1980.

Stretz, L., and Koenig, R.: "Offgas Treatment for Radioactive Waste Incinerators," Department of Energy Nuclear Air Cleaning Conference, October 1980.

CHAPTER 19
SOILS INCINERATION

There are thousands of sites throughout the United States where soils are contaminated with hazardous or toxic organics and where these contaminants are threatening surface or subsurface water supplies. Cleanup of these sites is becoming a major priority of the public. The focus of activity has generally been on-site remediation. Public concern over the safety and aesthetics of site cleanup has made trucking wastes through neighborhoods to a disposal site less and less acceptable. Likewise, there are fewer places for waste disposal. There are many on-site remediation methods being developed, and thermal treatment is one option.

Thermal treatment is usually not the least costly alternative, but it is one of the most acceptable and permanent techniques available. Compared to land disposal, it offers limited liability, immediate destruction, and mobility, minimizing the impact on local neighborhoods.

Thermal systems that have been developed for or used for site remediation are included in this chapter.

CONTAMINANTS

No sites have been found to be truly homogeneous. Some have a scattering of buried or exposed drums contaminated with toxic or hazardous organic or inorganic components. Many sites include ponds with significant hazardous constituents, with neither drums nor other containers, and some sites have no visible contamination but have underground wastes. Each site has its own unique characteristics.

Sites with contaminated soils often contain other materials which may also be contaminated. For instance, trees, tree stumps, rocks, and vegetation will likely be tainted, and their destruction or treatment must be included in a remediation plan.

REGULATORY ISSUES

The Resource Conservation and Recovery Act (RCRA) was the first major legislation enacted to provide complete control of the disposition of hazardous waste. This act defines a hazardous waste and regulates the transport, storage, and disposal of such wastes, as discussed in Chap. 3.

Subsequent regulatory legislation addressed the cleanup of waste accumulations and sites contaminated with hazardous wastes. These sites are potential sources of contamination of surface and subsurface water sources, and they represent a threat to plants, wildlife, and human populations. Often wastes accumulate on land where ownership is not clear or where no party can be readily identified as responsible.

The Comprehensive Environmental Response, Compensation and Liability Act (CERCLA) established a procedure (included in Superfund) for determining the party responsible for cleanup of a site and, further, for evaluating remedial procedures. The Superfund Amendments and Reauthorization Act (SARA) focuses on actual site cleanup. This legislation also established a *national priorities list* (NPL) which identifies the sites of greatest concern. This list is continuously updated and currently includes over 800 sites.

At present each site requires its own permits including RCRA, Toxic Substances Control Act (TSCA) (where PCBs are present), state, and local permits. Exceptions to this are NPL sites, which are not required to be permitted under RCRA or TSCA, according to Section 121 of SARA. At these sites the EPA regional administrator sets the applicable standards, which may require compliance with the provisions of incinerator permitting under RCRA without the necessity of obtaining an actual permit. Typically, the permit process takes from 6 to 18 months or more.

MOBILITY

Remedial incineration systems can be permanent, mobile, or portable. A permanent system is one that is installed on-site and that will have a projected life of at least 5 years, with no anticipated salvage value.

Mobile systems are brought to a site and are removed at the conclusion of the cleanup. They normally include all the equipment and subsystems necessary for operation of the facility. These include electric power generation equipment, a fuel supply, and equipment to collect wastewater and dispose of it. These systems may be designed for use on trailers. If they are skid-mounted, they will be designed for use on their skids, which can be removed from the trailer chassis. A mobile system is normally used for cleanups requiring 6 months or less.

Transportable equipment differs from mobile equipment in that it requires a significant installation effort. This equipment is provided in modular components and must be assembled before use. A process water supply will be sought on-site; the wastewater discharge will be disposed of on-site also, although water or wastewater discharge treatment facilities may be required. Transportable systems are designed so that they can be dismantled, removed, and reinstalled at another site at a future date. Such equipment is designed for a site cleanup effort in the range of 6 months to 5 years.

The majority of technologies proposed for site cleanup are transportable systems.

When evaluating portable units, one must recognize that the specific technology must carry with it all its own requirements, including utilities, laboratory facilities, personnel stations, and the process equipment itself. This could require 10 to 30 trucks, many acres of land, and 1½ to 3 months for setup. The portability of any thermal site cleanup system is subject to these technical constraints in addition to permitting constraints.

ROTARY KILN SYSTEMS

One of the first commercial cleanups of a contaminated site by incineration was at Sydney Mines near Tampa, Florida. The kiln system was designed, constructed, and operated by Ensco Pyrotech of Little Rock, Arkansas.

The equipment layout on site is shown in Fig. 19.1. There are five major elements of the system:

- Rotary kiln incinerator
- Afterburner/liquids incinerator
- Waste heat boiler
- Emissions control system/prime mover
- Process control/laboratory facility

The kiln, or solids module, consists of a trailer-mounted rotary kiln, solids preparation and charging equipment, a supplemental fuel burner, a combustion air fan, and an air discharge system. The charging system includes a belt conveyer, and kiln charging hopper with an integral ram feeder. During operation, contaminated solids are placed on the belt conveyer and are subsequently fed to the charging hopper. An intermittently operated ram feed mechanism located in the feed chute pushes the material into the kiln. The rotary kiln is lined with castable refractory for its initial 6 ft of length. The balance of the kiln is lined with refrac-

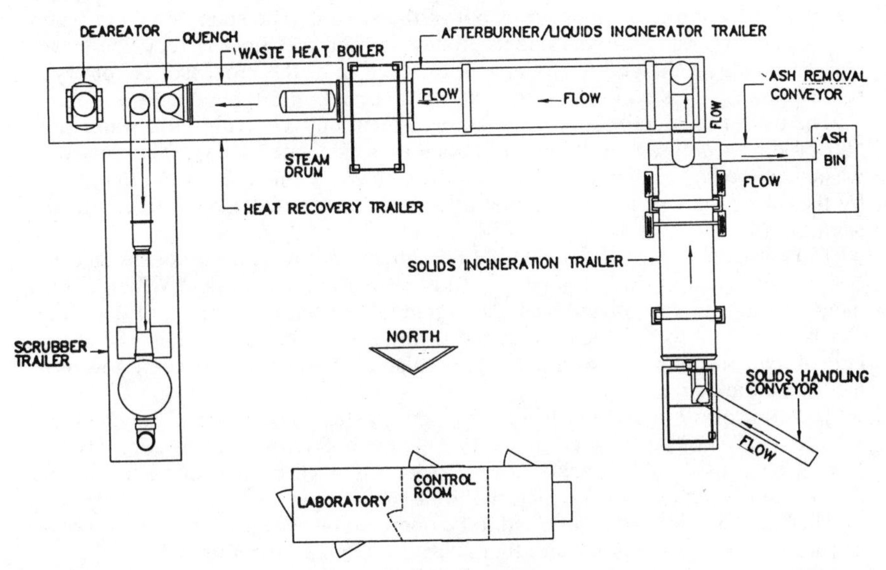

FIGURE 19.1 Transportable rotary kiln incineration system.

tory brick. Residuals generated from incineration are discharged to the end breeching where they fall into a dry ash discharge hopper. A water-cooled screw conveyer subsequently carries the ash to a storage bin where samples are taken for analysis. If residual organics are found in the ash, the entire ash load is recycled through the incinerator. If not, then the ash is taken to a landfill.

At Sydney Mines it was found that in the temperature range of 1300 to 1400°F organic contaminants volatilize from the soil. Some of these organics will then be destroyed in the kiln; the remaining contaminants are carried within the gas stream to downstream equipment.

To reduce the carryover of residual soil in the gas stream, cyclones were installed between the kiln and the secondary combustion chamber. The cyclones were found to remove over 70 percent of the airborne residual in the exhaust gas.

The secondary combustion chamber, or afterburner, is a refractory-brick-lined horizontal cylinder sized to provide a residence time of 2 s at a temperature of 2200°F at full waste load. The volatiles and products of combustion exiting the kiln pass through the afterburner where a supplemental fuel burner raises the gas temperature to destroy the organics.

A trailer-mounted fire-tube boiler is provided to generate the steam required by the exhaust gas cleaning system. A boiler feedwater treatment package is included within this transportable facility to increase system reliability. The deaerator and water polisher included in this package reduce boiler fouling and subsequent boiler maintenance. Excess steam can be discharged through a silenced steam vent. During normal operation more steam is generated than can be used by the system.

Steam and water are injected through an ejector nozzle system which creates a negative pressure, or draft. This draft is sufficient to draw gases from the kiln, through the cyclones, afterburner, waste heat boiler, quench elbow, and quench tank. In addition to producing the draft required as a prime mover within the system, the ejector system creates small atomized water particles which adsorb particulate matter from the gas stream. The saturated gas is directed to a mist eliminator where entrained water drops out of the stream. The spent scrubber water is directed to a lamella-type sludge processor. The overflow is recycled to the ejector while the sludge that collects at the bottom of the processor is conveyed to a container. This containerized sludge is fed back into the incinerator.

The use of an ejector scrubber requires steam but relatively little water flow and produces only a few hundred gallons of wastewater a day. The cost of a waste heat boiler to generate the required steam was found to be more than offset by the advantages of low water flow, decreased wastewater discharge, and the savings of an induced-draft fan.

The contaminants at the Sydney Mines site did not have a significant chloride content, and acid gas was not generated in measurable quantities. Where chlorinated organics are found and acid gas is generated from the burning of these materials, a packed tower could be added to the system downstream of the scrubber. A caustic solution would be recirculated through the packed tower to neutralize the acid gases.

The control room and the laboratory occupy the same trailer. Analyzers and recorders monitor and display flue gas constituents (carbon monoxide, oxygen, and nitrogen oxides). The laboratory is primarily used to provide chemical and heating value analyses of the material to be incinerated.

These five modules are designed to be operated on their trailers. Other mobile or transportable systems require that skids be removed from the trailer.

Other on-site incineration systems by other manufacturers utilize vertical afterburner (secondary combustion chamber) sections. There is additional cost and scheduling for installation of the vertical system; however, residuals carryover is minimized.

CONVEYER FURNACE

Shirco Infrared Systems of Dallas (presently marketed by ECOVA, located in Seattle), Thermal Process Systems (TPS) in Michigan, and NASS (National Applied Scientific Systems) of Dallas have each developed an enclosed conveyer incineration system for on-site incineration and have sold a number of them to cleanup contractors. A Shirco system operated by Haztech, Inc., was used for cleanup of a portion of the Peak Oil site in Brandon, Florida. The system is shown in schematic in Fig. 19.2. It is a variation of the infrared sludge incinerator described in Chap. 15.

The primary chamber of the infrared system is constructed of eight modular sections bolted to a support skid. The modules include feed and discharge sections and six 8-ft-long powered sections containing infrared heating elements.

A conveyer belt passes under a series of agitators which gently turn the feed to expose additional surface to the air and to hot combustion gases. When conveyed feed reaches the discharge end of the furnace, it drops into a transverse screw, then into an inclined ash screw conveyer for transport to a container for disposal. The discharge system is equipped with water sprays, using spent scrubber water, to cool the ash as it exits the furnace. Ash collected from the feed section and

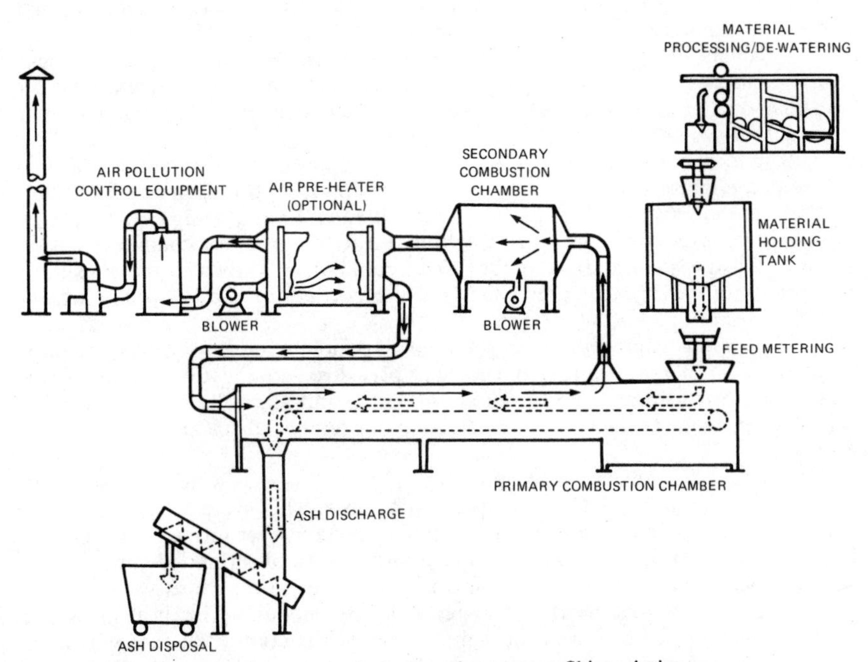

FIGURE 19.2 Conveyer furnace on-site incineration system, Shirco design.

hopper bottoms is transported to a discharge container by a longitudinal screw conveyer mounted on the furnace skid.

As feed passes through the chamber, it is exposed to radiant heat from silicon carbide heating elements, grouped into two thermal control zones. Each control zone is powered by a 500-kVA heating element power center, utilizing 480-V 60-Hz electric power.

The primary chamber is capable of maintaining process temperatures up to 1850°F. Feed residence time varies from 10 to 180 min by adjusting the belt speed. Either an oxidizing or a reducing (oxygen-deficient) atmosphere can be maintained in the unit.

Gases from the primary chamber pass through a crossover flue to the secondary combustion chamber (afterburner). The afterburner is fabricated in two sections, each skid-mounted with axles, which are bolted together at the site.

Process heat in the afterburner is provided by a system of four natural gas or propane burners. An array of silicon carbide heating elements placed downstream of the burners increase gas turbulence. The afterburner is sized to accommodate 2-s gas residence time at 2300°F at the maximum feed rate.

An internal insulated flue carries exhaust gases from the secondary combustion chamber (afterburner) to the scrubber system. Exhaust gases are split by manifold into two processing trains, each of which contains a quench section and an adjustable venturi. One side of the manifold can be closed for processing low gas flow rates associated with lightly contaminated soils or a low soils throughput. The gases are then passed through a common packed-tower scrubber with a closed-loop scrubber liquor recirculation system.

An induced-draft (ID) fan is located on the discharge side of the packed tower to exhaust the primary and secondary combustion chambers. A fiberglass-reinforced plastic exhaust stack is provided with ports for stack sampling and continuous emissions monitoring equipment.

In the event of a power failure or failure of the ID fan, an emergency bypass damper and stack will exhaust gas from the discharge of the secondary combustion chamber. The damper position can be manually controlled to provide proper draft during system shutdown.

The main control panel, off-gas analyzer, and motor control center are contained in a climate-controlled van. The control panel includes process controllers, alarm and status lights, and data recording and monitoring equipment. A temperature recorder is provided for each of the two primary chamber heating zones, plus the secondary chamber. The cumulative kilowatthour consumption for each primary chamber zone is also recorded for use in optimizing energy consumption.

Additional instrumentation on the control panel includes monitoring equipment for chamber pressures, burner gas pressure, combustion air pressure, and water flow rate and pressure. Status lights and alarm lights are provided for major process parameters such as chamber temperatures and burner operations.

The conveyer incinerator designed by Thermal Process Systems features direct firing of fossil fuel into the conveyer chamber. This system, shown in Fig. 19.3, includes a primary combustion chamber (the conveyer furnace) discharging into a thermal mixing unit which creates a high degree of turbulence prior to entering the secondary combustion chamber (afterburner). From the afterburner the gas stream is directed to the flue gas cleaning module. The hot gases are quenched and are then passed through a series of gas control devices which remove particulate and acid gases from the exhaust.

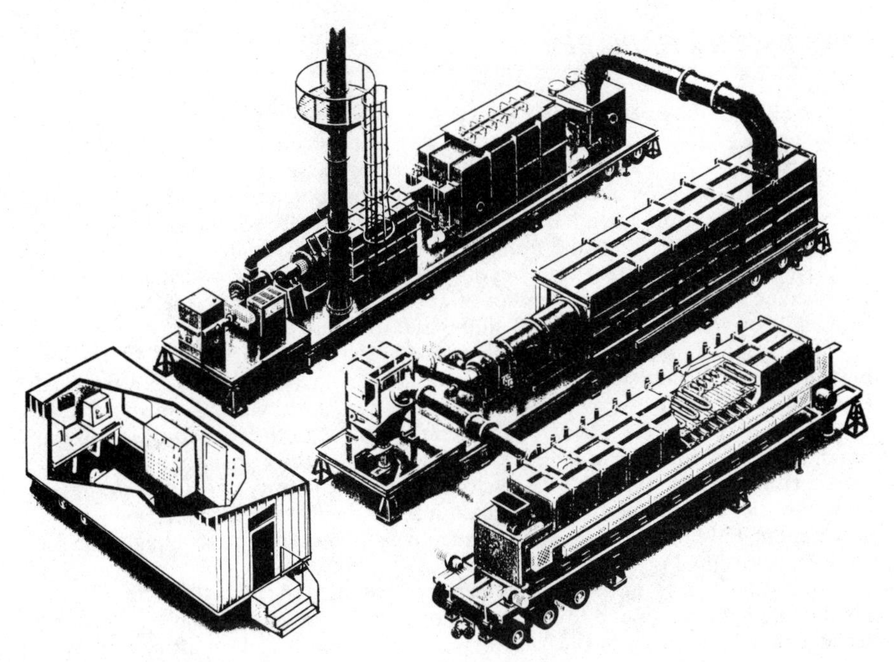

FIGURE 19.3 TPS conveyer furnace on-site incineration system.

The NASS system, described in Chap. 15 as applied to sludge incineration, utilizes fossil fuel as an indirect source of heat for the conveyer chamber. The conveyer incinerator (primary combustion chamber) module is shown in Fig. 19.4. This system also contains afterburning equipment and an exhaust gas cleaning system.

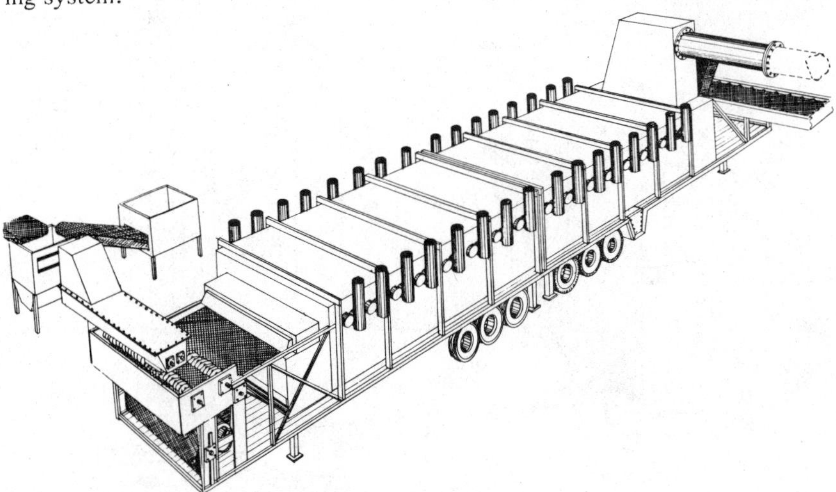

FIGURE 19.4 NASS on-site conveyer incineration system.

CIRCULATING FLUID BED

The circulating fluid bed incineration system, developed by Ogden Environmental Systems, Inc., is used for site cleanup as a transportable unit. Components of the system are transported to a site where they are assembled and erected.

The circulating fluid bed concept is distinct from conventional fluidized beds. Conventional reactors have a fixed bed depth and operate within a narrow range of gas velocities (between the minimum and maximum fluidization velocities), from 1.5 to 4.5 ft/s. At velocities beyond 5 ft/s the reactor freeboard and the off-gas become entrained with carryover of unburned particles from the bed. At velocities below 1.2 ft/s, the bed will slump and lose fluidization.

In the circulating bed concept, illustrated in the schematic in Fig. 19.5, combustible waste is introduced into the bed along with recirculated bed material from the hot cyclone. A high air/gas velocity (from 15 to 20 ft/s) elutriates both the bed and the combustible waste, which rises through the reaction zone to the top of the combustion chamber (freeboard) and passes through a hot cyclonic collector. Hot gas passes through the cyclone while the majority of solids drop to the bottom of the cyclone and are reinjected into the bed of the furnace.

Uncirculated flue gases pass to a convective gas cooler, then to a baghouse for removal of residual particulate.

Feed is introduced in the leg between the cyclone and the bed of the reactor. Solid or sludge waste is fed to the system through a feeding bin where a metering screw conveys waste from the bin to the feed leg. For liquid or slurry wastes, a pump is used to meter the waste feed from a stirred tank to the reactor. No atomizers or specialty nozzles are required for introduction of fluid wastes into the sand bed. The waste feed rate is automatically adjusted to maintain a preset oxygen concentration in the flue gas.

Lime can be added to the waste feed through a lime metering system to neutralize acid-generating constituents of the waste. As shown in Fig. 19.5, lime and

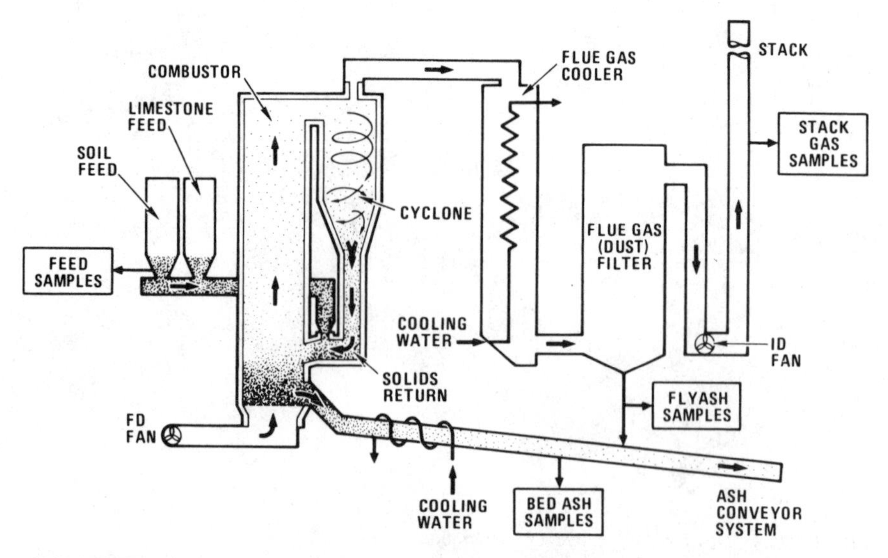

FIGURE 19.5 Circulating fluid bed schematic.

makeup sand are both added to the reactor through the waste feed metering screw.

The operating temperature is normally 1600°F although the system can withstand temperatures up to 2000°F on a continuous basis.

A combustion air fan provides air to the bed for fluidization and oxidation. The furnace draft is maintained by an induced-draft fan downstream of the cyclone.

Retention time of material within the system is controlled by controlling the discharge from the ash cooler. Cyclone bottom ash discharges to the reactor, and part of this ash flow is removed from the system through a water-cooled ash conveyer. By increasing the speed of this conveyer, additional material is removed from the furnace system, and the residence time within the system is likewise controlled. For instance, by lowering the conveyer discharge rate, less material is discharged from the system and the solids retention time is increased.

Flue gas exiting the cyclone passes through a conventional exhaust gas treatment system which removes particulate and other undesirable constituents from the gas stream.

A major feature of the circulating fluid bed system is its ability to control the residence time of wastes to well over 15 s. By increasing the residence time over that of conventional incineration systems, the temperature required for destruction can be substantially reduced. Generally, destruction of organics has been found to occur at temperatures below 1600°F in this system, rather than at the +2000°F temperatures required in conventional afterburners. This lower temperature translates to lower supplemental fuel requirements, lower (insignificant) quantities of nitrogen oxides produced, less refractory maintenance, less severe bed eutectics, etc.

ELECTRIC PYROLYZER

This system has been developed by the Westinghouse Electric Corporation for the processing of soils contaminated with organics. It has been built in a pilot size, as a fully mobile facility.

The waste requires minimal processing for introduction into this system. Process feed must pass through a 2-in grizzly (2-in open mesh). Soil can have a moisture content of up to 25 percent (by weight) before drying is necessary. Solid waste is dropped, by gravity, through the reactor, or pyrolyzer. Liquid feed is injected into the reactor.

The electric pyrolyzer process promotes the release of volatiles from the surface of the soil or other material. As waste is dropped into the unit, it passes through a high-temperature zone where the majority of the organics volatilize. The soil, or other solid waste, drops to the bottom of the unit, which is maintained at a high enough temperature to keep the soil and other inorganics in a molten state.

Supplemental electrodes within the melt ensure that its temperature will be sustained at a relatively high level, uniform throughout the melt. Any metals present will be found in their elemental form or as a salt, and they will be removed from the melt on a continuous basis from an appropriately placed tap. Other taps are located at other levels of the reactor to provide a means for discharge of slag and other materials generated by the process. The tapped materials fall into a water bath where they immediately cool. The cooled residual has the appearance of dark glass.

Any organics that have not volatilized as the soil drops through the reactor will be destroyed within the melt. With the presence of a melt, sizing of the soil particles and moisture content are not critical. The surface area of the soil particle is a factor in the evaporation of organics from the particle surface, but if volatilization does not occur as a surface phenomenon, the melt will provide the medium for destruction.

The electrodes generate a temperature on the order of 4000°F. The melt is maintained at a much lower temperature. The chemistry of the melt can be controlled by additives such as lime, salts, or other compounds. By adjusting the melt constituents neutralization of acid gas components of the waste occurs, and the properties of the slag can be controlled.

The off-gas from the reactor passes through a cyclone where the majority of particulate that may be elutriated into the gas stream is removed. A baghouse removes the balance of particulate matter. A wet scrubber placed downstream of the baghouse will remove any halogenated or sulfonated (acid) gases that were not neutralized within the reactor.

The pilot unit, with a capacity of from 5 to 10 tons/day of soil, is mounted on a single trailer with utility equipment on a second trailer. The equipment trailer requires a pad, but the utility trailer can be placed on level ground.

HIGH-TEMPERATURE FLUID WALL REACTOR

The high-temperature fluid wall reactor (HTFWR) has been developed by the Huber Corporation for the treatment of soils contaminated with organic materials. This system is designed as a transportable unit that can be assembled and installed in a number of weeks.

Contaminated soil is dropped, by gravity, through a metered screw feeder into a reactor. The organic contaminants volatilize from the surface of the soil and dissociate into their elements. The heated soil falls through a post reactor zone where it begins to cool and then falls into a solid waste bin, where it is collected for ultimate disposal. An induced-draft fan downstream of the reactor generates the draft required in the reactor to provide a draw for the gases produced.

A cyclone followed by a baghouse removes particulate from the exhaust gas stream. If halogenated or sulfonated contaminants are present in the waste or soil, the exhaust gas can be diverted to a caustic scrubber for acid neutralization. Activated carbon filters help ensure the capture of any organics that may be present in the gas stream as a result of operating upsets or other causes.

The heart of the reactor is a vertical graphite cylinder. It is surrounded by a set of electrodes, which in turn are encased within a cooled and insulated jacket. The electrodes provide the thermal energy required to heat the core to radiant temperatures. Temperatures in the range of 4000 to 5000°F are developed.

A nitrogen purge is introduced within the porous graphite core to act as a blanket, or fluid wall. It tends to prevent the soil from coming in contact with the graphite surface. The graphite is sensitive to plugging and fouling, and by the use of the nitrogen purge the core is kept clean and its pores are kept open. The nitrogen will also displace any air to prevent oxidation of the graphite tube.

The heat flux developed within the core by the energy radiated from the graphite rapidly heats the surface of the falling soil. Contaminants present are typically adsorbed onto the surface of the soil. These organic contaminants will volatilize under the intense temperature level present. At the high temperature generated within the core, the organic contaminants released from the soil will dissociate into their elemental constituents (carbon, hydrogen, oxygen, and, when present in the feed, nitrogen, chlorine, and other elements).

The soil itself does not have sufficient residence time within the reactor to see an increase in temperature of more than a few hundred degrees. The energy required for this process, therefore, only has to be sufficient to volatilize the surface contaminant components of the soil particle, not to heat the soil.

The reactor requires significant soils preparation. The process is sensitive to the surface area of the soil particle, and for effective contaminant release the soil must be sized to pass through a 32-mesh screen. Soil moisture content is also a factor in effective unit operation. The soil moisture must be reduced to less than 8 percent (by weight) to aid in the grinding of the soil and to promote effective volatilization of soil contaminants in the reactor itself.

SITE INCINERATION SYSTEMS SUPPLIERS

As more funding becomes available for site remediation, additional cleanup contractors and equipment vendors are entering the field. Table 19.1 lists vendors and contractors active in site cleanups by incineration as of 1990.

TABLE 19.1 Site Cleanup Incinerator Contractors and Manufacturers

Company	Activity
AET (Applied Environmental Technology) See U.S. Waste Thermal Processing	
Allis Chalmers (Boliden Allis) P.O. Box 512 Milwaukee, Wisconsin 53201	Kiln manufacturer 414/475-3862
Bell Lumber & Pole Co. 778 1st Street, NW St. Paul, Minnesota 55112	Site cleanup 612/633-4334
Canonic Engineers 6557 South Reveric Parkway Suite 155 Englewood, Colorado 80111	Site cleanup 303/790-1747
CE Raymond 650 Warrenville Road Lisle, Illinois 60532	Kiln manufacturer 312/971-2500
Chemical Waste Management, Inc. 3003 Butterfield Road Oakbrook, Illinois 60521	Contract disposal site cleanup 312/218-1500
Consertherm, Inc. 489 Sullivan Avenue South Windsor, Connecticut 06074	Kiln manufacturer 203/289-1588
Detoxco 2700 Ygnacio Valley Road Walnut Creek, California	Kiln manufacturer 415/930-7997

TABLE 19.1 Site Cleanup Incinerator Contractors and Manufacturers (Continued)

Ecova Corporation 15555 NE 33rd Redmond, Washington 98052	Site cleanup/conveyer furnace manufacturer 206/883-1900
Ensco/Pyrotech 333 Executive Court Little Rock, Arkansas 72205	Contract disposal 501/223-4100
Environmental Elements, Inc. (ENELCO) 3700 Koppers Street Baltimore, Maryland 21227	Kiln manufacturer 301/368-6742
Ford, Bacon & Davis 375 Chipeta Way Salt Lake City, Utah 84108	Kiln manufacturer 801/583-3773
Fuller Company P.O. Box 2040 Bethlehem, Pennsylvania 18001	Kiln manufacturer 215/264-6011
GA Technologies See Ogden Environmental Systems	
GDC Engineering 1822 Neosho Avenue Baton Rouge, Louisiana 70802	Site cleanup 504/383-8556
GSX Chemical Services, Inc. P.O. Box 210799 Columbia, South Carolina 29210	Contract disposal 803/798-2993
Harmon Environmental Services, Inc. 2075-A West Park Drive Stone Mountain, Georgia 30087	Site cleanup 404/469-3077
Haztech Inc. (Division of Westinghouse) 5280 Panola Industrial Blvd. Decatur, Georgia 30035	Site cleanup 404/981-9332
Heritage Environmental Services, Inc. 7901 West Morris Street Indianapolis, Indiana 46231	Site cleanup 317/243-0811
Industronics See Conserthern, Inc.	
International Technology Corporation 17461 Derian Avenue Irvine, California 92714	Site cleanup 714/261-644
International Waste Energy Systems 2150 Kienlen Avenue St. Louis, Missouri 63121	Kiln manufacturer 314/389-7273
J.M. Huber Corp. P.O. Box 2831 Borger, Texas 79008	Equipment manufacturer/ site cleanup 806/274-6331
John Zink Services 4401 S. Peoria Ave. Tulsa, Oklahoma	Equipment manufacturer 918/747-1371

TABLE 19.1 Site Cleanup Incinerator Contractors and Manufacturers (Continued)

Kennedy Van Saun P.O. Box 500 Danville, Pennsylvania 17821	Kiln manufacturer 717/275-3050
Kimmins Environmental Services 1501 2nd Avenue Tampa, Florida 33605	Site cleanup 813/248-3878
M&S Engineering & Manufacturing Co. 95 Rye Street Broad Brook, Connecticut 06016	Kiln manufacturer 203/627-9396
McGill Incorporated 5800 West 68th Street Tulsa, Oklahoma 74157	Kiln manufacturer 918/445-2431
NASS, Inc. 4215 San Gabriel Drive Dallas, Texas 75229	Conveyer furnace manufacturer 214/350-6274
Ogden Environmental Services P.O. Box 85178 San Diego, California 92138	Equipment manufacturer/ site cleanup 619/455-2383
OH Materials P.O. Box 551 Findlay, Ohio 45840	Site cleanup 416/423-3526
Pyrotech See Ensco	
Rollins Environmental Services, Inc. P.O. Box 73877 Baton Rouge, Louisiana 70897	Contract disposal 504/778-1242
Rollins Environmental Services, Inc. P.O. Box 609 Deer Park, Texas 77536	Contract disposal 713/479-6001
Rollins Environmental Services, Inc. 1 Rollins Plaza Wilmington, Delaware 19899	Contract disposal 302/479-2951
Ross Incineration 394 Giles Grafton, Ohio 44044	Contract disposal 216/748-2171
Reidel Environmental Services, Inc. 4611 N. Channel Ave. Portland, Oregon 97217	Site cleanup 503/286-4656
SCA Chemical Services, Inc. 1000 E. 111th Street Chicago, Illinois 60628	Contract disposal 312/660-7200
Shirco Infrared Systems, Inc. See Ecova	
Supratec, Inc. See NASS	
Thermal Process Systems (TPS) 12068 Market Street Livonia, Michigan 48150	Conveyer furnace manufacturer 313/591-1000

TABLE 19.1 Site Cleanup Incinerator Contractors and Manufacturers (Continued)

Thermall P.O. Box 1776 Peapack, New Jersey 07977	Kiln manufacturer 201/234-1776
Thermodynamics Corporation P.O. Box 69 Bedford Hills, New York 10507	Site cleanup 914/666-6066
Trade Waste Incineration 7 Mobile Avenue Sauget, Illinois 62201	Contract disposal 618/271-2804
Tricil, Ltd. 89 Queensway West, Suite 711 Mississauga, Ontario L5B 2V2	Contract disposal 416/270-8280
Tyger Construction Company P.O. Box 5684 Spartanburg, South Carolina 29304	Site cleanup 803/585-8381
UMA (Taciuk Processor) 210-2880 Glenwood Trail Calgary, Alberta T2C 2E7	Equipment manufacturer 403/279-8080
Universal Energy International (UEI) 10 Drawer 5180 Little Rock, Arkansas 72225	Equipment manufacturer 501/666-2446
U.S. Waste Thermal Processing 11090 Rose Avenue Fontana, California 92335	Site cleanup/equipment manufacturer 302/378-8888
Vesta Technology, Ltd. 2501 E. Commercial Blvd. Ft. Lauderdale, Florida 33308	Site cleanup 305/776-0330
Waste Management See Chemical Waste Management	
Waste Tech Services 445 Union Blvd. Lakewood, Colorado 80228	Site cleanup 303/987-1790
Westinghouse Waste Technology Services P.O. Box 286 Madison, Pennsylvania 15663	Equipment manufacturer/ site cleanup 412/722-5600
Weston Weston Way West Chester, Pennsylvania 19380	Site cleanup 215/692-3030

BIBLIOGRAPHY

Brunner, C. R.: *Site Cleanup by Incineration*, HMCRI, Silver Spring, Md., 1988.

Brunner, C. R., and Erickson, F.: "Site Remediation by Incineration/Thermal Treatment," presented at the International Conference on Incineration of Hazardous, Radioactive and Mixed Waste, San Francisco, May 1988.

Hill, M.: "Mobile Infrared Incinerators for Soil Decontamination," presented at the Air Pollution Control Association 80th Annual Meeting, New York, June 21, 1987.

CHAPTER 20
MISCELLANEOUS WASTE DESTRUCTION PROCESSES

There are a number of waste destruction systems which do not utilize burning processes. Some of these systems are described in this chapter.

MOLTEN SALT INCINERATION

The molten salt process has been developed by Atomics International (division of Rockwell International Corporation) to dispose of a wide variety of wastes: solids, liquids, and sludges. It is not a true incineration process because flame is not produced. But it is an oxidation process and a recombinant process where wastes are oxidized and/or chemically altered to innocuous substances.

Soluble alkali salts are used as the bed material. A single salt or a mixture of salts can be used. The more common salts in use are sodium chloride (NaCl), sodium sulfate (Na_2SO_4), sodium phosphate (Na_3PO_4), sodium carbonate (Na_2CO_3), and corresponding calcium salts.

The salt bed is heated to fluidization, its temperature a function of the salt material utilized. Waste is prepared before feeding, if solid waste, by shredding and is fed directly into the molten salt bed, if liquid or sludge. Typical temperatures of the bath are in the range of 1500 to 1800°F.

A schematic diagram of a molten salt incinerator is shown in Fig. 20.1. Waste components dissolve within the melt, producing an off-gas. The off-gas will contain carbon dioxide, moisture (steam), and oxygen and nitrogen from the air supply. It will also contain particulate matter, salt, and other components generated within the melt. After particulate removal the gas is discharged to the atmosphere.

The volume of the melt increases as waste is added to the system, and the reactor must be tapped periodically. Table 20.1 shows typical combustion products within the melt for particular waste components. Many of these compounds can be separated from the salt, and the salt can be recycled to the reactor for reuse, as shown in Fig. 20.1.

After startup, the waste stream may have sufficient heat content to maintain the salt bed in a hot, fluid condition without the addition of supplemental fuel.

With regard to waste destruction, the salt acts as a dispersing medium for both the waste being processed and the air used in the reaction. The salt acts as a cat-

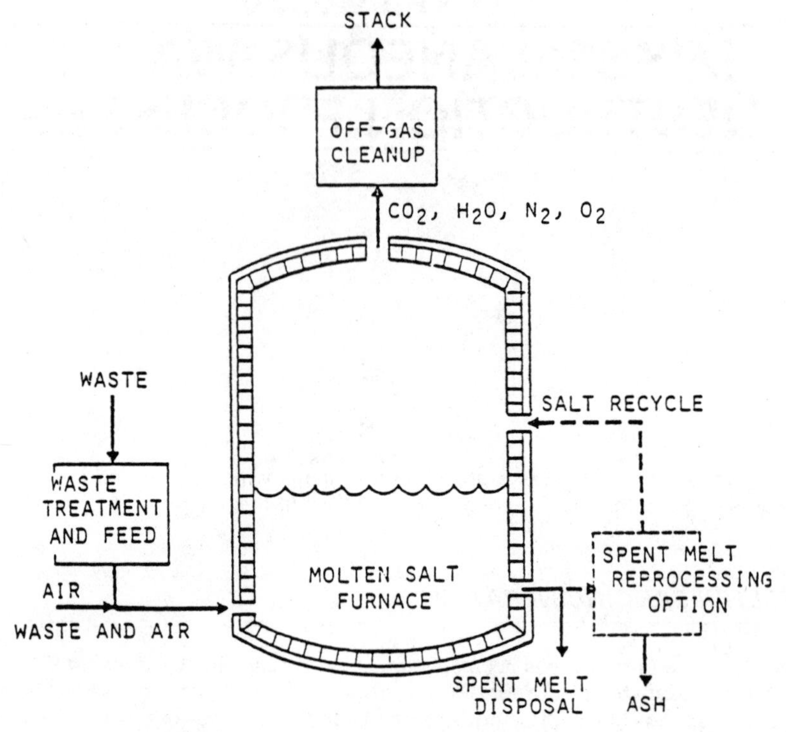

FIGURE 20.1 Molten salt reactor. *Source:* Rockwell International, Energy Systems Group, Canoga Park, Calif.

TABLE 20.1 Molten Salt Combustion Process Chemistry

Salt: Sodium Carbonate
Oxidizing Agent: Air

Element in Feed	Combustion Product
Carbon	carbon dixoide
Hydrogen	steam
Chlorine, fluorine	sodium chloride, fluoride
Phosphorus	sodium phosphate
Arsenic	sodium arsenate
Sulfur silicon	sodium sulfate
Silicon (glass)	sodium silicate
Iron (stainless steel)	iron oxide
Silver (photo film)	silver metal

Source: Rockwell International, Energy Systems Group, Canoga Park, Calif.

alyst for oxidation reactions and accelerates the destruction of organic materials while preventing the emission of acidic materials by neutralization. The bed acts as a heat sink for absorbing and distributing the heat generated by oxidation of the waste components. Further, ash generated and other noncombustible materials generated by the waste are physically retained within the melt.

In general, the temperatures developed within the salt melt are not high enough to produce significant quantities of nitrogen oxides.

Figure 20.2 is a schematic of a generalized molten salt waste destruction system. A startup heater within the reactor brings the salt to molten temperature. Compressed air is used for solids feeding, i.e., pneumatic conveying. A portion of the compressed air supply is heated to help provide the heat necessary to sustain the molten salt temperature.

Off-gas is exhausted through a quench or spray cooler where its temperature is reduced below 300°F. At this temperature the gas can pass through electrostatic or fabric filters for particulate removal prior to discharge through induced-draft blowers feeding an exhaust stack. High-efficiency particulate air filters (HEPA filters) may or may not be necessary, depending on the particulate loading and size distribution.

Liquid and sludge waste can be taken directly to the incinerator and charged beneath the melt surface. Solid waste, shown as item 18 in Fig. 20.2, is shredded to uniform size and is then transferred to a waste feeder. Shredded solid waste, residue from the particulate filter, fresh salt, and coke are conveyed, through a single pipeline, to a solids feeder which in turn feeds the furnace.

As waste is added, the melt level increases and overflows to a drain cart. When the ash and other inerts build up to 20 percent of the melt, as sampled in

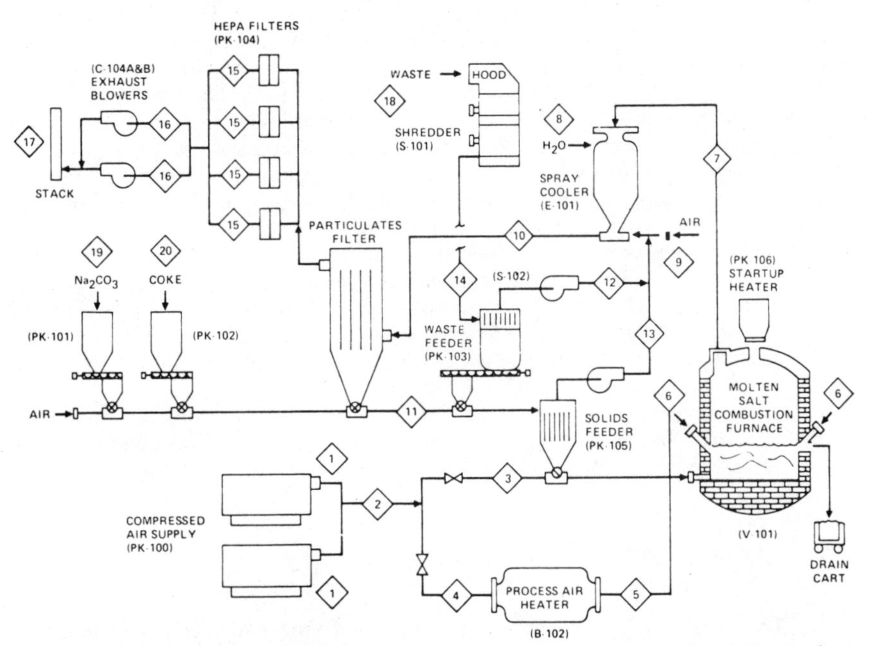

FIGURE 20.2 Molten salt system flow schematic. *Source:* Rockwell International, Energy Systems Group, Canoga Park, Calif.

the melt overflow, the melt must be tapped and fresh salt added. The purpose of the coke is to provide a source of carbon, i.e., combustion, to aid in startup and to act as a supplemental fuel when the waste heating value is insufficient.

Spurious emissions or leakages from the waste feeder and solid feeder are collected and passed through the off-gas particulate removal system to provide essentially "zero discharge" from the system.

FLASH DRYING

Flash drying is a system used for disposal of various sludge wastes. Sludge waste is dried and can then be burned or disposed of by another method or within the flash drying system. Figure 20.3 is a schematic of a flash drying system. Its principal elements are a hot gas heater, sludge mixer, cage mill, cyclone collector, vapor fan, and dry product conveyer.

The operation of this system is as follows:

Wet sludge is blended with a small quantity of dried sludge in the mixer to improve its transportability. This mixed sludge is fed to a windbox on the cage mill where hot gas, from 1000 to 1400°F (depending on the nature of the sludge), contacts the sludge. Moisture begins to evaporate into steam. The mixture of sludge, hot gas, and steam is ground together in the cage mill. By the time the sludge leaves the mill, it is virtually dry, with only 8 to 10 percent moisture. The dried sludge is pneumatically conveyed to a cyclone where separation of dry sludge from gas and moisture vapor occurs, at approximately 300°F.

A dry sludge is discharged by the dry product conveyer. This material can be burned to produce the hot gas required for initiation of the process, or the dry product can be incinerated in another furnace or taken away for other uses. The furnace shown in Fig. 20.3 utilizes conventional fuel for generating hot gas; the dry sludge is removed from the process as a dry material.

The spent gases from the cyclone must be deodorized and discharged. In this design a vapor fan blows these gases from the cyclone to a vertical heat exchanger (recuperator) within the furnace. The spent gas picks up heat from the recuperator and is heated further, within the furnace, to 1400°F. At this temperature, with an appropriate residence time, the odor components will destruct. The hot gas will pass over the tubes in the recuperator, losing some of its heat to the incoming gas. The gas continues to another heat exchanger, a preheater, which heats combustion air prior to its injection into the furnace, adjacent to the supplemental fuel burner. The spent gas will exit the preheater and will discharge the stack at approximately 500°F.

The flash drying process is a rapid process. Drying is practically instantaneous in the cage mill. In the cyclone, separation resulting from the flashing of the hot moisture occurs in 6 to 10 s.

THE CARVER-GREENFIELD PROCESS

The Carver-Greenfield process is a method of drying moisture-laden liquid or sludge through heating and evaporation. Figure 20.4 illustrates a typical system.

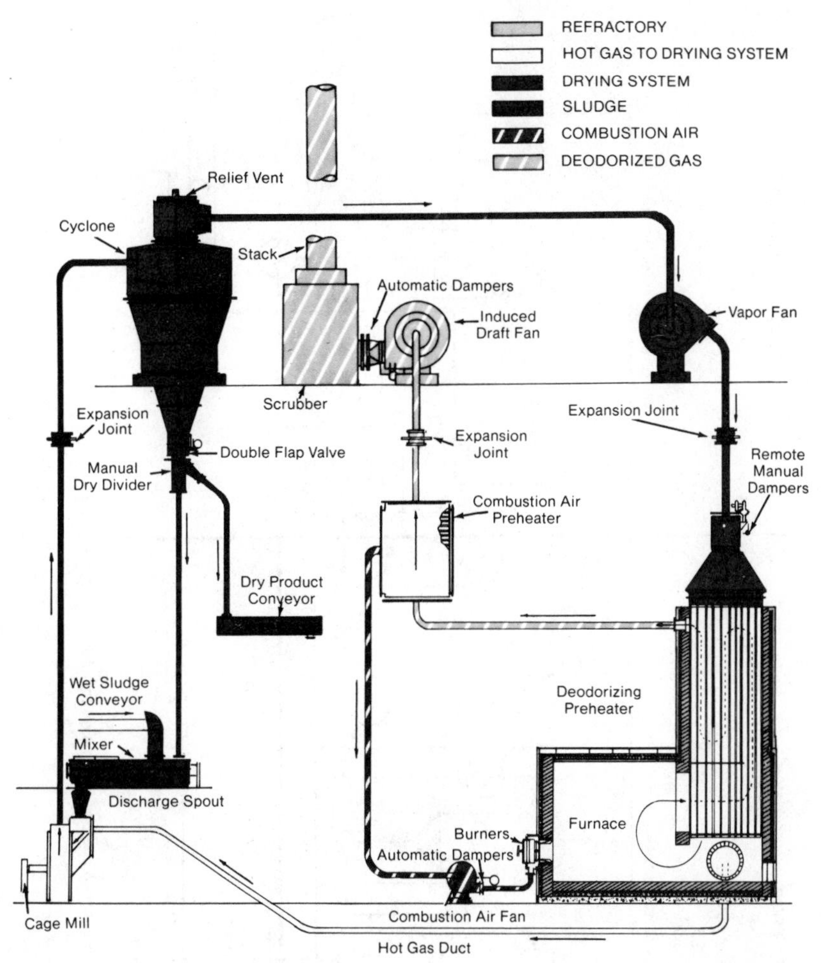

FIGURE 20.3 Flash dryer system. *Source:* C-E Raymond, Combustion Engineering, Inc., Chicago.

After the feed is passed through a grinder to give it a uniform consistency, it is pumped to a fluidizing tank where an oil is added to it. The feed-oil mixture passes through a series of evaporators where the moisture is evaporated, leaving the solids suspended in the oil as a fluid slurry. Four stages of evaporation are shown in Fig. 20.4; however, the number of stages required is a function of the feed properties, and as few as two stages may be sufficient for some materials.

The feed is pumped through a heat exchanger prior to each stage of evaporation and ultimately is discharged to a centrifuge. The centrifuged solids contain less than 1 percent of the original oil added to the feed and less than 2 percent by weight of moisture. The dry solids can be burned to provide steam for the evaporators, or, as with flash dryers, they can be utilized external to the system.

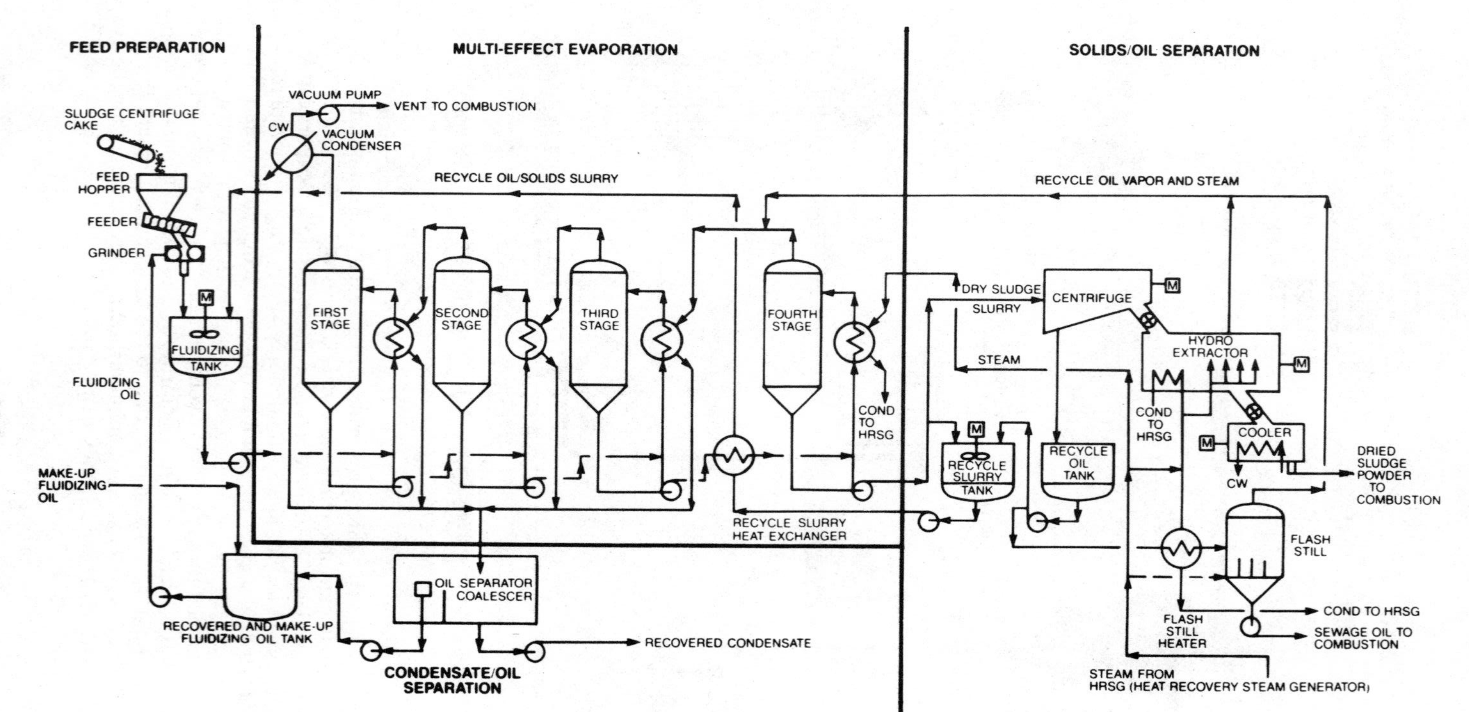

FIGURE 20.4 Carver-Greenfield process. *Source:* Foster-Wheeler/Carver-Greenfield, Livingston, N.J.

If additional oil removal is required, a hydroextractor can be used, as shown in Fig. 20.4. This is a steam-operated stripper where higher-temperature superheated steam is used to boil off the residual oil within the dried sludge.

The balance of the equipment used in this process is involved in the capture and recovery of the carrier oil, its reuse, and the generation of steam.

Petroleum-based oils, animal oils, or vegetable oils can be used in this process, depending on the nature of the sludge and its means of ultimate disposal.

The off-gases, including spent steam, from this system are normally burned in the steam boiler as combustion air. The only atmospheric discharge is the stack from the steam boiler. The only other discharge is the dried product.

WET AIR OXIDATION

Wet air oxidation is a process whereby a sludge will release bound water upon application of heat. Sludges which are colloidal gels, such as organic sludges, are composed of minute particles bound by sheaths which contain water as well as sludge solids. Upon application of heat the sheath will dissolve, releasing the bound water within the sludge particle. Upon release of bound water the sludge solids can be dewatered or can be further processed.

Thermal conditioning is a low-pressure wet air oxidation process where the end product, thermal conditioned sludge, is dewaterable. High-pressure wet air oxidation produces a sludge that is both dewaterable and oxidized.

Referring to Fig. 20.5, raw sludge is first ground to uniform consistency. A low-pressure pump feeds this sludge to a high-pressure positive-displacement pump which develops sufficient pressure to pump to the reactor. For thermal conditioning the reactor pressure will be 175 to 200 lb/in^2 gauge, and for high-pressure oxidation the reactor pressure can be as high as 1200 to 2400 lb/in^2 gauge.

Compressed air is injected into the sludge feed. Air is required for both thermal conditioning and wet air oxidation to provide turbulence within the sludge stream, increasing heat-transfer efficiency through the heat exchanger(s). It is also required for oxidation of some of the malodorous gaseous compounds released with the release of bound water. High-pressure wet air oxidation, however, requires additional air for oxidation of the sludge solids. Depending on the amount of air injected into the sludge, and system pressure, the conditioned sludge can be partially oxidized or can be fully oxidized to a sterile ash.

Steam is injected within the reactor to provide the required reactor pressure and residence time. Temperatures range from 300 to 650°F, and normal residual time within the reactor is from 10 to 30 min.

The hot sludge exiting the reactor passes through one or a series of heat exchangers where most of the available heat is transferred to the entering raw sludge to reduce the total system heating requirements. Raw sludge is pumped through the heat-exchanger tubes, and the hot conditioned sludge passes through the shell side of the heat exchanger.

From the heat exchanger the conditioned sludge goes to a separator, or storage tank, where a quiescent period is provided. Gases are released, and these gases are highly odorous. They are normally incinerated by direct flame or catalytic incineration before discharge to the atmosphere.

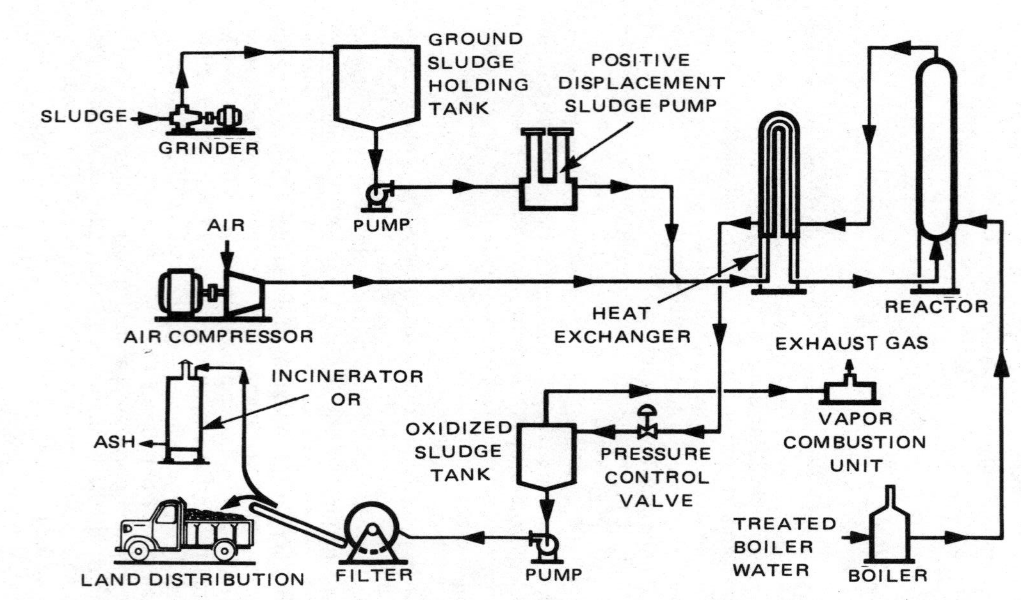

FIGURE 20.5 Flow sheet for high-pressure/high-temperature wet air oxidation. *Source:* ZIMPRO, Rothschild, Wis.

When a low-pressure thermal conditioning process is used, relatively little air is added to the system and the resultant sludge has approximately the same heating value (volatile content) as the raw sludge. Steam (or in some cases hot water) must be provided to maintain this reaction.

The high-pressure oxidation process, where relatively large quantities of air are introduced into the sludge stream, can produce a sterile ash, where the initial volatile content is completely oxidized and the resultant sludge has no heating value. Other than for initial startup, this process will often generate enough heat by oxidation of the volatiles to maintain itself without steam (or hot water) injection.

BIBLIOGRAPHY

Keller, F.: U.S. Patent 3,805,715, April 23, 1974.

"Municipal, Industrial Waste Dehydration System," *Resource Recovery and Energy Review*, (by magazine staff), January 1976.

Sommers, L., and Curtis, E.: "Wet Air Oxidation," *Journal of the Water Pollution Control Federation*, November 1977.

Swets, D., Pratt, L., and Metcalf, C.: "Thermal Sludge Conditioning in Kalamazoo," *Proceedings of the American Society of Civil Engineers*, March 1974.

Teletzke, G.: "Components of Sludge and Its Wet Air Oxidation Products," *Journal of the Water Pollution Control Federation*, June 1967.

CHAPTER 21
ENERGY RECOVERY

Incineration produces heat which can often be reclaimed and utilized. In one step, i.e., incineration, a waste can be destroyed and energy reclaimed by converting the waste heat to fuel-saving energy. Techniques for determining the energy available from waste heat will be presented in this chapter.

RECOVERING HEAT

Steam is used for incinerator heat recovery far more frequently than hot water or hot air (gas) generation. Steam is more versatile in its application, and 1 lb of steam contains significantly more energy than 1 lb of water or air. In general, while hot water is normally of use only for building heat during winter months or can be used in limited quantities for feedwater heating, steam can be used for process requirements and for equipment loads, which are year-round loads. Further, steam can be converted to hot water or used for air heating when these needs arise.

The calculations presented in this chapter are for steam generation.

APPROACH TEMPERATURE

With t the temperature of the heated medium (steam or hot water), t_i the temperature of the entering flue gas, and t_o the temperature of the flue gas exiting the boiler (see Fig. 21.1), the heat available can be calculated.

For any heat exchanger there is an approach temperature t_x. This temperature is the difference between the temperatures of the heated medium (t) and of the exiting flue gas (t_o). Therefore,

$$t_x = t_o - t$$

The more efficient the heat exchanger, the lower the approach temperature. The larger the heat exchanger, the lower t_x until, in the extreme case, with an infinitely large heat exchanger, t_x will be zero and the steam (or hot water) will be at the same temperature as the exiting flue gas.

In practice, the approach temperature of a waste heat boiler is on the order of 100°F for efficient and 150°F for standard, economical construction.

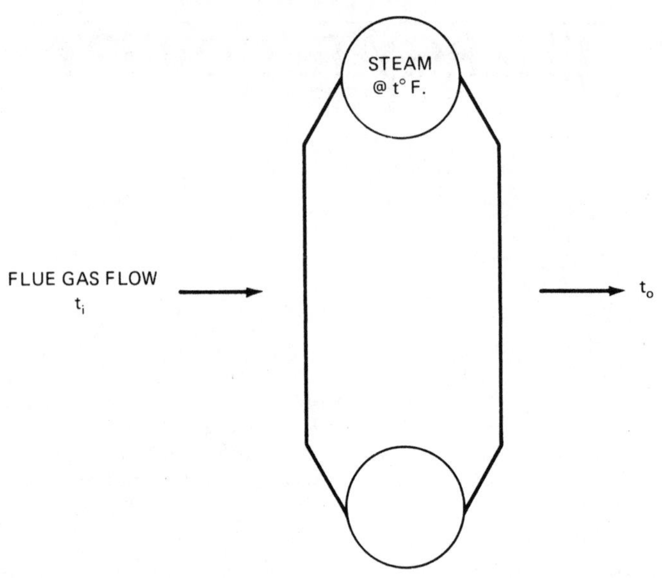

FIGURE 21.1 Waste heat boiler.

AVAILABLE HEAT

The heat available in exhaust or flue gas is equal to that heat at the boiler inlet less the gas heat content at the boiler outlet.

With Q the heat available from the flue gas stream, in Btu per pound, note the following:

$$Q = W(h \ @ \ t_i - h \ @ \ t_o)$$

The flue gas will have a dry and a wet component. Considering the dry gas component to have the properties of air (W_{dg}, h_a) and W_m the moisture component, this equation becomes

$$Q = W_{dg}(h_{ai} - h_{ao}) + W_m(h_{mi} - h_{mo})$$

The inlet temperature t_i is the temperature of the incinerator outlet. The outlet temperature of the heat exchanger (t_o) is defined by the approach temperature (t_x) and the temperature of the heated medium (t):

$$t_o = t_x + t$$

The enthalpy at the outlet of the heat exchanger must be evaluated at t_o.

Example. Consider a gas flow at 1400°F of 15,000 lb/h of dry gas plus 2000 lb/h of moisture. Let the available heat be calculated for the generation of saturated steam at 100, 200, and 400 lb/in^2 absolute with a 150°F approach temperature (note Table 4.1 for enthalpy values):

Inlet condition:
$$t_i = 1400°\text{F}$$
$$h_a = 341.85 \text{ Btu/lb}$$
$$h_{mi} = 1719.82 \text{ Btu/lb}$$
$$W_{\text{dg}} = 15{,}000 \text{ lb/h}$$
$$W_m = 2000 \text{ lb/h}$$

Outlet condition:

With $p = 100 \text{ lb/in}^2$ absolute, $t = 328°\text{F}$
$$t_o = 150 + 328 = 478°\text{F}$$
By interpolation, $h_{ao} = 101$ Btu/lb
$$h_{mo} = 1249 \text{ Btu/lb}$$
$$Q = 15{,}000(341.85 - 101) + 2000(1719.82 - 1249)$$
$$= 4.554 \text{ MBtu/h}$$

With $p = 200 \text{ lb/in}^2$ absolute, $t = 382°\text{F}$
$$t_o = 150 + 382 = 532°\text{F}$$
By interpolation, $h_{ao} = 115$ Btu/lb
$$h_{mo} = 1274 \text{ Btu/lb}$$
$$Q = 15{,}000(341.85 - 115) + 2000(1719.82 - 1274)$$
$$= 4.294 \text{ MBtu/h}$$

With $p = 400 \text{ lb/in}^2$ absolute, $t = 445°\text{F}$
$$t_o = 150 + 445 = 595°\text{F}$$
By interpolation, $h_{ao} = 130$ Btu/lb
$$h_{mo} = 1304 \text{ Btu/lb}$$
$$Q = 15{,}000(341.85 - 130) + 2000(1719.82 - 1305)$$
$$= 4.007 \text{ MBtu/h}$$

These calculations are summarized in Table 21.1. Steam temperature is listed in Table 21.2. The inlet is the total heat in the flue gas, related to 60°F:

$$15{,}000 \times 341.85 + 2000 \times 1714.82 = 8.567 \text{ MBtu/h}$$

The column Δt is the difference in flue gas temperatures entering and leaving the boiler. The efficiency noted is the available heat divided by the total heat in the flue gas entering the boiler.

Of significance is the relationship between available heat and Δt, the temperature difference of the flue gas across the boiler. The available heat is proportional to Δt.

For example, Q at $\Delta t = 805$ versus $\Delta t = 922$:

TABLE 21.1

Inlet, MBtu/h	°F	Steam pressure, psia	Steam tempera-ture, °F	Flue gas tempera-ture, °F	Δt, °F	Available heat, MBtu/h	Effi-ciency, %
8.567	1400	100	328	478	922	4.554	53
8.567	1400	200	382	532	868	4.294	50
8.567	1400	400	445	595	805	4.007	47

$$Q \text{ @ } 805 = \frac{805}{922} \times 4.554 = 4.0 \text{ MBtu/h}$$

Q at $\Delta t = 868$ versus $\Delta t = 922$:

$$Q \text{ @ } 868 = \frac{868}{922} \times 4.554 = 4.3 \text{ MBtu/h}$$

By comparing these values to the calculated values for Q in Table 21.1, it is clear that the available heat Q is directly proportional to Δt, the temperature loss in the flue gas.

STEAM GENERATION

Given the heat availability, the amount of steam that can be generated can be calculated. Figure 21.2 shows typical flow through a waste heat boiler producing steam. Makeup water temperature is raised to feedwater temperature by steam flow from the boiler and by the heat contained in return condensate. Condensate is returned to the deaerator.

Besides raising the feedwater temperature prior to injection into the boiler, the deaerator acts to help release dissolved oxygen from feedwater. Additional

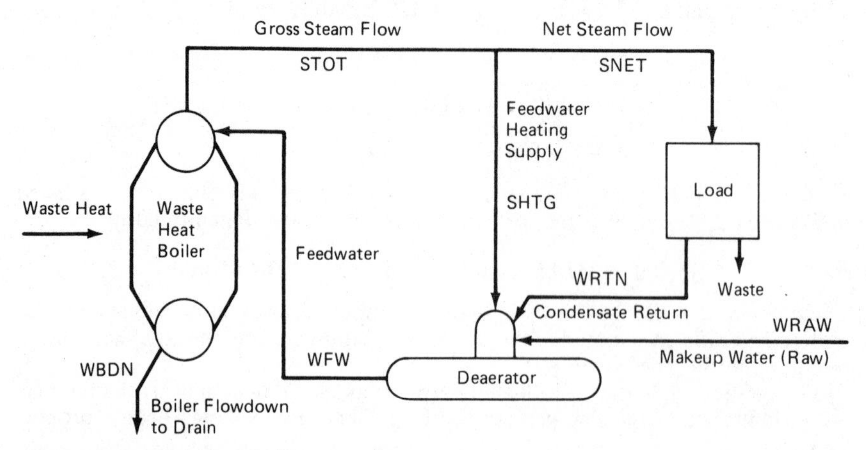

FIGURE 21.2 Waste heat boiler, steam flow.

feedwater treatment is usually employed to reduce, or prevent, scaling and corrosion of boiler surfaces. Water softeners are used to remove most of the calcium and magnesium hardness from raw water. Chemical addition is also used, typically as follows:

- *Sodium sulfite*. This is an oxygen-scavenging chemical that chemically removes the dissolved-oxygen component not removed in the deaerator. Hydrazine is another oxygen scavenger that is used in high-pressure (over 1200 lb/in^2 absolute) boiler applications.
- *Amine*. There are a number of amines in use for feedwater treatment. They are used for boiler pH or alkalinity control. Excess alkalinity (pH greater than 11) will result in accelerated scale buildup while low pH (below 6) can cause excessive boiler tube corrosion. Normally boiler water pH is maintained in the range of 8.0 to 9.5, slightly alkaline.
- *Phosphates*. This treatment is used to precipitate residual calcium and magnesium hardness remaining in feedwater after softening. Certain phosphates will act as dispersants, preventing adhesion of the precipitate to tube walls.

These chemicals will form a sludge, or mud, which will accumulate in the lower drum of a boiler. The boiler water must have a blowdown on a regular basis to prevent a buildup of mud within the boiler. This blowdown will normally represent from 2 to 5 percent of the boiler steam generation.

CALCULATING STEAM GENERATION

By using the steam tables (Table 21.2), calculations will be performed for obtaining steam, makeup, blowdown, and feedwater flows.

With an approach temperature of 150°F, generating 100 lb/in^2 absolute steam, dry and saturated (h_{stm} = 1187 Btu/lb, h_{bdn} = 298 Btu/lb), the heat available (from Table 21.1) is 4.554 MBtu/h. For this illustration, blowdown is 4 percent of the feedwater flow, feedwater is provided to the boiler at 220°F (h_{fw} = 188 Btu/lb), and 20 percent of the steam generation is returned as condensate at 170°F (h_{ret} = 138 Btu/lb). In addition, raw water enters the deaerator at 60°F (h_{mu} = 27 Btu/lb), and radiation loss from the boiler is 1 percent of the total boiler input.

The heat available for generating steam is the heat lost by the flue gas less the heat lost by boiler radiation:

$$\text{Waste heat} = 4.554 \text{ MBtu/h} - 0.01 \times 4.554 \text{ MBtu/h} = 4.508 \text{ MBtu/h}$$

Referring to Fig. 21.2,

$$\text{Waste heat} = \text{heat in steam} + \text{heat in blowdown} - \text{heat in feedwater}$$

$$\text{Heat in blowdown} = \text{WBDN} \times h_{bdn}$$

$$\text{WBDN} = 0.04 \times \text{WFW}$$

$$\text{Heat in blowdown} = 0.04 \times 298 \times \text{WFW} = 11.92\text{WFW}$$

$$\text{Heat in feedwater} = \text{WFW} \times h_{fw}$$

$$= 188\text{WFW}$$

TABLE 21.2 Saturated Steam Properties

Pressure, lb/in^2 absolute	Temperature, °F	Pressure, lb/in^2 absolute	Temperature, °F
14.7	212	45	274
15	213	50	281
16	216	55	287
17	219	60	293
18	222	65	298
19	225	70	303
20	228	75	308
21	231	80	312
22	233	90	320
23	235	100	328
24	238	125	344
25	240	150	358
26	242	175	371
27	244	200	382
28	246	250	401
29	248	300	417
30	250	350	432
32	254	400	445
34	258	450	456
36	261	500	467
38	264	600	486
40	267	700	503

Source: Ref. 21-1.

$$\text{Heat in steam} = \text{STOT} \times h_{\text{stm}} = \text{STOT} \times 1187 \text{ Btu/lb}$$

$$\text{STOT} = \text{WFW} - \text{WBDN} = \text{WFW} - 0.04\text{WFW} = 0.96\text{WFW}$$

$$\text{Heat in steam} = 0.96\text{WFW} \times 1187 = 1139.52\text{WFW}$$

$$\text{Waste heat} = 4.508 \text{ MBtu/h} = 11.92\text{WFW} + 1139.52\text{WFW} - 188\text{WFW}$$

$$= 963.44\text{WFW}$$

Therefore, $\text{WFW} = 4679 \text{ lb/h}$

and $\text{STOT} = 0.96 \times 4679 = 4492 \text{ lb/h}$

also $\text{WBDN} = 0.04 \times 4679 = 187 \text{ lb/h}$

To calculate steam required for feedwater heating, makeup, and condensate return flows, a material balance and a heat balance must be performed around the deaerator.

Material (flow) balance:

$$\text{WFW} = \text{SHTG} + \text{WRTN} + \text{WRAW}$$

From above: $\text{WRTN} = 0.2 \times \text{STOT} = 0.2 \times 4492 = 898 \text{ lb/h}$

$$\text{WFW} = 4679 \text{ lb/h}$$

Therefore,

$$\text{SHTG} = \text{WFW} - \text{WRTN} - \text{WRAW} = 4679 - 898 - \text{WRAW} = 3781 - \text{WRAW}$$

Heat balance:

$$\text{WFW} \times h_{fw} = \text{SHTG} \times h_{stm} + \text{WRTN} \times h_{ret} + \text{WRAW} \times h_{mu}$$

Therefore,

$$4679 \times 188 = (3781 - \text{WRAW}) \times 1187 + 898 \times 138 + \text{WRAW} \times 27$$

$$\text{WRAW} = 3218 \text{ lb/h}$$

$$\text{SHTG} = 3781 - 3218 = 563 \text{ lb/h}$$

Where condensate is not returned to the deaerator, i.e., to the boiler system, the steam required for feedwater heating will increase:

Material balance:

$$\text{SHTG} = \text{WFW} - \text{WRAW} = 4679 - \text{WRAW}$$

Heat balance:

$$4679 \times 188 = (4679 - \text{WRAW}) \times 1187 + \text{WRAW} \times 27$$

Therefore,

$$\text{WRAW} = 4030 \text{ lb/h}$$

$$\text{SHTG} = 649 \text{ lb/h}$$

In general, with no separate source of heat for feedwater heating (such as returned condensate), 12 to 15 percent of generated steam is required.

Table 21.3 summarizes the above calculations.

The net flow of steam (SNET) is that quantity of steam available for useful work. As can be seen, use of condensate return for feedwater heating increases the quantity of steam available for a load.

WATERWALL SYSTEMS

Larger incinerators dedicated to destruction of refuse or other paper-type waste materials are often designed with "waterwall" construction, as described in

TABLE 21.3

Avail. heat, MBtu/h	STOT, lb/h	WBDN, lb/h	WFW, lb/h	WRET, lb/h	WRAW, lb/h	SHTG, lb/h	WNET, lb/h
4.554	4492	187	4679	898	3218	563	3929
4.554	4492	187	4679	0	4030	649	3843

Chap. 13. These installations can be provided with a variety of features including the following:

- *Convection boiler section*. Boiler tubes are placed perpendicular to the flow of gas as it exits the incinerator. A major portion of available heat is captured by these tubes, producing saturated steam.
- *Economizer*. This is used to heat feedwater by extracting heat from gases as they leave the convection boiler section.
- *Superheater*. A tubular section is normally placed upstream of the convection section. Hot incinerator gases superheat steam generated from the convection section of the boiler.
- *Air preheater*. This is used in lieu of, or directly downstream of, the economizer. It produces heated combustion air from the relatively low-temperature gas flow at this location.

Calculations of steam generation from each of these sections are a complex task and will not be detailed here. Tables 13.8 and 13.9 indicate steam generation for typical waterwall incinerators for a variety of waste quality.

To calculate available heat by the methods of this chapter, the exit gas temperature (the temperature of flue gas exiting the boiler sections and entering the air emissions control system) can be assumed to be in the range of 350 to 550°F.

REFERENCE

21-1. Keenan, J., and Keyes, F.: *Thermodynamic Properties of Steam*, John Wiley & Sons, New York, 1957.

BIBLIOGRAPHY

Boyen, J.: *Practical Heat Recovery*, 1st ed., John Wiley & Sons, New York, 1978.

Brown, D.: "Energy Recovery from Industrial Waste," *Waste Age Magazine*, April 1981.

Energy from Municipal Solid Wastes, Report of the U.S. House of Representatives, June 1980.

Energy Utilization: Municipal Waste Utilization, U.S. Department of Energy, March 27, 1981.

Freeman, H.: "Pollutants from Waste-to-Energy Conversion Systems," *Environmental Science and Technology*, November 1978.

Hasselriis, F.: "Thermal Oxidizers Convert Industrial Waste into Valuable Energy," *Power*, February 1982.

"Heat Recovery," *Compressed Air*, (by staff) July 1980.

Lasday, S.: "Production of Energy Pellets from Biomass," *Industrial Heating*, November 1979.

Materials and Energy from Municipal Waste, Office of Technological Assessment, 1979.

Merrill, R.: *Disposing of Hazardous Wastes in Industrial Boilers*, USEPA, March 1981.

"Recovering the Energy in Wastes," *Mechanical Engineering*, May 1981.

Rinaldi, G.: *An Evaluation of Emission Factors for Waste to Energy Systems*, USEPA, 1979.

Ross, A.: "The Chemical Control of Air Heater Corrosion and Plugging," American Society of Mechanical Engineers, Winter Annual Meeting, November 1973.

"Specifying and Operating Reliable Waste Heat Boilers," *Chemical Engineering*, (by staff) August 13, 1979.

Stevens, J.: *Energy Recovery and Emissions from Municipal Waste Incineration*, Ontario Air Pollution Control Association, September 1977.

Stricker, G.: "Twin Incineration Systems with Heat Exchangers Start Up Smoothly," *Chemical Processing*, September 1977.

CHAPTER 22

AIR POLLUTION CONTROL

Air discharges from incinerators include particulate matter, gases, odor, and noise. In this chapter equipment available for air pollution control will be discussed. Chapter 23 addresses the control of acid gas discharges.

THE CONTROL PROBLEM

There are numerous types and sizes of air pollution control equipment and systems on the market today. They range from the unsophisticated, a series of baffles to separate large particulate matter from a gas stream, to the highly sophisticated, high-energy water scrubbing devices utilizing alkalai to clean gas streams of solid, liquid, and gaseous pollutants.

As presented in Chap. 7 and referenced in Chaps. 2 and 3, incinerators will necessarily produce air pollutants which must be removed from the gas stream prior to atmospheric discharge. These pollutants must be removed to provide discharge levels below those mandated by government statutes. In many cases incinerator discharges must be free of additional components in order to safeguard downstream equipment from excessive wear from erosion (physical wear of materials) and/or corrosion (chemical degradation of materials).

The generation of pollutants is a function of the following factors:

- Waste composition
- Charging rate
- Method of charge
- Furnace type
- Furnace design
- Burning conditions (three T's, i.e., temperature, turbulence, time)
- Excess air introduced

The more common pollutants discharged from incinerators which must be substantially removed from the gas stream include the following:

- Particulate matter
- Chlorides
- Sulfur oxides

Other pollutants, as discussed in Chap. 7, may not be present in incinerator discharges to the extent that they warrant particular attention. If they are present in any significant amounts, however, they will often be removed by the systems installed for control of the three pollutants identified above, or by control of the combustion process.

PARTICLE SIZE

The efficiency of any control device will vary as a function of particle catch size. Particulate matter is normally measured by mean physical size in units of one-millionth of a meter, or a micron, represented by μ.

Particle size cannot be calculated; it can be measured, and estimates can only be inferred from prior data based on measurement. As would be expected, therefore, data on particle sizes are rare. They are a function of waste properties, furnace design, and furnace operation.

Table 22.1 lists particle size data measured at the outlet of a multiple-hearth incinerator burning thermally conditioned sewage sludge. It lists the particulate size and loading both before and after the gas scrubbing system.

TABLE 22.1 Sewage Sludge Incineration, Airborne Particle Size

	Percentage by weight less than indicated size	
	Location 1[a]	Location 2[b]
Particulate loading,		
gr/st ft^3	1.88	0.01
lb/h	217.01	1.48
lb/ton	52.08	0.36
Size distribution, μ		
18.7	37.9	100.0
11.7	30.6	98.0
8.0	16.4	94.9
5.4	6.6	93.4
3.5	2.6	92.8
1.8	1.6	83.3
1.1	0.9	67.7
0.76	0.1	54.6

[a]Measured at incinerator outlet without controls, burning 100 tons/day wet sludge cake.
[b]Measured after venturi and tray scrubber with a total pressure drop of 30 in WC.
Source: Envisage Environmental Laboratory, Cleveland, Ohio, "Tests of the Cleveland Southerly Wastewater Treatment Center Multiple-Hearth Incinerators, May 1981."

Listed in Table 22.2 are particle size and emissions data from each of two incinerators burning municipal refuse.

Note that with sludge or refuse over half the particulate is larger than 30 μ. The scrubber, as shown in Table 22.1, removes practically all the particulate matter over 5 μ in diameter.

Table 22.3 lists chemical elements found in a typical particulate catch after organic (carbonaceous) components present were removed.

TABLE 22.2 Refuse Incineration, Airborne Particle Size

	Unit no. 1 (250 tons/day)	Unit no. 2 (120 tons/day)
Particle specific gravity, lb/ft^3	2.70	3.77
Particle bulk density, lb/ft^3	30.9	9.4
Loss on ignition at 1400°F, %	8.2	30.4
Size distribution (% by weight less than indicated size), μ:		
30	40.4	50.0
20	34.6	45.0
15	31.1	42.1
10	26.8	38.1
8	24.8	36.3
6	22.3	33.7
4	19.2	30.0
2	14.6	23.5
Particulate emission rate		
lb/ton	24.6	9.1
lb/h	256.3	45.5

Source: Ref. 22-1.

TABLE 22.3 Inorganic Component Analysis: Refuse Burning Incinerator

Component	Weight Percent
SiO_2	53.0
CaO	14.8
MgO	9.3
Al_2O_3	6.2
Na_2O	4.3
TiO_2	4.2
K_2O	3.5
Fe_2O_3	2.6
P_2O_5	1.5
ZnO	0.4
BaO	0.2

Source: Ref. 22-2.

Particle diameters and their general properties and characteristics are shown in Fig. 22.1. Some of the information in this figure can be used to calculate the minimum size chamber required for settling of particulate. For example:

A 30-μ particle has a settling velocity of approximately 5 cm/s, equivalent to $5 \div 2.54 \div 12 = 0.164$ ft/s. If it passes through a 2-ft-high chamber at 20 ft/s,

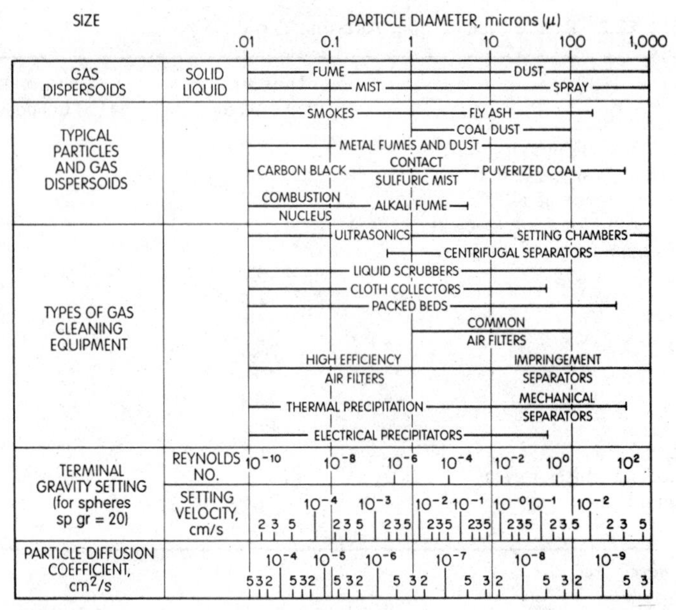

FIGURE 22.1 Particle size data. *Source:* Ref. 22-3.

the chamber length would have to be long enough to provide sufficient time for settling:

$$\frac{2 \text{ ft high}}{0.164 \text{ ft/s}} = 12.2 \text{ s to settle}$$

$$20 \text{ ft/s velocity} \times 12.2 \text{ s} = 244 \text{ ft long}$$

The time necessary for settling of a 30-μ particle would require a chamber length of 244 ft, assuming no turbulence, which would increase settling velocity.

This calculation indicates the impracticality of utilizing chambers for removal of small-size particulate matter. As previously noted, half of the particulate generated from incineration processes may be less than 30 μ in diameter. The smaller the diameter, the larger the settling time, therefore, the larger the settling chamber required. For the above example of a 244-ft-long chamber, the size of such equipment makes their use of questionable value.

Of interest is the fact that particles greater than 10 μ can be seen as individual particles by the naked eye.

Below 10 μ, although the eye cannot distinguish individual particles, there is a net effect of particle size on atmospheric clarity. As shown in Fig. 22.2, particulate in the range of 0.3 to 1.0 μ will severely affect visual clarity.

REMOVAL EFFICIENCY

The efficiency of a control device is simply the weight of material removed divided by the weight of that material entering the device. It can be calculated for

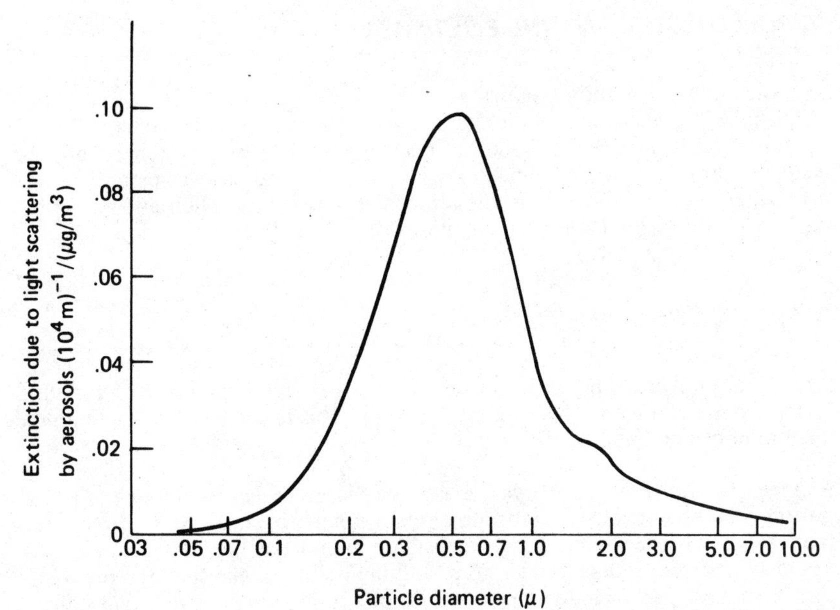

FIGURE 22.2 Efficiency of light scattering as a function of particle diameter. *Source:* Ref. 22-4.

a specific size of particulate or can be based on total particulate loading. For example, given the particle size information in Table 22.1:

Total flow into control devices, 217.01 lb/h for particles less than 1.1 μ:

$$0.9\% \times 217.01 \text{ lb/h} = 1.95 \text{ lb/h}$$

Total flow out of control device, 1.48 lb/h for particles less than 1.1 μ:

$$67.7\% \times 1.48 \text{ lb/h} = 1.00 \text{ lb/h}$$

Removal efficiency of particles less than 1.1 μ:

$$\frac{1.95 - 1.00}{1.95} = 48.72\%$$

Removal efficiency of total particulate load:

$$\frac{217.01 - 1.48}{217.01} = 98.32\%$$

As would be expected, in this case the removal efficiency of small particulate matter is significantly less than the removal of larger particles—less than 50 percent of the 1.1-μ particles are removed, while over 98 percent of the total particulate input is removed.

AIR POLLUTION CONTROL EQUIPMENT

The choice of air pollution equipment depends not just on the emission quality and quantity but also on conditions external to the incinerator system. For instance, if water is not readily available on a site, wet scrubbing systems cannot be considered; limited sewer facilities may preclude the use of electrostatic precipitators where the precipitator soot loading is sewered; space limitations will impact choice of a baghouse for air pollution control; and so on.

EQUIPMENT SELECTION

There are a number of industries where air emissions control has been found to be effective with only one type of control device. This is particularly true with the incineration of municipal waste:

- *Municipal solid waste*. In the vast majority of large-scale installations of refuse burning equipment, electrostatic precipitators have been found to provide effective and reliable gas cleaning. Particulate matter has been the more critical of emissions from refuse incinerators, and the ability of electrostatic precipitators to remove particles down to fractions of a micron is unique. Other control devices have been used, including wet scrubbers and baghouses, but their efficiency and wet discharge (in the case of scrubbing equipment) and reliability (in the case of fabric filters) have not been always found to be satisfactory. With the growing concern for control of other emissions in the off-gas, such as acid gases and organics, dry scrubbing systems have been developed. These and associated control systems will be discussed in Chap. 23.
- *Municipal sludge*. The incineration of sewage sludge requires not only the control of particulate but also the removal of gaseous components from the exhaust stream. In general, a sludge incinerator does not operate at the higher temperatures that are developed within a refuse incinerator. As a result, additional unburned organics may be present in the gas stream exiting a sludge incinerator as compared to the exhaust of a refuse incinerator. Wet scrubbers have been found to be effective in removal of gaseous as well as particulate components from sludge incinerator exhaust. In the majority of installations these scrubbers are high-energy devices, requiring in excess of 30 in WC differential; however, their water flow requirement is usually not a limitation. Sewage sludge incinerators are located at wastewater treatment plants where vast quantities of water are available.

Other incinerator usage may be required to provide wet scrubbing if, for instance, acid components are generated and the waste is considered hazardous. In general, however, except for the above two instances, the variability of industrial waste streams precludes generalization of the type of control equipment to provide. Industrial applications each require their own analysis.

The design of air emissions control equipment for a particular installation requires more than selection of the appropriate type of device; flues, dampers, and air moving equipment must also be selected and sized. Sizing of ancillary equipment, and the control device itself, is in large measure a function of particle size distribution. As noted previously in this chapter, particle size cannot be calculated. Estimates can be made from previous measurements, such as the distribution of particle size listed in Tables 22.1 and 22.2. There is no certainty, however, that these data are representative of any other installation. The data are based on grab sampling, where a discrete number of samples have been obtained and each of these samples has been analyzed. At another time of day, this distribution could be different. If there were any change in waste quality, such as in moisture content or heating value (which normally will vary from one lot to another), the size distribution will most likely change. These tables can be used as guides to relative magnitudes of particulate in off-gas, but they cannot necessarily be used as a design basis. Accurate size distribution can be determined, unfortunately, only by full-scale testing of actual equipment. Control device designers and manufacturers will normally be required to make their own estimates of size data to determine equipment sizing, based on their own experience and related empirical factors. An incinerator designer is usually not able to provide this information.

SETTLING CHAMBERS

As discussed previously, settling chambers (expansion chambers), which are basically long, boxlike structures, are ineffective for all but the heaviest, i.e., largest, particles. They are normally not used where the particulate size is 40 μ or less.

Some attempts have been made to utilize gravity settling by decreasing the height of the chamber, i.e., by placing horizontal plates in the direction of flow. With the previous calculations, if the vertical drop were reduced from 2 ft to 1 in, maintaining the same velocity of the gas stream (20 ft/s) a length of 1/24 of 260, or less than 11 ft, would be required for settling of larger particles.

In general, settling chambers are not practical for use as pollution control devices other than as part of a larger system to help remove larger particle sizes from the gas stream.

DRY IMPINGEMENT SEPARATORS

Impingement separators are essentially a series of baffles placed in the gas stream. The relatively high inertia of particulate within the gas stream will maintain their direction of flow while the gaseous component of the stream will change its direction of flow around a baffle. Particulate matter will drop, and the gas flow will continue through the process stream.

This method of particulate removal can be effective for larger particles, above 15 μ, but smaller particulate matter will continue to flow with the gas stream.

DRY CYCLONIC SEPARATORS

The cyclone is an inertial separator. As shown in Fig. 22.3, gas entering the cyclone forms a vortex which eventually reverses and forms a second vortex leaving the cyclonic chamber. Particulate, because of their inertia, tend to move toward the outside wall. They will drop from this wall, the sides of the cyclone, to an external receiver for ultimate disposal.

Cyclones will remove larger-size particulate matter from the gas stream (greater than 15 μ), but will have negligible effect on smaller particles. A cyclone collector is often placed before another control device such as an electrostatic precipitator or a baghouse. The cyclone will remove larger particles from the gas stream, and in many cases the cyclone solids discharge is tied directly to

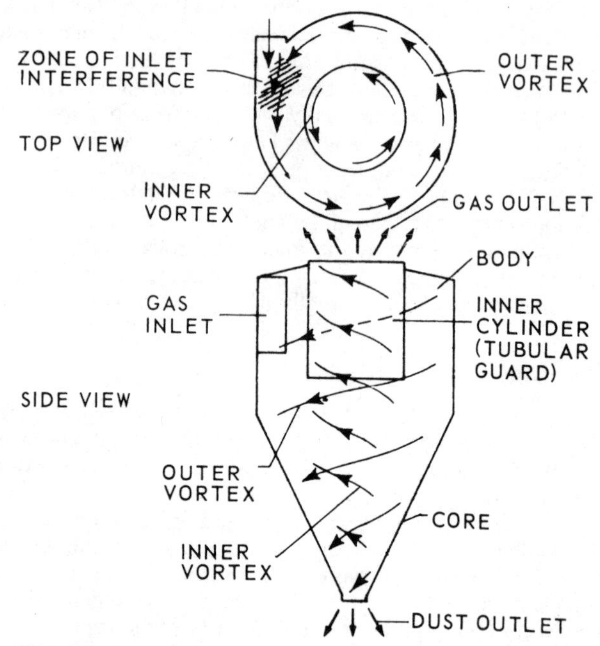

FIGURE 22.3 Conventional reverse-flow cyclone. *Source:* Ref. 22-5.

the incinerator, feeding collected particulate back to the furnace for complete burnout. Removal of these larger particles from the gas stream results in increased efficiency for the downstream control equipment, which now has only the smaller-sized particles to deal with. Of note is that the larger particles are normally the greatest weight component of particulate in a gas stream. For instance, 85 percent of the emissions weight may be over 15 μ, with only 15 percent below 15 μm. Use of a cyclonic collector, which has a good removal efficiency for larger particles, will allow provision of less efficient downstream control devices while providing a high overall removal efficiency, based on the total particulate size loading.

With D_{pc} the particle cut size, the ratio D_p/D_{pc} is the actual particle size D_p related to the particle cut size. The cut size is the diameter of those particles collected by a particular piece of equipment with 50 percent efficiency. Figure 22.4 shows the relationship between particle size and collection efficiency. Collection efficiency drops off rapidly as the particle size decreases.

One danger in the use of cyclonic separators is the reluctance of collected particulate to drop from the cyclone walls. This condition, agglomeration, may occur if the dust is fibrous, sticky, or hygroscopic (water-absorbing) or if the gas stream contains too much particulate matter (100 g/ft^3 is a practical maximum, with 7000 gr/lb).

WET COLLECTION

Wet devices for particulate removal utilize two mechanisms. First, particles are wetted by contact with a liquid droplet. Second, the wetted particles impinge on

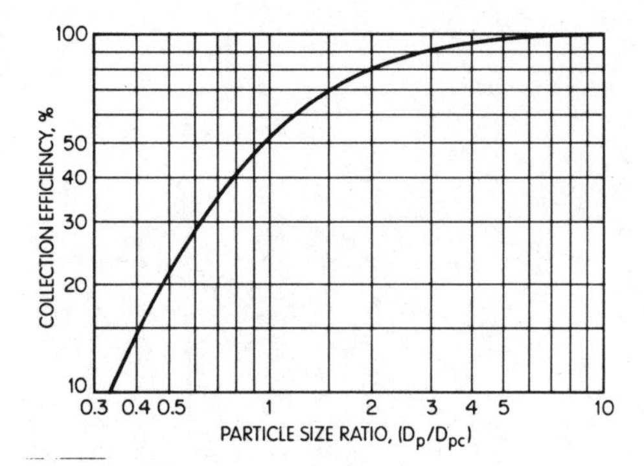

FIGURE 22.4 Cyclone efficiency vs. particle size ratio. *Source:* Ref. 22-6.

a collecting surface where they are subsequently removed from that surface by a liquid flush.

One of the simplest devices for wet collection of particles is wetting a collecting baffle. Particulate is continually washed from the baffle, or wall, preventing agglomeration and presenting a continual clean liquid surface to aid in catch and removal.

LIQUID SPRAYING

A spray directed along the path of dust or other particulate matter will impinge upon these particles with an efficiency directly proportional to the number of droplets and to the force imparted to the droplets. This results in a droplet size range; i.e., the smaller the size of the individual droplets, the greater their number. However, the smaller the droplet particle size, the less impingement, or inertial force, is associated with them. The optimum water droplet particle size is approximately 100 μ. Above 100 μ there are too few particles, and below 100 μ the droplets have insufficient inertia.

The mechanism of diffusion promotes deposition of dust particles on water droplets. Diffusion, or brownian movement, is that property of materials of different diameters to intermingle although the materials may be at rest, much as natural gas diffusing within a contained room although the air within that room is at rest. Diffusion helps the particulate and droplets come in contact and is inversely proportional to the size of water and/or solid particulate. The smaller the particle or droplet size, the more rapid the diffusion or the quicker the wetting process will be.

Spray chambers at times are utilized as a low-cost means of removing heavier particulate from a gas stream. Water is sprayed at rates of 3 to 8 gal/min per 100 ft^3/min of gas flow, and the heavier particulate matter is wetted and drops out of the gas stream.

CYCLONE COLLECTORS

Cyclone collectors are basically dry cyclones provided with a water spray to promote collection and to remove collected particulate.

One type of wet cyclonic collector is shown in Fig. 22.5. The equipment is a

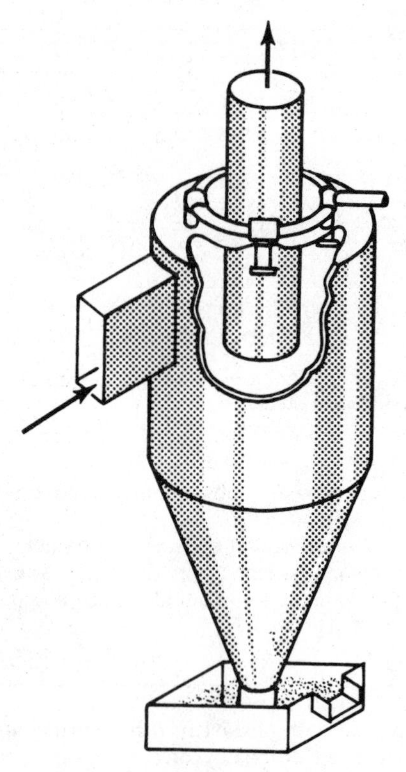

FIGURE 22.5 Conventional cyclone converted to a scrubber. *Source:* Ref. 22-5.

conventional dry cyclone, shown in Fig. 22.3, with water sprays installed near the gas entrance. Wetting of the particulate matter occurs almost immediately upon entering the cyclone, and the spray water washes the cyclone walls clean.

Figure 22.6 shows a cyclonic collector where gases enter tangentially from the bottom of the unit. The placement of sprays within the scrubber promotes a cyclonic effect as the gas rises to the scrubber exit. As with the other cyclones, water collects from the walls of the scrubber and exits from the bottom of the unit.

Cyclone collectors require 4 to 10 gal/min of water per 1000 ft^3/min of entering gas flow. Their pressure drop can be as low as 2 or as high as 8 in WC.

Dimensions of a typical wet cyclone are shown in Fig. 22.7.

WET CORROSION

Gases exiting an incinerator will contain sulfide and halogen components from the waste feed. These components, in addition to organic residuals, can result in significant acid formation in the gas scrubbing liquid. The use of nonmetallic chambers or nonmetallic-coated carbon steel is a growing method of corrosion control. These materials, such as flake polyester, fiberglass, neoprene, and epoxy, have good anticorrosion properties; however, they are temperature-

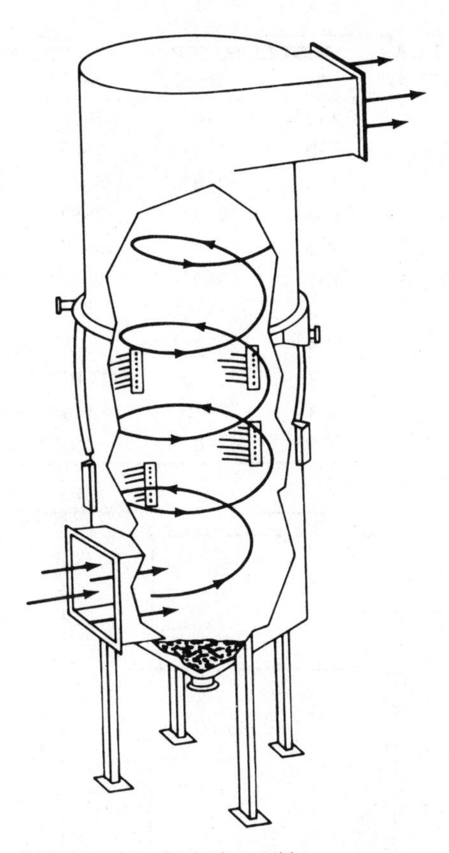

FIGURE 22.6 Cyclonic scrubber.

sensitive. In most cases these materials cannot withstand temperatures above 160 to 200°F. Use of such materials must include provisions for emergency quenching to provide cooling if the normal cooling water supply is lost.

Table 22.4 lists the resistance of selected alloys to incinerator scrubber solutions having relatively high acidity. Also listed is the resistance of these alloys to corrosion when fabricated as a fan (induced-draft fan) downstream of the scrubber. Note that stainless steel (316L, 310, 446, and USS 18-18-2) has poor corrosion resistance to acid attack.

MIST ELIMINATORS

Wet gas cleaning equipment floods a gas with scrubbing liquid. Water droplets usually are carried off by the gas stream exiting the scrubber. A mist eliminator is a stationary piece of equipment that removes most of the entrained water droplets from the gas. This is desirable to reduce the plume exiting the stack (excess water in off-gas will appear as a white plume), and will reduce the moisture collected at the induced-draft fan and other downstream equipment. Decreasing this liquid accumulation will also decrease attendant corrosion.

SATURATED GAS VOLUME (ACFM)*	MIST ELIM. SEP.		CYCLONE SEP.	
	A	B	A	B
2,500	2′-2″	7′-8″	3′-0″	8′-0″
5,000	3′-0″	8′-11″	4′-0″	10′-0″
10,000	4′-3″	10′-6″	5′-4″	15′-1″
25,000	6′-8″	14′-0″	7′-8″	20′-6″
50,000	9′-5″	17′-3″	10′-4″	27′-4″
75,000	11′-7″	19′-9″	12′-3″	31′-9″
100,000	13′-4″	21′-11″	13′-10″	37′-0″
150,000	16′-4″	25′-8″		
200,000	18′-10″	28′-10″	NOT RECOMMENDED	
300,000	23′-1″	33′-10″	(Parallel Units	
500,000	29′-9″	41′-3″	can be used)	

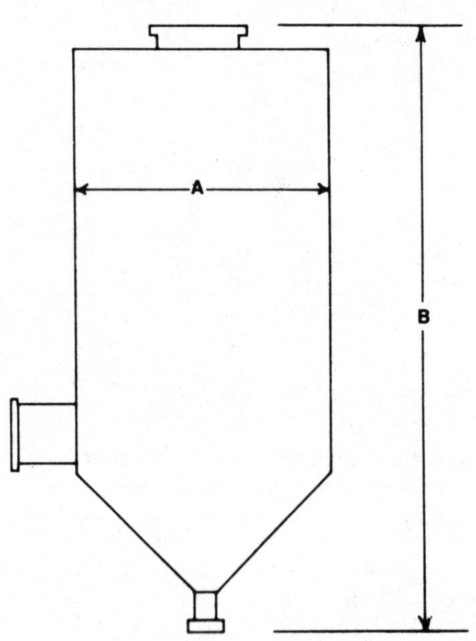

FIGURE 22.7 Wet cyclone. *Source:* Koch Engineering Co., Inc., Pollution Control Div., New York.

The most common mist eliminator in use with scrubbing equipment is the chevron type, Fig. 22.8. The moisture-laden gas flow travels through a series of baffles. Gas exits the baffles and the heavier moisture particles, because of their inertia, impact the chevron-shaped baffles and fall to the bottom of the chamber, leaving the gas stream.

Figure 22.9 is a diagrammatic representation of an impingement tray tower which functions as a mist eliminator. The wet gas stream passes through a perforated plate and then impacts on a series of small plates, impingement plates. As

TABLE 22.4 Evaluation of Alloys for Incinerator Scrubber

Alloy	Corrosion Results	
	Scrubber Solutions	Fan Deposits
Ti6Al-4V	good resistance	good resistance
Hastelloy C	good	good
Inconel 625	good	good
Hastelloy F	good	–
Hastelloy C-276	good	–
Hastelloy G	good	–
Ti75A	good	–
S-816	good	pitted
Carpenter 20	pitted	pitted, SCC
Incoloy 825	pitted	pitted
Incoloy 800	–	pitted
316L	pitted, SCC	pitted, SCC
310	pitted	–
446	pitted	–
Inconel 600	trenches	–
Inconel 601	trenches	–
Armco 22-13-5	pitted	pitted, SCC
USS 18-18-2	pitted	pitted, SCC
Type 304	pitted, SCC	pitted, SCC

SCC: Stress corrosion cracking.
Source: Ref. 22-7.

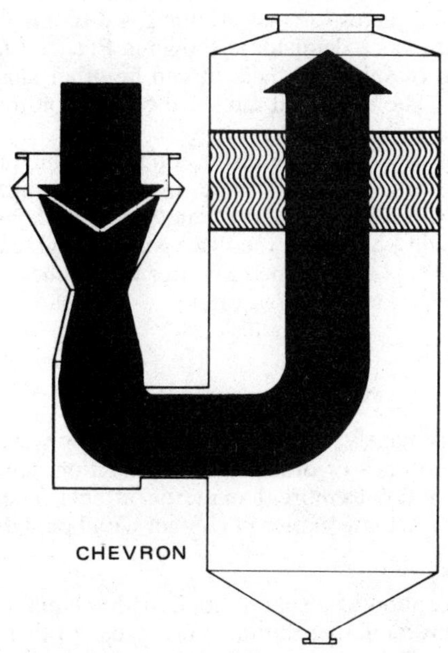

FIGURE 22.8 Mist eliminator. *Source:* Peabody-Galion Co., Princeton, N.J.

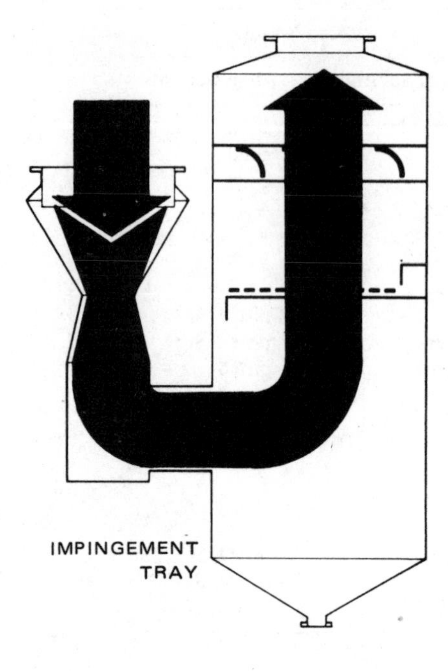

FIGURE 22.9 Mist eliminator. *Source:* Peabody-Galion Co., Princeton, N.J.

with the chevron scrubber, inertial forces allow the gas to flow around the impingement plates while the entrained moisture falls to the bottom of the tower. The use of impingement plates for gas scrubbing is discussed later in this chapter.

A packed tower used as a demister is shown in Fig. 22.10. Moisture-laden gas passes through a bed of spherical balls (it can be other shapes) which tends to allow passage of gas. The entrained moisture collects on the packing and eventually drops to the bottom of the tower.

There are many other types of demister equipment, unique to particular manufacturers. They typically have a pressure drop in the range of 1 to 3 in WC at rated flow. They are often used as part of another item of gas cleaning (or scrubbing) equipment, as will be seen in the discussion of high-energy scrubbing systems. Dimensions of a typical cyclone separator with a chevron-type demister are listed in Fig. 22.7 for various gas flow rates.

GAS SCRUBBING

The wet collectors previously discussed were basically washing devices used in conjunction with centrifuges or other inertial collection devices. Scrubbing of a gas stream, however, involves more than inertial effects. The mechanisms for gas scrubbing (bringing particulate matter in contact with liquid droplets) includes the following:

- *Interception*. Interception of a solid particle with a liquid particle occurs when the two particles have relative motion toward each other and are within one radius of each other. This is the radius of the smaller of the particles.
- *Gravitational force*. Gravitational force causes a particle, when passing an obstacle, to fall from its streamline and settle upon the surface of the obstacle.

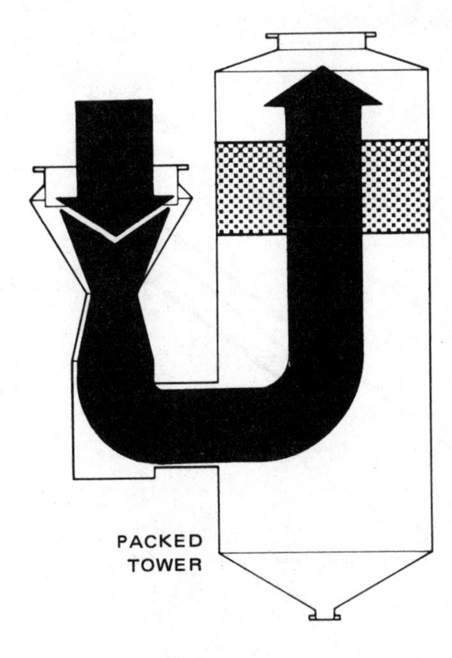

FIGURE 22.10 Mist eliminator. *Source:* Peabody-Galion Co., Princeton, N.J.

- *Impingement*. Impingement occurs when an object, placed in the path of a particle-laden gas stream, causes the gas to flow around the obstacle. Larger particles will tend to continue in a straight path because of their inertia and may impinge on the obstacle and be collected.
- *Contraction and expansion*. Contraction of a gas stream will tend to produce condensation of the moisture within the stream. High turbulence in a contracted area will result in good contact between solid particulate and the liquid particles. The dust-laden liquid particles will have the same velocity as the rest of the gas stream, and then passing through an area of expansion, these particles will continue their direction of flow while the balance of the gas stream can be directed to flow in another direction. In effect, this process of contraction and expansion produces good separation of particulate matter from the gas stream.

The above mechanisms are normally all present in gas scrubbing equipment at the same time.

The effectiveness of a scrubbing system is usually directly related to the pressure drop across the scrubber. The higher the pressure drop, the greater the turbulence and mixing and, therefore, the more effective the scrubbing action. This feature is illustrated by the graph in Fig. 22.11. For a 2-μ size particle, for instance, a pressure differential of 8 in WC will have a removal efficiency of 95 percent, whereas a 35-in WC differential will result in almost total (99.9 percent) removal from the gas stream.

Scrubbing systems are differentiated by system pressure drop. A low-energy system is normally defined as one producing less than 12 in WC for particulate removal, whereas high-energy systems will have significantly higher pressure differential, from 20 to over 60 in WC. Likewise, medium-energy systems produce 7 to 20 in WC differential.

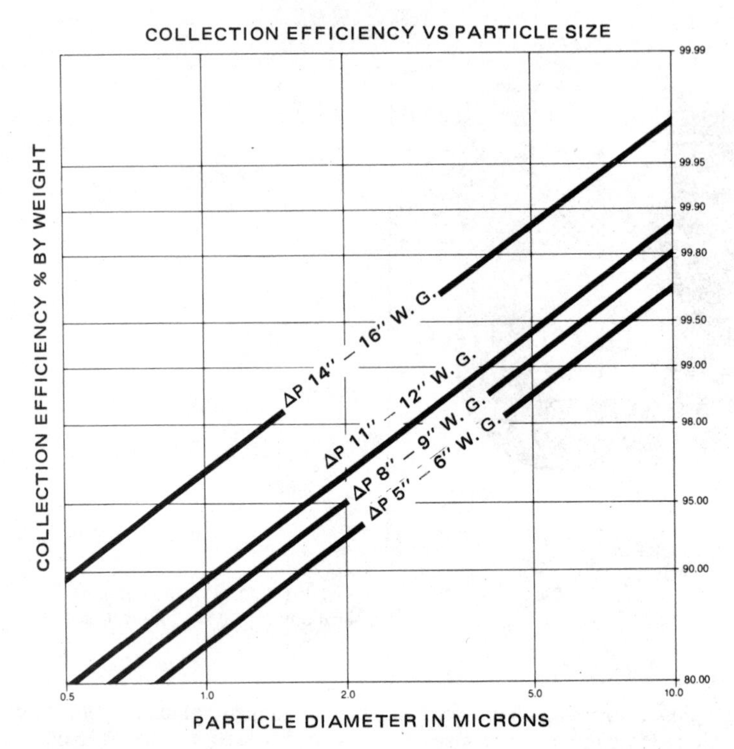

FIGURE 22.11 Collection efficiency vs. particle size. *Source:* Joy Manufacturing Co., Western Precipitator Div., Los Angeles.

CYCLONIC SCRUBBER

A cyclonic scrubber is shown in Fig. 22.12. The damper (referred to as a *spin damper*) creates contraction within the entering gas stream. Normally water is injected immediately upstream of the spin damper. The flow expands within the cyclonic chamber, and the expansion and contraction effect plus the inertial effects within the cyclonic tower scrub the gas and separate the dust-laden water from the gas stream.

This scrubber is termed *low-energy*. The pressure drop across the entire unit ranges from 6 to 12 in WC. The spin damper position can usually be controlled to provide a lesser or greater inlet velocity to the cyclone. This control also provides the necessary adjustment when passing low flow versus high or rated flow. The water consumption will vary from 5 to 15 gal per 1000 ft^3 of gas.

VENTURI SCRUBBER

Venturi scrubbers are widely used, where water is readily available, as high-efficiency, high-energy gas cleaning devices. The heart of this system is a venturi

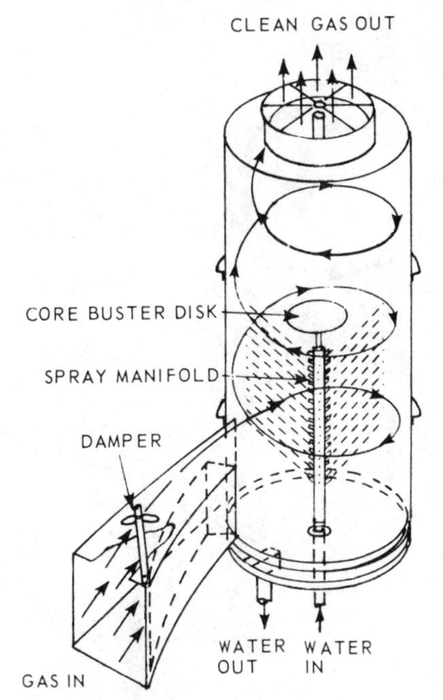

FIGURE 22.12 Cyclonic scrubber. *Source:* Pease Anthony Construction Corp., Division of Chemical Construction Corp.

throat where gases pass through a contracted area, reaching velocities of 200 to 600 ft/s, and then pass through an expansion section. From the expansion section the gas enters a large chamber for separation of particles or for further scrubbing. Water is injected at the venturi throat or just upstream of the venturi section. Figure 22.13 illustrates a venturi scrubber where water is injected at its throat. For this design, 5 to 7 gal per 1000 ft^3 of gas is required.

An adjustable venturi throat is shown in Fig. 22.14. Two throat flaps are illustrated. However, it is often designed with a single flap. Water is injected into a precooler section immediately before the throat. The throat area is adjustable and is normally controlled (manually or automatically) to maintain a desired pressure drop. Note that in this illustration a vane or chevron demister (water separator) is included in the outlet chamber to remove entrained moisture particles.

A flooded-disk (or plumb-bob) scrubber design is illustrated in Fig. 22.15. The conical plug is positioned to increase or decrease the gas flow area, similar to the throat flap design. Typical dimensions for this unit are given for a range of gas flows entering the system, through flange A. Water is injected above the venturi section, tangentially to the inlet gas flow.

A serious problem associated with venturi sections is the erosive effect of the gas-liquid mixture passing through the throat. As indicated, the throat velocity is extremely high, creating the potential for wear of the throat surface through the mechanism of erosion. In addition, the corrosive effect of the wet, acidic gas flow is heightened by its turbulent action.

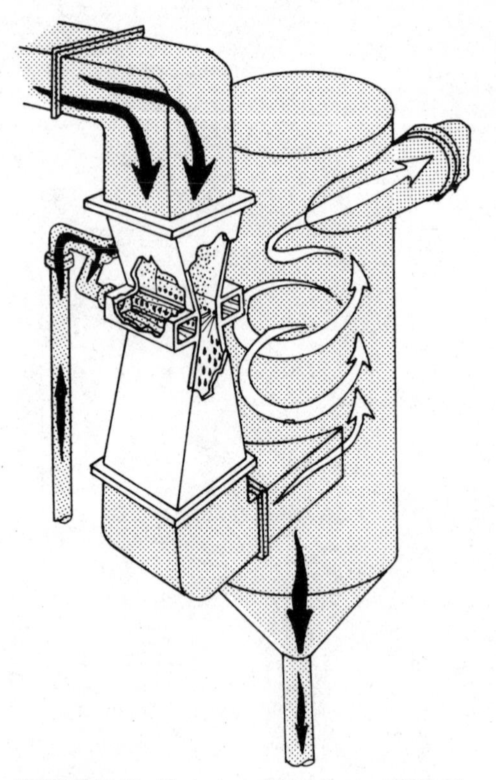

FIGURE 22.13 Venturi scrubber. *Source:* UOP Air Correction Div., Des Plaines, Ill.

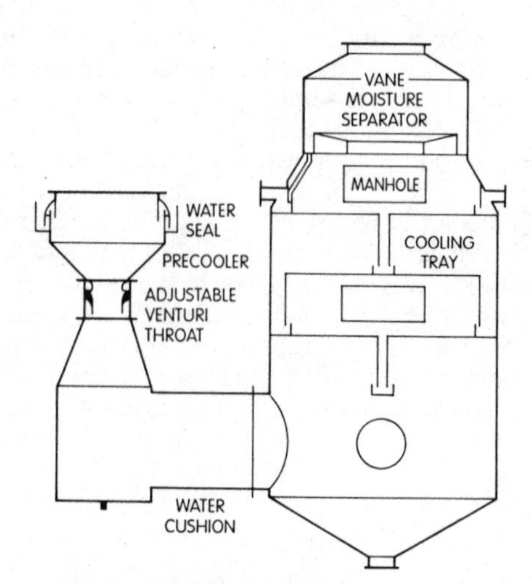

FIGURE 22.14 Venturi scrubber.

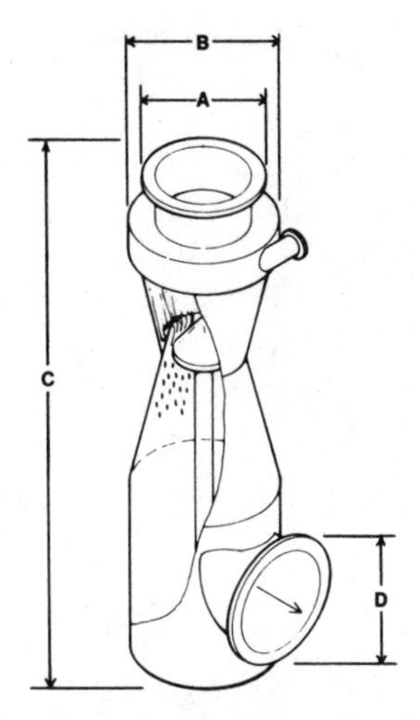

Saturated Gas Volume (ACFM)	A	B	C	D
2,500	1′-0″	3′-6″	6′-9″	1′-1″
5,000	1′-4″	3′-10″	7′-11″	1′-6″
10,000	1′-10″	4′-4″	8′-10″	2′-0″
25,000	3′-0″	5′-6″	10′-4″	3′-3″
50,000	4′-2″	7′-2″	13′-7″	4′-5″
75,000	5′-2″	8′-2″	14′-10″	5′-7″
100,000	5′-11″	8′-11″	16′-3″	6′-0″
150,000	7′-3″	10′-3″	18′-3″	7′-6″
350,000	11′-0″	13′-6″	25′-0″	11′-0″

FIGURE 22.15 Flooded disk scrubber.

TRAY SCRUBBERS

Impingement tray scrubbers are essentially perforated plates with target baffles. Tray scrubbers have no large gas-directing baffles but are simply perforated plates within a tower, usually immediately downstream of a venturi. Figure 22.16 illustrates a typical impingement plate. A water level is maintained above the trays (there are usually two or more trays). The geometric relationship of the tray thickness, hole size, and spacing, as well as the impinger details, results in a high-efficiency device for the removal of small-size particulate, less than 5 μ.

As many as 300 openings are provided per square foot of tray area. The openings can be from $\frac{1}{16}$ to $\frac{3}{8}$ in in diameter. Gas flows up through the openings, while water is trying to flow, by gravity, counter to the gas flow. Highly effective turbulence with attendant mixing of, or scrubbing of, the solid particulate with the water effectively catches the small-micron particulate and removes them from the gas stream.

Tray scrubbers will have a pressure drop of 2 to 3 in WC per tray. Impingement plates will add another ½ to 1 in WC differential per tray.

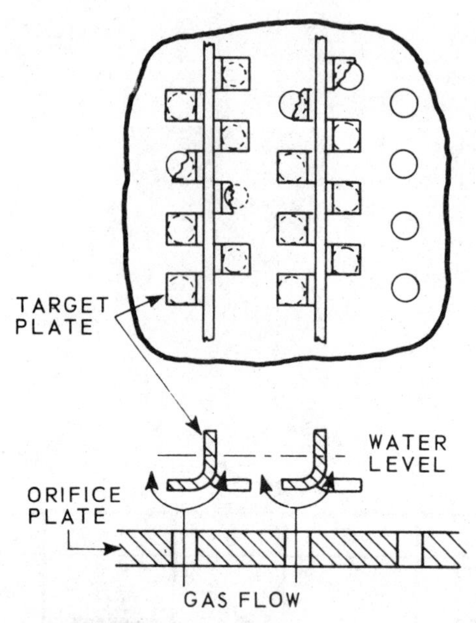

FIGURE 22.16 Arrangement of target plates in impingement scrubber. *Source:* Ref. 22-5.

SELF-INDUCED SCRUBBER

There are a number of scrubbers that employ a unique geometry, utilizing the gas flow to generate scrubbing action. For instance, in the unit shown in Fig. 22.17, the water level controls the scrubbing action of the gas. The higher the water

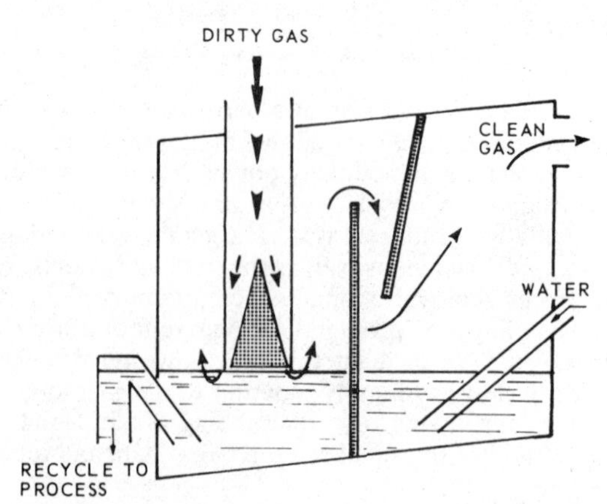

FIGURE 22.17 Self-induced spray scrubber. *Source:* Joy Manufacturing Co., Western Precipitator Div., Los Angeles.

level, the greater, or longer, the contact between the gas stream and the water particles. As the water level decreases, however, the gas discharge into the water bath will generate a surface effect, atomizing some of this water surface. The net effect of this system is to obtain relatively good efficiency while utilizing relatively low water flows and low gas differential pressure.

ELECTROSTATIC PRECIPITATOR

Electrostatic precipitators (ESPs) are effective devices for the removal of airborne particulate matter. The first commercial ESP was designed and constructed by Fredrick G. Cottrell earlier in this century, and they are sometimes referred to as *Cottrell process equipment*. In Europe they are called *electrofilters*.

An ESP (a typical unit is shown in Fig. 22.18 with dimensional information listed in Table 22.5) operates as follows:

- The gas stream passes through a series of discharge electrodes. These electrodes are negatively charged, in the range of 1000 to 6000 V. This voltage creates a corona around the electrode. A negative charge is induced in the particulate matter passing through the corona.
- A grounded surface, or collector electrode, surrounds the discharge electrode. Charged particulate will collect on the grounded surface, usually in the form of plate surfaces.
- Particulate matter will be removed from the collector surface for ultimate disposal.

Typical discharge electrode and collector plate designs are shown in Fig. 22.19. A variation of this design is the two-stage ESP, where the gas passes through a corona discharge prior to entering the collector plate area.

The ESP is extremely efficient in the collection of small-size particulate, down to the submicrometer range. It can be designed for temperatures as high as 750°F; however, its efficiency is sensitive to variation in temperature and flue gas humidity. As shown in Fig. 22.20, ESP efficiency can have substantial variance when it is operated beyond its design point.

Removal of particulate from the collecting surface is the key to the success of an ESP installation. If it is not removed, it will act as an insulator, preventing the required electrostatic action from occurring, reducing the ability of the ESP to function.

Various methods have been developed for removing particulate. The most common method is use of "rappers," members that are sequenced to "rap" each plate section at regular intervals. Particulate will fall off the plates by this action, collecting in a hopper or series of hoppers beneath the ESP for eventual final disposal. The ESP pictured in Fig. 22.18 is provided with four such collection hoppers.

Another method of cleaning the plates is to wet them. Wet electrostatic precipitators have been developed where sprays wet the incoming flue gas stream to a saturated or supersaturated condition. The electric charge is transferred to the liquid droplets, and the liquid charges, collects, and washes away the particulate from the gas stream.

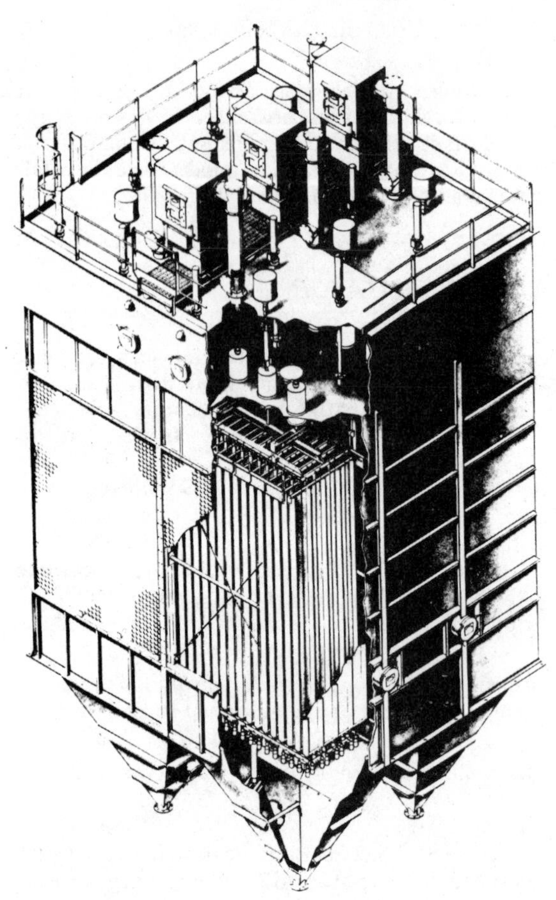

FIGURE 22.18 Parallel-plate precipitator. *Source:* Ref. 22-8.

TABLE 22.5 Typical Electrostatic Precipitator Dimensions

Capacity, ft^3/min	Number of modules	$W \times L \times H$, ft
12,000	5	10 × 4 × 18
20,000	8	10 × 6 × 19
25,000	11	10 × 8 × 19
35,000	14	10 × 10 × 20
40,000	17	10 × 12 × 20
50,000	21	10 × 14 × 21
60,000	25	10 × 17 × 22
70,000	29	10 × 20 × 23
80,000	33	10 × 22 × 23
100,000	43	10 × 19 × 24

Source: Compiled from Ref. 22-8.

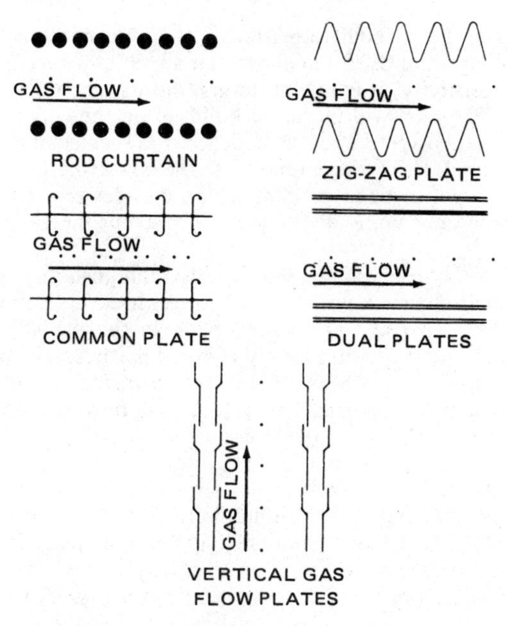

FIGURE 22.19 Some special electrodes used in electrical precipitators. *Source:* Ref. 22-8.

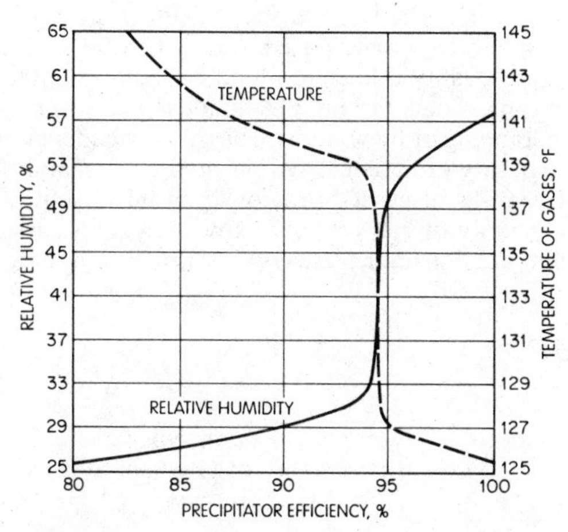

FIGURE 22.20 Curves showing effects of variation of the gas stream temperature and humidity on efficiency of a specific electrical precipitator installation. *Source:* Ref. 22-9.

The resistivity of the particulate matter is a significant parameter in ESP design and in the ability of an ESP to collect that specific material. If a particle, or dust, has a high resistivity, it is unable to give up its electric charge to the collecting electrode. The dust will therefore build up on the collector, acting as an insulating layer. As this layer increases in depth, the surface of the dust layer will develop a significant electric charge relative to the collector. This charge will generate a back discharge, or back corona, which can destroy the collecting electrode. Eventually the insulating dust layer will prevent the ESP from collecting particulate.

With too small a resistivity a dust will readily relinquish its negative charge to the collector and will assume a positive charge. With the collecting electrode at a positive potential, the particle is repelled back to the gas stream. In the gas stream, which is saturated with negatively charged particles, the dust will pick up a negative charge again and will eventually return to the collector plate and be repelled. The low-resistivity particle, therefore, will be successively repelled by the collecting electrode and will not be collected. It will pass through the ESP system.

Electrical resistivity is measured in units of ohm-centimeters. The optimum resistivity for particulate matter to be effectively collected within an ESP is from 10,000 to 10,000,000,000 $\Omega \cdot$ cm. The resistivity of most materials varies significantly with temperature, and the use of an ESP may very well mandate the temperature range of collection to that temperature range where the resistivity is within acceptable limits.

With the proper sensitivity a dust particle will relinquish part of its charge to the collecting electrode. The rate at which the charge dissipates increases as the dust layer builds up. When the weight of the collected dust exceeds the electrostatic force available to hold the layer to the collector, the dust particles will fall off under their own weight or will be jarred loose when the collectors are rapped.

Another particulate quality associated with resistivity is the tendency to agglomerate, to form a hard or a tarlike mass impossible to remove from the collector electrodes by rapping or by water washing. The tendency to agglomerate is influenced by the quantity of moisture within the gas stream and by the temperature of the gas stream, as discussed previously in this chapter.

Collection is a function of gas velocity as well as of the other factors noted. The velocity through the collector plates will normally range from 2 to 4 ft/s.

FABRIC FILTERS

Fabric filters, or baghouses, are prevalent in all types of industrial applications. They are essentially a series of permeable bags which allow the passage of gas but not particulate matter. They are effective for particle sizes in the submicron range.

Filter fabrics will usually be woven with relatively large spaces, in excess of 50 μ. The filtering process, therefore, is not just simple sieving, since particles less than 1 μ are caught. Filtering occurs as a result of impaction, diffusion, gravitational attraction, and electrostatic forces generated by interparticle friction. The dust layer itself acts as a filter medium.

As dust collects on a bag filter surface, the resistance of the filter increases. In addition to this dust mat resistance there is a clean cloth resistance which is a function of the type of cloth fiber and its weave. Clean cloth resistance is measured by an ASTM permeability test procedure: Permeability is measured as that air volume, in standard cubic feet per minute, that will produce a pressure differential of 0.5 in WC across 1 ft^2 of cloth area. The usual values of permeability are 7 to 70 $ft^3/(min \cdot ft^2)$.

Typical filter fabrics in common use are listed in Table 22.6. Operating temperatures for continuous (long) and intermittent (short) duty are provided as well as information on flammability, permeability, and resistance to corrosive attack.

The filtering ratio of a fabric is the ratio of gas flow, in cubic feet per minute, to filter area, in square feet. The more efficient the filter cloth, the greater the filtering ratio. Figure 22.21 shows the filtering ratio for various materials as a function of dust loading, grains of particulate per cubic foot of gas. Particulate from an incinerator would most closely resemble carbon black or diatomaceous earth, curves 2 and 5, respectively.

A major feature of a baghouse is its ability to discharge collected particulate. A number of different methods have been developed for particulate removal, including the following:

- *Shaker mechanism* (Fig. 22.22). An eccentric rod physically shakes a bag section, and the falling particles drop to the bottom of a silo, by gravity, for eventual disposal. The shaker motor is sequenced to operate in conjunction with operation of the fresh air dampers. Fresh air is admitted to that damper section that has its bags shaken, or agitated. The fresh, clean air aids in discharging the collected dust.
- *Compressed air* (Fig. 22.23). A blast of compressed air is directed into the inside of each bag, discharging the dust accumulated on the external surface of the bags. Wire retainers are provided to help reinforce the bags against the abrupt action of the air blast.
- *Repressurization*. The filter sections are independent of each other. Through a series of inlet and exhaust valves, as shown in Fig. 22.24, the dirty gas flow passes through the inside of the bags. On a timing sequence the flow will be reversed to pressurize the outside of the bags of a specific baghouse section. Under external pressure the internal dust loading will fall from the bag surfaces to a collection hopper for disposal.
- *Sonic cleaning*. A source of intense sound tuned to the resonant frequency of the bags will create sympathetic vibrations of the bags. Under this vibration the dust will fall from the bags for collection and eventual removal.

A typical baghouse assembly is shown in Fig. 22.25, with dimensions and operating parameters listed in Table 22.7.

ODOR EMISSIONS

Odor from incinerators can best be controlled by an alteration of the burning process. In general, odors emanating from combustion processes are composed of unburned hydrocarbons and, to a lesser degree, inorganic compounds. By proper combustion equipment design and operation, the generation of odor can be effectively reduced or eliminated.

Unfortunately, waste streams are normally not consistent in chemistry or quality. As waste quality changes, the efficiency of the burning process for that

TABLE 22.6 Filter Fabric Characteristics

	Operating exposure, °F					Resistance[b]			
Fiber	Long	Short	Supports combustion	Air permeability,[a] $ft^3/(min \cdot ft^2)$	Composition	Abrasion	Mineral acids	Organic acids	Alkali
Cotton	180	225	yes	10–20	Cellulose	G	P	G	G
Wool	200	250	no	20–60	Protein	G	F	F	P
Nylon	200	250	yes	15–30	Polyamide	E	P	F	G
Orlon	240	275	yes	20–45	Polyacrylonitrile	G	G	G	F
Dacron	275	325	yes	10–60	Polyester	E	G	G	G
Polypropylene	200	250	yes	7–30	Olefin	E	E	E	E
Nomex	425	500	no	25–54	Polyamide	E	F	E	G
Fiberglass	550	600	yes	10–70	Glass	P-F	E	E	P
Teflon	450	500	no	15–65	Polyfluoroethylene	F	E	E	E

[a] At 0.5 in WC.
[b] P = poor, F = fair, G = good, E = excellent.
Source: Ref. 22-10.

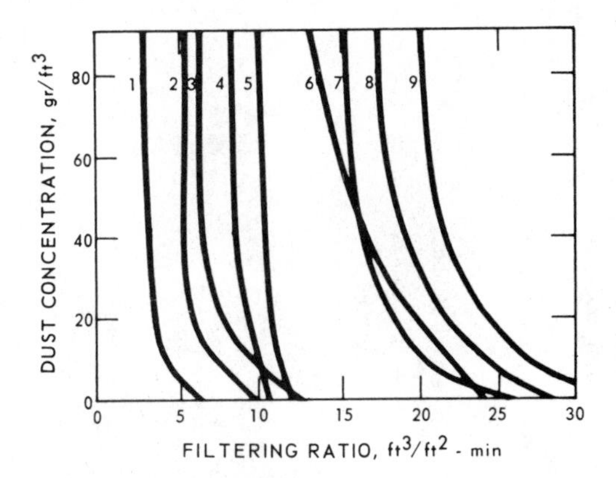

KEY

1 MAGNESIUM TRISILICATE
2 CARBON BLACK
3 STARCH DUST
4 RESINOX
5 DIATOMACEOUS EARTH
6 KAOLIN
7 CEMENT OR LIMESTONE DUST
8 COAL DUST
9 LEATHER BUFFING DUST

FOR NUMBERS 1-99.94 - 99.99% PASSING 325 MESH FOR NUMBERS 7 AND 8 . 95% PASSING 200 MESH. NUMBER 9.60 MESH AVERAGE

FIGURE 22.21 Typical performance of fabrics on a variety of dusts. *Source:* Ref. 22-11.

waste will change and burning will have to be adjusted to reduce the possibility of odor generation. Changes in the combustion process may not be practical with a waste that changes in character within a short time. Additional mechanisms of odor control may be required.

A number of methods of odor control, including incineration, are currently in use, as described below.

Dilution

The principle of dilution is to reduce the odor concentration to a level where the odor is no longer detectable or, if detectable, is no longer considered noxious. Methods of dilution include provision of tall stacks to disperse odor by the time the odorous compounds reach a populated surface. Another method of diluting odorous discharges is to increase the temperature of the gas exiting the stacks, thereby increasing its outlet velocity. Also increasing the velocity of discharge from the stack by decreasing the stack outlet diameter will promote dilution. These latter methods, increasing the gas exit velocity, will increase the time until the odorous compounds return to earth, at which time they will be increasingly dilute.

Yet another method of odor control by dilution is the addition of clean air to the odorous gas exiting the stack. Clean air can be added by providing a flue damper just upstream of the induced-draft fan or by a barometric damper where there is no such fan in the system.

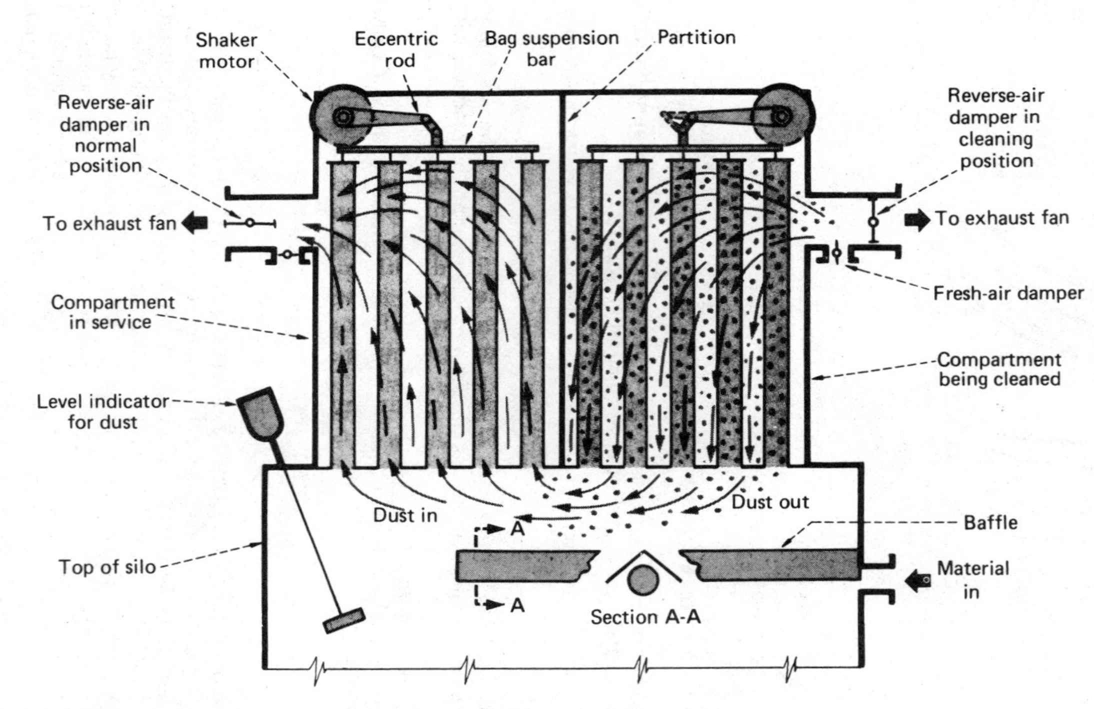

FIGURE 22.22 Baghouse with shaker mechanism. *Source:* Ref. 22-12.

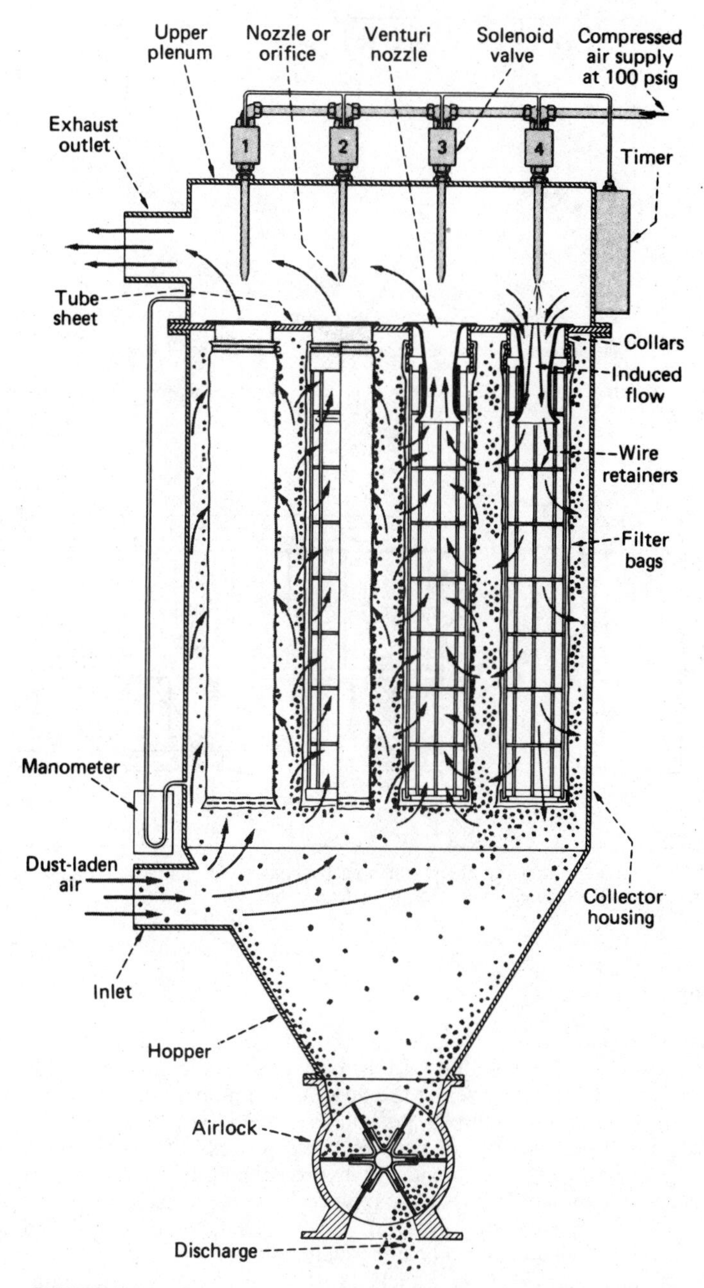

FIGURE 22.23 Compressed air utilization in baghouse. *Source:* Ref. 22-13.

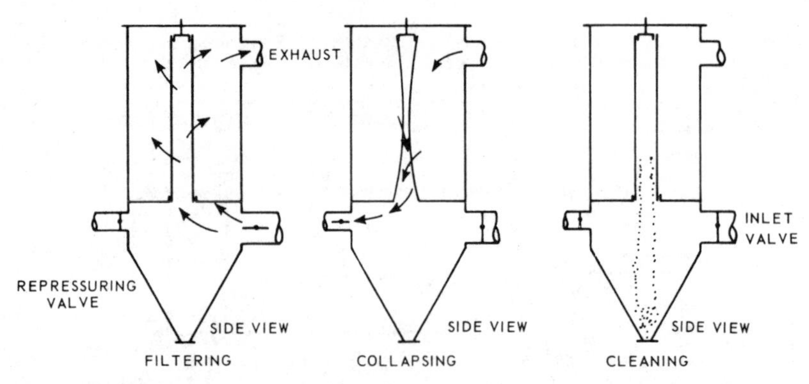

FIGURE 22.24 Reverse air flexing to clean dust collector bags by repressurizing. *Source:* Ref. 22-5.

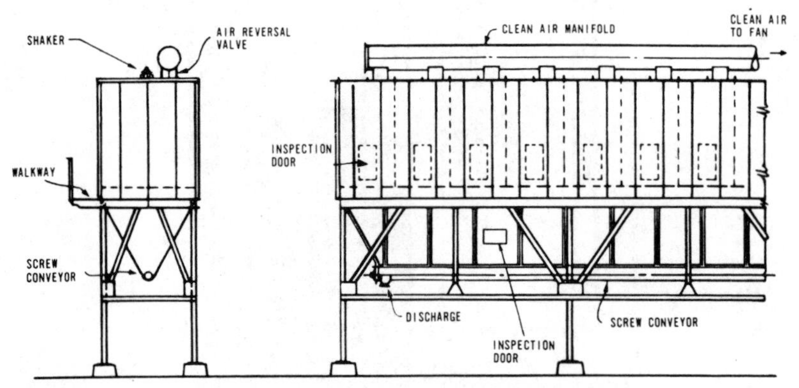

FIGURE 22.25 Fully automatic compartmented baghouse with hopper discharge screw conveyer. *Source:* Buell Engineering Co., Cleveland, Ohio.

Absorption

In this method of odor control the odorous constituent is absorbed into a solution by either chemical bond or solubility. In the case of solubility an odorous gas may be absorbed by a liquid. However, if the odorous component of the gas does not condense and dissolve, the odorous gas will eventually be released.

Chemical absorption is generally an oxidative process. As noted previously, many odors are unburned, i.e., unoxidized, hydrocarbons. By mixing or scrubbing with chemicals with the ability to release oxygen, such as potassium permanganate ($KMNO_4$), sodium hypochlorite (NaOCl), caustic (NaOH), or chlorine dioxide (ClO_2), the odorous constituent can be effectively oxidized and lose its odor characteristic. Figure 22.26 shows a typical packed tower used for odor absorption.

Adsorption

A number of materials have the physical property of high adsorptive capacity, the ability to adsorb or attract gases or vapors onto their surface. Adsorptive media

TABLE 22.7 Typical Baghouse Parameters

Capacity, ft³/min		Number of bags	Cloth, ft²	Number of hoppers	A, ft	B, ft	Weight, lb
5:1 Ratio	12:1 Ratio						
4,400	10,700	40	888	1	23	6	6,900
5,300	12,800	48	1,066	2	24	7	7,800
6,200	14,900	56	1,243	2	25	8	8,500
7,100	17,100	64	1,421	2	27	9	9,600
8,000	19,200	72	1,598	2	28	10	10,300
8,900	21,300	80	1,776	2	29	11	11,100
9,800	23,400	88	1,954	3	30	12	12,000
10,700	25,600	96	2,131	3	31	14	12,700
11,500	27,700	104	2,309	3	32	15	13,500
12,400	29,800	112	2,486	3	33	16	14,300
13,300	32,000	120	2,664	3	34	17	15,000
14,200	34,100	128	2,842	4	35	18	16,200
15,100	36,200	136	3,019	4	36	19	17,000
16,000	38,400	144	3,197	4	37	20	17,800
16,900	40,500	152	3,374	4	39	21	18,600
17,800	42,600	160	3,552	4	40	22	19,400
18,600	44,800	168	3,730	5	41	23	20,300
19,500	46,900	176	3,907	5	42	25	21,100
20,400	49,000	184	4,085	5	43	26	21,900
21,300	51,100	192	4,262	5	44	27	22,700

Source: Buell Engineering Co., Cleveland, Ohio; *A*, length; *B*, width.

commonly used include activated carbon, silica gel, aluminum oxide, and magnesium silicate. The particles of each of these materials have extremely high area-to-weight ratios, providing high retentive capacity.

Activated carbon is the most widely used adsorbent. One reason for this is that it has a low affinity for moisture. Moisture will not compete with contaminants in searching out the carbon surface. Other adsorbents will attract moisture and will have, therefore, a short useful life in wet gas environments.

Another property of an adsorbent material is its ability to regenerate. Normally, with application of heat, the contaminant will be released, and the adsorbent will be reactivated for reuse. Reactivation may also require sparging of the spent adsorbent with steam.

Fume Incineration

The ultimate odor destruction technique is burning. Separate fume incinerators are often utilized for odor destruction when other techniques are not effective. Destruction of odor by incineration requires the use of supplemental fuel, and with current energy costs and availability this technique is avoided whenever possible. The proper use of fume incinerators or afterburners, i.e., provision of the appropriate temperature and residence time, will completely destroy odor, whether organic or inorganic. A discussion of fume or gas burning equipment is included in Chap. 17.

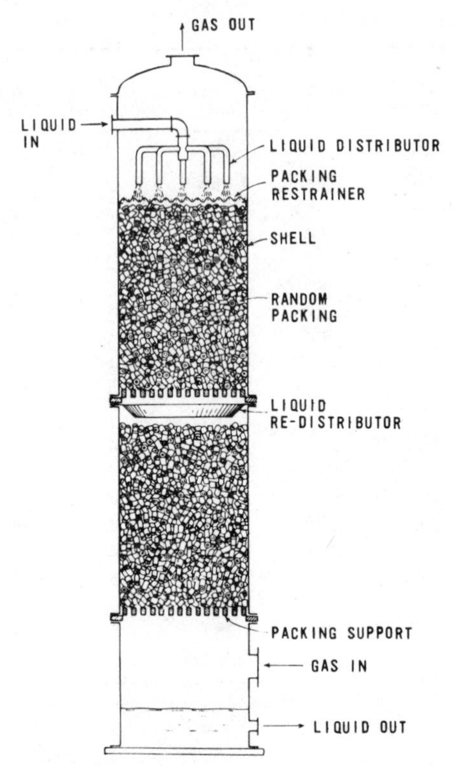

FIGURE 22.26 Schematic diagram of a packed tower. *Source:* Ref. 22-14.

Other Techniques

A number of other methods of odor control have been developed which have application to specific odor types.

Masking is a technique where one odor is added to another to mask its unpleasantness. If an inappropriate masking agent is applied or if the original odor changes in characteristic, the resultant odor can be far worse than the original odor.

Catalytic oxidation utilizes a catalyst such as palladium, rhodium, or platinum to induce oxidation at lower temperatures than direct flame oxidation. The catalyst can be sensitive to the constituents of the odor and can easily foul or permanently lose its potency with unanticipated odor sources. Catalytic incineration is discussed in a previous chapter.

Techniques for odor control from incineration processes are normally either fume incineration (afterburner) or packed tower (absorber).

Dilution is not a positive means of odor control, and it is dependent on atmospheric conditions, which are highly variable. Adsorption requires lower temperatures than those generated by incineration systems in order to maintain the adsorption property of the active media.

Scrubbing with a reagent has been used for odor and gas emissions control. Table 22.8 lists some of these reagent systems.

TABLE 22.8 Gas Absorption Systems of Commercial Importance

			Degree of Commercial Importance		
Solute	Solvent	Reagent	High	Moderate	Low
CO_2, H_2S	water		X	–	–
CO_2, H_2S	water	monoethanolamine	X	–	–
CO_2, H_2S	water	diethanolamine	X	–	–
CO_2, H_2S	water	triethanolamine	–	–	X
CO_2, H_2S	water	diaminoisopropanol	–	–	X
CO_2, H_2S	water	methyldiethanolamine	–	–	X
CO_2, H_2S	water	K_2CO_3, Na_2CO_3	X	–	–
CO_2, H_2S	water	NH_3	–	X	–
CO_2, H_2S	water	NaOH, KOH	–	X	–
CO_2, H_2S	water	K_3PO_4	–	X	–
HCl, HF	water		X	–	–
HCl, HF	water	NaOH	X	–	–
Cl_2	water		X	–	–
SO_2	water		–	–	X
SO_2	water	NH_3	–	X	–
SO_2	water	Xylidine	–	X	–
SO_2	water	Dimethylaniline	–	X	–
SO_2	water	$Ca(OH)_2$, oxygen	–	–	X
SO_2	water	Aluminum hydroxide-sulfate	–	X	–
NH_3	water		X	–	–
NO_2	water		X	–	–
HCN	water	NaOH	X	–	–
CO	water	Copper ammonium salts	X	–	–

Source: Ref. 22-15.

HEPA FILTERS

HEPA (high-efficiency particulate air) filters are extremely efficient filters that were originally developed for control of particulate from nuclear energy facilities. They can remove over 99.97 percent of particles 0.3 μ and greater.

The standard test for evaluating removal efficiency is the DOP (dioctylphalate) smoke test. DOP particles are generated in a standard size of 0.3 μ. The percentage of removal of DOP smoke with a filter differential pressure drop of 1.0 in WC is defined as the filter efficiency.

A typical HEPA filter, as shown in Fig. 22.27, is constructed of a glass fiber mat pleated to increase its unit surface area. The filter is mounted in a frame, and a series of frames are mounted in a filter band to provide the required flow capacity. Table 22.9 lists typical HEPA filter sizes, and Fig. 22.28 shows a series of curves relating flow to clean filter resistance.

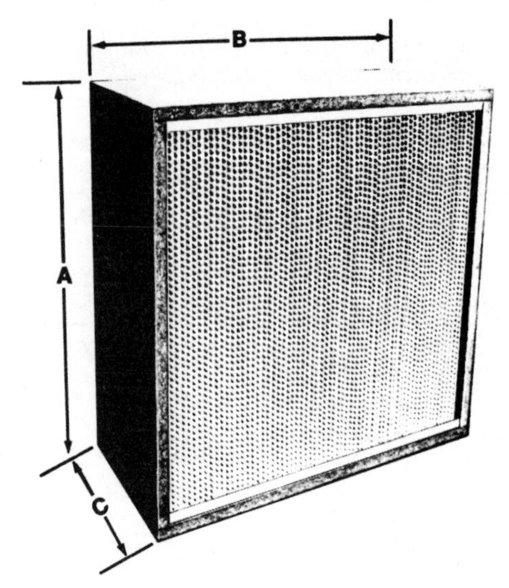

FIGURE 22.27 HEPA filter. *Source:* American Air Filter Co., Louisville, Ky.

TABLE 22.9 Typical HEPA Filter Sizes

Rated value at 1 in WG, st ft^3/min	Actual dimensions (less gaskets), in		
	A	*B*	*C*
900	24	36	$5\frac{7}{8}$
1230	24	48	$5\frac{7}{8}$
1550	24	60	$5\frac{7}{8}$
1900	24	72	$5\frac{7}{8}$
750	30	24	$5\frac{7}{8}$
925	30	30	$5\frac{7}{8}$
1150	30	36	$5\frac{7}{8}$
1550	30	48	$5\frac{7}{8}$
1975	30	60	$5\frac{7}{8}$
2350	30	72	$5\frac{7}{8}$
900	36	24	$5\frac{7}{8}$
1150	36	30	$5\frac{7}{8}$
1400	36	36	$5\frac{7}{8}$
1900	36	48	$5\frac{7}{8}$
2350	36	60	$5\frac{7}{8}$
2850	36	72	$5\frac{7}{8}$

Source: American Air Filter Co., Louisville, Ky.

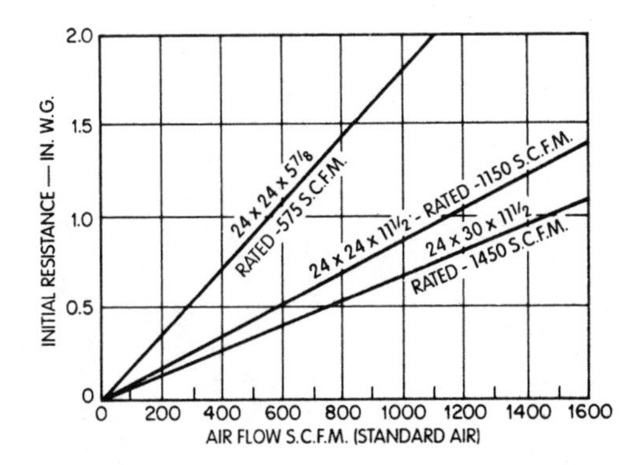

FIGURE 22.28 Typical HEPA filter characteristics. *Source:* American Air Filter Co., Louisville, Ky.

SUMMARY TABLES

As a concise guide to the types of air pollution control equipment available and their performance, Tables 22.10 through 22.13 have been included in this chapter.

Table 22.10 summarizes the advantages and disadvantages of the equipment discussed in this chapter. Typical collection efficiencies of control equipment as a function of particle size are listed in Table 22.11. Table 22.12 relates collection efficiency to gas stream components, and Table 22.13 is an overview of incinerator air pollution control devices.

TABLE 22.10 Advantages and Disadvantages of Collection Devices

Collector	Advantages	Disadvantages
Gravitational	Low pressure loss, simplicity of design and maintainance.	Much space required. Low collection efficiency.
Cyclone	Simplicity of design and maintenance	Much head room required.
	Little floor space required	Low collection efficiency of small particles.
	Dry continuous disposal of collected dusts	Sensitive to variable dust loadings and flow rates.
	Low to moderate pressure loss.	–
	Handles large particles.	–
	Handles high dust loadings.	–
	Temperature independent.	–
Wet collectors	Simultaneous gas absorption and particle removal.	Corrosion, erosion problems.
		Added cost of wastewater treatment and reclamation.
	Ability to cool and clean high-temperature, moisture-laden gases.	Low efficiency on submicron particles.
		Ineffective for hydrophobic materials.
	Corrosive gases and mists can be recovered and neutralized.	Contamination of effluent stream by liquid entrainment.
	Reduced dust explosion risk	Freezing problems in cold weather.
	Efficiency can be varied	Reduction in buoyancy and plume rise.
		Water vapor contributes to visible plume under some atmospheric conditions.
Electrostatic precipitator	99+ percent efficiency obtainable	Relatively high initial cost.
	Very small particles can be collected	Precipitators are sensitive to variable dust loadings or flow rates.
	Particles may be collected wet or dry	Resistivity causes some material to be economically uncollectable.

	Pressure drop and power requirement are small compared to other high-efficiency collectors.	Precautions are required to safeguard personnel from high voltage.
	Maintenance is nominal unless corrosive or adhesive materials are handled.	Collection efficiencies can deteriorate gradually and imperceptibly.
	Few moving parts	–
	Can be operated at high temperatures (550° to 850°F.).	–
Fabric filtration	Dry collection possible	Sensitivity to filtering velocity.
	Decrease of performance is noticeable	High-temperature gases must be cooled to 200° to 550°F.
	Collection of small particles possible	Affected by relative humidity (condensation).
	High efficiencies possible	Susceptibility of fabric to chemical attack.
Afterburner, direct flame	High removal efficiency of submicron odor-causing particulate matter.	High operational cost. Fire hazard.
	Simultaneous disposal of combustible gaseous and particulate matter.	Removes only combustibles.
	Direct disposal of non-toxic gases and wastes to the atmosphere after combustion.	–
	Possible heat recovery	–
	Relatively small space requirement	–
	Simple construction	–
	Low maintenance	–
Afterburner, catalytic	Same as direct flame afterburner	High initial cost.
	Compared to direct flame: reduced fuel requirements, reduced temperature, insulation requirements, and fire hazard.	Catalysts subject to poisoning. Catalysts require reactivation.

Source: Ref. 22-5, p. xxiv.

TABLE 22.11 Average Collection Efficiencies of Gas Cleaning Equipment

Equipment Type	Percentage Efficiency at		
	50 μ	5 μ	1 μ
Inertial collector	95	16	3
Medium-efficiency cyclone	94	27	8
Low-resistance cellular cyclones	98	42	13
High-efficiency cyclone	96	73	27
Impingement scrubber	98	83	38
Self-induced spray deduster	100	93	40
Void spray tower	99	94	55
Fluidized bed scrubber	>99	99	60
Irrigated target scrubber	100	97	80
Electrostatic precipitator	>99	99	86
Irrigated electrostatic precipitator	>99	98	92
Flooded disk scrubber, low energy	100	99	96
Flooded disk scrubber, medium energy	100	99	97
Venturi scrubber, medium energy	100	>99	97
High-efficiency electrostatic precipitator	100	>99	98
Venturi scrubber, high energy	100	>99	98
Shaker-type fabric filter	>99	>99	99
Reverse-jet fabric filter	100	>99	99

Source: Ref. 22-13.

TABLE 22.12 Average Control Efficiency of Air Pollution Control (APC) Systems

APC Type	APC System Removal Efficiency (wt %)								
	Mineral Particulate	Combustible Particulate	Carbon Monoxide	Nitrogen Oxides	Hydrocarbons	Sulfur Oxides	Hydrogen Chloride	Polynuclear Hydrocarbons	Volatile Metals
None (flue settling only)	20	2	0	0	0	0	0	10	2
Dry expansion chamber	20	2	0	0	0	0	0	10	0
Wet bottom expansion chamber	33	4	0	7	0	0	10	22	4
Spray chamber	40	5	0	25	0	0.1	40	40	5
Wetted wall chamber	35	7	0	25	0	0.1	40	40	7
Wetted, close-spaced baffles	50	10	0	30	0	0.5	50	85	10
Mechanical cyclone (dry)	70	30	0	0	0	0	0	35	0
Medium-energy wet scrubber	90	80	0	65	0	1.5	95	95	80
Electrostatic precipitator	99	90	0	0	0	0	0	60	90
Fabric filter	99.9	99	0	0	0	0	0	67	99

Source: Ref. 22-16, p. 260.

TABLE 22.13 Maximum Demonstrated Collection Efficiency of Incinerator Control Equipment

Collection Device	Collection Efficiency, Percent
Settling chamber	35
Wetted baffles	53
Cyclones	75 to 80
Impaction scrubbers (with pressure drop less than ten inches of water)	94 to 96
Electrostatic precipitators	99+
Bag filters	99+

Source: Ref. 22-17.

REFERENCES

22-1. Hagerty, D., and Pavoni, J.: *Solid Waste Management*, Van Nostrand Reinhold Company, New York, 1973.

22-2. Niessen, W., and Sarofin, A.: "Incinerator Air Pollution: Facts and Speculation," National Incinerator Conference, American Society of Mechanical Engineers, 1970.

22-3. First, M., and Dreiner, P.: "Concentrations of Particulates Found in Air," *Industrial Hygiene & Occupational Medicine*, 5:387, 1952.

22-4. Pueschel, R., and Noll, K.: "Visibility and Aerosol Size Frequency," *Journal of Applied Meteorology*, vol. 6, 1967.

22-5. *Control Techniques for Particulate Pollutants*, National Air Pollution Control Administration, AP-51, January 1969.

22-6. *Air Pollution Manual*, American Industrial Hygiene Association, 1972.

22-7. Miller, P.: *Corrosion Studies in Municipal Incinerators*, USEPA SW-72-3-3, 1972.

22-8. Marchello, J., and Kelly, J.: *Gas Cleaning for Air Quality Control*, Marcel Dekker, Inc., New York, 1975.

22-9. Coulter, R.: "Smoke, Dust, Fumes Closely Controlled in Electric Furnaces," *Iron Age*, 173:107–110, January 1954.

22-10. Friedrick, H.: "Primer on Fabric Dust Collection," *Air Engineering*, vol. 9, May 1967.

22-11. Hersey, H.: "Reverse-Jet Filters," *Industrial Chemist*, 31:138, March 1955.

22-12. Kraus, M.: "Baghouses," *Chemical Engineering*, April 9, 1979, pp. 94–106.

22-13. Lapple, C.: *Interim Report: Stack Contamination—200 Areas*, HDC-611, Government Printing Office, August 6, 1948.

22-14. Treybal, R.: *Mass Transfer Operations*, McGraw-Hill, New York, 1955.

22-15. Kohl, R.: "Gas Absorption," *Chemical Engineering*, 66:12–127, 1959.

22-16. Niessen, W.: *Combustion and Incineration Processes*, 1st ed., Marcel Dekker, Inc., New York, 1978.

22-17. Kaiser, E.: "Incinerators to Meet New Air Pollution Standards," Presented at the Mid-Atlantic Section Meeting of the Air Pollution Control Association, New York, April 1967.

BIBLIOGRAPHY

"Air Pollution Control," *Power*, June 1961.

Balakrishnan, N., Cheng, C., and Patel, M.: "Emerging Technologies for Air Pollution Control," *Pollution Engineering*, November 1979.

Battelle Memorial Institute: *Process Modifications for Control of Particulate Emissions from Stationary Combustion*, USEPA 650, October 1974.

Brink, J.: "Air Pollution Control with Fiber Mist Eliminators," *Canadian Journal of Air Pollution Control*, June 1963.

Controlling Airborne Particulates, Report of the National Academy of Sciences, 1980.

Cross, F.: *Air Pollution Odor Control Primer*, Technomic, Lancaster, Pa. 1973.

Danielson, J.: *Air Pollution Engineering Manual*, County of Los Angeles, Calif., Air Pollution Control District, AP-40, May 1973.

"Fabric Filters, Dry Scrubbers," *Power*, January 1982.

Grant, A., Tailor, T., and Powers, J.: "Odor Control Incineration," *Chemical Processing*, February 1972.

Greiner, G.: "Latest Developments in Air Pollution Control Technology," *Specifying Engineer*, September 1981.

Heaney, F.: "Air Pollution Controls at Braintree Incinerator," *Journal of the Air Pollution Control Association*, August 1972.

Industrial Guide for Air Pollution Control, USEPA 625/6-78-004, July 1978.

Kaplan, N., and Maxwell, M.: "Removal of Sulfur Dioxide from Industrial Waste Gases," *Chemical Engineering*, October 17, 1977.

Lapple, C.: "Processes Use Many Collection Types," *Chemical Engineering*, 58:145–151, May 1951.

Lasater, R., and Hopkins, J.: "Removing Particulates from Stack Gases," *Chemical Engineering*, October 17, 1977.

Pierce, R.: "Estimating Acid Dewpoints in Stack Gases," *Chemical Engineering*, November 4, 1977.

Strauss, W.: *Industrial Gas Cleaning*, 2d ed., Pergamon Press, Elmsford, N.Y., 1980.

Teller, A.: "Absorption with Chemical Reaction," *Chemical Engineering*, 67:111–124, July 1, 1960.

CHAPTER 23
ACID GAS CONTROL

A significant proportion of industrial waste streams derive from processes where chlorine is a major constituent of the product. Biomedical wastes have also a significant chlorine component, which results from the extensive use of chlorinated plastics in hospitals. To a lesser extent sulfur is found in many wastes. Both chlorine and sulfur generate an acid discharge in the incinerator exhaust, and under state and federal statutes many of these occurrences of acid gas discharges must be controlled. In addition to the reduction of chlorine and sulfur components of the exhaust gas stream, the reduction of nitrogen oxides, which is also the subject of regulatory control, will be discussed.

REGULATORY REQUIREMENTS FOR ACID GAS REMOVAL

Many state regulations require control of acid gas discharges. This can generally be met with a scrubber, a high stack to encourage dispersion, or a combination of these features. TSCA (see Chap. 3) requires all incinerators permitted under its requirements to include acid gas scrubbing. The RCRA incinerator regulations (discussed in Chap. 3) mandate acid gas scrubbing to reduce any incinerator discharge in excess of 4 lb of hydrogen chloride to a lesser figure. In this chapter techniques of acid gas control will be discussed.

WET SCRUBBING

The wet scrubbing process reduces the temperature of the gas stream and removes the acidic component in one system. It normally consists of a quench, to reduce the temperature of the off-gas, a venturi scrubber, for particulate control, followed by a tray tower or a packed tower. An alkali solution is circulated in the tower to wash acidic components from the gas stream.

Generally, caustic soda (NaOH) or lime is used for acid absorption from incinerator exhaust gas. Removal of acidic gas components is subject to the following equations:

- Caustic scrubbing:

$$\begin{array}{ccccccccc} 64.06 & & 16.00 & & 80.00 & & 142.04 & & 18.02 \\ SO_2 & + & \tfrac{1}{2}O_2 & + & 2NaOH & \rightarrow & Na_2SO_4 & + & H_2O \\ 1.00 & & & & 1.25 & & 2.22 & & 0.28 \end{array} \tag{23.1}$$

As shown, 1 lb of SO_2 requires 1.25 lb of NaOH for neutralization and produces 2.22 lb of Na_2SO_4 and 0.28 lb of H_2O.

Likewise, for HCl scrubbing with caustic soda:

$$\begin{array}{ccccccc} 36.46 & & 40.00 & & 58.44 & & 18.02 \\ HCl & + & NaOH & \rightarrow & NaCl & + & H_2O \\ 1.00 & & 1.10 & & 1.61 & & 0.49 \end{array} \tag{23.2}$$

And 1 lb of HCl requires 40 lb of NaOH for complete neutralization, producing 1.61 lb of NaCl and 0.49 lb of H_2O.

- Lime. The use of lime (CaO) requires the slaking of lime into a concentrated liquid, calcium hydroxide, $Ca(OH)_2$, per Eq. (23.3):

$$\begin{array}{ccccc} 56.08 & & 18.02 & & 74.10 \\ CaO & + & H_2O & \rightarrow & Ca(OH)_2 \\ 1.00 & & 0.32 & & 1.32 \end{array} \tag{23.3}$$

The $Ca(OH)_2$ produced combines with the acidic component of the gas stream as follows:

$$\begin{array}{ccccccc} 64.06 & & 74.10 & & 120.14 & & 18.02 \\ SO_2 & + & Ca(OH)_2 & \rightarrow & CaSO_3 & + & H_2O \\ 1.00 & & 1.16 & & 1.88 & & 0.28 \end{array} \tag{23.4}$$

From Eq. (23.4), 1.16 lb of $Ca(OH)_2$ is required for neutralization of 1 lb of SO_2. With 1 lb of lime required for 1.32 lb of $Ca(OH)_2$, 1.53 lb of lime is required for this reaction. $CaSO_3$ is generated from this reaction at the rate of 1.88 lb/lb SO_2, and 0.28 lb of H_2O is likewise produced.

The neutralization of HCl is as follows, using lime as the reagent:

$$\begin{array}{ccccccc} 72.92 & & 74.10 & & 110.98 & & 36.04 \\ 2HCl & + & Ca(OH)_2 & \rightarrow & CaCl_2 & + & 2H_2O \\ 1.00 & & 1.01 & & 1.52 & & 0.49 \end{array} \tag{23.5}$$

Neutralization of 1 lb of HCl requires 1.01 lb of $Ca(OH)_2$, or 0.77 lb of lime, CaO. For each pound of HCl neutralized, 1.52 lb of $CaCl_2$ and 0.49 lb of H_2O are generated.

To a lesser degree, the use of lime as a reagent results in the absorption of carbon dioxide, as follows:

$$CO_2 + Ca(OH)_2 \rightarrow CaCO_3 + H_2O \tag{23.6}$$

This carbonation of lime occurs due to the relatively high carbon dioxide content of the flue gas.

The above equations are for the stoichiometric or ideal quantities of alkali. The actual use of alkali will be in excess of these figures, because alkali will also scrub carbon dioxide out of the gas stream, because mixing or scrubbing and contact of reagent and solution is never ideal, and because the distribution of acid gases in the exhaust is not consistent.

In the selection of lime versus caustic for scrubbing, the cost of each and the solubility of the salts generated by each should be considered. The cost of sodium hydroxide is over 10 times that of calcium hydroxide (lime). However, calcium salts are much less soluble than those of sodium. For instance, at 200°F the solubility of sodium sulfate is 43 g/100 mL while the solubility of calcium sulfate is only 0.2 g/100 mL. When calcium hydroxide is used for acid gas cleaning, a white calcium sulfate precipitate is formed which, if unchecked, will settle out in tanks, idle piping, and treatment ponds.

DRY SCRUBBING SYSTEMS

Dry scrubbing systems used to remove acid gas components operate as sorbent systems rather than as the washing systems inherent in wet scrubber designs. Wet systems operate with relatively high pressure requirements, utilizing high-horsepower fans. Their ability to remove small particulate matter from the gas stream (below 2 μ) is limited, and carbonaceous aerosols are hydrophobic and may not be effectively captured by a wet system. The major problem inherent in a wet scrubbing system, however, is the need for relatively large freshwater supplies and the generation of a wastewater. The wastewater may itself be classified as a hazardous waste.

There are a number of types of dry scrubbing systems in use on municipal solid waste incinerators and on hazardous waste incineration facilities. These systems have been relatively common in Europe during the past five years and recently have been adapted to United States installations. The more common of these are discussed subsequently.

DRY SCRUBBER OPERATION

In a dry scrubber, lime is used to neutralize acid gases while either lime or caustic can be used in a wet scrubber. Collection and removal occur in the same device with a wet scrubber. A dry scrubber requires a number of major subsystems, i.e., units for neutralization, collection, and removal.

A generic dry scrubbing system is shown in Fig. 23.1. Incinerator off-gas is normally reduced in temperature to the range of 400 to 600°F, often by passing the hot gas stream through a waste heat boiler. This temperature range allows the use of conventional materials (carbon steel, etc.) within the absorber.

Lime is injected as a calcium hydroxide slurry into the absorber, in the range of 5 to 50 percent calcium hydroxide to water. The reactions are basically those described in Eqs. (23.4) and (23.5) for the removal of sulfur dioxide and HCl from the gas stream. The slurry strength is controlled through a dilution waterline.

The absorbent (lime slurry) is injected into the absorber (spray dryer) as a finely atomized spray, producing droplets in the range of 30- to 100-μ diameter. The absorber is sized to provide up to 12-s gas residence time. With this reten-

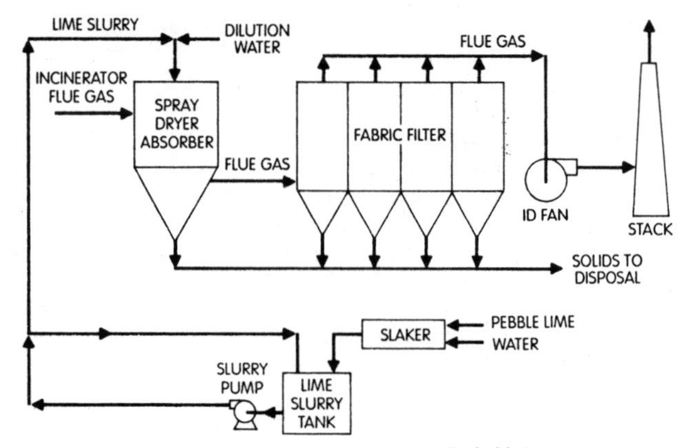

FIGURE 23.1 Spray dryer system. *Source:* Ref. 23-1.

tion time and with appropriate absorber geometry, the lime slurry is evaporated and does not come in contact with the reactor walls. The temperature of the gas exiting the absorber is normally maintained in the range of 250 to 300°F.

Flue gases exiting the absorber are passed through a collection device, either an electrostatic precipitator or, as shown in this illustration, a baghouse. Solids in the gas stream form a cake on the surface of the bags, and the growth of these solids provides additional opportunity for acid removal in the system. Use of an electrostatic precipitator requires that all acids be removed in the absorber, and the absorber has to be designed appropriately for this increased loading.

Some designs utilize a dry injection system. Water is injected upstream of the absorber, with dry lime (not lime slurry) injected in the absorber. A wet-dry (lime slurry) system has been found to be more difficult to maintain than a dry system. Lime tends to cake up in piping and nozzles and on surfaces exposed to the slurry. The major advantage of such a system, however, is that it requires less equipment than a dry system, which includes a separate quench chamber, or cooling tower.

FLAKT

Flakt is one of a number of manufacturers that have developed dry scrubbing systems for use on municipal and hazardous waste incinerators for acid gas control. Two of their designs are the dry system and the wet-dry system. Both utilize a baghouse for particulate and lime collection and disposal.

The dry system, shown in Fig. 23.2, accepts flue gas from the incinerator in the range of 490 to 520°F. Quench water is atomized by nozzles in the gas cooler to provide a uniform reduction of temperature of the gas stream to the range of 230 to 280°F. The injected water is completely vaporized in this section.

From the gas cooler the stream enters the bottom of a cyclonic absorber tower, or dry scrubber. Larger particles are removed from the gas stream by inertial effect. Above the point of gas entrance a dry lime powder is injected into the scrubber with the aid of compressed air. The lime nozzle is directed downward, counter to the upward direction of the exhaust gas stream. The gas, now

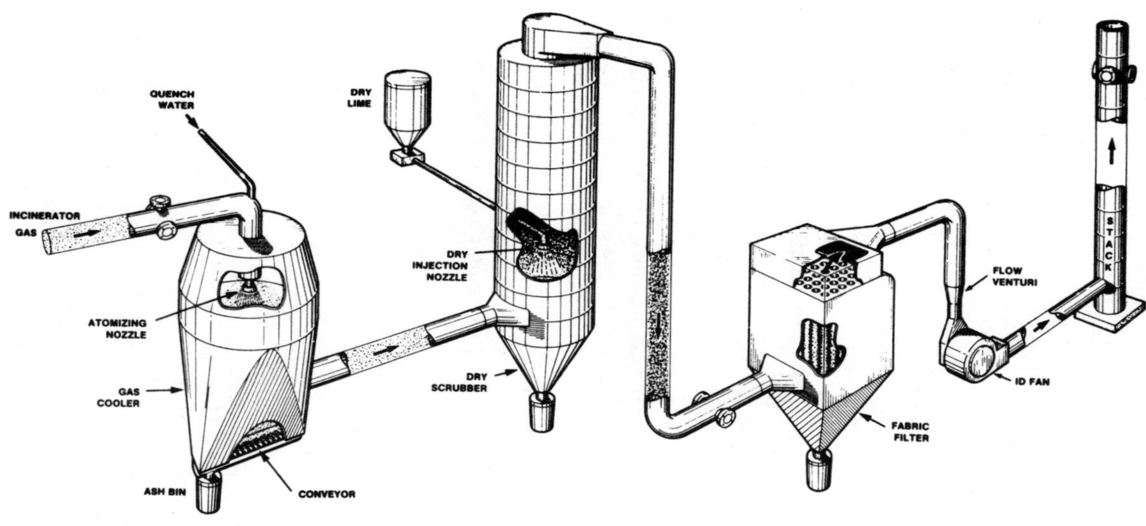

FIGURE 23.2 Flakt's dry system. *Source:* Ref. 23-2.

cooled and conditioned with lime and containing dry lime reaction products, is directed to a baghouse (fabric filter) for collection, further growth and acid removal on the bag surfaces, and eventual disposal.

In the wet-dry system, shown in Fig. 23.3, hot off-gases exiting the incinerator enter a scrubber directly. Lime is injected into the scrubber as a slurry, through a unique nozzle arrangement that produces a fine, well-atomized spray. As with the dry system, water is evaporated completely. No entrained moisture is present in the gas stream exiting the scrubber.

From the wet-dry scrubber, the gas stream passes through a fabric filter for particulate collection and removal.

In general, baghouses used in these systems are designed to withstand a temperature of 480°F, and the gas entering the baghouse is limited to a maximum sustained temperature of 280°F. Bags are fabricated of a chemically inert material, such as Teflon conglomerates, with collection occurring on the outside of the bags. Flakt utilizes compressed air to clean the bags, pulsed to provide continual cleaning during continual use.

These systems can operate with fabric filter "ash" recirculation. This ash contains unreacted hydrated lime which is mixed with fresh lime in either slurry or dry form before reinjection into its associated scrubber.

Table 23.1 demonstrates the effectiveness of dry scrubbing in the removal of hydrogen chloride from the gas stream. Removal efficiency is affected by temperature, and differences between wet-dry and dry systems may not be significant.

Sulfur dioxide removal is indicated in Table 23.2. It appears that temperature has a greater effect on sulfur dioxide removal than on the removal of HCl, and at 285°F there is no difference in performance of the wet vs. the wet-dry system.

In removal of either HCl or sulfur dioxide, recycling appears to have no effect on performance.

The removal of dioxins and dibenzofurans is demonstrated in Tables 23.3 and 23.4, respectively, based on a series of tests on the Flakt scrubber. Efficiencies at all test points are extremely high. In many cases, the level of dioxins or furans was undetectable.

TELLER SYSTEM

One of the Teller dry scrubber systems utilizes lime for acid neutralization plus a separate absorbent. Teller also markets a system similar to the Flakt lime injection scrubber. Hot incinerator exhaust gas enters the inlet of a quench reactor, shown in Fig. 23.4. This is a cyclone where larger particles drop out of the gas stream through inertial action. Rising through the reactor, an alkali solution is sprayed into the swirling gas stream and fully wets it. The alkali (a lime slurry or sodium carbonate solution) neutralizes the acid component of the gas stream and quenches any sparklers within the gas flow. The reactor is designed to promote neutralization of the acid gas components within 1 s by formation of an alkaline mist within the reactor. The gas residence time within the reactor is a nominal 7 s.

The neutralized gas stream leaves the quench reactor at its adiabatic temperature (normally from 150 to 180°F) and passes through the dry venturi (Fig. 23.5). A highly crystalline inert material with 3- to 20-μ-diameter particle size is in-

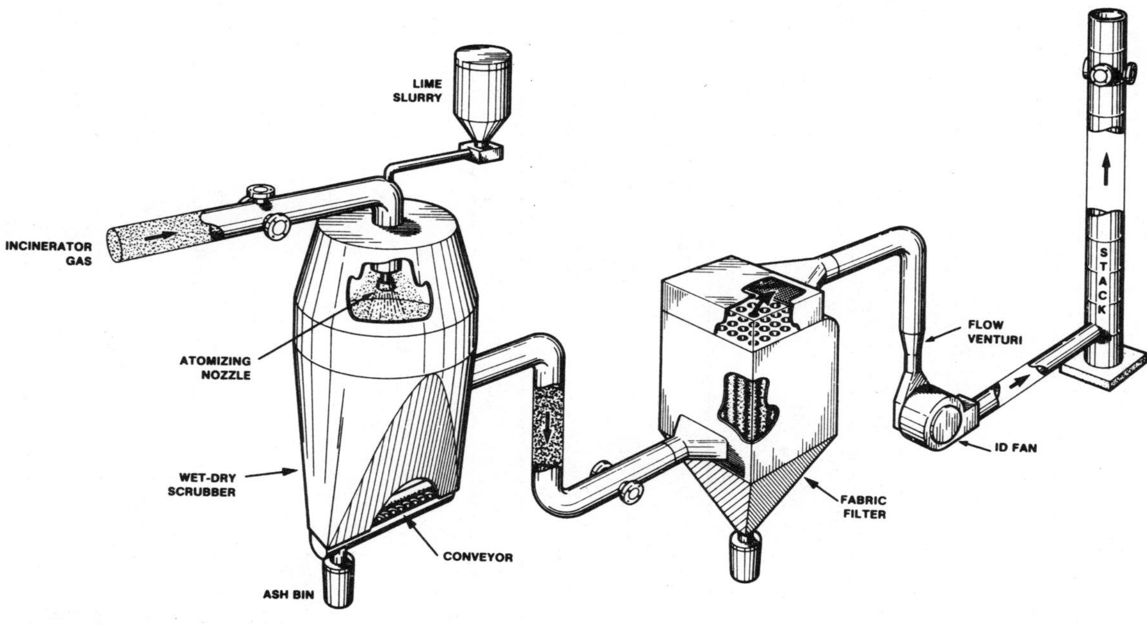

FIGURE 23.3 Flakt's wet-dry system. *Source:* Ref. 23-2.

TABLE 23.1 Hydrogen Chloride Concentrations and Removal Efficiencies Corrected to 8% Oxygen in the Gas Stream

	Dry system				Wet-dry system	
Flue gas temp. @ baghouse inlet, °F	230	260	285	>400	285	285 + recycle
Inlet, ppmw	423	464	475	392	366	470
Outlet, ppmw	7	9	29	91	29	42
Efficiency, %	98	98	94	77	92	91

Note: ppmw = parts HCl per million parts of gas by weight.
Source: Ref. 23-2.

TABLE 23.2 Sulfur Dioxide Concentrations and Removal Efficiencies Corrected to 8% Oxygen in the Gas Stream

	Dry system				Wet-dry system	
Flue gas temp. @ baghouse inlet, °F	230	260	285	>400	285	285 + recycle
Inlet, ppmw	119	118	99	117	106	106
Outlet, ppmw	4	10	41	83	35	43
Efficiency, %	96	92	58	29	67	60

Note: ppmw = parts HCl per million parts of gas by weight.
Source: Ref. 23-2.

TABLE 23.3 Polychlorinated Dibenzo-*P*-Dioxin Concentrations and Removal Efficiencies Corrected to 8% Oxygen in the Gas Stream

	Dry system				Wet-dry system	
Flue gas temp. @ baghouse inlet, °F	230	260	285	>400	285	285 + recycle
Inlet, $ng/(N \cdot m^3)$	580	1400	1300	1030	1100	1300
Outlet, $ng/(N \cdot m^3)$	0.2	ND	ND	6.1	ND	0.4
Efficiency, %	>99.9	>99.9	>99.9	99.4	>99.9	>99.9

ND = not detected.
Source: Ref. 23-2.

TABLE 23.4 Polychlorinated Dibenzofuran Concentrations and Removal Efficiencies Corrected to 8% Oxygen in the Gas Stream

	Dry system				Wet-dry system	
Flue gas temp. @ baghouse inlet, °F	230	260	285	>400	285	285 + recycle
Inlet, $ng/(N \cdot m^3)$	300	940	1000	560	660	850
Outlet, $ng/(N \cdot m^3)$	2.3	ND	1.0	1.2	ND	0.9
Efficiency, %	99.3	>99.9	99.9	99.8	>99.9	99.9

ND = not detected.
Source: Ref. 23-2.

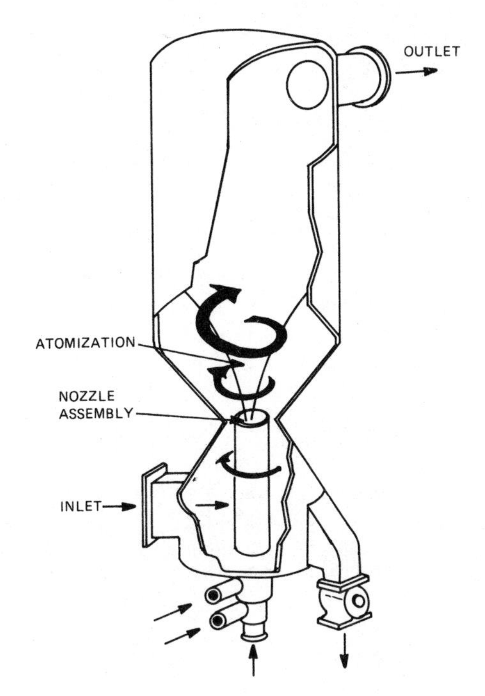

FIGURE 23.4 Upflow quench reactor.

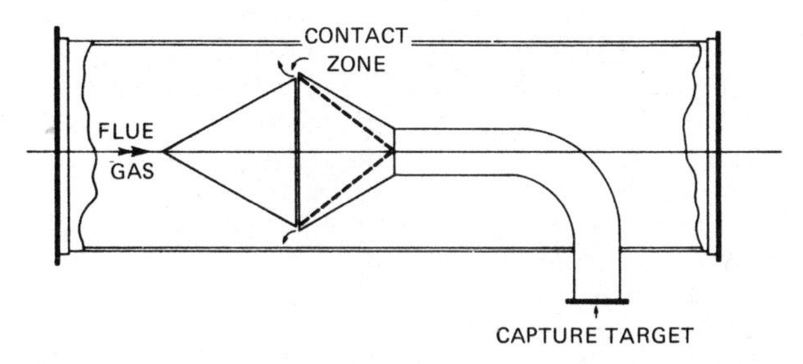

FIGURE 23.5 Teller dry scrubber.

jected into the gas stream through the venturi. Talc can be used, or waste product fines can be used from various manufacturing industries.

Within the venturi the generated turbulence will tend to complete the neutralization process. In addition, the injected powder will adsorb particulate matter within the gas stream and will also act as a catalyst in promoting agglomeration of particulate matter. Particles down to the submicrometer range have been found to agglomerate to particles of 10-μ diameter and greater.

A baghouse is placed downstream of the dry venturi to catch the particles in the gas stream. In measuring particulate removal efficiencies after the baghouse, 95 to 99 percent of the particulate matter exiting the incinerator is removed, in addition to over 90 percent of the HCl present and over 65 percent of the sulfur dioxide in the exhaust gas. Teller estimates that use of an ESP in lieu of a baghouse would result in a drop of particulate removal efficiency to approximately 90 percent. This is due to the growth and residence of particles on fabric filter bags, which are not present in an ESP.

NITROGEN OXIDES REMOVAL

Nitrogen oxides enter the exhaust gas stream as a product of combustion, as described in Chap. 7. They will also be introduced into the gas stream where nitrogen is a component of the fired waste. While the combustion process can be controlled to reduce the nitrogen oxides generated, nitrogen in the waste stream will generate nitrogen oxides that are not as readily reduced within the incinerator.

The reduction of nitrogen oxide formation requires control of combustion temperature and heat release. The high temperatures associated with supplemental fuel burners should be reduced by increasing the heat transfer to surrounding surfaces. Highly turbulent burners with high heat release rates should be avoided if nitrogen oxide generation is to be curtailed.

The following techniques and systems have been found effective in the control of nitrogen oxides once formation has occurred:

- *Noncatalytic reduction*. In the narrow temperature range of 1600 to 1650°F, the following reactions will occur relatively effectively:

$$4NH_3 + 4NO + O_2 \rightarrow 4N_2 + 6H_2O$$

$$4NH_3 + 2NO_2 \rightarrow 3N_2 + 6H_2O$$

  The introduction of ammonia will convert the nitrogen oxides present to diatomic nitrogen and water vapor. In this temperature window the NO_x level will be reduced by 60 percent or more. Above a temperature of 1600°F the ammonia will begin to oxidize and itself produces NO_x. Below 1600°F the reaction becomes too slow to be practical.

  This process becomes less effective with the presence of sulfur oxides and hydrogen chloride. Ammonium sulfate, ammonium sulfite, and ammonium chloride will be generated, and these salts can foul boiler preheater or other relatively low-temperature surfaces. They will also cause a visible plume, which is difficult to control.

- *Selective catalytic reduction* (SCR). The above reactions can be promoted and encouraged with the use of a catalyst. Platinum, palladium, vanadium oxide, and titanium oxide catalysts can reduce nitrogen oxides by 80 percent, and at a lower temperature and a wider temperature range than noncatalytic reduction.

  Ammonia is injected into the exhaust gas upstream of the catalyst bed at a temperature between 500 and 800°F. Over 80 percent of the nitrogen oxides present will be converted to diatomic nitrogen and water vapor. The phenomenon of ammonia slippage, or breakthrough, whereby ammonia will pass through the catalyst bed and combine with sulfur oxides and hydrogen chloride to form troublesome salts in the exhaust, does not occur to any significant degree in SCR systems.

  Where sulfur is present in the exhaust, platinum and palladium-family catalysts are not used. They will readily lose their effectiveness (will be poisoned) by sulfur oxides.

REFERENCES

23-1. Kroll, P., and Williamson, P.: "Application of Dry Flue Gas Scrubbing to Hazardous Waste Incineration," *Journal of the Air Pollution Control Association*, 36 (11), November 1986.

23-2. The National Incinerator Testing and Evaluation Program: Air Pollution Control Technology, Report EPS 3/UP/2, Environment Canada, Ottawa, September 1986.

BIBLIOGRAPHY

Brunner, C. R.: *Hazardous Air Emissions from Incineration*, Chapman & Hall, 2d ed., New York, 1986.

Ellis, T. D., Diemer, R. B., Jr., and Brunner, C. R.: *Industrial Hazardous Waste Incineration, AIChE Short Course Text*, Amer. Institute of Chemical Engineers, New York, 1987.

Feindler, K.: "Long-Term Results of Operating TA Luft Acid Gas Scrubbing Systems," ASME Solid Waste Processing Division, *Proceedings of the 1986 National Waste Processing Conference*, Denver, 1986.

Flynn, B., Hsieh, H., and Curtis, D.: "Effect of Product Recycle on Dry Acid Gas Emission Control," ASME Solid Waste Processing Division, *Proceedings of the 1984 National Waste Processing Conference*, Orlando, Fla.

Hurst, B., and White, C.: "Thermal DeNox: A Commercial Selective Noncatalytic NO_x Reduction Process for Waste-to-Energy Applications," ASME Solid Waste Processing Division, *Proceedings of the 1986 National Waste Processing Conference*, Denver.

Kiang, Y.: "The Formation of Nitrogen Oxides in Hazardous Waste Incineration," ASME Solid Waste Processing Division, *Proceedings of the 1982 National Waste Processing Conference*, New York.

Russell, S., and Roberts, J.: "Oxides of Nitrogen: Formation and Control in Resource Recovery Facilities," ASME Solid Waste Processing Division, *Proceedings of the 1984 National Waste Processing Conference,* Orlando, Fla.

CHAPTER 24
COMPREHENSIVE DESIGN EXAMPLE

The analytical methods developed in this text will be tied together in this chapter. Design parameters will be developed for an incinerator system burning solid and liquid waste. Refractory design will also be included.

EXAMPLE

A multipurpose incinerator is required to burn plant trash (rubbish) and an organic liquid. The trash is normally contaminated with chlorides in relatively small quantities. The organic liquid, cycloheptane (C_7H_{14}), contains trace amounts of solid contaminants. The facility must be designed for destruction of 2000 lb/h of rubbish and 475 lb/h of liquid cycloheptane, either burning concurrently or firing the trash alone.

Future plant expansion may result in the burning of solid organic powders. Additional capacity shall be obtained by operating additional hours during the week.

CRITERIA

It will be assumed that the wastes charged are hazardous and that the incinerator will have to comply with RCRA requirements for hazardous wastes containing chlorides. Not knowing the specific nature of the waste stream with regard to parameters of destruction, the following criteria will be used:

- A temperature zone of 2200°F should be provided with a retention time of 2 s.
- A minimum of 3 percent oxygen should be found in the off-gas stream.
- A maximum particulate loading of 0.03 gr/dry st ft^3 should be allowed.

Additional criteria shall include the following:

- A maximum external temperature of 200°F at the surface of the primary chamber
- A temperature limit of 250°F at the surface of the secondary chamber

INCINERATOR EQUIPMENT

The incinerator system must handle solid waste and/or liquid waste. A rotary kiln with an afterburner or a modular combustion unit (excess air incinerator) could be used. The provision that powders be incinerated in the future requires that turbulence be provided within the incinerator. Without sufficient turbulence to agitate the powder waste and expose the powder surfaces to air and heat, the powder may not burn out properly. The rotary kiln appears to be the favorable option because it provides turbulence to the solid waste charge.

A rotary kiln will be used for solid waste incineration. The off-gas will pass through an afterburner (secondary combustion chamber) where the required temperature and residence time will be maintained.

Fuel oil (no. 2 grade) will be used to obtain the required temperature if the waste will not maintain combustion. The liquid waste stream will be fired in the afterburner.

SOLID WASTE

The incinerator must burn 2000 lb/h of rubbish, type 1 waste, as noted in Table 8.2. The total heat release is as follows:

$$6500 \text{ Btu/lb} \times 2000 \text{ lb/h} = 13.0 \text{ MBtu/h}$$

Assuming a kiln heat release of 25,000 Btu/(h · ft^3), from Chap. 17, the required volume can be calculated:

$$\frac{13{,}000{,}000 \text{ Btu/h}}{25{,}000 \text{ Btu/(h} \cdot \text{ft}^3)} = 520 \text{ ft}^3$$

With a 4:1 ratio of length to diameter, following the methods of Chap. 17,

$$L = 4D$$

$$V = \frac{\pi D^2 L}{4} = \frac{\pi D^2}{4}(4D) = \pi D^3$$

$$D = \left(\frac{V}{\pi}\right)^{1/3} = \left(\frac{520}{\pi}\right)^{1/3} = 5.491 \text{ ft}$$

$$L = 4D = 4 \times 5.491 = 21.964 \text{ ft}$$

The kiln inside dimensions are, therefore:

$$D = 5 \text{ ft } 6 \text{ in inside diameter}$$

$$L = 22 \text{ ft } 0 \text{ in inside length}$$

For an external temperature of 200°F with an internal temperature of 1600°F a refractory will be selected. The kiln is continually rotating, and convection heat transfer can be best approximated as a vertical wall. The kiln surface is oxidized steel. Using the resistance graph, Fig. 5.5, a resistance R of 5.3 is required. From Chap. 5:

$$x = RK$$

Using a dense castable, $K = 5$, over 2 ft of refractory is required. Since 2 ft of refractory is not practical (too large), a two-wall refractory system will be used, insulating castable against the kiln surface and 6 in of firebrick adjacent to the hot stream. Following the methods of Chap. 5,

$$R = R_1 + R_2 = \frac{X_1}{K_1} + \frac{X_2}{K_2}$$

with

$$X_1 = \text{6-in firebrick}$$

$$K_1 = 10.0 \text{ Btu} \cdot \text{in/(ft}^2 \cdot {}^\circ\text{F} \cdot \text{h) for firebrick}$$

and

$$K_2 = 1.3 \text{ Btu} \cdot \text{in/(ft}^2 \cdot {}^\circ\text{F} \cdot \text{h) for insulating castable}$$

To find the thickness of castable required X_2,

$$X_2 = \left(R - \frac{X_1}{K_1}\right)K_2$$

$$= \left(5.3 - \frac{6}{10.0}\right) \times 1.3 = 6.11 \text{ in}$$

Therefore the kiln shall be lined with 7 in of insulating castable protected from the gas stream by 6 in of firebrick.

From the resistance graph the heat loss Q from the kiln, which will have a surface temperature of 200°F, will be 263 Btu/(h · ft²).

The internal kiln dimensions are 5-ft 6-in diameter by 22 ft 0 in long, and with a total lining of 13 in the kiln external dimensions are as follows:

Diameter: 5 ft 6 in + 2 × 13in = 7 ft 8 in

Length: 22 ft + minimum of wall diameter (13 in) = 23 ft 1 in *minimum*

Heat loss through the kiln wall, per the resistance graph, is 263 Btu/(h · ft²), as noted above. Now that the external kiln size is determined, the total heat loss can be calculated.

Kiln external cylindrical area:

$$\pi DL = \pi \times (7 \text{ ft } 8 \text{ in}) \times (23 \text{ ft } 1 \text{ in}) = 556 \text{ ft}^2$$

Heat loss:

$$556 \text{ ft}^2 \times 263 \text{ Btu/(h} \cdot \text{ft}^2) = 0.15 \text{ MBtu/h}$$

LIQUID WASTE

Cycloheptane will be fired in the secondary chamber, or afterburner, at its supply rate of 475 lb/h. To determine the heating value of cycloheptane, the methods of Chap. 8 will be used.

Using the heat of formation method, with Tables 8.3 and 8.5:

$$\begin{array}{llllll} -37.5 & & & -94.1 & -68.3(g) & \\ C_7H_{14} & + 10.5O_2 & + 3.3197(10.5)N_2 \rightarrow & 7CO_2 & + 7H_2O & + 34.857N_2 \\ 98.21 & & & 44.01 & 18.02 & 28.02 \\ 1.00 & & & 3.1368 & 1.2843 & 9.9449 \end{array}$$

$$\begin{aligned} \Delta H_f\ H_2O(g)&: \quad -68.3 \times 7 = -478.1 \\ \Delta H_f\ CO_2&: \quad -94.1 \times 7 = -658.7 \\ \Delta H_f\ C_7H_{14}&: \quad 37.5 \times 1 = \underline{37.5} \\ \Sigma\Delta H_f &= -1099.3 \end{aligned}$$

$$\frac{1}{98.21 \text{ g/mol}} \times (-1099.3 \text{ kcal/mol}) \times 1802 = -20{,}170 \text{ Btu/lb (exothermic)}$$

Using the hydrocarbon approximation:

$$\frac{10.5 \text{ lb} \cdot \text{mol } O_2}{\text{mol } C_7H_{14}} \times \frac{\text{mol } C_7H_{14}}{98.21 \text{ lb}} \times \frac{184{,}000 \text{ Btu}}{1 \text{ lb} \cdot \text{mol } O_2} = 19{,}672 \text{ Btu/lb}$$

These two heating values are relatively close:

$$\frac{20{,}170 - 19{,}672}{20{,}170} \times 100 = 2.5\%$$

The more rigorous number, 20,170 Btu/lb, will be used for further calculations.

The generation of off-gas is calculated from the above equation, noting that the second line of values beneath the equation is the normalized quantity of each component:

$$\frac{3.1368 \text{ lb } CO_2}{\text{lb } C_7H_{14}} + \frac{9.9449 \text{ lb } N_2}{\text{lb } C_7H_{14}} = \frac{13.0817 \text{ lb DG}}{\text{lb } C_7H_{14}}$$

$$\frac{13.0817 \text{ lb DG}}{\text{lb } C_7H_{14}} \times \frac{\text{lb } C_7H_{14}}{20.170 \text{ kBtu}} = \frac{6.486 \text{ lb DG}}{10 \text{ kBtu}}$$

$$\frac{1.2843 \text{ lb } H_2O}{\text{lb } C_7H_{14}} \times \frac{\text{lb } C_7H_{14}}{20.170 \text{ kBtu}} = \frac{0.637 \text{ lb } H_2O}{10 \text{ kBtu}}$$

The size of the afterburner will be calculated using an external low-pressure atomizing burner, as described in Chap. 16. This burner is utilized because of the presence of solids in the liquid waste stream and the low-pressure compressed air required. With a heat release of 60,000 Btu/(h · ft^3) and 475 lb/h of cycloheptane, the *minimum* chamber size is calculated as follows:

$$475 \text{ lb/h} \times 20{,}170 \text{ Btu/lb} = 9.581 \text{ MBtu/h}$$

$$\frac{9{,}581{,}000 \text{ Btu/h}}{60{,}000 \text{ Btu/(h} \cdot \text{ft}^3)} = 159.683 \text{ ft}^3$$

For a 4-ft-diameter chamber the required length can be calculated:

Area: $$\frac{\pi}{4}(4)^2 = 12.566 \text{ ft}^2$$

Length: $$\frac{159.683 \text{ ft}^3}{12.566 \text{ ft}^2} = 12.708 \text{ ft}$$

Use a length of 12 ft 9 in.

As noted above, this is the *minimum* chamber size to take care of the heat release of the cycloheptane. Additional fuel may be required to bring the 1600°F gas stream leaving the kiln to the required 2200°F.

The afterburner refractory will be selected on the basis of 2200°F inside temperature and 250°F external temperature. Using an outside surface of steel, with conduction averaged as a vertical wall, from the resistance graph (Fig. 5.5), the required wall thickness is calculated as follows:

$$R = \frac{X}{K} = 4.7 \text{ h} \cdot \text{ft}^2 \cdot {}^\circ\text{F/Btu} \qquad \text{(from resistance graph)}$$

Using insulating castable,

$$K = 1.9 \text{ Btu} \cdot \text{in/(h} \cdot \text{ft}^2 \cdot {}^\circ\text{F)}$$

$$X = KR = 1.9 \times 4.7 = 8.93 \text{ in}$$

A thickness of 9.0 in will be used.

The heat loss, again from the resistance graph, is 418 Btu/(h · ft^2).

INCINERATOR CALCULATIONS

For rubbish, from Table 8.3, the ash is 10 percent of the as-received weight, and moisture is 25 percent of the as-received weight. For the mass balance, Table 24.1, the wet feed is 2000 lb/h. The moisture content is 25 percent, which is 500 lb/h. The dry feed is the difference between wet feed and moisture, or 1500 lb/h.

The ash content is 10 percent of the wet feed, or 200 lb/h, and the volatile content is therefore 1500 − 200 = 1300 lb/h. This is 200 ÷ 1500 = 13.3 percent of the dry weight. From the tabular data in Chap. 8, type 1 waste has an as-received heating value of 6500 Btu/lb. This calculates to 6500 × 2000 ÷ 1300 = 10,000 Btu/lb volatile.

Dry gas and moisture generation will be taken as 7.50 lb/10 kBtu and 0.51 lb/10 kBtu, respectively, as noted in Chap. 13 for refuse.

The mass balance is computed using 150 percent excess air for burning the solid waste.

In the heat balance, 120 Btu/lb is assumed for ash losses, and the kiln radiation loss has been calculated for a 200°F external kiln temperature, 0.15 MBtu/h. The radiation loss related to heat release is 0.15 ÷ 13 = 1.2 percent.

The temperature developed within the kiln, to obtain the 12.61 MBtu/h released, is calculated from the methods of Chap. 4, as follows:

TABLE 24.1 Mass Flow

	Kiln	Afterburner	Kiln w/ afterburner	12% CO_2
Wet feed, lb/h	2000	475		
Moisture, %	25			
lb/h	500			
Dry feed, lb/h	1500	475		
Ash, %	13.3			
lb/h	200			
Volatile, lb/h	1300	475		1775
Volatile htg. value, Btu/lb	10,000	20,170		
MBtu/h	13.00	9.58	22.58	
Dry gas, lb/10 kBtu	7.50	6.49		
lb/h	9751	6217		15,968
Comb. H_2O, lb/10 kBtu	0.51	0.64		
lb/h	663	613		1276
Dry gas + comb. H_2O, lb/h	10,414	6830		17,244
100 air, lb/h	9114	6355	15,469	15,469
Total air fraction	2.5	1.1		2.29
Total air, lb/h	22,785	6991	29,776	35,424
Excess air, lb/h	13,671	636	14,307	19,955
Humidity/dry gas (air), lb/lb	0.01	0.01		
Humidity, lb/h	228	67	295	
Total H_2O, lb/h	1391	680	2071	
Total dry gas, lb/h	23,422	6853	30,275	35,923

Temperature		MBtu/h
1700°F	$425.08 \times 23{,}422 + 1890.11 \times 1.391 =$	12.59
x		13.05
1800°F	$453.24 \times 23{,}422 + 1948.02 \times 1.391 =$	13.33

$$x = 1700 + (1800 - 1700)\frac{13.05 - 12.59}{13.33 - 12.59} = 1762°\text{F}$$

The off-gas from the solid waste must be raised to 2200°F, and 63.78 gal/h fuel oil is required, from the heat balance, to obtain this heat increase.

At this point the gas volume will be evaluated. Using Table 4.5, at 2200°F the gas flow is calculated as follows:

Dry gas: $67.0\ \text{ft}^3/\text{lb} \times 30{,}275\ \text{lb/h} \times \text{h}/3600\ \text{s} = 563\ \text{ft}^3/\text{s}$

Moisture: $107.9\ \text{ft}^3/\text{lb} \times 2071\ \text{lb/h} \times \text{h}/3600\ \text{s} = 62\ \text{ft}^3/\text{s}$

Total $= 625\ \text{ft}^3/\text{s}$

For a 2-s retention time a volume of $2 \times 625 = 1250\ \text{ft}^3$ is required. This volume is 8 times greater than that minimum volume necessary for firing the liquid waste stream, cycloheptane as initially calculated. It is also twice the volume required for the rotary kiln at the head of the system.

A reasonably sized chamber would have an internal diameter of 10 ft and a length of 16 ft 6 in. As calculated previously, the wall thickness is 9 in. The external diameter would be therefore 10 ft + 2(9 in) = 11 ft 6 in.

The radiation and convection loss from the kiln is also as noted previously, 418 Btu/(h · ft^2).

The total heat loss from the afterburner is calculated as follows:

$$A = \pi DL = \pi(11.5 \text{ ft})(16.5 \text{ ft}) = 596 \text{ ft}^2$$

Heat transfer: $596 \text{ ft}^2 \times 418 \text{ Btu/(h} \cdot \text{ft}^2) = 0.25 \text{ MBtu/h}$

The percent oxygen, as noted previously, must be at least 2 percent within the afterburner. The total gas flow, from the heat balance, is 30,275 lb/h dry gas plus 2071 lb/h H_2O, a total of 32,346 lb/h wet gas flow. The oxygen component of this flow is the sum of the excess oxygen injected in the primary chamber (kiln) plus the excess oxygen injected into the afterburner:

Kiln: $13{,}671 \text{ lb/h excess air} \times 0.2315 \text{ lb } O_2\text{/lb air}$

$= 3165 \text{ lb/h oxygen}$

Afterburner: $8220 \text{ lb/h total air} \times \dfrac{0.1 \text{ excess air}}{1.1 \text{ total air}}$

$= 636 \text{ lb/h excess air} \times 0.2315 \text{ lb } O_2\text{/lb air}$

$= 147 \text{ lb/h oxygen}$

The total oxygen in the gas flow is therefore

$$3165 + 147 = 3.312 \text{ lb/h } O_2$$

The percentage of oxygen is

$$\frac{3312 \text{ lb/h } O_2}{32{,}346 \text{ lb/h gas flow}} \times 100 = 10.24\% \text{ oxygen, by weight}$$

This is well in excess of the minimum oxygen requirement, 2 percent of the gas flow.

The above calculations, as noted in the first column of Tables 24.1 and 24.2, the mass flow and heat balance sheets, represent burning solid waste, using no. 2 fuel oil in the afterburner to reach the desired temperature. The second column is used to calculate combustion characteristics of the waste liquid, cycloheptane. Cycloheptane will be used in lieu of fuel oil if it has sufficient heating value to raise the temperature of the kiln off-gas to at least 2200°F.

The mass flow sheet calculates the air feed and products of combustion of cycloheptane. The heat losses when burning cycloheptane are inserted in the heat balance sheet, second column.

The third column of the mass flow and heat balance sheets represents the conditions within the afterburner when burning cycloheptane in the afterburner and trash in the kiln. The temperature developed within the afterburner is 2272°F, more than the 2200°F minimum required for complete destruction of the combustible components within the off-gas.

The total off-gas flow is slightly less when burning cycloheptane (30,275 + 2071 = 32,346 lb/h) than when burning fuel oil (30,764 + 1941 = 32,705 lb/h) in

TABLE 24.2 Heat Balance

	Kiln	Afterburner	Kiln w/ afterburner
Cooling air wasted, lb/h			
°F			
Btu/lb			
MBtu/h			
Ash, lb/h	200		
Btu/lb	120		
MBtu/h	0.02		
Radiation, %	1.2	2	
MBtu/h	0.15	0.25	
Humidity, lb/h	228	67	
Correction (@970 Btu/lb), MBtu/h	−0.22	−0.06	
Losses, total, MBtu/h	−0.05	0.19	0.14
Input, MBtu/h	13.00	9.58	22.58
Outlet, MBtu/h	13.05	9.39	22.44
Dry gas, lb/h	23,422	6853	30,275
H_2O, lb/h	1391	680	2071
Temperature, °F	1762		2272
Desired temp., °F	2200		2272
MBtu/h	16.34		
Net, MBtu/h	3.54[a]		
Fuel oil, air fraction	1.1		
Net Btu/gal	55,507		
gal/h	63.78		
Air, lb/gal	114.64		
lb/h	7312		
Dry gas, lb/gal	115.12		
lb/h	7342		
H_2O, lb/gal	8.62		
lb/h	550		
Dry gas with fuel oil, lb/h	30,764		30,275
H_2O with fuel oil, lb/h	1941		2071
Air with fuel oil, lb/h	30,097		29,776
Outlet, MBtu/h	21.96[a]		22.58
Reference t, °F	60	60	60

[a]Including 0.25 heat loss in the afterburner.

the afterburner. The flue gas flow parameters are calculated for each of these two cases in Table 24.3, the flue gas discharge table.

PARTICULATE EMISSIONS

Allowable particulate emissions are related to a 12 percent CO_2 discharge standard, i.e., 0.03 gr/dry st ft^3 corrected to 12 percent CO_2. This correction is calculated as follows:

Trash: Assume $C_6H_{10}O_5$ (cellulose)

Cycloheptane: C_7H_{14}

TABLE 24.3 Flue Gas Discharge

	Kiln	Kiln w/ afterburner
Inlet, °F	2200	2272
Dry gas, lb/h	30,764	30,275
Heat, MBtu/h	21.96	22.44
Btu/lb dry gas	714	741
Adiabatic t, °F	178	179
H_2O saturation, lb/lb dry gas	0.6008	0.6279
lb/h	18,483	19,260
H_2O inlet, lb/h	1941	2071
Quench H_2O, lb/h	16,542	17,189
gal/min	33	34
Outlet temp., °F	120	120
Raw H_2O temp., °F	70	70
Sump temp., °F	148	149
Temp. diff., °F	78	79
Outlet, Btu/lb dry gas	111.65	111.65
MBtu/h	3.43	3.42
Req'd. cooling, MBtu/h	18.95	19.06
H_2O, lb/h	242,949	241,266
gal/min	486	483
Outlet, ft^3/lb dry gas	16.515	16.515
ft^3/min	8468	8443
Fan press., in WC	15	15
Outlet, actual, ft^3/min	8792	8766
Outlet, H_2O/lb dry gas	0.08128	0.08128
H_2O, lb/h	2500	2493
Recirc. (ideal), gal/min	33	34
Recirc. (actual), gal/min	265	275
Cooling H_2O, gal/min	486	483

Equilibrium equation:

$$\begin{array}{l} \qquad\qquad\qquad\qquad\qquad\qquad\qquad\qquad 44.01 \qquad\qquad\quad 28.2 \qquad\qquad 32.00 \\ C_6H_{10}O_5 + C_7H_{14} + 16.5xO_2 + 16.5(3.3197)xN_2 \rightarrow 13CO_2 + 12H_2O + 54.775xN_2 + 16.5(x - 1) \\ \qquad\qquad\qquad\qquad\qquad\qquad\qquad\qquad 572.13 \qquad\qquad 1534.80 \qquad 528x - 528 \end{array}$$

With the unit molecular weight above the dry gas components and the total molecular weight beneath the components, the total dry gas flow is

$$572.13 + 1534.80x + 528.00x - 528.00 = 44.13 + 2062.80x$$

The CO_2 fraction present is 0.12 (12 percent CO_2):

$$0.12 = \frac{572.13}{44.13 + 2062.80x}$$

$$x = 2.290$$

This calculation indicates that 229 percent total air, or 129 percent excess air, will produce a CO_2 concentration of 12 percent of the exiting dry gas stream. The actual dry gas flow must be calculated for this condition in order to determine allowable emissions.

The mass flow sheet, Table 24.1, column 4, is calculated for 129 percent excess air, or 12 percent CO_2. The total dry gas flow is 35,923 lb/h.

Emissions, as stated above, are 0.03 gr/dry st ft^3 at 12 percent CO_2, and 1 dry ft^3 of air (dry gas) at standard conditions weighs 0.075 lb:

$$(35{,}923 \text{ lb/h}) \times \left(\frac{\text{dry st ft}^3}{0.075 \text{ lb}}\right) \times \left(\frac{0.03 \text{ gr}}{\text{dry st ft}^3}\right) \times \left(\frac{\text{lb}}{7000 \text{ gr}}\right)$$

$$= 2.05 \text{ lb/h emissions}$$

CONTROL EQUIPMENT

Air pollution control equipment must be provided to reduce particulate emissions to an acceptable level. From Table 7.6 the particulate discharge from a multiple-chamber incinerator is approximately 7 lb/ton of charge. For the solid waste incinerated in this example,

$$2000 \frac{\text{lb}}{\text{h}} \times 7 \frac{\text{lb}}{\text{ton}} \times \frac{\text{ton}}{2000 \text{ lb}} = 7 \frac{\text{lb}}{\text{h}}$$

Allowable emissions are 2.05 lb/h.

Required collection efficiency:

$$\frac{7 - 2.05}{7} \times 100 = 71\%$$

Examining Table 22.12 shows that a low-efficiency wet scrubber has a particulate removal efficiency of up to 90 percent. This is more than adequate to provide the 71 percent removal efficiency required. A pressure drop of 12 in WC will be assumed for the scrubber. The ID fan will be sized for an additional 3 in WC drop through the exhaust gas system.

PSD EVALUATION

As noted in Chap. 2, when emissions exceed *de minimis* values, PSD procedures are triggered. For particulate matter the trigger is 25 tons/year. For this example emissions of 2.05 lb/h of particulate are anticipated. On a yearly basis, with 5000 h/year operations,

$$2.05 \frac{\text{lb}}{\text{h}} \times 5000 \frac{\text{h}}{\text{year}} \times \frac{\text{ton}}{2000 \text{ lb}} = 5.13 \frac{\text{tons}}{\text{year}}$$

This is below the *de minimis* value for particulates.

The nitrogen oxide *de minimis* value is 40 tons/year. Using Fig. 7.2,

For the kiln: 13.00 MBtu/h heat release
150 percent excess air
1700°F

Generation of NO_x is 0.3 lb/MBtu.

In 1 year:

$$13.00 \frac{\text{MBtu}}{\text{h}} \times \frac{0.3 \text{ lb}}{\text{MBtu}} \times 5000 \frac{\text{h}}{\text{year}} \times \frac{\text{ton}}{2000 \text{ lb}} = 9.75 \frac{\text{tons}}{\text{year}}$$

For the afterburner: 9.58 MBtu/h heat release
10 percent excess air
2216°F

Generation of NO_x is 0.2 lb/MBtu.

In 1 year:

$$9.58 \text{ MBtu/h} \times 0.2 \text{ lb/MBtu} \times 5000 \text{ h/year} \times \text{ton}/2000 \text{ lb} = 4.79 \text{ tons/year}$$

Total NO_x generation:

$$9.75 + 4.79 = 14.54 \text{ tons/year } NO_x$$

This is less than the 40 tons/year *de minimis* value.

The carbon monoxide *de minimis* value is 100 tons/year. Examining Table 7.3, which is not directly applicable to trash burning (Table 7.3 applies to pure hydrocarbon compounds, whereas trash contains oxygen as an elemental component), indicates an order of magnitude of CO generation. At 2200°F, 50 percent excess air, 3×10^{-6} lb of CO is generated per pound of stoichiometric air (15,469 lb/h):

$$3 \times 10^{-6} \text{ lb CO/lb air} \times 15{,}469 \text{ lb air/h} \times 5000 \text{ h/year} \times \text{ton}/2000 \text{ lb}$$
$$= 0.12 \text{ ton/year CO}$$

This value of CO is insignificant compared to the *de minimis* value, 100 tons/year.

From the above calculations a PSD analysis will not be triggered by the quantities of particulate, NO_x, and/or CO generated by this system.

CONCLUSION

The previous analysis resulted in a determination of the following equipment items and parameters:

- Use of a rotary kiln, 7 ft 8 in outside diameter by 23 ft 1 in long lined with 7-in insulating firebrick and 6 in of firebrick.
- An afterburner with 11 ft 6 in outside diameter by 16 ft 6 in long lined with 9 in of insulating castable refractory.
- Use of a wet scrubber with 71 percent (minimum) collection efficiency.

- At 5000 h of operations per year this system will not require a PSD new-source review.
- The solid waste incinerated will not generate sufficient heat to maintain 2200°F in the afterburner. Fuel oil must be provided, 63.78 gal/h, for 2200°F operation.
- The liquid waste has sufficient heating value to replace all the fuel oil otherwise required to maintain 2200°F.
- Makeup water at a temperature of 70°F is required at a flow of 486 gal/min to cool the incinerator off-gas.
- A recirculation flow of 275 gal/min is required for adiabatic cooling (quencher flow). For an assumed pump head of 30 lb/in² gauge, the pump horsepower required is calculated as follows, assuming 85 percent pump efficiency:

$$\frac{\dfrac{275\text{ gal}}{\text{min}} \times \dfrac{8.34\text{ lb}}{\text{gal}} \times \dfrac{30\text{ lb}}{\text{in}^2} \times \dfrac{\text{ft}^3}{62.4\text{ lb}} \times \dfrac{144\text{ in}^2}{\text{ft}^2}}{33{,}000\,\dfrac{\text{ft}\cdot\text{lb}}{\text{min}\cdot\text{hp}} \times 0.85\text{ efficiency}} = 5.66\text{ hp}$$

  A 7.5-hp recirculation pump motor is required.
- Combustion air for the kiln is 22,785 lb/h (22,785 ÷ 0.075 ÷ 60 = 5063 st ft³/min). Assuming a fan pressure of 8 in WC, with 75 percent efficiency, the fan motor horsepower calculation is as follows:

$$\frac{\dfrac{5063\text{ ft}^3}{\text{min}} \times 8\text{ in} \times \dfrac{62.4\text{ lb}}{\text{ft}^3} \times \dfrac{\text{ft}}{12\text{ in}}}{33{,}000\,\dfrac{\text{ft}\cdot\text{lb}}{\text{min}\cdot\text{hp}} \times 0.75\text{ efficiency}} = 8.51\text{ hp}$$

  A 10-hp fan motor is required.
- Combustion air required in the afterburner is 6991 lb/h (1554 st ft³/min) when burning liquid waste and 7312 lb/h (1625 st ft³/min) when burning fuel oil. Using the larger flow value, the fan horsepower required, at 12 in WC fan pressure with 80 percent efficiency, is as follows:

$$\frac{\dfrac{1625\text{ ft}^3}{\text{min}} \times 12\text{ in} \times \dfrac{62.4\text{ lb}}{\text{ft}^3} \times \dfrac{\text{ft}}{12\text{ in}}}{33{,}000\,\dfrac{\text{ft}\cdot\text{lb}}{\text{min}\cdot\text{hp}} \times 0.80\text{ efficiency}} = 3.84\text{ hp}$$

  Use a 5-hp fan motor.
- The induced-draft fan must provide a maximum of 8792 ft³/min gas flow at 15 in WC. The required horsepower, at 80 percent efficiency, is

$$\frac{\dfrac{8792\text{ ft}^3}{\text{min}} \times 15\text{ in} \times \dfrac{62.4\text{ lb}}{\text{ft}^3} \times \dfrac{\text{ft}}{12\text{ in}}}{33{,}000\,\dfrac{\text{ft}\cdot\text{lb}}{\text{min}\cdot\text{hp}} \times 0.80\text{ efficiency}} = 25.98\text{ hp}$$

  A 30-hp fan motor (minimum) is required.

- The scrubber system will produce a net spent water discharge which must be drained. To calculate this discharge:

	Fuel oil	Liquid waste
Entering in flue gas, lb/h	1,941	2,071
In scrubber, lb/h	242,949	241,266
Total entering, lb/h	244,890	243,337
Exiting in flue gas	2,500	2,493
To drain, lb/h	242,390	240,844
gal/min	485	482
Exiting temperature, °F	178	179

CHAPTER 25
METRIC ANALYSIS

Throughout this book the calculations were performed with customary (United States or English) units, i.e., foot-pound-second-degree Fahrenheit. Conventions derived from these units were also used, such as gallon for liquid volume and horsepower for the rate of doing work. In this chapter SI units will be used and demonstrated. The more important tables in this text have been converted to SI, and the comprehensive calculations of Chap. 24 will be repeated in SI units.

SI UNITS

The SI units derive from an international treaty, the Metric Convention, that resulted from a meeting in France in 1872 that attempted to standardize measurement throughout the world. *Le Système International d'Unités*, abbreviated in English as SI units, is a body of standards of measurement based on the meter, kilogram, second, ampere, kelvin, mole, and candela.

The relationships between SI units and customary units of interest in this text are listed in Table 25.1. A number of other tables have been included in this chapter for convenience in performing calculations in SI (metric) units, as follows:

Table 25.2 Temperature Conversion
Table 25.3 Enthalpy, Air, and Moisture
Table 25.4 Properties of Air–Water Vapor Mixture
Table 25.5 Typical Combustion Parameters for Stoichiometric Burning
Table 25.6 No. 2 Fuel Oil
Table 25.7 Natural Gas
Table 25.8 Specific Volume

EXAMPLE

This example is similar to the example presented in Chap. 24. Material and heat balances will be in SI units; heat-transfer and chemical calculations will be converted from U.S. units.

TABLE 25.1 SI Units

atm × 101.324	= kPa
Btu × 0.252	= kcal
Btu × 1.0551	= kJ
Btu/ft^2 × 11.357	= kJ/m^2
Btu/ft^3 × 8.899	= $kcal/m^3$
Btu/ft^3 × 37.256	= kJ/m^3
Btu/gal × 66.5786	= $kcal/m^3$
Btu/gal × 0.2788	= kJ/L
Btu · in/(h · ft^2 · °F) × 0.5193	= kJ/(h · m · °C)
Btu/lb × 0.5556	= kcal/kg
Btu/lb × 2.326	= kJ/kg
(°F − 32)/1.8	= °C
ft × 0.3048	= m
ft^3 × 0.02832	= m^3
ft^2 × 0.09290	= m^2
ft^2 · h · °F/Btu × 0.0489	= m^2 · h · °C/kJ
ft/min × 0.00508	= m/s
ft^3/min × 1.6992	= m^3/h
ft^3/lb × 0.062434	= m^3/kg
gal × 3.785	= L (liter)
gal × 0.003785	= m^3
gal/min × 0.2271	= m^3/h
gr × 64.7989	= mg (milligram)
gr/ft^3 × 2288.097	= mg/m^3
hp × 0.7457	= kW
hp × 745.6999	= W
in × 25.4	= mm
inHg × 3.3769	= kPa
inH_2O × 0.2488	= kPa
mmHg × 133.3224	= Pa
lb × 0.4536	= kg
lb/Btu × 0.4299	= kg/kJ
lb/ft^3 × 16.017	= kg/m^3
lb/gal × 0.1198	= kg/L
lb/ton × 0.5	= kg/t
lb/in^2 × 6.8947	= kPa
lb/in^2 × 68.966	= mb (millibar)
ton × 0.9072	= t

A multipurpose incinerator is required to burn plant trash (rubbish) and an organic liquid. The trash is normally contaminated with chlorides in relatively small quantities. The organic liquid, cycloheptane (C_7H_{14}), contains trace amounts of solid contaminants. The facility must be designed for destruction of 907 kg/h of rubbish and 215 kg/h of liquid cycloheptane, either burning concurrently or firing the trash alone.

Future plant expansion may result in the burning of solid organic powders. Additional capacity will be obtained by operating additional hours during the week.

TABLE 25.2 Temperature Conversion

Fahrenheit	Degree[a]	Celsius	Fahrenheit	Degree	Celsius	Fahrenheit	Degree	Celsius
32.0	0	-17.8	131.0	55	12.8	230.0	110	43.3
33.8	1	-17.2	132.8	56	13.3	231.8	111	43.9
35.6	2	-16.7	134.6	57	13.9	233.6	112	44.4
37.4	3	-16.1	136.4	58	14.4	235.4	113	45.0
39.2	4	-15.6	138.2	59	15.0	237.2	114	45.6
41.0	5	-15.0	140.0	60	15.6	239.0	115	46.1
42.8	6	-14.4	141.8	61	16.1	240.8	116	46.7
44.6	7	-13.9	143.6	62	16.7	242.6	117	47.2
46.4	8	-13.3	145.4	63	17.2	244.4	118	47.8
48.2	9	-12.8	147.2	64	17.8	246.2	119	48.3
50.0	10	-12.2	149.0	65	18.3	248.0	120	48.9
51.8	11	-11.7	150.8	66	18.9	249.8	121	49.4
53.6	12	-11.1	152.6	67	19.4	251.6	122	50.0
55.4	13	-10.6	154.4	68	20.0	253.4	123	50.6
57.2	14	-10.0	156.2	69	20.6	255.2	124	51.1
59.0	15	-9.4	158.0	70	21.1	257.0	125	51.7
60.8	16	-8.9	159.8	71	21.7	258.8	126	52.2
62.6	17	-8.3	161.6	72	22.2	260.6	127	52.8
64.4	18	-7.8	163.4	73	22.8	262.4	128	53.3
66.2	19	-7.2	165.2	74	23.3	264.2	129	53.9
68.0	20	-6.7	167.0	75	23.9	266.0	130	54.4
69.8	21	-6.1	168.8	76	24.4	267.8	131	55.0
71.6	22	-5.6	170.6	77	25.0	269.6	132	55.6
73.4	23	-5.0	172.4	78	25.6	271.4	133	56.1
75.2	24	-4.4	174.2	79	26.1	273.2	134	56.7
77.0	25	-3.9	176.0	80	26.7	275.0	135	57.2
78.8	26	-3.3	177.8	81	27.2	276.8	136	57.8
80.6	27	-2.8	179.6	82	27.8	278.6	137	58.3
82.4	28	-2.2	181.4	83	28.3	280.4	138	58.9
84.2	29	-1.7	183.2	84	28.9	282.2	139	59.4
86.0	30	-1.1	185.0	85	29.4	284.0	140	60.0
87.8	31	-0.6	186.8	86	30.0	285.8	141	60.6
89.6	32	0.0	188.6	87	30.6	287.6	142	61.1
91.4	33	0.6	190.4	88	31.1	289.4	143	61.7
93.2	34	1.1	192.2	89	31.7	291.2	144	62.2
95.0	35	1.7	194.0	90	32.2	293.0	145	62.8
96.8	36	2.2	195.8	91	32.8	294.8	146	63.3
98.6	37	2.8	197.6	92	33.3	296.6	147	63.9
100.4	38	3.3	199.4	93	33.9	298.4	148	64.4
102.2	39	3.9	201.2	94	34.4	300.2	149	65.0
104.0	40	4.4	203.0	95	35.0	302.0	150	65.6
105.8	41	5.0	204.8	96	35.6	303.8	151	66.1
107.6	42	5.6	206.6	97	36.1	305.6	152	66.7
109.4	43	6.1	208.4	98	36.7	307.4	153	67.2
111.2	44	6.7	210.2	99	37.2	309.2	154	67.8
113.0	45	7.2	212.0	100	37.8	311.0	155	68.3
114.8	46	7.8	213.8	101	38.3	312.8	156	68.9
116.6	47	8.3	215.6	102	38.9	314.6	157	69.4
118.4	48	8.9	217.4	103	39.4	316.4	158	70.0
120.2	49	9.4	219.2	104	40.0	318.2	159	70.6
122.0	50	10.0	221.0	105	40.6	320.0	160	71.1
123.8	51	10.6	222.8	106	41.1	321.8	161	71.7
125.6	52	11.1	224.6	107	41.7	323.6	162	72.2
127.4	53	11.7	226.4	108	42.2	325.4	163	72.8
129.2	54	12.2	228.2	109	42.8	327.2	164	73.3

TABLE 25.2 Temperature Conversion (Continued)

Fahrenheit	Degree"	Celsius	Fahrenheit	Degree	Celsius	Fahrenheit	Degree	Celsius
329.0	165	73.9	581	305	152	2147	1175	635
330.8	166	74.4	590	310	154	2192	1200	649
332.6	167	75.0	599	315	157	2237	1225	663
334.4	168	75.6	608	320	160	2282	1250	677
336.2	169	76.1	617	325	163	2327	1275	691
338.0	170	76.7	626	330	166	2372	1300	704
339.8	171	77.2	635	335	168	2417	1325	718
341.6	172	77.8	644	340	171	2462	1350	732
343.4	173	78.3	653	345	174	2507	1375	746
345.2	174	78.9	662	350	177	2552	1400	760
347.0	175	79.4	671	355	179	2597	1425	774
348.8	176	80.0	680	360	182	2642	1450	788
350.6	177	80.6	689	365	185	2687	1475	802
352.4	178	81.1	698	370	188	2732	1500	816
354.2	179	81.7	707	375	191	2777	1525	829
356.0	180	82.2	716	380	193	2822	1550	843
357.8	181	82.8	725	385	196	2867	1575	857
359.6	182	83.3	734	390	199	2912	1600	871
361.4	183	83.9	743	395	202	2957	1625	885
363.2	184	84.4	752	400	204	3002	1650	899
365.0	185	85.0	770	410	210	3047	1675	913
366.8	186	85.6	788	420	216	3092	1700	927
368.6	187	86.1	806	430	221	3137	1725	941
370.4	188	86.7	824	440	227	3182	1750	954
372.2	189	87.2	842	450	232	3227	1775	968
374.0	190	87.8	860	460	238	3272	1800	982
375.8	191	88.3	878	470	243	3317	1825	996
377.6	192	88.9	896	480	249	3362	1850	1010
379.4	193	89.4	914	490	254	3407	1875	1024
381.2	194	90.0	932	500	260	3452	1900	1038
383.0	195	90.6	977	525	274	3497	1925	1052
384.8	196	91.1	1022	550	288	3542	1950	1066
386.6	197	91.7	1067	575	302	3587	1975	1079
388.4	198	92.2	1112	600	316	3632	2000	1093
390.2	199	92.8	1157	625	329	3677	2025	1107
392	200	93	1202	650	343	3722	2050	1121
401	205	96	1247	675	357	3767	2075	1135
410	210	99	1292	700	371	3812	2100	1149
419	215	102	1337	725	385	3857	2125	1163
428	220	104	1382	750	399	3902	2150	1177
437	225	107	1427	775	413	3947	2175	1191
446	230	110	1472	800	427	3992	2200	1204
455	235	113	1517	825	441	4037	2225	1218
464	240	116	1562	850	454	4082	2250	1232
473	245	118	1607	875	468	4127	2275	1246
482	250	121	1652	900	482	4172	2300	1260
491	255	124	1697	925	496	4217	2325	1274
500	260	127	1742	950	510	4262	2350	1288
509	265	129	1787	975	524	4307	2375	1302
518	270	132	1832	1000	538	4352	2400	1316
527	275	135	1877	1025	552	4397	2425	1329
536	280	138	1922	1050	566	4442	2450	1343
545	285	141	1967	1075	579	4487	2475	1357
554	290	143	2012	1100	593	4532	2500	1371
563	295	146	2057	1125	607	4577	2525	1385
572	300	149	2102	1150	621	4622	2550	1399

TABLE 25.2 Temperature Conversion (Continued)

Fahrenheit	Degree[a]	Celsius	Fahrenheit	Degree	Celsius	Fahrenheit	Degree	Celsius
4667	2575	1413	5702	3150	1732	6692	3700	2038
4712	2600	1427	5747	3175	1746	6737	3725	2052
4757	2625	1441	5792	3200	1760	6782	3750	2066
4802	2650	1454	5837	3225	1774	6827	3775	2079
4847	2675	1468	5882	3250	1788	6872	3800	2093
4892	2700	1482	5927	3275	1802	6917	3825	2107
4937	2725	1496	5972	3300	1816	6962	3850	2121
4982	2750	1510	6017	3325	1829	7007	3875	2135
5027	2775	1524	6062	3350	1843	7052	3900	2149
5072	2800	1538	6107	3375	1857	7097	3925	2163
5117	2825	1552	6152	3400	1871	7142	3950	2177
5162	2850	1566	6197	3425	1885	7187	3975	2191
5207	2875	1579	6242	3450	1899	7232	4000	2204
5252	2900	1593	6287	3475	1913	7277	4025	2218
5297	2925	1607	6332	3500	1927	7322	4050	2232
5342	2950	1621	6377	3525	1941	7367	4075	2246
5387	2975	1635	6422	3550	1954	7412	4100	2260
5432	3000	1649	6467	3575	1968	7457	4125	2274
5477	3025	1663	6512	3600	1982	7502	4150	2288
5522	3050	1677	6557	3625	1996	7547	4175	2302
5567	3075	1691	6602	3650	2010	7592	4200	2316
5612	3100	1704	6647	3675	2024	7637	4225	2329
5657	3125	1718						

[a]Example: 153°F = 67.2°C
153°C = 307.4°F

TABLE 25.3 Enthalpy, Air, and Moisture

Temp., °C	kJ/kg Relative to 0°C	
	Air	H_2O
50	49.03	2539.81
75	74.24	2623.70
100	99.48	2679.90
125	124.81	2735.33
150	150.29	2790.48
175	175.85	2845.39
200	201.52	2900.39
225	227.35	2955.61
250	253.35	3011.06
275	279.47	3066.96
300	305.77	3123.11
325	332.19	3179.91
350	358.81	3236.99
375	396.06	3294.51
400	412.55	3352.60
425	439.66	3411.38
450	466.96	3470.44
475	494.40	3530.15
500	522.02	3590.51
525	549.77	3650.81
550	577.70	3712.28
575	605.75	3773.74
600	633.95	3836.49
650	690.64	3955.24
700	747.62	4018.43
750	804.90	4083.98
800	862.38	4149.22
850	920.12	4213.82
900	978.31	4280.54
950	1036.74	4415.10
1000	1095.69	4549.81
1050	1155.00	4687.11
1100	1214.66	4826.99
1150	1274.63	4969.29
1200	1334.92	5111.70
1250	1395.49	5257.16
1300	1456.34	5403.04
1350	1517.43	5548.42
1400	1578.78	5700.07
1500	1701.92	6002.44

Source: Table 4.1.

TABLE 25.4 Properties of Saturated Air–Water Vapor Mixture

Temperature, °C	Relative to 0°C		
	Humidity, kg/kg DA[a]	Enthalpy, kJ/kg DA	Volume, m^3/kg DA
15	0.0106	42.09	0.830
16	0.0114	44.95	0.834
17	0.0122	47.91	0.838
18	0.0130	50.99	0.842
19	0.0138	54.19	0.846
20	0.0147	57.52	0.850
21	0.0157	61.00	0.854
22	0.0168	64.63	0.858
23	0.0178	68.40	0.863
24	0.0189	72.34	0.867
25	0.0201	76.44	0.872
26	0.0214	80.74	0.876
27	0.0228	85.23	0.881
28	0.0242	89.91	0.886
29	0.0257	94.80	0.893
30	0.0273	99.91	0.896
31	0.0290	105.26	0.902
32	0.0307	110.87	0.907
33	0.0326	116.74	0.913
34	0.0346	122.87	0.918
35	0.0367	129.29	0.924
36	0.0389	136.03	0.930
37	0.0412	143.10	0.937
38	0.0437	150.51	0.943
39	0.0455	158.26	0.950
40	0.0490	166.40	0.957
41	0.0519	174.97	0.964
42	0.0550	183.95	0.972
43	0.0567	193.39	0.976
44	0.0608	203.30	0.985
45	0.0652	213.70	0.995
46	0.0690	224.68	1.004
47	0.0731	236.23	1.013
48	0.0773	248.37	1.023
49	0.0808	261.17	1.030
50	0.0866	274.63	1.042
51	0.0916	288.86	1.053
52	0.0969	303.87	1.064
53	0.1026	319.76	1.076
54	0.1086	336.48	1.088
55	0.1149	354.18	1.101
56	0.1216	372.89	1.114
57	0.1288	392.75	1.128
58	0.1364	413.77	1.143
59	0.1444	435.98	1.158
60	0.1530	459.59	1.175
61	0.1622	484.73	1.192

TABLE 25.4 Properties of Saturated Air–Water Vapor Mixture (Continued)

	Relative to 0°C		
Temperature, °C	Humidity, kg/kg DA[a]	Enthalpy, kJ/kg DA	Volume, m^3/kg DA
62	0.1719	511.42	1.210
63	0.1822	539.80	1.230
64	0.1933	570.09	1.250
65	0.2051	602.29	1.272
66	0.2178	636.85	1.295
67	0.2313	673.77	1.320
68	0.2459	713.29	1.346
69	0.2614	755.68	1.374
70	0.2782	801.21	1.404
71	0.2964	850.63	1.436
72	0.3160	903.48	1.471
73	0.3372	960.81	1.508
74	0.3602	1,022.85	1.548
75	0.3851	1,090.24	1.592
76	0.4125	1,164.00	1.640
77	0.4424	1,244.58	1.691
78	0.4752	1,332.90	1.748
79	0.5113	1,430.02	1.810
80	0.5511	1,537.21	1.879
81	0.5957	1,656.90	1.955
82	0.6453	1,790.32	2.040
83	0.7010	1,940.01	2.135
84	0.7639	2,108.68	2.242
85	0.8352	2,300.13	2.363
86	0.9176	2,521.06	2.502
87	1.0129	2,776.69	2.663
88	1.1242	3,075.90	2.851
89	1.2564	3,428.90	3.074
90	1.4140	3,852.32	3.340
91	1.6098	4,376.69	3.669
92	1.8550	5,032.35	4.080
93	2.1702	5,876.64	4.609
94	2.5896	6,999.49	5.313
95	3.1730	8,558.75	6.290
96	4.0676	10,955.74	7.789
97	5.5756	14,989.91	10.301
98	8.6556	23,232.55	15.469
99	18.7300	50,187.64	32.327

[a]DA = Dry Air

Source: Table 4.5.

TABLE 25.5 Typical Combustion Parameters for Stoichiometric Burning

kJ	kg Air/ 10 MJ	kg Dry gas/10 MJ	kg H_2O/ 10 MJ	Element fractions				
				C	H_2	S	O_2	N_2
10,000	1.3361	1.2086	0.1274	0.165	0.031	0.01	0.012	0.781
12,000	1.6172	1.4724	0.1448	0.208	0.036	0.01	0.017	0.730
14,000	1.8984	1.7362	0.1622	0.251	0.040	0.01	0.021	0.678
16,000	2.1796	2.0000	0.1796	0.294	0.044	0.01	0.025	0.627
18,000	2.4608	2.2637	0.1970	0.337	0.049	0.01	0.029	0.576
20,000	2.7419	2.5275	0.2144	0.380	0.053	0.01	0.033	0.524
22,000	3.0231	2.7913	0.2318	0.423	0.057	0.01	0.037	0.473
24,000	3.3043	3.0551	0.2492	0.466	0.062	0.01	0.041	0.421
26,000	3.5854	3.3188	0.2666	0.509	0.066	0.01	0.045	0.370
28,000	3.8666	3.5826	0.2840	0.552	0.070	0.01	0.049	0.319
30,000	4.1478	3.8464	0.3014	0.595	0.074	0.01	0.053	0.267
32,000	4.4290	4.1102	0.3188	0.638	0.079	0.01	0.058	0.216
34,000	4.7101	4.3739	0.3362	0.681	0.083	0.01	0.062	0.164
36,000	4.9913	4.6377	0.3536	0.724	0.087	0.01	0.066	0.113
38,000	5.2725	4.9015	0.3710	0.767	0.092	0.01	0.070	0.062
40,000	5.5536	5.1653	0.3884	0.810	0.096	0.01	0.074	0.010

Source: Table 8.6.

TABLE 25.6 No. 2 Fuel Oil, 38,949 kJ/L, 13.05 kg/L

Total air:	1.1	1.2	1.3
kg air/L	13.734	14.982	16.231
kg dry gas/L	13.791	15.039	16.288
kg H_2O/L	1.032	1.048	1.065
Temperature, °C		Heat available, kJ/L	
100	35,075	34,926	34,778
150	34,269	34,060	33,847
200	33,463	33,211	32,910
250	32,650	32,310	31,965
300	31,827	31,420	31,009
350	30,995	30,521	30,042
400	30,007	29,452	28,891
450	29,297	28,685	28,069
500	28,431	27,749	28,391
550	27,555	26,802	26,045
600	26,668	25,844	25,015
650	25,772	24,875	23,974
700	24,875	23,906	22,932
750	23,969	22.926	21,879
800	23,054	21,937	20,816
850	22,130	20,939	19,744
900	21,197	19,931	18,661
950	20,259	18,917	17,571
1000	19,312	17,894	16,472
1050	18,359	16,865	15,367
1100	17,397	15,828	17,214
1150	16,435	14,787	13,134
1200	15,474	13,745	9,053
1300	13,523	11,639	6,791
1400	11,547	9,505	4,499
1500	9,552	7,352	2,187

CRITERIA

It will be assumed that the wastes charged are hazardous and that the incinerator will have to comply with RCRA requirements for hazardous wastes containing chlorides. Since the specific nature of the waste stream with regard to parameters of destruction is not known, the following criteria will be used:

- A temperature zone of 1200°C should be provided with a retention time of 2 s.
- A minimum of 3 percent oxygen should be present in the off-gas stream.
- A maximum particulate loading of 69 mg per normal cubic meter should be allowed.

Additional criteria include the following:

TABLE 25.7 Natural Gas, 37,256 kJ/m³, 0.80 kg/(N · m³)

Total air:	1.1	1.2	1.3
kg air/m³	12.09	12.67	13.25
kg dry gas/m³	11.40	11.98	12.56
kg H_2O/m³	1.65	1.65	1.67
Temperature, °C		Heat available, kJ/m³	
100	7639	7629	7602
150	7462	7444	7410
200	7286	7260	7217
250	7102	7068	7017
300	6926	6884	6825
350	6742	6692	6625
400	6525	6467	6393
450	6365	6299	6216
500	6173	6107	6016
550	5981	5899	5800
600	5780	5698	5592
650	5588	5490	5375
700	5380	5282	5159
750	5172	5074	4943
800	4972	4865	4727
850	4771	4649	4502
900	4563	4433	4278
950	4355	4217	4054
1000	4139	4000	3830
1050	3930	3784	3597
1100	3714	3560	3365
1150	3498	3336	3133
1200	3282	3111	2901
1300	2849	2663	2436
1400	2401	2214	1971
1500	1952	1750	1475

- A maximum external temperature of 93°C at the surface of the primary chamber
- A temperature limit of 121°C at the surface of the secondary chamber

INCINERATOR EQUIPMENT

The incinerator system must handle solid waste and/or liquid waste. A rotary kiln with an afterburner or a modular combustion unit could be used. The provision that powders be incinerated in the future requires that turbulence be provided within the incinerator. Without sufficient turbulence to agitate the powder waste and expose the powder surfaces to air and heat, the powder may not burn proper-

TABLE 25.8 Specific Volume

Temperature, °C	Air, m^3/kg	H_2O, m^3/kg
50	0.915	1.473
75	0.986	1.587
100	1.056	1.701
125	1.127	1.815
150	1.198	1.929
175	1.269	2.043
200	1.339	2.157
225	1.410	2.271
250	1.481	2.385
275	1.552	2.499
300	1.622	2.613
325	1.693	2.726
350	1.764	2.840
375	1.835	2.954
400	1.905	3.068
425	1.976	3.182
450	2.047	3.296
475	2.118	3.410
500	2.188	3.524
525	2.259	3.638
550	2.330	3.752
575	2.401	3.866
600	2.471	3.980
625	2.542	4.093
650	2.613	4.207
675	2.684	4.321
700	2.754	4.435
725	2.825	4.549
750	2.896	4.663
775	2.966	4.777
800	3.037	4.891
825	3.108	5.005
850	3.179	5.119
875	3.249	5.233
900	3.320	5.347
925	3.391	5.460
950	3.462	5.574
975	3.532	5.688
1000	3.603	5.802
1050	3.745	6.030
1100	3.886	6.258
1150	4.028	6.486
1200	4.169	6.714
1250	4.311	6.941
1300	4.452	7.169
1350	4.594	7.397
1400	4.735	7.625
1450	4.877	7.853
1500	5.018	8.081

TABLE 25.8 Specific Volume (Continued)

Temperature, °C	Air, m^3/kg	H_2O, m^3/kg
1550	5.159	8.308
1600	5.301	8.536
1650	5.442	8.764
1700	5.584	8.992
1750	5.725	9.220
1800	5.867	9.448
1850	6.008	9.675
1900	6.150	9.903
1950	6.291	10.131
2000	6.433	10.359
2050	6.574	10.587
2100	6.716	10.815
2150	6.857	11.042

Source: Table 4.4.

ly. The rotary kiln appears to be the favorable option because it provides turbulence to the solid waste charge.

A rotary kiln will be used for solid waste incineration. The off-gas will pass through an afterburner (secondary combustion chamber) where the required temperature and residence time will be maintained.

Fuel oil (no. 2 grade) will be used to obtain the required temperature if the waste will not maintain combustion. The liquid waste stream will be fired in the afterburner.

SOLID WASTE

The incinerator must burn 907 kg/h of rubbish, type 1 waste, as noted in Table 8.2 (6500 Btu/lb or 15,119 kJ/kg). The total heat release is as follows:

$$15{,}119 \text{ kJ/kg} \times 907 \text{ kg/h} = 13{,}712{,}933 \text{ kJ/h}$$

Assuming a kiln heat release rate of 931,400 kJ/(h · m³) [25,000 Btu/(h · ft³)], from Chap. 12, the required volume can be calculated:

$$\frac{13{,}712{,}933 \text{ kJ/h}}{931{,}400 \text{ kJ/(h} \cdot \text{m}^3)} = 14.723 \text{ m}^3$$

With a 4:1 ratio of length to diameter, by the methods of Chap. 12

$$L = 4D$$

$$V = \frac{\pi D^2 L}{4} = \frac{\pi D^2}{4}(4D) = \pi D^3$$

$$D = \frac{V}{\pi} \exp \frac{1}{3} = \frac{14.723}{\pi} \exp \frac{1}{3} = 1.673 \text{ m}$$

$$L = 4D = 4 \times 1.673 = 6.692 \text{ m}$$

The kiln inside dimensions are, therefore,

$$D = 170\text{-cm inside diameter}$$

$$L = 670\text{-cm inside length}$$

For an external temperature of 93°C with an internal temperature of 871°C, a refractory will be selected. The kiln is continually rotating, and convection heat transfer can be best approximated as a vertical wall. The kiln surface is oxidized steel. Using the resistance graph, Fig. 5.5, we see that a resistance R of 5.3 is required. From Chap. 5,

$$x = RK$$

Using a dense castable, $K = 5$, over 61 cm (2 ft) of refractory is required. A 61-cm thickness of refractory is not practical (too large), and a two-wall refractory system will be used, insulating castable against the kiln surface and 15 cm (6 in) of firebrick adjacent to the hot stream. Following the methods of Chap. 5,

$$R = R_1 + R_2 = \frac{X_1}{K_1} + \frac{X_2}{K_2}$$

with

X_1 = 15-cm (6-in) firebrick

K_1 = 5.19 kJ/(m · °C · h) [10.0 Btu · in/(ft^2 · °F · h)] for firebrick

and

K_2 = 0.68 kJ/(m · °C · h) [1.3 Btu · in/(ft^2 · °F · h)] for insulating castable

To find the thickness of castable required X_2:

$$X_2 = \left(R - \frac{X_1}{K_1}\right) \times K_2$$

$$= \left(5.3 - \frac{6}{10.0}\right) \times 1.3 = 6.11 \text{ in } (15.5 \text{ cm})$$

Therefore the kiln should be lined with 16 cm (7 in) of insulating castable protected from the gas stream by 15 cm (6 in) of firebrick.

From the resistance graph the heat loss Q from the kiln, which will have a surface temperature of 93°C (200°F), will be 2987 kJ/(h · m^2) [263 Btu/(h · ft^2)].

The internal kiln dimensions are 170-cm (5 ft 6 in) diameter by 670 cm (22 ft 0 in) long and with a total lining of 33 cm (13 in). The kiln external dimensions are:

Diameter: 170 cm + 2 × 33 cm = 236 cm

Length: 670 cm + minimum of wall diameter (33 cm) = 703 cm *minimum*

Heat loss through the kiln wall, per the resistance graph, is 2987 kJ/(h · m^2) [263 Btu/(h · ft^2)], as noted above. Now that an external kiln size has been determined, the total heat loss can be calculated:

Kiln external cylindrical area:

$$\pi DL = \pi \times 2.34 \text{ m} \times 7.03 \text{ m} = 51.679 \text{ m}^2$$

Heat loss:

$$51.679 \text{ m}^2 \times 2987 \text{ kJ/(h} \cdot \text{m}^2) = 154{,}365 \text{ kJ/h}$$

LIQUID WASTE

Cycloheptane will be fired in the secondary chamber, or afterburner, at its supply rate of 215 kg/h (475 lb/h). To determine the heating value of cycloheptane, the methods of Chap. 8 will be used.

By the heat of formation method, with Tables 8.3 and 8.5:

$$\begin{array}{lllll} -37.5 & & -94.1 & -68.3(g) & \\ C_7H_{14} + 10.5O_2 + 3.3197(10.5)N_2 \rightarrow & & 7CO_2 \quad + & 7H_2O \quad + & 34.857N_2 \\ 98.21 & & 44.01 & 18.02 & 28.02 \\ 1.00 & & 3.3168 & 1.2843 & 9.9499 \end{array}$$

$$\begin{aligned} \Delta H_f \; H_2O \; (g){:} \quad & -68.3 \times 7 = -478.1 \\ \Delta H_f \; CO_2{:} \quad & -94.1 \times 7 = -658.7 \\ \Delta H_f \; C_7H_{14}{:} \quad & +37.5 \times 1 = \underline{+37.5} \\ & \Sigma \Delta H_f = -1099.3 \end{aligned}$$

$$\frac{1}{98.21 \text{ g/mol}} \times (-1099.3 \text{ kcal/mol}) \times 1802 = -20{,}170 \text{ Btu/lb}$$

$$[-46{,}915 \text{ kJ/kg}] \text{ (exothermic)}$$

Using the hydrocarbon approximation gives

$$\frac{10.5 \text{ lb} \cdot \text{mol } O_2}{\text{mol } C_7H_{14}} \times \frac{\text{mol } C_7H_{14}}{98.21 \text{ lb}} \times \frac{184{,}000 \text{ Btu}}{1 \text{ lb} \cdot \text{mol } O_2} = 19{,}672 \text{ Btu/lb } (45{,}757 \text{ kJ/kg})$$

These two heating values are relatively close:

$$\frac{46{,}915 - 45{,}757}{46{,}915} \times 100 = 2.5\%$$

The more accurate number, 46,915 kJ/kg, will be used for further calculations.

The generation of off-gas is calculated from the above equation, noting that the second line of values beneath the equation is the normalized quantity of each component:

$$\frac{3.1368 \text{ kg } CO_2}{\text{kg } C_7H_{14}} + \frac{9.9449 \text{ kg } N_2}{\text{kg } C_7H_{14}} = \frac{13.0817 \text{ kg DG}}{\text{kg } C_7H_{14}}$$

$$\frac{13.0817 \text{ kg DG}}{\text{kg } C_7H_{14}} \times \frac{\text{kg } C_7H_{14}}{46{,}915 \text{ kJ}} = \frac{2.788 \text{ kg DG}}{10 \text{ MJ}}$$

$$\frac{1.2843 \text{ kg } H_2O}{\text{kg } C_7H_{14}} \times \frac{\text{kg } C_7H_{14}}{6915 \text{ kJ}} = \frac{0.274 \text{ kg } H_2O}{10 \text{ MJ}}$$

The size of an afterburner will be calculated for an external low-pressure atomizing burner, as described in Chap. 14. This burner is utilized because of the presence of solids in the liquid waste stream and the low-pressure compressed air required. With a heat release of 2,235,360 kJ/m³ (60,000 Btu/ft³) and 215 kg/h of cycloheptane, the *minimum* chamber size is calculated as follows:

$$215 \text{ kg/h} \times 46{,}915 \text{ kJ/kg} = 10{,}086{,}725 \text{ kJ/h}$$

$$\frac{10{,}086{,}725 \text{ kJ/h}}{2{,}235{,}360 \text{ kJ/(h} \cdot \text{m}^3)} = 4.512 \text{ m}^3$$

For a 122-cm-diameter chamber the required length is as follows:

Area: $\pi/4\ (122)^2 = 11{,}690 \text{ cm}^2 = 1.169 \text{ m}^2$
Length: $4.512 \text{ m}^3/1.169 \text{ m}^2 = 3.860 \text{ m}$

Use a length of 390 cm.

As noted above, this is the *minimum* chamber size to take care of the heat release of the cycloheptane. Additional fuel may be required to bring the 871°C gas stream leaving the kiln to the required 1200°C.

The afterburner refractory will be selected on the basis of 1200°C inside temperature and 121°C external temperature. Using an outside surface of steel, with conduction averaged as a vertical wall, from the resistance graph (Fig. 5.5) the required wall thickness is calculated as follows:

$$R = \frac{X}{K} = 4.7 \text{ h} \cdot \text{ft}^2 \cdot {}^\circ\text{F/Btu} \qquad \text{(from resistance graph)}$$

$$= 0.23 \text{ h} \cdot \text{m}^2 \cdot {}^\circ\text{C/kJ}$$

Using insulating castable gives

$$K = 1.9 \text{ Btu} \cdot \text{in/(h} \cdot \text{ft}^2 \cdot {}^\circ\text{F)} = 0.989 \text{ kJ/(h} \cdot \text{m} \cdot {}^\circ\text{C)}$$

$$X = KR = 0.989 \text{ kJ/(h} \cdot \text{m} \cdot {}^\circ\text{C)} \times 0.23 \text{ h} \cdot \text{m}^2 \cdot {}^\circ\text{C/kJ} = 0.227 \text{ m}$$

A thickness of 23 cm will be used.

The heat loss, again from the resistance graph, is 418 Btu/(h · ft²) [4747 kJ/(h · m²)].

INCINERATOR CALCULATIONS

For rubbish, from Table 8.3, the ash is 10 percent of the as-received weight, and moisture is 25 percent of the as-received weight. For the mass balance, Table 25.8, the wet feed is 907 kg/h. The moisture content is 25 percent, which is 226.75

kg/h. The dry feed is the difference between the wet feed and moisture, or 680.25 kg/h.

The ash content is 10 percent of the wet feed, or 90.7 kg/h, and the volatile content is therefore 680.25 − 90.7 = 589.55 kg/h. This is 90.7 ÷ 680.25 = 13.3 percent of the dry weight. From the tabular data in Chap. 8, type 1 waste has an as-received heating value of 15,119 kJ/kg (6500 Btu/lb). This calculates to 15,119 kJ/kg × 907 kg/h ÷ 589.55 kg/h = 23,260 kJ/kg volatile.

Dry gas and moisture generation will be taken as 3.22 kg/10 MJ (7.50 lb/10 kBtu) and 0.219 kg/10 MJ (0.51 lb/10 kBtu), respectively, as noted in Chap. 13, for refuse.

The mass balance is computed using 150 percent excess air for burning the solid waste.

In the heat balance, 270 kJ/kg is assumed for ash losses, and the kiln radiation loss has been calculated previously in this chapter for a 93°C external kiln temperature as 0.154 MJ/h. The radiation related to heat release is 0.154 ÷ 13.713 = 1.1 percent.

The temperature developed within the kiln, to obtain the 13.713 MJ/h released, is calculated from the methods of Chap. 4 as follows:

Temperature		MJ/h
900°C	978.31 × 10,605 + 4280.54 × 630 =	13.072
x		13.674
950°C	1036.74 × 10,605 + 4415.10 × 630 =	13.776

$$x = 900 + (950 - 900)\frac{13.674 - 13.072}{13.776 - 13.072} = 943°\text{C}$$

The off-gas from the solid waste must be raised to 1200°C, and 250 L/h fuel oil is required, from the heat balance, to obtain this heat increase.

At this point the gas volume will be evaluated. Using Table 25.8 at 1200°C, the gas flow is calculated as follows:

Dry gas: 4.169 m^3/kg × 30,275 kg/h × h/3600 s = 15.87 m^3/s

Moisture: 6.714 m^3/kg × 2071 kg/h × h/3600 s = 1.75 m^3/s

Total = 17.62 m^3/s

For a 2-s retention time a volume of 2 × 17.62 = 1250 m^3 is required.

This volume is 8 times greater than the minimum volume necessary for firing the liquid waste stream, cycloheptane as initially calculated. It is also twice the volume required for the rotary kiln at the head of the system.

A reasonable chamber would have an internal diameter of 300 cm and a length of 503 cm. As calculated previously, the wall thickness is 23 cm. The external diameter would be, therefore, 300 cm + (2 × 23 cm) = 346 cm.

The radiation and convection loss from the kiln is also as noted previously, 4747 kJ/(h · m^2).

The total heat loss from the afterburner is calculated as follows:

$$A = \pi DL = \pi(3.46 \text{ m})(5.03 \text{ m}) = 54.68 \text{ m}^2$$

Heat transfer: 54.68 m^2 × 4747 kJ/(h · m^2) = 0.260 MJ/h

The oxygen, as noted previously, must be at least 3 percent within the afterburner. The total gas flow, from the heat balance, is 13,707 kg/h dry gas plus 936 kg/h H_2O, a total of 14,643 kg/h wet gas flow. The oxygen component of this flow is the sum of excess oxygen injected in the primary chamber (kiln) and the excess excess oxygen injected into the afterburner:

Kiln: $\quad 6189 \text{ kg/h excess air} \times 0.2315 \text{ kg } O_2/\text{kg air} = 1433 \text{ kg/h } O_2$

Afterburner: $\quad 3164 \text{ kg/h total air} \times \dfrac{0.1 \text{ excess air}}{1.1 \text{ total air}}$

$$= 288 \text{ kg/h excess air} \times 0.2315 \text{ kg } O_2/\text{kg air}$$

$$= 66.67 \text{ kg/h } O_2$$

The total oxygen in the gas flow is therefore

$$1433 + 67 = 1500 \text{ kg/h } O_2$$

The percentage of oxygen is

$$\frac{1500 \text{ kg/h } O_2}{14{,}643 \text{ kg/h gas flow}} \times 100 = 10.24\% \ O_2 \text{ by weight}$$

This is well in excess of the minimum oxygen requirement, 3 percent of the gas flow.

The above calculations, as noted in the first column of Tables 25.9 and 25.10, the mass flow and heat balance sheets, represent burning solid waste by using no. 2 fuel oil in the afterburner to reach the desired temperature. The second column is used to calculate combustion characteristics of the waste liquid, cycloheptane. Cycloheptane will be used in lieu of fuel oil if it has sufficient heating value to raise the temperature of the kiln off-gas to at least 1200°C.

The mass flow sheet calculates the air feed and products of combustion of cycloheptane. The heat losses from burning cycloheptane are inserted in the heat balance sheet, second column.

The third column of the mass flow and heat balance sheets represents the conditions within the afterburner when cycloheptane is burned in the afterburner at 1230°C, more than the 1200°C minimum required for complete destruction of the combustible components within the off-gas.

The total off-gas flow is slightly less for burning cycloheptane (13,707 + 936 = 14,643 kg/h) than fuel oil (14,053 + 888 = 14,941 kg/h) in the afterburner. The flue gas flow parameters are calculated for each of these two cases in Table 25.11, the flue gas discharge table.

PARTICULATE EMISSIONS

Allowable particulate emissions are related to a 12 percent CO_2 discharge standard, that is, 69 mg per normal cubic meter corrected to 12 percent CO_2. This correction is calculated as follows:

TABLE 25.9 Mass Flow

	Kiln	Afterburner	Kiln w/ afterburner	12% CO_2
Wet feed, kg/h	907	215		
Moisture, %	25			
kg/h	226.75			
Dry feed, kg/h	680.25	215		
Ash, %	13.3			
kg/h	90.7			
Volatile, kg/h	589.55	215		805
Volatile htg. value, kJ/kg	23,260	46,915		
MJ/h	13.713	10.087	23.80	
Dry gas, kg/10 MJ	3.22	2.79		
kg/h	4416	2814		7230
Comb. H_2O, kg/10 MJ	0.219	0.275		
kg/h	300	277		577
Dry gas+comb. H_2O, kg/h	4716	3091		7807
100% Air, kg/h	4126	2876	7002	7002
Total air fraction	2.5	1.1		2.29
Total air, kg/h	10,315	3164	13,479	16,035
Excess air, kg/h	6189	288	6479	9033
Humidity/dry gas (air), kg/kg	0.01	0.01		
Humidity, kg/h	103	29	132	
Total H_2O, kg/h	630	306	936	
Total dry gas, kg/h	10,605	3102	13,707	16,263

Source: Table 25.1.

Trash: Assume $C_6H_{10}O_5$ (cellulose)

Cycloheptane: C_7H_{14}

Equilibrium equation:

$$C_6H_{10}O_5 + C_7H_{14} + 16.5xO_2 + 16.5(3.3197)xN_2 \rightarrow \overset{44.01}{\underset{572.13}{13CO_2}} + 12H_2O + \overset{28.2}{\underset{1534.80}{54.775x}} + \overset{32.00}{\underset{528x\,-\,528}{16.5(x-1)}}$$

With the unit molecular weight above the dry gas components and the total molecular beneath the components, the total dry gas flow is

$$572.13 + 1534.80x + 528.00x - 528.00 = 44.13 + 2062.80x$$

The CO_2 fraction present is 0.12 (12 percent CO_2):

$$0.12 = \frac{572.13}{44.13 + 2062.80x}$$

$$x = 2.290$$

This calculation indicates that 229 percent total air, or 129 percent excess air, will produce a CO_2 concentration of 12 percent of the exiting dry gas stream. The actual dry gas flow must be calculated for this condition in order to determine allowable emissions.

TABLE 25.10 Heat Balance

	Kiln	Afterburner	Kiln w/ afterburner
Cooling air wasted, kg/h			
°C			
kJ/lb			
MJ/h			
Ash, kg/h	90.7		
kJ/kg	465		
MJ/h	0.042		
Radiation, %	1.1	2	
MJ/h	0.151	0.202	
Humidity, kg/h	103	13	
Correction @ 2256 kJ/kg, MJ/h	−0.232	−0.029	
Losses, total, MJ/h	−0.039	0.173	0.134
Input, MJ/h	13.713	10.087	23.800
Outlet, MJ/h	13.674	9.914	23.588
Dry gas, kg/h	10,605	3102	13,707
H_2O, kg/h	630	306	936
Temperature, °C	943		1230
Desired temp., °C	1200		1230
MJ/h	17.377		
Net, MJ/h	3.876[a]		
Fuel oil, air fraction	1.1		
Net, MBtu/(h · L)	15,474		
L/h	250		
Air, kg/L	13.734		
kg/h	3434		
Dry gas, kg/L	13.791		
kg/h	3448		
H_2O, kg/L	1.032		
kg/h	258		
Dry gas with fuel oil, kg/h	14,053		13,707
H_2O with fuel oil, kg/h	888		936
Air with fuel oil, kg/h	13,749		13,479
Outlet, MJ/h	23.299[a]		23.800
Reference t, °C	0	0	0

[a]Including 0.173 MJ/h heat loss in the afterburner.
Source: Table 24.2.

The mass flow sheet, Table 25.9, column 4, is calculated for 129 percent excess air, that is, 12 percent CO_2. The total dry gas flow is 16,263 kg/h.

Emissions, as stated above, are 69 mg per normal cubic meter at 12 percent CO_2, and 1 dry ft^3 of air (dry gas) at standard conditions weighs 1.201 kg:

$$16{,}263 \text{ kg/h} \times \frac{\text{N} \cdot \text{m}^3}{1.201 \text{ kg}} \times 69 \text{ mg/(N} \cdot \text{m}^3) \times \frac{\text{kg}}{1{,}000{,}000 \text{ mg}}$$

$$= 0.93 \text{ kg/h emissions}$$

TABLE 25.11 Flue Gas Discharge

	Kiln	Kiln w/afterburner
Inlet, °C	1200	1230
Dry gas, kg/h	14,053	13,707
Heat, MJ/h	23.299	23.800
kJ/kg dry gas	1658	1736
Adiabatic *t*, °C	81	82
H_2O saturation, kg/kg dry gas	0.5957	0.6453
kg/h	8371	8845
H_2O inlet, kg/h	888	936
Quench H_2O, kg/h	7483	7909
L/m	125	132
Outlet temp., °C	49	49
Raw H_2O temp., °C	21	21
Sump temp.,[a] °C	64	65
Temp. diff., °C	43	44
Outlet, kJ/kg dry gas	261.17	261.17
MJ/h	3.670	3.580
Req'd. cooling,[b] MJ/h	19.629	20.220
H_2O, kg/h	109,055	109,785
L/m	1818	1830
Outlet, m^3/kg dry gas	1.030	1.030
m^3/min	241	235
Fan press., kPa	3.732	3.732
Outlet, m^3/min	250	244
Outlet, kg H_2O/kg dry gas	0.0808	0.0808
H_2O, kg/h	1135	1108
Recirc. (ideal), L/m	125	132
Recirc. (actual), L/m	998	1055
Cooling H_2O, L/m	1818	1830

[a]Sump temperature is the adiabatic temperature divided by 1.26.
[b]Required cooling is Req'd. cooling, MJ/h, × 0.2389/Temp. diff.
Source: Table 24.3.

CONTROL EQUIPMENT

Air pollution control equipment must be provided to reduce particulate emissions to an acceptable level. From Table 7.6 the particulate discharge from a multiple-chamber incinerator is approximately 7 lb/ton (3.5 kg/t) of charge. For the solid waste incinerated in this example:

$$907 \text{ kg/h} \times 3.5 \text{ kg/t} \times \frac{\text{t}}{1000 \text{ kg}} = 3.17 \text{ kg/h}$$

Allowable emissions are 0.93 kg/h. The required collection efficiency is

$$\frac{3.17 - 0.93}{3.17} \times 100 = 71\%$$

Examining Table 22.12, we see that a low-efficiency wet scrubber has a particulate removal efficiency of up to 90 percent. This is more than adequate to provide the 71 percent removal efficiency required. A pressure drop of 3 kPa will be assumed for the scrubber. The ID fan will be sized for an additional 0.75-kPa drop through the exhaust gas system.

CONCLUSION

The previous analysis resulted in a determination of the following equipment items and parameters:

- Use a rotary kiln with 236-cm outside diameter by 703 cm long lined with 16-cm insulating firebrick and 15 cm of firebrick.
- Use an afterburner with 168-cm outside diameter by 413 cm long lined with 23-cm insulating castable refractory.
- Use a wet scrubber with 71 percent (minimum) collection efficiency.
- The solid waste incinerated will not generate sufficient heat to maintain 1200°C in the afterburner. Fuel oil must be provided, 250 L/h, for 1200°C operation.
- The liquid waste has sufficient heating value to replace all the fuel oil otherwise required to maintain 1200°C.
- Makeup water at a temperature of 21°C is required at a flow of 1818 L/m to cool the incinerator off-gas.
- A recirculation flow of 998 L/m is required for adiabatic cooling (quencher flow). For an assumed pump head of 207 kPa, the pump horsepower required is calculated as follows, assuming 85 percent pump efficiency:

$$\frac{998 \text{ L/m} \times 207 \text{ kPa}}{44{,}746 \text{ L} \cdot \text{kPa/(min} \cdot \text{hp)} \times 0.85 \text{ efficiency}} = 5.43 \text{ hp}$$

  A 7.5-hp recirculation pump is required.
- Combustion air for the kiln is 10,315 kg/h (10,315 ÷ 1.201 m^3/kg ÷ 60 = 316 N · m^3/min). Assuming a fan pressure of 2 kPa, with 75 percent efficiency, the fan motor horsepower calculation is as follows:

$$\frac{143 \text{ N} \cdot \text{m}^3\text{/min} \times 2 \text{ kPa}}{44.715 \text{ N} \cdot \text{m}^3 \cdot \text{kPa/(min} \cdot \text{hp)} \times 0.75 \text{ efficiency}} = 8.53 \text{ hp}$$

  A 10-hp fan motor is required.
- Combustion air required in the afterburner is 3164 kg/h (43.9 N · m^3/m) when burning liquid waste and 3434 kg/h (47.7 N·m^3/m) when burning fuel oil. Using the larger flow value, the fan horsepower required, a 3-kPa fan pressure with 80 percent efficiency, is as follows:

$$\frac{47.7 \text{ N} \cdot \text{m}^3\text{/min} \times 3 \text{ kPa}}{44.715 \text{ N} \cdot \text{m}^3 \cdot \text{kPa/(min} \cdot \text{hp)} \times 0.75 \text{ efficiency}} = 4.00 \text{ hp}$$

  Use a 5-hp fan motor.

- The induced-draft fan must provide a maximum of 250 N · m^3/m gas flow at 3.732 kPa. The required horsepower, at 80 percent efficiency, is

$$\frac{250 \text{ m}^3/\text{min} \times 3.732 \text{ kPa}}{44.715 \text{ N} \cdot \text{m}^3 \cdot \text{kPa}/(\text{min} \cdot \text{hp}) \times 0.75 \text{ efficiency}} = 26.08 \text{ hp}$$

  A 30-hp fan motor (minimum) is required.
- The scrubber system will produce a net spent water discharge which must be drained. To calculate this discharge:

	Fuel oil	Liquid waste
Entering in flue gas, kg/h	888	936
In scrubber, kg/h	109,055	109,785
Total entering, kg/h	109,943	110,721
Exiting in flue gas, kg/h	1,135	1,108
To drain, kg/h	108,808	109,613
L/m	1,813	1,827
Exiting temperature, °C	64	65

GLOSSARY

abrasion: The wearing away of surface material by the scouring action of moving solids, liquids, or gases.

absorption: The penetration of one substance into or through another.

activated carbon: A highly absorbent form of carbon used to remove odors and toxic substances from gaseous emissions or to remove dissolved organic material from wastewater.

acute LC(50): A concentration of a substance, expressed as parts per million parts of medium, that is lethal to 50 percent of the test population of animals under specified test conditions.

acute toxicity: Any poisonous effect produced within a short time, usually up to 24 to 96 h, resulting in severe biological harm and often death.

adhesion: Molecular attraction which holds the surfaces of two substances in contact, such as water and rock particles.

adsorption: The attachment of the molecules of a liquid of gaseous substance to the surface of a solid.

aerosol: A particle of solid or liquid matter that can remain suspended in the air because of its small size.

afterburner: A device that includes an auxiliary fuel burner and combustion chamber to incinerate combustible gas contaminants.

agricultural solid waste: The solid waste that is generated by the rearing of animals and the producing and harvesting of crops or trees.

air classifier: A system that uses a forced air stream to separate mixed material according to size, density, and aerodynamic drag of the pieces.

air-cooled wall: A refractory wall with a lane directly behind it through which cool air flows.

air emissions: For stationary sources, the release or discharge of a pollutant by an owner or operator into the ambient air either by means of a stack or as a fugitive dust, mist, or vapor as a result inherent to the manufacturing or forming process.

air heater: A heat exchanger through which air passes and is heated by a medium of a higher temperature, such as hot combustion gases in metal tubes.

air pollutant: Dust, fumes, smoke and other particulate matter, vapor, gas, odorous substances, or any combination thereof. Also any air pollution agent or combination of such agents, including any physical, chemical, biological, radioactive substances, or matter which is emitted into or otherwise enters the ambient air.

air pollution: The presence of any air pollutant in sufficient quantities and of such characteristics and duration as to be, or likely to be, injurious to health or welfare, animal or plant life, or property, or as to interfere with the enjoyment of life or property.

ambient air: That portion of the atmosphere external to buildings to which the general public has access.

animal waste: The high organic waste that is generated by the breeding, maintenance, use, and slaughter of animals.

ash: Inorganic residue remaining after ignition of combustible substances determined by definite prescribed methods.

autogenous (autothermic) combustion: The burning of wet organic material where the moisture content is at such a level that the heat of combustion of the organic material is sufficient to vaporize the water and maintain combustion at a particular temperature; no auxiliary fuel is required except for startup of the process.

baffles: Deflector vanes, guides, grids, grating, or similar devices constructed or placed in air or gas flow systems, flowing water, or slurry systems to effect a more uniform distribution of velocities; absorb energy; divert, guide, or agitate fluids; and check eddies.

bagasse: An agricultural waste material consisting of the dry pulp residue that remains after juice is extracted from sugar cane or sugar beets.

baghouse: An air pollution abatement device used to trap particulates by filtering gas streams through large fabric bags usually made of cloth or glass fibers.

barometric damper: A hinged or pivoted plate that automatically regulates the amount of air entering a duct, breeching, flue connection, or stack, thereby maintaining a constant draft within an incinerator.

batch-fed incinerator: An incinerator that is periodically charged with waste; one charge is allowed to burn out before another is added.

battery wall: A double or common wall between two incinerator combustion chambers; both faces are exposed to heat.

bilge oil: Waste oil which accumulates, usually in small quantities, in the lower spaces in a ship, just inside the shell plating, and usually mixed with larger quantities of water.

biodegradable: Any substance that decomposes chemically and/or physically through the action of microorganisms.

biological waste: Waste derived from living organisms.

biomass: The amount of living matter in a given unit of the environment.

blowdown: The minimum discharge of recirculating water for the purpose of discharging materials contained in the process, the further buildup of which would cause concentrations or amounts exceeding limits established by best engineering practice.

blower: A fan used to force air or gas under pressure.

bottom ash: The solid material that remains on a hearth or falls off the grate after thermal processing is complete.

breeching: A passage that conducts the products of combustion to a stack or chimney.

bridge wall: A partition between chambers over which the products of combustion pass.

British thermal unit: The amount of heat required to raise the temperature of one pound of water one degree Fahrenheit.

bulky waste: Large items of solid waste such as household appliances, furniture, large auto parts, trees, branches, stumps, and other oversize wastes whose large size precludes or complicates their handling by normal solid waste collection, processing, or disposal methods.

bunker "C" oil: A general term used to indicate a heavy viscous fuel oil.

burial ground: A disposal site for unwanted radioactive materials that uses earth or water for a shield.

burning area: The horizontal projection of a grate, a hearth, or both.

burning hearth: A solid surface to support the solid fuel or solid waste in a furnace and upon which materials are placed for combustion.

burning rate: The volume of solid waste incinerated or the amount of heat released during incineration. The burning rate is usually expressed in pounds of solid waste per square foot of burning area per area or in British thermal units per cubic foot of furnace volume per hour.

butterfly damper: A plate or blade installed in a duct, breeching, flue connection, or stack that rotates on an axis to regulate the flow of gases.

cake: The solids discharged from a dewatering apparatus.

calcination: The process of heating a waste material to a high temperature without fusing in order to effect useful change (e.g., oxidation, pulverization).

carbonaceous matter: Pure carbon or carbon compounds present in the fuel or residue of a combustion process.

carbon sorption: The process in which a substance (the sorbate) is brought into contact with a solid (the sorbent), usually activated carbon, and held there by either chemical or physical means.

carcinogenic: Capable of causing the cells of an organism to react in such a way as to produce cancer.

catalytic combustion system: A process in which a substance is introduced into an exhaust gas stream to burn or oxidize vaporized hydrocarbons or odorous contaminants; the substance itself remains intact.

caustic soda: Sodium hydroxide (NaOH), a strong alkaline substance used as an acid neutralizer.

centrifugal collector: A mechanical system using centrifugal force to remove aerosols from a gas stream or to dewater sludge.

chain grate stoker: A stoker with a moving chain as a grate surface. The grate consists of links mounted on rods to form a continuous surface that is generally driven by a shaft with sprockets.

chamber: An enclosed space inside an incinerator.

charge: The amount of solid waste introduced into a furnace at one time or, in the steel industry, the addition of iron and steel scrap or other materials into the top of an electric arc furnace.

charging chute: An overhead passage through which waste materials drop into an incinerator.

charging gate: A horizontal, movable cover that closes the opening on a top-charging furnace.

charging hopper: An enlarged opening at the top of a charging chute.

checker work: A pattern of multiple openings through which the products of combustion pass to accelerate the turbulent mixing of gases.

chute-fed incinerator: An incinerator that is charged through a chute that extends above it.

clamshell bucket: A vessel used to hoist and convey materials; it has two jaws that clamp together when it is lifted by specially attached cables.

classification: The separation and rearrangement of waste materials according to composition (e.g., organic or inorganic), size, weight, color, shape, and the like, using specialized equipment.

clinkers: Hard, sintered, or fused pieces of residue formed in a furnace by the agglomeration of ash, metals, glass, and ceramics.

coal refuse: Waste products of coal mining, cleaning, and coal preparation operations and containing coal, matrix material, clay, and other organic and inorganic material.

codisposal: The technique in which sludge is combined with other combustible materials (e.g., refuse, refuse-derived fuel, coal), to form a furnace feed with a higher heating value than the original sludge.

coffin: A thick-walled container (usually lead) used for transporting radioactive materials.

cold drying hearth: A surface upon which unheated waste materials are placed to dry or to burn. Hot combustion gases are then passed over the materials.

combustibles: Materials that can be ignited at a specific temperature in the presence of air to release heat energy.

combustion: The production of heat and light energy through a chemical process, usually oxidation.

combustion air: The air used for burning a fuel.

combustion gases: The mixture of gases and vapors produced by burning.

commercial waste: All types of solid wastes generated by stores, offices, restaurants, warehouses, and other nonmanufacturing activities, excluding residential and industrial wastes.

compost: A relatively stable mixture of decomposed organic waste materials, generally used to fertilize and condition the soil.

conduction: The transfer of heat by physical contact between substances.

conical burner: A hollow, cone-shaped combustion chamber with an exhaust vent at its point and a door at its base through which waste materials are normally charged; air is delivered to the burning waste inside the cone. Also called a *teepee burner*.

conservation: The protection, improvement, and use of natural resources according to principles that will ensure their highest economic or social benefits.

construction and demolition waste: The waste building materials, packaging, and rubble resulting from construction, remodeling, and demolition operations on pavements, houses, commercial buildings, and other structures.

consumer waste: Materials used and discarded by the buyer, or consumer, as opposed to wastes created and discarded in-plant during the manufacturing process.

contaminant: Any physical, chemical, biological, or radiological substance in water.

continuous-feed incinerator: An incinerator into which solid waste is charged almost continuously to maintain a steady rate of burning.

controlled air incinerator: An incinerator excess or starved air with two or more combustion chambers in which the amounts and distribution of air are controlled.

corrosion: The gradual wearing away of a substance by chemical action.

curie: A measure of radioactivity.

curtain wall: A refractory construction or baffle that deflects combustion gases downward.

cyclone separator: A separator that uses a swirling airflow to sort mixed materials according to the size, weight, and density of the pieces.

dairy waste: The waste generated by dairy plants in their processing of milk to produce cream, butter, cheese, ice cream, and other dairy products.

decibel: A unit of sound measurement.

decontamination/detoxification: Processes which will convert pesticides into nontoxic compounds or the selective removal of radioactive material from a surface or from within another material.

deep-well injection: Disposal of raw or treated hazardous wastes by pumping them into deep wells for filtration through porous or permeable subsurface rock and then containment within surrounded layers of impermeable rock or clay.

dewatering: A physical process which removes sufficient water from sludge so that its physical form is changed from that of a fluid to that of slurry or damp solid.

dispersion technique: The use of dilution to attain ambient air quality levels including any intermittent or supplemental control of air pollutants varying with atmospheric conditions.

domestic waste or household waste: Solid waste, comprised of garbage and rubbish, which normally originates in the residential private household or apartment house.

downpass: A chamber or gas passage placed between two combustion chambers to carry the products of combustion downward.

draft: The difference between the pressure within an incinerator and that in the atmosphere.

drag conveyer: A conveyer that uses vertical steel plates fastened between two continuous chains to drag material across a smooth surface.

drop arch: A form of construction that supports a vertical refractory furnace wall and serves to deflect gases downward.

drying hearth: A solid surface in an incinerator upon which wet waste materials, liquids, or waste matter that may turn to liquid before burning is placed to dry or to burn with the help of hot combustion gases.

duct: A conduit, usually metal or fiberglass, round or rectangular in cross section, used for conveyance of air.

dump: A site used to dispose of solid wastes, or other containerized wastes, without environmental controls.

dust: Fine grain particles light enough to be suspended in air.

effluent: Waste materials, usually waterborne, discharged into the environment, treated or untreated.

electrostatic precipitator: An air pollution control device that imparts an electric charge to particles in a gas stream, causing them to collect on an electrode.

elutriation: The reduction of the concentration of an impurity on a solid by repeated washings.

emission rate: The amount of pollutant emitted per unit of time.

encapsulation: The complete enclosure of a waste in another material in such a way as to isolate it from external effects such as those of water or of air.

environment: Water, air, land, and all plants and human and other animals living therein, and the interrelationships which exist among them.

environmentally persistent waste: Any waste which, if exposed to a natural environment, remains hazardous for an extended time.

evaporation: The physical transformation of a liquid to a gas at any temperature below its boiling point.

evase stack: An expanding connection on the outlet of a fan or in an airflow passage to convert kinetic energy to static pressure.

exhaust system: The system comprised of a combination of components which provides for enclosed flow of exhaust gas from the furnace exhaust port to the atmosphere.

explosive wastes: Wastes which are unstable and may readily undergo violent chemical change or explode.

extraction test procedure: A series of laboratory operations and analyses designed to determine whether, under severe conditions, a solid waste, stabilized waste, or landfill material can yield a hazardous leachate.

federal register: Daily publication that is the official method of notice of executive branch actions, including proposed and final regulations.

feedlot waste: High concentrations of animal excrement that result from raising large numbers of animals on a relatively small, confined area of land.

filter: A porous device through which a gas or liquid is passed to remove suspended particles or dust.

firebrick: Refractory brick made from fireclay.

fireclay: A sedimentary clay containing only small amounts of fluxing impurities, high in hydrous aluminum silicate and capable of withstanding high temperatures.

fire point: The lowest temperature at which an oil vaporizes readily enough to burn at least 5 s after ignition.

fixed carbon: The ash-free carbonaceous material that remains after volatile matter is driven off a dry solid waste sample.

fixed grate: A grate without moving parts, also called a *stationary grate*.

flammable waste: A waste capable of igniting easily and burning rapidly.

flash drying: The process of drying a wet organic material by passing through a high-temperature zone at such a rate that the water is rapidly evaporated but the organic material, protected by the boiling point of water, is not overheated.

flash point: The minimum temperature at which a liquid or solid gives off sufficient vapor to form an ignitable vapor-air mixture near the surface of the liquid or solid.

flue: Any passage designed to carry combustion gases and entrained particulates.

flue-fed incinerator: An incinerator that is charged through a shaft that functions as a chute for charging waste and as a flue for carrying products of combustion.

flue gas: The products of combustion, including pollutants, emitted to the air after a production process or combustion takes place.

fluidized bed combustion: Oxidation of combustible material within a bed of solid, inert (noncombustible) particles which under the action of vertical hot airflow will act as a fluid.

fly ash: The airborne combustion residue from burning fuel.

food waste: The organic residues generated by the handling, storage, sale, preparation, cooking, and serving of foods, commonly called *garbage*.

forced draft: The positive pressure created by the action of a fan or blower which supplies the primary or secondary combustion air in an incinerator.

fossil fuel: Natural gas, petroleum, coal, and any form of solid, liquid, or gaseous fuel derived from such materials for the purpose of creating useful heat.

fouling: The impedance to the flow of fluids or heat that results when material accumulates in flow passages or on heat-absorbing surfaces in an incinerator or other combustion chamber.

fuel: Any material which is capable of releasing energy or power by combustion or other chemical or physical means.

fugitive emissions: Emissions other than those from stacks or vents.

fume: Solid particles under 1 μm in diameter, formed as vapors condense or as chemical reactions take place.

furnace: A combustion chamber; an enclosed structure in which heat is produced.

furnace arch: A nearly horizontal structure that extends into a furnace and serves to deflect gases.

fusion point: The temperature at which a particular complex mixture of minerals (ash) can flow under the weight of its own mass.

garbage: Solid waste resulting from animal, grain, fruit, or vegetable matter used or intended for use as food.

grain loading: The rate at which particles are emitted from a pollution source, in grains per cubic foot of gas emitted, 7000 gr = 1 lb.

grate: A piece of furnace equipment used to support solid waste or solid fuel during the drying, igniting, and burning process.

grease: A group of substances including fats, waxes, free fatty acids, calcium and magnesium soaps, mineral oils, and certain other nonfatty materials.

guillotine damper: An adjustable plate, used to regulate the flow of gases, installed vertically in a breeching.

hardness: A characteristic of water, imparted by salts of calcium, magnesium, and iron, that causes curdling of soap, deposition of scale in boilers, damage in some industrial processes, and sometimes objectionable taste.

hazardous waste: A waste, or combination of wastes, that may cause or significantly contribute to an increase in mortality or an increase in serious irreversible, or incapacitating reversible illness or that pose a substantial present or potential hazard to human health or the environment when improperly treated, stored, transported, disposed of, or otherwise managed.

hearth: The bottom of a furnace on which waste materials are exposed to the flame.

heat balance: An accounting of the distribution of the heat input and output of an incinerator or boiler, usually on an hourly basis.

heavy metals: Metallic elements, such as mercury, chromium, lead, cadmium, and arsenic, with high atomic weights and which tend to accumulate in the food chain.

hot drying hearth: A surface upon which waste materials are placed to dry or to burn. Hot combustion gases first pass over the materials and then under the hearth.

hydrocarbon: Any of a vast family of compounds containing carbon and hydrogen in various combinations, found especially in fossil fuels.

ignitability: The characteristic property of liquids having a flash point of less than 140°F; nonliquids liable to cause fires through friction, absorption of moisture, spontaneous chemical change, or retained heat from manufacturing, or are liable, when ignited, to burn so vigorously and persistently as to create a hazard; ignitable compressed gas; oxidizers.

ignition arch: A refractory furnace arch or surface located over a fuel bed to radiate heat and to accelerate ignition.

ignition temperature: The lowest temperature of a fuel at which combustion becomes self-sustaining.

incineration: An engineered process using controlled flame combustion to thermally degrade waste materials.

incinerator stoker: A mechanically operable moving-grate arrangement for supporting, burning, or transporting solid waste in a furnace and discharging the residue.

induced draft: The negative pressure created by the action of a fan, blower, or other gas-moving device located between an incinerator and a stack.

industrial waste: Unwanted materials produced in or eliminated from an industrial operation.

infiltration air: Air that leaks into the chambers of ducts of an incinerator.

inorganic matter: Chemical substances of mineral origin, not containing carbon-to-carbon bonding.

inorganic refuse: Solid waste composed of matter other than plant, animal, and certain carbon compounds, e.g., metals and glass.

institutional solid waste: Solid waste generated by educational, health care, correctional, and other institutional facilities.

insulated wall: A furnace wall on which refractory material is installed over insulation.

isokinetic sampling: Sampling in which the linear velocity of the gas entering the sampling nozzle is equal to that of the undisturbed gas stream at the sample point.

junk: Unprocessed, discarded materials that are usually not suitable for reuse or recycling, e.g., rags, paper, toys, metal, furniture.

kraft paper: A comparatively coarse paper noted for its strength and used primarily as a wrapper or packaging material.

landfill: A land disposal site employing an engineered method of disposing of wastes on land that minimizes environmental hazards by spreading wastes in thin layers, compacting the wastes to the smallest practical volume, and applying cover materials at the end of each operating day.

lantz process: A destructive distillation technique in which the combustible components of solid waste are converted to combustible gases, charcoal, and a variety of distillates.

lethal dose: The quantity of a substance which is fatal to 50 percent of the population on which it is tested.

lime: Any of a family of chemicals consisting essentially of calcium oxide or hydroxide made from limestone (calcite).

liner: The material used on the inside of a furnace wall to ensure that a chamber is impervious to escaping gases; the material used on the inside of a sanitary landfill to ensure that the basin is impervious to fluids.

macroencapsulation: The isolation of a waste by embedding it in, or surrounding it with, a material which acts as a barrier to water or air.

magnetic separation: The process by which a permanent magnet or electromagnet is used to attract magnetic materials away from mixed waste.

masking: Blocking out one sight, sound, or smell with another.

metal cleaning wastes: Any cleaning compounds, rinse waters, or any other waterborne residues derived from cleaning any metal process equipment including, but not limited to, boiler tube cleaning, boiler fireside cleaning, and air preheater cleaning.

microencapsulation: The isolation of a waste from external effects by mixing it with a material which then cures or converts it to a solid, nonleaching barrier.

milled refuse: Solid waste that has been mechanically reduced in size.

mining waste: Residues which result from the extraction of raw materials from the earth.

mist: Liquid particles, measuring 40 to 500 μm in diameter, that are formed by condensation of vapor.

mixing chamber: A chamber usually placed between the primary and secondary combustion chambers and in which the products of combustion are thoroughly mixed by turbulence that is created by increased velocities of gases, checkerwork, or turns in the direction of the gas flow.

mixture: Any combination of two or more chemical substances if the combination does not occur in nature and is not, in whole or in part, the result of a chemical reaction.

modular combustion unit: One of a series of incinerator units designed to operate independently and can handle small quantities of solid waste.

movable grate: A grate with moving parts.

multiple-chamber incinerator: An incinerator that consists of two or more chambers, arranged as in-line or retort types, interconnected by gas passage parts or flues.

municipal incinerator: A privately or publicly owned incinerator primarily designed and used to burn residential and commercial solid waste within a community.

municipal solid wastes: Garbage, refuse, sludges, and other discarded materials resulting from residential and nonindustrial operations and activities.

mutagenicity: The property of a substance or mixture of substances which, when it interacts with a living organism, causes the genetic characteristics of the organism to change and its offspring to have a decreased life expectancy.

natural draft: The negative pressure created by the height of a stack or chimney and the difference in temperature between flue gases and the atmosphere.

natural gas: A natural fuel containing methane and hydrocarbons that occurs in certain geologic formations.

neutralization: The chemical process in which the acidic or basic characteristics of a fluid are changed to those of water.

nitrogenous wastes: Animal or plant residues that contain large amounts of nitrogen.

nonferrous: Metals that contain no iron, e.g., aluminum, copper, brass, and bronze materials.

odor threshold: The lowest concentration of an airborne odor that a human being can detect.

offal: The viscera and trimmings of a slaughtered animal removed from the carcass.

office wastes: Discarded materials that consist primarily of paper waste including envelopes, ledgers, and brochures.

oil: Oil of any kind or in any form including, but not limited to, petroleum, fuel oil, sludge, oil refuse, and oil mixed with wastes other than dredged spoil.

on-site incinerator: An incinerator that burns solid waste on the property used by the generator thereof.

opacity: Degree of obscuration of light; e.g., a window has zero opacity, while a wall has 100 percent opacity.

open burning: The combustion of any material without control of combustion air or combustion temperature, without containment of the combustion reaction in an enclosure, and/or without provision of a stack or a vent for discharge of combustion products.

open-hearth furnace: A long, wide, shallow reverberatory furnace used to produce steel from cast or pig iron.

open-pit incinerator: A burning apparatus that has an open top and a system of closely spaced nozzles that place a stream of high-velocity air over the burning zone.

organic matter: Chemical substance comprised mainly of carbon, covalently bonded. May have its origin in animal or plant life, coal, petroleum, or laboratory synthesis.

organic nitrogen: Nitrogen combined in organic molecules such as protein, amines, and amino acids.

Orsat: An apparatus used to analyze flue gases volumetrically by dissolving the constituent gases selectively in various solvents.

oscillating-grate stoker: A stoker whose entire grate surface oscillates to move the solid waste and residue over the grate surface.

overfire air: Air under control as to quantity and direction, introduced above and beyond a fuel bed by induced or forced draft.

oxidant: A substance containing oxygen that reacts chemically to release some of or all its oxygen component to another substance.

oxidation: The addition of oxygen to a compound.

oxide: A compound of two elements, one of which is oxygen.

packed tower: A pollution control device that forces dirty gas through a tower packed with crushed rock, wood chips, or other packing while liquid is sprayed over the packing material. Pollutants in the gas stream either dissolve in or chemically react with the liquid.

paper: The term for all kinds of matted or felted sheets of fiber laid down on a fine screen from a water suspension. Specifically, as one of the two subdivisions of the general term, paper refers to materials that are lighter in basis weight, thinner, and more flexible than paperboard, the other subdivision.

particulate matter: Any material, except water in uncombined form, that is or has been airborne and exists as a liquid or a solid at standard conditions.

particulates: Fine liquid or solid particles such as dust, smoke, mist, fumes, or smog, found in the air or in emissions.

pathogen: Organism capable of causing disease.

pathogenic waste: Discarded materials that contain organisms capable of causing disease.

permeability: The property of a solid material which allows fluid to flow through it.

pH: A measure of the acidity (0, most acid) or alkalinity (14, most alkaline) of a liquid or solid on a scale of 0 to 14.

pig: A container, usually lead, used to ship or store radioactive materials.

plastics: Nonmetallic compounds that result from a chemical reaction and are molded or formed into rigid or pliable structural material.

plume: Visible emissions from a flue or chimney.

pollutant: Dredged spoil, solid waste, incinerator residue, sewage, garbage, sewage sludge, munitions, chemical wastes, biological materials, radioactive materials, heat, wrecked or discarded equipment, rock, sand, cellar dirt, and industrial, municipal, and agricultural waste discharged into water.

pollution: The presence of matter or energy whose nature, location, or quantity produces undesired environmental effects. Also, the artificial or human-introduced alteration of the chemical, physical, biological, and radiological integrity of water.

porosity: The ratio of the volume of pores of a material to the volume of its mass.

pour point: The lowest temperature at which an oil will flow or can be poured under specified conditions of test.

precipitators: Air pollution control devices that collect particles from an emission by mechanical or electrical means.

primary combustion air: The air admitted to a combustion system when the fuel is first oxidized.

primary pollutant: A pollutant emitted directly from a polluting stack.

primary standard: A national air emissions standard intended to establish a level of air quality that, with an adequate margin of error, will protect public health.

process waste: Any designated toxic pollutant which is inherent to or unavoidable resulting from any manufacturing process, including that which comes into direct contact with or results from the production or use of any raw material, intermediate product, finished product, by-product, or waste product.

proximate analysis: The analysis of a fuel to determine (on a percentage basis) how much moisture, volatile matter, fixed carbon, and ash the sample contains; usually the fuel's heat value is also obtained.

putrescible: A substance that can rot quickly enough to cause odors and attract flies.

pyrolysis: The chemical decomposition of organic matter through the application of heat in an oxygen-deficient atmosphere.

quench tank: A water-filled tank used to cool incinerator residues or hot materials during industrial processes.

radioactive: Substances that emit rays either naturally or as a result of scientific manipulation.

rated incinerator capacity: The number of tons of solid waste that can be processed in an incinerator per 24-h period when specified criteria prevail.

reactivity: The tendency to create vigorous reactions with air or water, tendency to explode, to exhibit thermal instability with regard to shock, ready reaction to generate toxic gases.

reciprocating grate stoker: A stoker with a bed of bars or plates arranged so that alternate pieces, or rows of pieces, reciprocate slowly in a horizontal sliding mode and act to push the solid waste along the stoker surface.

recycled material: A material that is used in place of a primary, raw, or virgin material in manufacturing a product and consists of materials derived from postconsumer waste, industrial scrap, material derived from agricultural wastes and other items, all of which can be used in the manufacture of new products.

refractory material: Nonmetallic substances used to line furnaces because they can endure high temperatures and resist abrasion, spalling, and slagging.

refractory wall: A wall made of heat-resistant material.

refuse: All solid materials which are discarded as useless.

refuse-derived fuel: The combustible, or organic, portion of municipal waste that has been separated out and processed for use as fuel.

residual oil: A general term used to indicate a heavy viscous fuel oil.

residual wastes: Those solid, liquid, or sludge substances from human activities in the urban, agricultural, mining, and industrial environments remaining after collection and necessary treatment.

residue: Solid or semisolid materials such as, but not limited to, ash, ceramics, glass, metal, and organic structures remaining after incineration or processing.

residue conveyer: Generally a drag or flight conveyer used to remove incinerator residue from a quench through to a discharge point.

resource conservation: Reduction of the amounts of wastes generated, reduction of overall consumption, and utilization of recovered resources.

resource recovery: The extraction of useful materials or energy from waste.

retort-type incinerator: A multiple-chamber incinerator in which the gases travel from the end of the ignition chamber, then pass through the mixing and combustion chambers.

reverberatory furnace: A furnace in which the fuel is not in direct contact with the charge (waste) but the heating effect is basically generated by reflection down from a refractory roof.

rocking-grate stoker: A stoker with a bed of bars or plates on axles; when the axles are rocked in a coordinated manner, the solid waste is lifted and advanced along the surface of the grate.

rotary kiln stoker: A cylindrical, inclined device that rotates, thus causing the solid waste to move in a slow cascading and forward motion.

sanitary landfill: A method of disposing of refuse on the land without creating nuisances or hazards to the public health or safety, by utilizing the principles of engineering to confine the refuse to the smallest practical area, to reduce it to the smallest practical volume, and to cover it with a layer of earth at the conclusion of each day's operation, or more frequently as may be necessary.

scrubbing: The removal of impurities from a gas stream by spraying of a fluid.

secondary burner: A burner installed in the secondary combustion chamber of an incinerator to maintain a minimum temperature and to complete the combustion of incompletely burned gas.

secondary combustion air: The air introduced above or below the fuel (waste) bed by a natural, induced, or forced draft.

secondary pollutant: A pollutant formed in the atmosphere by chemical changes taking place between primary pollutants and other substances present in the air.

secondary standard: A national air quality standard that establishes that ambient concentration of a pollutant that, with an adequate margin of safety, will protect the public welfare (all parts of the environment other than human health) from adverse impacts.

settling chamber: Any chamber designed to reduce the velocity of the products of combustion and thus to promote the settling of fly ash from the gas stream before it is discharged to the next process or to the environment.

settling velocity: The velocity at which a given dust will fall out of dust-laden gas under the influence of gravity only; also called *terminal velocity*.

sewage sludge: A semiliquid substance consisting of settled sewage solids combined with varying amounts of water and dissolved materials.

shredder: A machine used to break up waste materials into smaller pieces by cutting or tearing.

siftings: The fine materials that fall from a fuel bed through its grate openings during incineration.

silo: A storage vessel, generally tall relative to its cross section, for dry solids; materials are fed into the top and withdrawn from the bottom through a control mechanism.

slag: The more or less completely fused and vitrified matter separated during the reduction of metal from its ore.

slagging: Destructive chemical action that forms slag on refractory materials subjected to high temperatures; also a molten or viscous coating produced on refractory materials by ash particles.

sliding damper: A plate normally installed perpendicular to the flow of gas in a breeching and arranged to slide across it to regulate the flow.

sludge: Any solid, semisolid, or liquid waste generated from a municipal, commercial, or industrial wastewater treatment plant, water supply treatment plant, or air pollution control facility, or any other such waste having similar characteristics and effects.

slurry: A pumpable mixture of solids and fluid.

smoke: Particles suspended in air after incomplete combustion of materials containing carbon.

solid waste: Any garbage, refuse, sludge from a waste treatment plant, water supply treatment plant, or air pollution control facility, and other discarded material, including solid, liquid, semisolid, or contained gaseous material resulting from industrial, commercial, mining, and agricultural operations, and from community activities.

soot: Carbon dust formed by incomplete combustion.

source: Any building, structure, facility, or installation from which there is or may be the discharge of pollutants.

source reduction: Reduction of the amount of materials entering the waste stream by voluntary or mandatory programs to eliminate the generation of waste.

spray chamber: A chamber equipped with water sprays that cool and clean the combustion products passing through it.

stack: Any chimney, flue, vent, roof monitor, conduit, or duct arranged to discharge emissions to the ambient air.

stationary source: Any building, structure, facility, or installation which emits or may emit any air pollutant.

stoichiometric combustion: Combustion with the theoretical air quantity.

stoker: A mechanical device to feed solid fuel or solid waste to a furnace.

street refuse: Solid waste picked up when streets and sidewalks are swept manually and mechanically, wastes from public waste receptacles, and material removed from catch basins.

syneresis: The process whereby a colloidal gel releases bound water by the application of heat.

teratogenic: Affecting the genetic characteristics of an organism so as to cause its offspring to be misshapen or malformed.

theoretical air: The quantity of air, calculated from the chemical composition of a waste, that is required to burn the waste completely.

thermal efficiency: The ratio of heat used to total heat generated.

threshold dose: The minimum application of a given substance to produce a measurable effect.

toxicity: The degree of danger posed by a substance to animal or plant life.

toxic substance: A chemical or mixture that may present an unreasonable risk of injury to health or to the environment.

transuranium: Nuclide having an atomic number greater than that of uranium (i.e., greater than 92).

traveling grate stoker: A stoker that is essentially a moving chain belt carried on sprockets and covered with separated, small metal pieces called keys.

trommel: A perforated, rotating, horizontal cylinder that may be used in resource recovery facilities to break open trash bags, to remove glass and such small items as stone and dirt, and to remove cans from incinerator residue.

tuyeres: Openings or ports in a grate through which air can be directed to improve combustion.

ultimate analysis: The chemical analysis of a solid, liquid, or gaseous fuel.

underfire air: Forced or induced combustion air (quantity and direction are controlled) introduced under a grate to promote burning within a fuel bed.

unit-suspended wall: A furnace wall or panel that is hung from a steel structure.

unreclaimable residues: Residual materials of little or no value remaining after incineration.

vapor: The gaseous phase of substances that are liquid or solid at atmospheric temperature and pressure, e.g., steam.

vapor plume: The stack effluent consisting of flue gases made visible by condensed water droplets or mist.

vitrification: A process whereby high temperatures effect permanent chemical and physical change in a ceramic body.

volatile: Any substance that evaporates at a low temperature.

volatility: The property of a substance or substances to convert to vapor or gas without chemical change.

waste: Unwanted materials left over from manufacturing processes, or refuse from places of human or natural habitation.

waste reduction: The prevention or restriction of waste generation at its source by redesigning products or the patterns of production and consumption.

wastewater: Water carrying dissolved or suspended solids from homes, farms, businesses, institutions, and industries.

waterwall incinerator: An incinerator whose furnace walls consist of vertically arranged metal tubes through which water passes and absorbs the radiant energy from burning solid waste.

windbox: A chamber below a furnace grate or surrounding a burner, through which air is supplied under pressure to burn the fuel.

wood residue: Bark, sawdust, slabe, chips, shavings, mill trim, and other wood products derived from wood processing and forest management operations.

zero gas: A gas containing less than one part per million parts of sulfur dioxide.

BIBLIOGRAPHY

The Clean Air Act, USEPA, 42USC.
Code of Federal Regulations—40, Protection of Environment, Revised July 1, 1978.
Common Environmental Terms: A Glossary, USEPA, November 1979.
Resource Conservation and Recovery Act, USEPA, May 1980.

Index

ABOUT THE AUTHOR

Calvin R. Brunner, P.E., is a consulting engineer based in Reston, Virginia. He has over 20 years of experience in the incineration field, specializing in the design, operation, and evaluation of incineration systems for industrial installations, remediation sites, resource recovery facilities, hospitals, and wastewater treatment plants. Mr. Brunner is the author of several textbooks and numerous articles and studies on incineration and related topics. He has developed specialty computer programs for the analysis of thermal disposal systems. He also presents seminars throughout the country on the design and application of incineration systems.